ADVANCES IN CHEMORECEPTION

Volume I

COMMUNICATION
BY CHEMICAL SIGNALS

ADVANCES IN CHEMORECEPTION

Edited by

JAMES W. JOHNSTON, JR.
Schools of Medicine and Dentistry
Georgetown University
Washington, D.C. 20007

DAVID G. MOULTON
Monell Chemical Senses Center
University of Pennsylvania
Philadelphia, Pennsylvania 19103
and Veterans Administration Hospital
University & Woodland Avenues
Philadelphia, Pennsylvania 19104

AMOS TURK
Department of Chemistry
The City College of the City University of New York
New York, New York 10031

Volume I

COMMUNICATION BY CHEMICAL SIGNALS

APPLETON-CENTURY-CROFTS
Educational Division
MEREDITH CORPORATION
New York

ISBN 978-1-4684-7157-1   ISBN 978-1-4684-7155-7 (eBook)
DOI 10.1007/978-1-4684-7155-7

390-48350-8

CONTRIBUTORS

JOHN E. BARDACH
School of Natural Resources
University of Michigan
Ann Arbor, Michigan

P. Z. BEDOUKIAN
Technical Director
Compagnie Parento, Inc.
40 Ashley Road
Hastings-on-Hudson, New York

F. H. BRONSON
The Jackson Laboratory
Bar Harbor, Maine

GORDON M. BURGHARDT
Department of Psychology
University of Tennessee
Knoxville, Tennessee

COLIN G. BUTLER
Rothamsted Experimental Station
Harpenden, Herts.
England

TRYGG ENGEN
Walter S. Hunter Laboratory of
  Psychology
Brown University
Providence, Rhode Island

A. GABBA
Instituto di Entomologia Agraria
  dell'Università di Pavia
Pavia, Italy

J. LE MAGNEN
Laboratoire de Physiologie des
  Sensibilitiés Chimiques et Régulations
  Alimentaires de l'E.P.H.E.
Collège de France
Paris, France

R. MYKYTOWYCZ
CSIRO Division of Wildlife Research
Canberra, Australia

M. PAVAN
Instituto di Entomologia Agraria
  dell'Università di Pavia
Pavia, Italy

ALISTAIR M. STUART
North Carolina State University at
  Raleigh
Raleigh, North Carolina

BRUNHILD STÜRCKOW
Department of Biology
Northeastern University
Boston, Massachusetts

JOHN H. TODD
Woods Hole Oceanographic Institute
Woods Hole, Massachusetts

W. K. WHITTEN
The Jackson Laboratory
Bar Harbor, Maine

EDWARD O. WILSON
The Biological Laboratories
Harvard University
Cambridge, Massachusetts

DISCUSSANTS

MICHAEL E. MASON
International Foods and Flavors, Inc.
1515 Highway 36
Union Beach, New Jersey

LOUIS M. ROTH
U. S. Army Natick Laboratories
Natick, Massachusetts

N. T. WERTHESSEN
Office of Naval Research
Boston, Massachusetts

PREFACE

Research on the chemical senses has been growing at a remarkable rate over the last decade. This growth has greatly expanded our understanding of the electrical properties and ultrastructure of chemosensory organs, of the role of chemoreception in the control of behavior, of the organization of higher centers in the chemosensory pathways, and of the chemical constituents of mixtures of biologic significance. But one area where advances have been especially impressive is concerned with the properties of pheromones and substances with similar biologic effects.

Pheromones are compounds, produced by certain animals, which have the effect of inducing one or more specific responses within members of the same or closely related species. Some of these substances—the "primer" pheromones—act on the endocrine system, probably through the central nervous system. The pregnancy block induced in mice by the odor of strange males is a striking example. Others, such as "signalling" or "releaser" pheromones, elicit an immediate behavioral response. Sex attractants are prominent examples of this group.

While the chemical identity of these compounds is known, in a relatively few cases (and these, mainly in insects) we now have extensive information about their impact on the receiving organism at the receptor, physiologic and behavioral levels. As a consequence, it is becoming increasingly clear that pheromones and related substances are exploited more intensively and by a far wider range of species than is generally assumed. They can transmit what is often surprisingly precise and complex information that can exert a powerful control over the relationship between an animal and its external environment.

The study of communication by chemical signals currently absorbs the activities of workers from a diverse range of disciplines, including zoology, perfumery, physiology, organic chemistry, electron microscopy, and psychology. Since this variety also reflects the diversity of approaches characteristic of the broader study of chemoreception, it is an appropriate topic for the first of a continuing series of monographs on chemoreception. It was only after the outline of this book had been prepared and authors had agreed to contribute that it became clear that many of them had not met each other or discussed their own work with more than a small group of workers in this area. At this stage the advantages of bringing them together began to emerge. A meeting was arranged and took place in Auburn, Massachusetts on June 27 and 28, 1968. A selection of the discussions which took place at that meeting is included in this book. The meeting was arranged by Clark University Biology Department, and sponsored by the U.S. Army Research Office.

THE EDITORS

vii

CONTENTS

Contributors   v

Preface   vii

1 INTRODUCTION   1
   Edward O. Wilson

2 DEFINING "COMMUNICATION"   5
   Gordon M. Burghardt

3 PURITY, IDENTITY, AND QUANTIFICATION OF PHEROMONES   19
   P. Z. Bedoukian

4 CHEMICAL COMMUNICATION IN INSECTS: BEHAVIORAL
   AND ECOLOGIC ASPECTS   35
   Colin G. Butler

5 THE ROLE OF CHEMICALS IN TERMITE COMMUNICATION   79
   Alistair M. Stuart

6 RESPONSES OF OLFACTORY AND GUSTATORY RECEPTOR CELLS IN INSECTS 107
   Brunhild Stürckow

7 RESEARCHES ON TRAIL AND ALARM SUBSTANCES IN ANTS   161
   A. Gabba and M. Pavan

8. CHEMICAL COMMUNICATION IN FISH   205
   John E. Bardach and John H. Todd

9 CHEMICAL PERCEPTION IN REPTILES   241
   Gordon M. Burghardt

10 THE ROLE OF PHEROMONES IN MAMMALIAN REPRODUCTION   309
    W. K. Whitten and F. H. Bronson

11 THE ROLE OF SKIN GLANDS IN MAMMALIAN COMMUNICATION   327
    R. Mykytowycz

12 MAN'S ABILITY TO PERCEIVE ODORS   361
    Trygg Engen

    GENERAL DISCUSSION (CHAPTERS 8-12)   385

    COMMUNICATION BY CHEMICAL SIGNALS (CONCLUSION)   393
    J. Le Magnen

    Index   405

ADVANCES IN CHEMORECEPTION

Volume I

COMMUNICATION BY CHEMICAL SIGNALS

1

Introduction

EDWARD O. WILSON

The Biological Laboratories
Harvard University
Cambridge, Massachusetts 02138

Odors drift back and forth across the boundary of human consciousness. Nevei explicit, never freighted with exact information, they carry, above all, emotional meaning. Odors haunt, intrigue, repel, evoke, are redolent of. Unlike sights and sounds, they seldom instruct, command, guide, inform. It is appropriate that a poet such as Baudelaire could express what odors mean to us as incisively as any psychologist:

> *Some wardrobe, in a house long uninhabited,*
> *Full of the powdery odours of moments that are dead—*
> *At time, distinct as ever, an old flask will emit*
> *Its perfume; and a soul comes back to live in it.**

It seems natural that, given this character of human olfaction, biologists have found it difficult to approach the subject of chemical communication in a systematic fashion. The other obstacle is that odors are "invisible," which is another way of saying that we are microsmatic, generally inept at classifying and measuring odors. We have had to depend on machinery to do the job for us.

Yet in the past 10 years the study of chemical communication has been able to take shape as a discipline. Two developments, one conceptual and the other methodologic, have made this possible. The first is the concept of the releaser, or sign stimulus, as created by the ethologic school of animal behavior. A releaser, defined as simply as possible in the description of animal communication, is a discrete, relatively simple and specific signal produced by one animal that evokes a specific, and often complex, response on the part of another animal. The releaser and the

---

* Translation from the French by Edna St. Vincent Millay. Reprinted by permission of George Dillon and Washington Square Press.

response are adaptive with respect to both animals involved; and being adaptive, they tend to be genetically programmed (innate). What made the primitive physiologic models of ethology enticing at the outset was the notion they created that animal communication is isomorphic in nature: for one response there is one signal or set of signals, and vice versa, and it remains for the biologists to synthesize the signal to unlock the response. Such a view, held widely in the 1950's, was of course oversimplified, but it was also very heuristic. Those who worried about the analysis of chemical communication were bound to ask the question: if visual and auditory communication is so predominantly mediated by releasers, must not the same be true for chemical communication? Should we not be able to induce many behavior patterns by the simple presentation of appropriate synthetic chemicals? I believe it was a faith in this elementary proposition that brought many of us to the subject in the first place. Complexity and compromise in our explanations were to come later.

The methodologic advance that made the difference was organic microanalysis. In the late 1950's, when Butenandt and his associates isolated and identified the first sex attractant (bombykol), they had to extract 250,000 female moths to obtain on the order of 10 mg of pure pheromone; nothing less was feasible for structural identification. In the early 1960's gas chromatography came into general use, and it was then possible to analyze complex mixtures of odorants using less than milligram quantities. With the employment of the coupled gas chromatograph-mass spectrometer in the last several years, chemists acquired the ability to identify substances in microgram quantities or less. Since most pheromones are produced in nanogram or microgram quantities, or less, per individual animal, the new natural-products chemistry has made pheromone identification an all but routine matter.

The analysis of a communication system is tripartite: it considers the origin, the transmission, and the reception of the signals. And that covers a great deal of biology. With reference to the origin of a signal we must consider not only the chemistry of the pheromone but also its biosynthesis, its intracellular transport, and its secretion—topics that require modern methods of biochemistry and electron microscopy. In studying transmission we must employ the techniques of physical chemistry in order to estimate the shape and rate of spread of the "active space" within which the molecules are at or above threshold concentration. In studying reception we must consider chemosensory physiology, one of the most difficult and still inchoate topics of modern biology, and finally we must consider neurophysiology.

Additional complexity is added in that there is as yet no very efficient classification of chemical signals. The term pheromone applies only to signals functioning within a single species, but there also exist chemical releasers of comparable specificity and efficiency that operate between species. Some of the signals are mutually adaptive—for example those that operate in mutualism—and they are therefore part of a communication system in the strictest sense of the word. There also exist defensive secretions that repel rather than cripple or kill and can therefore be termed communicative in a looser sense. In fact, some of the volatile secretory substances of ants serve to cripple enemies, to repel them, as well as to alarm other members of the same colony. Perhaps the way out of the semantic difficulties is to refer to interspecific chemical signals as allomones and to the study of pheromones and allomones together as chemical semiotics.

The term *allomone* was suggested by W. L. Brown and T. Eisner in *The American Naturalist*, 102:188-191, 1968.

The Auburn Conference, the first to my knowledge ever specially convened at the international level, displayed the eclectic nature of chemical semiology. Chemists met with biologists of the widest range of persuasion—insect physiologists, mammalian endocrinologists, a herpetologist, an ichthyologist, an experimental psychologist, animal behaviorists, and zoologists and physiologists identified with various other research specialties. To an unusual degree the conferees were meeting each other for the first time.

A certain expected awkwardness and occasional semantic confusion in the discussion periods were more than made up for by the sense of excitement over being present at an early event in the development of what is quickly emerging as an important new biologic discipline. The papers to follow should convey to the reader these impressions, and also a sense that considerable coherence is already being attained in the subject. For the keynote in research on chemical communication is now the rapid discovery of new phenomena, and these phenomena exist all around us in rich abundance. The conference was marked by a common interest in bringing them to light and to deepen their analyses in all aspects of their origin, transmission, and reception.

②

Defining "Communication"*

GORDON M. BURGHARDT†

Department of Psychology
University of Tennessee
Knoxville, Tennessee 37916

INTRODUCTION 5
DEFINITIONS OF COMMUNICATION 6
CONCLUSION 16
SUMMARY 17
REFERENCES 17

INTRODUCTION

The study of "communication" has been of great interest recently to students of animal behavior, psychology, ethology, and related fields. Until recently I was quite sure that I knew what the area of animal communication entailed. Now I am not so sure. This change came about because of some of my research, discussed elsewhere in this book, which shows that newborn garter snakes will respond to chemical cues from those organisms which constitute the species-characteristic diet, such as worms, fish, and frogs. This response was considered to be an example of interspecific chemical communication, as evidenced by an invitation to present this work at the symposium, "Communication by Chemical Signals," on which the contents of this book were based. I had not interpreted this behavior as communicatory, and somehow the label fit uncomfortably, but I was unable to resolve the conflict on the basis of some current defini-

* Originally presented as "The Communication Gap" at the annual meeting of the Psychonomic Society in St. Louis, November 1, 1968.
† Supported, in part, by grant MH-15707 from the National Institute of Mental Health. I thank Benjamen B. Beck, Lori S. Burghardt, Michael G. Johnson, and William S. Verplanck for reading the manuscript and making many helpful suggestions.

tions of communication. This led to discussing the topic of chemical communication, animal communication, and communication in general with a large number of people, as well as searching the literature. I soon discovered that if I was confused, many other people were confused also, although they, perhaps, did not realize it. Indeed, many authors neglect even to give a definition so that one may know which usage is being employed. While it is often possible to determine usage from context, and while "intuitively everyone knows what communication is," when workers in various areas of "communication" gather they feel constrained to grapple with the need for precise meaning. But is this possible? I rapidly approached the point where "animal communication" seemed merely one of the latest "in" or "vogue" fields labeled with a phrase which very well might turn out to be meaningless in the sense of being a scientifically valid or useful concept.

The present effort is an attempt to avoid such a pessimistic conclusion. This chapter brings together the types of definitions implicitly or explicitly used most frequently, with a brief critical analysis of each. Because of the multitude of approaches, each of the various definitions will be stated in an overly simplified manner. These definitions consist of the various characteristics of communication that have been advanced in attempts to set it apart as a scientific term.* Then I will evaluate briefly each definition, using as a primary criterion its applicability to certain examples of behavior that are generally acknowledged to be or not to be communication at the animal level. Those arguments which state that communication involves only high order symbolic or complex cognitive functions, which probably do not exist in nonhuman animals, will not be dealt with; in other words, communication viewed as involving an uniquely human type of language is not under consideration here. Indeed, if this is what is meant by communication, then I submit that we do not need the concept since we have the term language, which is generally restricted to that which occurs in humans. (Bees are sometimes excepted. Whether or not they should be is an interesting question but inappropriate here.) On the other hand, I am not unmindful of the "communication" processes found within an organism involving, for instance, the various organs or systems (such as the nervous system). Even on the cellular lever, communication is referred to in the analysis of interactions between the nucleus and the cytoplasm as implied in the name of the mediator—"messenger RNA." However, to restrict communication considerations to organisms appears no more arbitrary and as necessary tactically as the similar restriction usually placed on behavior in psychology and ethology (Verplanck, 1957).

DEFINITIONS OF COMMUNICATION

The Act of Discrimination

Perhaps the most elementary and straightforward definition of communication is simply that it "is the discriminatory response of an organism to a stimulus." This

---

* Since the original presentation of this paper, the valuable volume edited by Sebeok (1968) has become available and has been referred to extensively in this revision.

definition was proposed by S. S. Stevens (1950). By this definition a pigeon pecking at a green key is communicating. It is clear that if we accept this definition, the term "communication" is quite unnecessary since the term "stimulus control" is perfectly adequate and does not entail the excess implications of the term communication.

Information Transfer

A definition, predicated on the concept of *information transmission* as the essential ingredient, is one commonly given initially by various scientists, whether physical, biological, or psychological. This is especially true of those for whom systems analysis, computers, and cybernetics have replaced the telephone switchboard as the model for behavioral study. This concept of communication has been greatly influenced by the writings of Shannon and Weaver (1949). Communication exists wherever a transfer of information is shown to occur. For example, Batteau (1968) sees communication as the "transformation of information from one carrier to another." Although clarity is not a major attribute of papers with this orientation, it appears that "carrier" is not synonomous with organism.* Indeed "any change in the physical world is a consequence of a message being written in it" (Batteau, 1968). Although couched in different language, it appears that this definition is essentially no different from the preceding one.† Consider the following example: If a person or animal detours around a tree or rock, he is indeed responding to information concerning the size, location, etc., of the object. While this example is rather crude, it demonstrates that information transfer is really not different from stimulus control, at least in the way it appears to be defined—although perhaps not in the way it is used. Therefore, information transfer is, by itself, an inadequate criterion for distinguishing communication. At best, "information" is merely a means of scaling stimulus control. A common procedure, especially in perception, is to quantify "physically" stimuli into bits with $\log_2 N$ alternatives. Although this gives an aura of preciseness, it is a mirage due to the utter lack of rigor possible in specifying the alternatives.

The Organism Restriction

Suppose we tried to rescue the preceding distinction by requiring that two *organisms* (plant or animal) be involved in either the stimulus control or information transfer. This is also inadequate since there seems to be no basic difference in stepping aside to avoid stepping on a snake, bumping into a tree, or stubbing your toe on a rock.

---

* The origins of mathematical communications theory in electrical engineering make it abundantly clear that living organisms are unnecessary for the concept of communication in this technical sense.
† The stimulus control approach may even have an advantage over the information model as used by some, such as Batteau. It has the advantage of a clearly defined criterion (discriminatory response) and the inclusion of an organism. The first sentence in Shannon and Weaver also exposes a rather unstable stage. "The word *communication* will be used here in a very broad sense to include all the procedures by which one mind may affect another" (Shannon and Weaver. 1949. p. 3).

The Animal Restriction

Perhaps, however, we can eliminate plants and restrict communication to events between *animals*. "In the most general sense, communication includes any stimulus arising from one animal and eliciting a response in another" (Scott, 1968). But here again there are certain problems. What about, for instance, the flowers that give off insect-attracting scents? What about the orchid whose flower parts mimic the visual sexual signals of female insects to the extent that the male attempts to copulate with the flowers and in the process polinates them (Wickler, 1968)? In other words, there are the symbiotic, commensal, and parasitic associations, the first of which Marler (1967) is willing to consider communicative. Although the label "communication" might not seem appropriate at first, it would seem that these examples involve stimuli which "work" differently than the rough bark of a tree which leads to its avoidance. In other words, they differ as they treat another organism, plant or animal, as an interbehavioral event, and not merely as a physical entity.

The Species-Member Restriction

Perhaps, however, we should qualify the statement to read as follows: Communication is the transfer of information between *species members*. In other words, communication is limited to intraspecific events. Diebold (1968) has utilized this distinction, as have Frings and Frings (1964). This is, however, a rather arbitrary restriction of communication, since interspecific phenomena that would come under most usages of communication are well known, such as the so-called threats and warning stimuli shown by one species to another species. Anyone owning a cat, dog, or almost any other pet also knows that this just won't do. Cross-fostering and imprinting experiments show that social signals can be given to normally inappropriate organisms. Bastian (1968), while not specifically eliminating interspecific behavior, would restrict communication to "interactions of a social nature." He goes on to state: "This restriction is surely warranted because the interactions of an animal with its social environment are most often marked by some degree of specialization and by relatively greater reciprocity of influence and concomitant fluidity than interactions with its nonsocial environment" (p. 576). The arguments of reciprocity and specialization will be dealt with below. Note that the implication of the definition is that communication is identical with *social behavior*, which Bastian recognizes. Again, what does the term communication add. Here we also need some clarity on what is "social" if restriction to intraspecific events is not sufficient. The best operational technique would call "social" all those instances where one animal treats another *as if it were* one of its own species. This is not only difficult to apply in practice, but it also breaks down upon consideration of cannibalism (predation on one's own species).

Two organizations appear to be needed as a minimum in communication. But this is not universally accepted. For instance, Sebeok (1965), who uses basically an information-transfer criterion based on encoding and decoding of messages, states that communication can take place within one organism. His example is echolocation in bats and porpoises. However, this, if looked at carefully, is quite unsatisfactory. The groping of a person in the dark for a light switch is exactly the same sort of phenomenon, and calling this "communication" certainly does not aid our analysis. As a matter of fact, an authority on echolocation made the following point in a recent volume on communication (Griffin, 1968):

> While the source of echoes may be the body of another animal, only passive physical reradiation of sound waves is involved rather than active reply by the second animal. Hence, echolocation does not properly fall within any reasonable definition of communication behavior, and its discussion in the present volume is justified only by its indirect relevance to the physiological and behavioral phenomena that may be important both in echolocation and in communication (p. 155).

Therefore, although communication does involve stimuli, and two organisms are needed at a minimum, we can conclude that an approach limited to an analysis of responses to stimuli alone is inadequate to distinguish communication. Let us then begin to look at another series of distinctions.

The Necessity of Reciprocity

One way of viewing communication is to define its occurrence only when *reciprocal responding* occurs between organisms. That is, A responds to B, B then responds to the response of A, and so forth. There are variations of this approach (e.g., Bastian, 1968), but they could all be characterized minimally as follows: Communication occurs when one organism (sender) acts as a stimulus for another organism (discriminative stimulus, transfer of information) and the second organism (receiver) responds, which response can itself act as a stimulus (signal) for the first organism. Let us look at an example: If a predator attacks an animal, the victim often responds. For instance, when a garter snake strikes a worm, the latter will at least squirm. The responses of the prey are more vigorous the more tightly the predator holds on. These are certainly reciprocal interactions, but to call them "communication" would vitiate any distinctive meaning of the term.

Conversely, if B does not respond to A rather quickly, this does not imply that no communication has taken place. The temporal factor alone may be such that the experimenter does not see the response. In other words, it may take place a week or a year later. Certainly, we refer to letters or memos as communications, although no response may ever be made. In addition, books, television shows, and commercials are all, I think, properly considered communications even though they do not necessarily elicit an overt response. The soap salesman is communicating with me although I may never buy the product. Of course, it is fallacious to argue here that no response is a response: "no response"—failure to respond—is not to be confused with no movement, passivity, or freezing which may be directly attributable to a stimulus. This comment is not meant to imply that responses by the receivers never occur or are unnecessary.

The Necessity of Specialization

The Frings, in their interesting little book on animal communication (1964), are aware of many of the points discussed above. They, too, are concerned with a definition of communication. The one they arrived at required the sender, the source of the message, to use some specialized structure or behavior to produce the signal.* Now, as one thinks about this definition, it does seem to be attractive. Wing markings, color patches, head bobbings, ritualized movements, and facial expressions can all be properly considered to be involved in communication. Are there some flaws in this definition? First of all, one might quibble with the reductionistic implications involved in the need for some specialized morphological structure. They do mention behavioral methods, but a specialized behavioral method is even more difficult to determine than a specialized structure. Nonetheless, let us assume that the criteria of specialized structures or methods are possible to ascertain.

Consider the response of snakes to chemical cues from their prey. On the basis of this definition, we can certainly exclude it as communication. In other words, the trailing by snakes of worm trails would not be communication because the worm is presumably not using a structure specialized for laying it, especially for snakes to follow. But I submit that the problem is not that easy. Let us consider the classical example of chemical communication, trail pheromones in ants. If there is anything which one would want to call chemical communication, this would certainly appear to be a strong contender, and usually it is so considered (Frings and Frings, 1964; Wilson, 1968). There are specialized structures and behavior involved in the depositing of the trail, and other worker ants respond to this trail, hence—communication. But looking at the behavior or structure of the sender is inadequate, as it is if we define the response as following the trail, and does not differentiate it as communicatory behavior. How can this statement be supported? Ants indeed follow a pheromone trail back to the nest. However, recently it has been shown (Watkins, Gehlbach, and Baldridge, 1967) that the small blind snake (*Leptotyphlops*) also follows pheromone trails of worker ants—it trails them right back to the nest, where it feeds upon the ant brood. Now, it seems reasonable to view the behavior of the blind snake as no more involving communication than the response of a garter snake to worm extracts or trails. Here, then, is a paradox.† It could be argued that if the ant trailing a worker ant's trail is considered communication and the snake trailing the identical trail is not, then the success of any attempt to define communication wholly on a physicalistic basis is impossible.‡ On the other hand, to consider both these examples (the snake trailing as well as the ant trailing the pheromone trail) as communication renders the concept of communication rather empty. For instance, the snake's orientation to prey cues is not basically different from a cow's orientation to its food source, grass, and to consider the latter communication only emphasizes the absurdity involved. But to deny communication in the pheromone trails of ants would eliminate not only chemical communication in general, but also many similar examples in the animal kingdom using other modalities as well.

* They further stated (p. 4) that the receiver be of the same species as the sender. This requirement has already been eliminated here as a necessity.
† The Frings' intraspecific requirement would allow them to resolve the problem. However, since the present author argues against the requirement, he obviously cannot use it.
‡ A good case against a model for psychology based on the model of the physical sciences has recently been made by Deese (1969), who has had impeccable S-R behaviorist credentials. Actually, the image of the physical sciences held by most "scientific" students of behavior is based on the outmoded Newtonian view, which is considerably different from modern theoretical physics (for example).

How, then, can we distinguish communication if specialized structures, mutual responding, limitations to intraspecific events, and other frequent distinctions are inadequate? Before answering this question, we should note the view that all predator-prey relationships be omitted from consideration as communication (Klopfer and Hatch, 1968). Then what remains with the above criteria will leave a satisfactory residue on the question of communication. However, I have already alluded to threats and warnings; for example, the stomping of skunks and hissing and rattling of snakes, which are associated with predator-prey relationships, but which certainly appear to be examples of communication.

The Appeal to Levels

It is clear that the biological world is composed of organisms that differ greatly in the complexity of their morphological, physiological, and behavioral characteristics. Therefore, the making of comparisons between different species always should be performed cautiously, especially when common mechanisms are implied. After stating this somewhat neglected truism, what more should we do? Considerably more according to Schneirla (1953) and, concerning communication, Tavolga (1968). Although complexity in biology is usually viewed as relative and continuous, Schneirla postulates a hierarchical series of levels of behavioral integration. These are not continuous, but are qualitatively different. The tendency has been to associate these levels with phylogenetic position. This is not always possible, and there is a regrettable propensity to ignore the behavioral differences within a group (i.e., fish, birds). The tendency becomes almost a rule in the study of comparative animal intelligence (e.g., Bitterman, 1965).

Another characteristic of the "levels" approach is that the behavioral level is usually tied to a specified degree of structural complexity. This opens the door to a reductionism not yet warranted by our knowledge of the physiological and anatomical correlates of behavior. An essential conservatism would be introduced into the study of behavior if we had to have a "reasonable" physiological mechanism before studying new behavioral phenomena in animals.

But given these caveats, how does the approach by levels contribute to the discussion of animal communication?* Briefly, there are six major levels of interorganismic interaction: vegetative, tonic, phasic, signal, symbolic, and language. The last three are said to comprise communication. The definition of communication, however, turns out to be only a more restricted version of Frings and Frings thesis (described above), and many of the same criticisms apply.

> "For communication to take place, the emitter most possess a specialized stimulus-producing mechanism (chemical, morphological, or behavioral). The stimulus must occupy a narrow portion of the available spectrum of the channel (frequency range, duration, patterning, chemical specificity). The receiver must possess specialized receptors and respond in a specific manner." (Tavolga, 1968, pp. 275-276.)

---

* Tavolga (1968) has attempted this task and should be referred to for a more complete explication.

The following of ant pheromones by the blind snake (which utilizes specialized chemical-sense organs) would have to be classified communication by this definition, while the following of worm "mucus" by a garter snake utilizing an equally specialized receptor would not be communication. In addition, the so-called qualitative levels are not that easy to discriminate. In other words, it is much easier to discuss them in the abstract than to deal with a specific example. I, at least, found the crucial distinction (as far as communication is concerned) between "phasic" and "signal" difficult to apply.

Another complication discussed briefly by Tavolga is that "the emitter and the receiver need not necessarily operate on the same level" (p. 276). This complicates the problem to the point where the use of this approach would eliminate the possibility of any precise meaning of the term communication. It is another bit of evidence that the concept has little scientific value.

The Multi-criterion Approach

Finally, we might conceive of communication as involving an injunctive definition similar to a definition of life (Lorenz, 1965). In other words, no one criterion is necessary and sufficient; certain combinations are required. While there may be some merit in that approach to communication, the parallel is far from exact. I have tried to make a list of criteria that, when taken in combination, would separate commonly accepted examples of communication from those not considered to be communication. I met with little success unless the position advocated below is made a necessary, but not necessarily a sufficient, criterion.

An Attempted Resolution

Is it, therefore, possible to define or to identify communication? Should we accept the conclusion that communication is a scientifically useless or redundant term? It will be argued here that it need not be this way if we make a certain criterion central to our definition. This criterion might be, at least on the surface, unacceptable to many scientists. However, at the present time there seems to be no other way of rescuing the term "communication." My position is that communication is indeed a valid term if we realize how we have, in fact, been using it.

The central criterion of the concept of communication must involve *intent* in some manner. This is implicit in the definitions of the Frings and of Tavolga, where the sender uses a specialized structure or method of behavior—specialized for what? They wanted to avoid the problem of purpose, and so they made a reductionistic jump that does involve "intent", but not so blatantly. Now intent, however, defined, characterizes all generally agreed upon examples of communication from protozoa to man. Must this mean vitalism and unscientific, unanalyzable types of arguments? Does it mean an incompatibility with a broad operational, experimental, and empirical approach to behavior? I do not think so. Intent can be looked at scientifically merely by considering that *it is to the real or perceived advantage of the signaler or the signaler's group for it to get its message across to whatever organism is involved.* In other words, it is adaptive. The signal mechanism has either been phylogenetically or ontogenetically "built" into the organism via natural selection in the first instance or probably via reinforcement in the other, acting, of course, upon the consequences of previous events. This crucial distinction between the "innate" and the "learned" has been recently elaborated on by Lorenz (1965) and Skinner (1966), although the present case does not depend on any given viewpoint in this controversial area. It does depend on the acceptance of both evolutionary and experiential determinants in behavior, although complex interactions certainly occur. In addition, the probabilistic nature of this approach must be emphasized. To be evolutionarily adaptive, communication must enhance survival value and reproductive success (Klopfer and Hatch, 1968). When identifying behavior as communicative, we are implying this. I submit that we must build our approach to communication on this conceptual base and not rely on concepts that are associated with biological and behavioral phenomena (such as signal characteristics, receiver mechanisms, reciprocity, and so forth), but that are by no means the most important parameters involved. In fact, this approach is based upon the use of a highly restricted teleology (see Ayala, 1968).

It is certainly empirically possible to determine whether it is to the signaler's advantage to get his message to another organism.* Obviously it needs to be received before a completed communicatory interchange can occur. Associated with this would be an energy expenditure by the signaler involving overt behavior and evolved morphological characteristics, such as colors, patterns, odors, and the like. Their use by the wrong animal, such as a predator, is irrelevant.† By this approach, we clearly see the blind snake's following of the pheromone trail is *not* communication. It is an unfortunate mistake from the sender's viewpoint. It is similar to tracing a murderer from clues left at the scene of the crime. Certainly the criminal was not attempting to communicate his identity to the authorities (psychoanalytic interpretations notwithstanding). This reliance on the "motivation," "intent," or as I would prefer, the *adaptive significance to the sender (or his group)* alone is intentional; a receiver, such as a predator, can be rewarded for responding to a signal at a distinct disadvantage to the sender.‡ For this reason it is not possible to completely accept the definition advanced by Klopfer and Hatch (1968, p. 32) in which communication "necessitates the existence of a code shared between two or more individuals" and is "mutually beneficial to its possessors." Emphasis on the signaler also neutralizes the problem encountered with the levels approach of having an act communicatory as far as either the signaler or receiver are concerned, but not both. Hockett and Altman (1968)

---

* In some cases, such as alarm cries, the sender might indeed be increasing his chance of capture, which may be maladaptive to the sender as an individual while conferring some advantage upon the signaler's "group," which is, by the way, not necessarily identical with all its species members. Consequently, the approach has room for altruistic behavior, however defined.
† "Intent" has been used in discussions of language as communication (Fearing, 1953). Although the usage in the present sense would include that covered by Fearing, the converse would not necessarily be true. However, his abstract definition would come close to encompassing that implied here. ". . . intent . . . refers to the fact that the act of producing content is *directed* rather than random or aimless, and implicitly or explicitly assumes future effects" (p. 76). Nonetheless, his usage is clearly "psychological" rather than "biological" or "ethological" as used here.
‡ That it might be an advantage to a species to have some of its senders picked off by predators (elimination of the less fit or excess numbers) is a legitimate point (witness the problems with ungulate herds in areas where predators have been eliminated by man). However, this consideration effectively moves the level of discourse to another sphere which is not appropriate here.

in contrast go even further and place all their chips on the receiver: "To count as communicative . . . a particular act of an animal in a setting must be, from the point of view of some receiver, one of a set of two or more acts any one of which *might* occur in just that perceived setting, but each of which precludes all the others" (p. 69). This is, in essence, a sophisticated statement of the information-transfer theory.

In complex behaviors, such as courtship, long chains may be involved where the response of B is simultaneously a stimulus for the original sender A. However, "reciprocity" of this type is not a necessary situation. Human communication, in this approach, does not contradict the above but involves merely the elaboration and extension of processes having the same ends.

The definition also allows us to make certain relevant distinctions that should be made. It is important to keep in mind Cherry's notion (1957) that to tell a man to jump off a bridge involves communication. To push him off the bridge is not communication. Therefore, a stimulus from A which has a direct effect on the responses of B, by dint of the physical or chemical force of the stimuli involved, is not communicating. This is a crucial distinction for the present argument. Obviously all signals, communicatory or not, are received through chemical or physical energy of some kind. However, a stimulus that is not intense enough to elicit the response directly, and yet does so, is qualitatively different from one that elicits the response passively as far as the receiver is concerned (such as a push), or, perhaps even reflexively (such as a startle response to a loud sound). Hockett and Altman (1968) also recognize this distinction and state the problem as follows: "To what extent is the communicative act (the signal) effective in itself, via direct energetic consequences, and to what extent only via triggering?" (p. 68). Another way to phrase the difference is that a communicatory signal is referring to something beyond itself (i.e., a future event) as shaped by prior events. This is implicit in the use of the term information. If this point were clearly emphasized, the distinction between information and stimulus control might be maintained, with the former being stimuli associated with communication as defined here.

Limitations and Extensions

It might be thought that certain instances would pose problems for this distinction. If a person is told to avoid a falling piano, or if he is pushed out of the way, what is the difference? The adaptive value is clear. However, the push itself contained no information. Indeed, he could have been pushed into the path of the descending ivories and associated structures. In neither case was information passed that would elicit a learned or evoked response. This example raises another point. The receiver of the verbal warning would presumably save his life or prevent serious injury by responding appropriately. This is clearly adaptive for the receiver, and since he is a member of the signaler's group, communication in the present sense is involved.

Let us apply the definition of communication to another phenomenon. We have heard recently of chemical interspecific or cross-specific communication where very potent noxious fumes are emitted by certain animals to ward off predators—especially in certain arthropods, as shown by Eisner and Meinwald (1966). Is this communication, as they claim? Is the spray of a skunk chemical communication? By the present approach, these are not examples of communication, but are much more closely akin to the pushing of a person off a bridge. The warning movements of a skunk and its body markings are involved in the communicatory phases of its defensive behavior. The actual defensive behavior which the animal is warning against is not communication. In the example of arthropods, only if we can find some warning stimuli associated with the animal before it gives the spray can we say communication occurs. If we only have the chemical spray, then we do not have chemical communication. Indeed, Eisner and Meinwald do refer to these sprays as weapons. To accept them as also legitimate communication would logically extend the concept of communication to the use of Mace (a chemical spray used by police) and tear gas. Eisner and Meinwald even mention a defensive spray in nasute termites which acts as an alarm pheromone for its own species. Here is an example where exactly the same stimulus can mediate a communicatory act and also serve as a weapon. It is legitimate, however, to ask which came first evolutionarily and which function was derived.

In the psychological learning experiment, such as the lever pressing of a rat in a Skinner box, adaptive behavior is certainly evidenced if the animal "learns" to get food, escape shock, or whatever. The question now arises as to whether the rat is communicating with the experimenter, as implied in the famous cartoon of the talking rats ("Boy, have I got this guy conditioned"). Perhaps experimental psychologists have really been studying animal communication intensively for several decades. If this is true, then what is new about the study of animal communication outside of the use of more species-characteristic signals and behavior? Both a little and a lot.

The rat pressing a bar is responding to an inanimate physical object on a level similar to the rocks and trees mentioned earlier. That a human is programming reinforcement dependencies is irrelevant. It is merely a way of controlling environmental events to an extent unobtainable in nature. Indeed, if some external agent is controlling events in the world at large, every adaptive response of an organism could be considered communication with God. The latter's friends here on earth (or the moon) would object, of course, since the recognized method of communication with the Deity is prayer. Since only selected anecdotal evidence can be adduced for the efficacy of prayer, the advantage of this rapprochement between religion and science is clear. The converse of the argument, however, is that the psychological experimenter is behaving like the God he generally disavows. The direction of this argument, by the way, is a good reason why lever pressing as communication should not be seriously argued in the author's presence!

A more serious claim for communication between experimenter and subject involves the effect of the experimenter bias, where the subject picks up inadvertent cues from the experimenter as to the correct response as in the case of Kluge Hans, the mathematical genius horse (Pfungst, 1911). Clearly this is adaptive for the subject, but not so for the experimenter (qua experimenter), and so communication in the present sense is not involved. In general, an *experimenter* cannot be a communicator unless such communication is part of the problem under study, such as training guide dogs for the blind.

Recently, the study of nonverbal communication in humans has been intensively studied (e.g., Mehrabian, 1969). While many gestural signals are "unconscious" or "unintentional," they can clearly be viewed as communication here. Intent viewed in the present sense as applying to the behavior of all organisms is not to be confused

with human "intentional" or "self-aware" behavior, which is at the present time in-
capable of being studied other than in humans (Verplanck, 1962). The present approach includes both unintentional and intentional behavior as communication if
they meet the proper requirements. Certainly, most persons who speak and write
fluently are not able to state the rules of grammar they are so beautifully obeying.

The last instance to be considered is the *decoy* or *mimic situation,* where one
organism possesses certain structures or performs behavior that characterizes a different
species, with the result that a third species behaves similarly to both. This area has
received a recent excellent treatment by Wickler (1968). It is with these instances
that a definition of communication is really put to the test. By the present definition,
all such examples, including protective camouflage, are communicatory if some advantage accrues to the organism by seeming to be what it is not. The alligator snapping
turtle who waves a worm-like appendage in his mouth to lure unsuspecting fish
certainly has a message to get across, as does the bird who acts as if her wing is broken
to lure the predator away from her nest. However, there is a group of mimicry situations that I think deserve classification as "communicatory mimicry." This occurs only
when the original relationship between the mimicked and the responder is communicatory. There are some fascinating examples of this type of mimicry. Lloyd (1965)
found that predacious female fireflies (*Photuris*) behave like females of a prey species
of firefly (*Photinus*). They mimic the flash responses of *Photinus* females and succeed
in attracting and devouring male *Photinus*. Since a communicatory relationship between male and female *Photinus* is mimicked, communicatory mimicry is involved.
Likewise, the previously mentioned flowers that mimic insect sexual releasers involve
communicatory mimicry. Animals that mimic warning stimuli of other organisms are
also in this category. Another probable example is that of harmless snakes that produce
what often appears to humans (a potential predator) as a good imitation of a rattlesnake by rapidly vibrating their tails. Also, the recent work on intraspecific mimicry
(automimicry) by Brower (1969) raises some interesting theoretical possibilities.

CONCLUSION

The definition tentatively proposed here is as follows: Communication is the
phenomenon of one organism producing a signal that, when responded to by another
organism, confers some advantage (or the statistical probability of it) to the signaler
or his group. I refuse, at this point, to get into further definitional arguments such
as what constitutes the "group." Nonetheless, I think that more thought about communication along these lines would be helpful. Also note that "purpose" in the psychological sense has been purposely ignored. Marler (1961) has adequately pointed
out the vagueness of the concept in relation to communication and has also excellently
discussed the various classes of communication signals and their relationships to human
language. Smith (1969) has made a preliminary classification of animal communicatory
signals.

The definition of communication advanced here is an example of the use of a
functional definition in the study of behavior. Many vexing areas of psychology
might be clarified and synthesized somewhat more satisfactorily if such an approach

were more generally employed. I do not claim originality for its application here since many people have mentioned "adaptive" in discussing communication. For instance, Scott (1968) distinguishes an important class of communication (based on his very general definition discussed above) as involving behavior having a "specialized signaling function" (p. 21). He uses bird song as an example. In the same volume, Bastian (1968) also attempts to deal with the "purpose" of communication by viewing behavior within its ecological context.

I feel that currently there is no possible way to justify scientifically the term "animal communication" unless we are willing to use a *restricted* teleology based upon the adaptive value of the behavior to the sender—based upon "reinforcement" dependencies operating during the early life of the organism, over the preceding generations, or both. Natural selection and conditioning are teleological in that they produce and maintain end-directed behavior—and this is exactly the type of behavior dealt with in communication.

SUMMARY

The problem explored is whether the concept of communication has any scientific validity. Various criteria which have been advanced including stimulus control, information, intraspecificity, reciprocity, and specialization are examined and rejected as uniquely essential. It is concluded that communication can be rescued as a useful term only if "intent" is included as a necessary aspect where intent is viewed in the context of the sender's adaptive behavior based upon prior phylogenetic and ontogenetic events. This notion is an essentially probabilistic functional one. It extends communication to areas such as mimicry where it is not generally applied, while excluding behaviors as noncommunicatory that are impossible to exclude with other approaches. This should enable researchers to concentrate on relevant variables.

REFERENCES

Ayala, F. J. 1968. Biology as an autonomous science. Amer. Sci., 56:207-221.
Bastian, J. 1968. Psychological perspectives. *In* Sebeok, T. A., ed. Animal Communication. Bloomington, Indiana University Press, pp. 572-591.
Batteau, D. W. 1968. The world as source; the world as sink. *In* Freedman, S.J., ed. The Neuropsychology of Spatially Oriented Behavior. Homewood, Ill., Dorsey Press, pp. 197-203.
Bitterman, M. E. 1965. Phyletic differences in learning. Amer. Psychol., 20:396-410.
Brower, L. P. 1969. Ecological chemistry. Sci. Amer., 220:22-29.
Cherry, C. 1957. On Human Communication. New York, John Wiley & Sons, Inc.
Deese, J. 1969. Behavior and fact. Amer. Psychol., 24:515-522.
Diebold, A. R. 1968. Anthropological perspectives. *In* Sebeok, T. A., ed. Animal Communication. Bloomington, Indiana University Press, pp. 525-571.
Eisner, T., and Meinwald, J. 1966. Defensive secretions of arthropods. Science, 153: 1341-1350.
Fearing, F. 1953. Toward a psychological theory of human communication. J. Personality, 22:71-88.

Frings, H., and Frings, M. 1964. Animal Communication. New York, Blaisdell Publishing Co.

Griffin, D. R. 1968. Echolocation and its relevance to communication behavior. *In* Sebeok, T. A., ed. Animal Communication. Bloomington, Indiana University Press, pp. 154-164.

Hockett, C. F., and Altman, S. A. 1968. A note on design features. *In* Sebeok, T. A., ed. Animal Communication. Bloomington, Indiana University Press, pp. 61-72.

Klopfer, P. H., and Hatch, J. J. 1968. Experimental considerations. *In* Sebeok, T. A., ed. Animal Communication. Bloomington, Indiana University Press, pp. 31-43.

Lloyd, J. F. 1965. Aggressive mimicry in *Photuris:* Firefly femmes fatales. Science, 149:653-654.

Lorenz, K. Z. 1965. Evolution and Modification of Behavior. Chicago, University of Chicago Press.

Marler, P. 1961. The logical analysis of animal communication. J. Theoret. Biol., 1:295-317.

——— 1967. Animal communication signals. Science, 157:769-774.

Mehrabian, A. 1969. Significance of posture and position in the communication of attitude and status relationships. Psychol. Bull., 71:359-372.

Pfungst, O. 1911. Clever Hans (The horse of Mr. von Osten). New York, Henry Holt and Company.

Schneirla, T. C. 1953. The concept of levels in the study of social phenomena. *In* Sherif, H., and Sherif, C., eds. Groups in Harmony and Tension. New York, Harper & Row, Publishers, pp. 52-75.

Scott, J. P. 1968. Observation. *In* Sebeok, T. A., ed. Animal Communication. Bloomington, Indiana University Press, pp. 17-30.

Sebeok, T. A. 1965. Animal communication. Science, 147:1006-1014.

——— 1968. Animal Communication. Bloomington, Indiana University Press.

Shannon, C. E., and Weaver, W. 1949. The Mathematical Theory of Communication. Urbana, University of Illinois Press.

Skinner, B. F. 1966. The phylogeny and ontogeny of behavior. Science, 153:1205-1213.

Smith, W. J. 1969. Messages of vertebrate communication. Science, 165:145-150.

Stevens, S. S. 1950. A definition of communication. J. Acoust. Soc. Amer., 22:689-690.

Tavolga, W. N. 1968. Fishes. *In* Sebeok, T. A., ed. Animal Communication. Bloomington, University of Indiana Press, pp. 271-288.

Verplanck, W. S. 1957. A glossary of some terms used in the objective science of behavior. Psychol. Rev., 64 (6), Part 2, pp. 1-42.

——— 1962. Unaware of where's awareness. *In* Erikson, C. W., et al., eds. Behavior and Awareness. Durham, North Carolina, Duke University Press, pp. 130-158.

Watkins, J. F., II, Gehlbach, F. R., and Baldridge, R. S. 1967. Ability of the blind snake. *Leptotyphlops dulcis,* to follow pheromone trails of army ants *Neivamyrmex nigrescens* and *N. opacithorax.* Southwest. Natural., 12:455-462.

Wickler, W. 1968. Mimicry in Plants and Animals. New York, McGraw-Hill Book Company.

Wilson, E. O. 1968. Chemical systems. *In* Sebeok, T. A., ed. Animal Communication. Bloomington, Indiana University Press, pp. 75-102.

3

Purity, Identity, and
Quantification of Pheromones

P. Z. BEDOUKIAN

Tech. Director, Compagnie Parento, Inc.
40 Ashley Road
Hastings-on-Hudson, N. Y. 13076

INTRODUCTION 19
PURITY 20
IDENTITY OF SUBSTANCES 21
STRUCTURAL ISOMERISM 22
CHEMICAL FACTORS IN CHEMORECEPTION STUDIES 30
PHYSICAL FACTORS IN CHEMORECEPTION STUDIES 31
MEASUREMENT OF QUANTITIES IN CHEMORECEPTION STUDIES 31
REFERENCES 33

INTRODUCTION

This chapter attempts to acquaint the reader with some of the pitfalls and difficulties which face the investigator of communication through chemoreception. An awareness of these problems is necessary in order to avoid the possibility of arriving at erroneous conclusions. Results of investigations which do not give details of the methods of purifiction, identification, and measurement of the chemical employed are open to criticism and doubt.

In the realm of chemoreception, we enter a somewhat bizarre area far beyond the usual concepts of purity, identity, or measurement. A "pure" substance is no longer sufficiently pure; positive identity is not assured by the generally accepted methods, nor are the quantities involved measurable by usual laboratory procedures.

Nevertheless, one need not be discouraged from attacking problems of chemoreception because some of the demands are too exacting. Although we may not be able to achieve absolute purity and identity, or measure infinitesimal quantities with absolute exactitude, we can at least show what the limitations are. The important point is to report as completely as possible the methods employed so that others can repeat the work and progress from there.

In this discussion, certain selected odorants will be used as models to illustrate the importance of purity, identity, and measurement of materials used in chemoreception studies.

PURITY

In an absolutely pure substance, every molecule must be identical with every other molecule, but since we cannot normally detect single molecules in a given sample, purity becomes a relative term. For practical reasons, each field of science has its own definition of purity. For example, concepts of purity differ when considering insecticides, catalysts, antibiotics, or odorants. As further advances are made in analytic techniques, standards of purity will undoubtedly become higher.

It is already recognized that in many areas, the elimination of fractional parts per million of impurities produces dramatic results. For example, in the field of electronics, silicon semiconductor technology in the United States could not be established until the impurity problem in silicon was brought to a manageable level. To achieve such limitations of impurities, of course, major efforts over a period of time are required.

After accepting the idea of relative purity in the field of olfaction and chemoreception, one is led to the study of the effects of various types of impurities to determine the suitability of certain chemicals for chemoreception problems.

The following examples may help to illustrate the problem. A sample of phenylethyl alcohol approved by a perfumer may be 97.5 percent pure; a sample he rejected may be of 99.5 percent purity. The perfume industry will rightfully accept the perfumer's choice as against the analytically purer material for reasons that are quite clear. Phenylethyl alcohol is used as an odorant and its olfactory acceptability is of paramount importance. Critical examination of the two samples may reveal that the rejected material, although analytically purer, contained impurities which were highly objectionable to the perfumer. On the other hand, the impurities found in the accepted product, though greater in quantity, were either odorless or blended so well with phenylethyl alcohol that they presented no objectionable features. Such occurrences are common in the fragrance industry.

Sometimes the situation is reversed. A perfumer may select a certain odorant simply because it contains a trace of an impurity which makes the product highly desirable from his point of view. The identity of the impurity, often resulting from certain manufacturing processes, may not have been established yet. The industry is so conscious of the importance of trace components in an odorant that judicious amounts of such components are sometimes added to a product to make it acceptable to the perfumer or to make it more desirable than a competitor's fragrance.

In applying these situations to chemoreception it can readily be seen that the presence of trace impurities may cause the experimenter to draw incorrect conclusions. The possibility that the unknown impurity in the tested material may be the active component should always be taken into consideration.

An instance may be cited where a powerful fly attractant was isolated from a plant. After considerable work, the attractant was purified and identified. It was extremely active and seemed to offer great commercial possibilities, but when the material was synthesized, it was found to be inactive. All the analytic data indicated that the synthetic material was identical with the natural material isolated. It was finally assumed that the natural material contained some trace impurity which accounted for its biologic activity but which was present in too small a quantity to be isolated and identified.

Perfumers can recall many cases of odorants obtained from various companies which could not be distinguished by analysis but which possessed such different olfactory qualities that some could not be used for the purpose intended. Even accepted grades exhibited marked differences. These variations may also be attributed to trace impurities.

The influence of trace components are often very puzzling, and in some cases appear to be due to synergistic effects (Mann, 1967). The importance of trace components is also proven by the fact that when certain materials are subjected to excessive purification, they may lose much of their odor or olfactory qualities. It should be mentioned that such problems can often be overcome. For example, in the case of phenylethyl alcohol, used in large volumes by the perfume industry, difficulties were encountered as new and improved processes were adopted by the industry. It was originally manufactured by the hydrolysis of phenylacetonitrile which was converted to phenylacetic acid whose ester was reduced to phenylethyl alcohol. By applying the Friedel-Crafts reaction, it was found more economic to manufacure this product by the interaction of benzene with ethylene oxide in the presence of aluminum chloride. The perfumer rejected this material because he considered it inferior to the alcohol obtained by the reduction of the phenylacetic ester. A critical examination of its impurities showed that the alcohol made by the new process did not contain traces of isonitrile which remained in the finished phenylethyl alcohol prepared from the phenylacetic ester. Introduction of traces of the nitrile satisfied the perfumers. On the other hand, certain impurities such as dibenzyl, inherent in the Friedel-Crafts process, were gradually eliminated and a superlative product of over 99.9 percent purity was obtained. Even at that stage, however, different cuts from an efficient fractionating column possess distinctly different odors, and it is necessary to combine various fractions to obtain the desired grades of phenylethyl alcohol (Carpenter, 1954, 1957).

The above examples illustrate the role of trace components or impurities in olfactory research, and should serve to alert the investigator of chemoreception studies.

IDENTITY OF SUBSTANCES

Strictly speaking, for two substances to be identical, all the molecules of one must be identical with those of the other. Since this is not in the realm of possibility, we must accept practical limitations imposed upon our concept of identity. For example, in the case of a substance which was isolated from a natural source and then synthesized, we have to consider the natural and synthetic samples identical if all the analytic, chemical, biologic, and olfactory tests fail to distinguish between them (Barnes et al., 1948).

Structural Isomerism

Different substances that have the same molecular formula are said to be isomers. Thus, ethyl alcohol and dimethyl ether both have the molecular formula C_2H_6O, but differ vastly in their properties. In some cases, there may be hundreds of isomers which will correspond to a given molecular formula. If a compound of known formula is to be identified, it is helpful first to characterize its functional groups.

We select as an example a compound whose formula is $C_{10}H_{22}O$. Many isomers have this composition. The number can be drastically reduced if, for instance, it is discovered that the compound is an alcohol. The latter class may include many individual members. If we subsequently find that by replacing the hydroxyl group with a hydrogen we obtain a compound that is a saturated acyclic terpene hydrocarbon, it becomes obvious that we are dealing with a saturated acyclic monoterpene alcohol. We are then restricted to a small number of possible isomers which are of interest to the student of olfaction or chemoreception.

It will be seen that we can have nine (I to IX) different isomeric alcohols meeting our requirements for a saturated acyclic monoterpene alcohol (Fig. 1). These compounds have actually been synthesized and it is not surprising to find that they possess different odors (Houlihan, 1961).

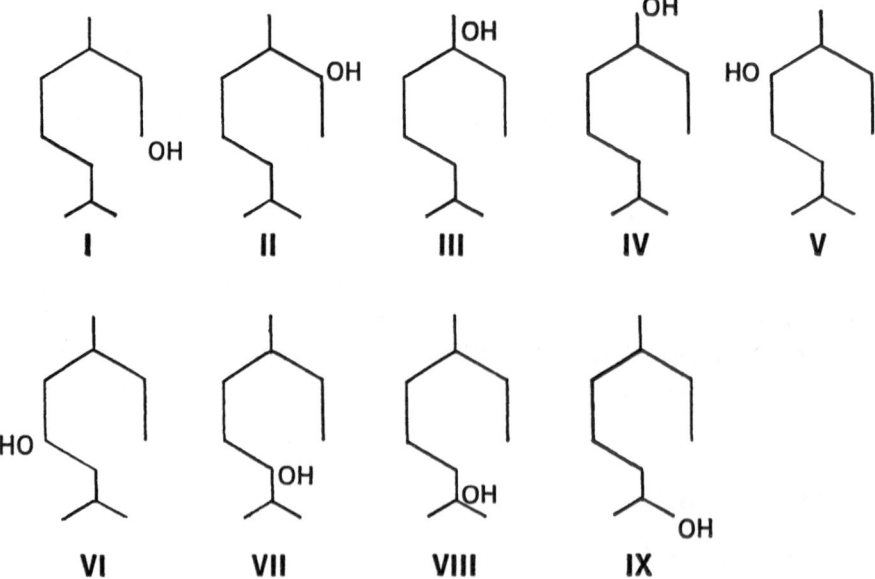

FIG. 1

All the isomers differ from each other in structure and generally show marked differences in their physical and organoleptic properties. We must thus narrow down the differences still further for studies in chemoreception or olfaction.

Geometric isomerism: double bonds

Geraniol can exist in two geometric isomeric forms (X, XI) (Fig. 2). The isomerism is due to the double bond between carbon atoms 2 and 3. No geometric isomerism is possible at carbon atoms 6 and 7 because the double bond is attached to two identical methyl groups. Compounds X (geraniol) and XI (nerol) are widely found in nature, and to date some 250 essential oils have been reported to contain geraniol and nerol (Gildemeister and Hoffmann, 1960).

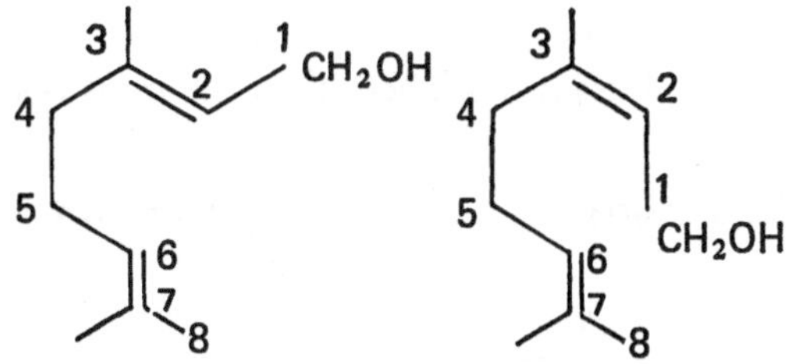

FIG. 3

This type of isomerism is quite common among terpenoids. The two compounds, geraniol and nerol, are notably unlike in all their physical and organoleptic properties. The former has an odor of geranium-rose, and the latter of a citrus-lemony rose.

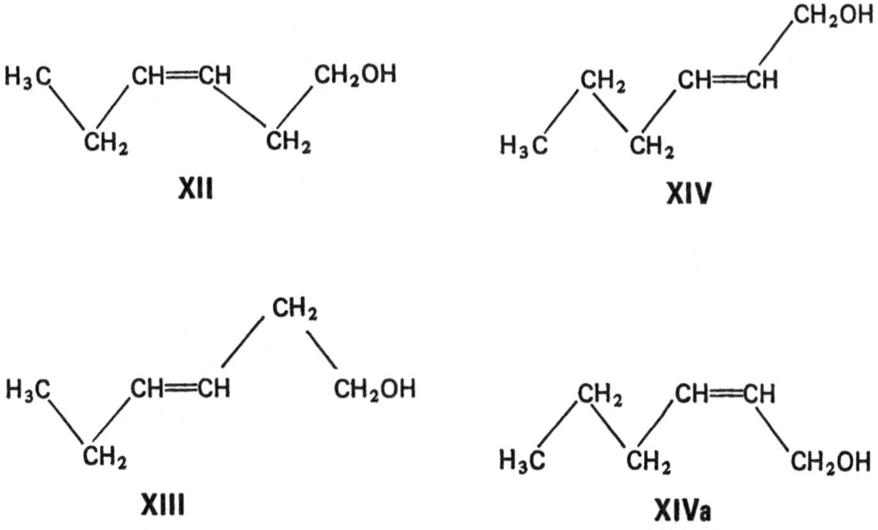

FIG. 2

It is interesting that the two geometric isomers have also been reported to differ markedly in their capacity to attract flies (Hopf, 1968).

Examples of geometric isomerism are legion, and little would be gained by enumerating more than a few representative ones to illustrate these differences.

Cis-3-hexon-1-ol (XII), an unsaturated alcohol occurring in grass and many fruits and vegetables, possesses a fresh, green-leafy odor, whereas the *trans* isomer (XIII) has a somewhat fatty-green, chrysanthemum character (Fig. 3). Experienced perfumers and flavorists can detect as little as 5 percent of the *trans* isomer mixed with the *cis* merely by smelling solutions of the mixture (Crombie and Harper, 1950; Bedoukian, 1963). The related unsaturated alcohol, *trans*-2-hexen-1-ol (XIV), has an entirely different smell. Although the preparation of *cis*-2-hexen-1-ol (XIVa) has not been reported, an impure sample prepared in the writer's laboratory appeared to have a greener note (Bedoukian, 1966).

Wide odor differences have been reported in large ring compounds possessing a double bond. Thus, the naturally occurring *cis*-civetone (XV) is reported (Stoll et al., 1948) to have a more pleasant musk odor than the *trans* isomer (XVI) (Fig. 4).

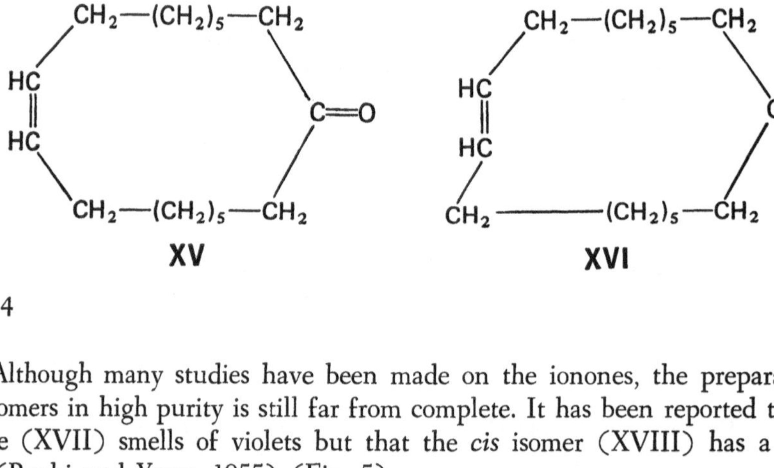

FIG. 4

Although many studies have been made on the ionones, the preparation of all the isomers in high purity is still far from complete. It has been reported that *trans*-α-ionone (XVII) smells of violets but that the *cis* isomer (XVIII) has a cedarwood odor (Buchi and Yang, 1955) (Fig. 5).

FIG. 5

Apparently, no chemoreception studies have been made on the ionones, nor have there been attempts to study the enantiomers of either *cis*- or *trans*-α-ionone. The role of geometric isomerism in chemoreception can be illustrated by the fact that the gypsy moth attractant, 10-acetoxy-1-hydroxy-*cis*-7-hexadecene (XIX) is of the *cis* form and that in contrast, the synthetically prepared *trans* isomer (XX) is a very poor attractant (Jacobson and Jones, 1962) (Fig. 6).

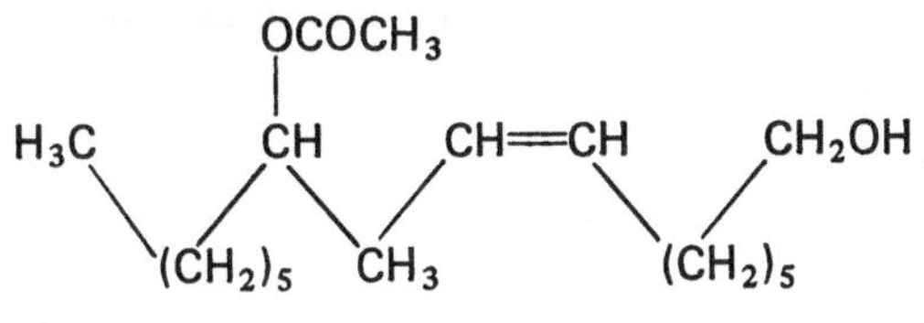

XIX

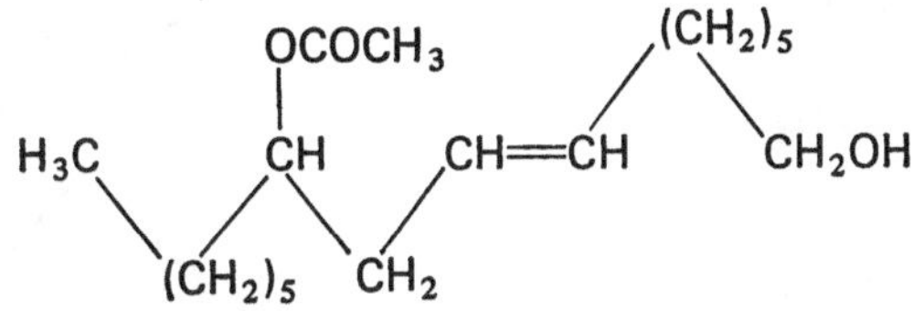

XX

FIG. 6

The position of the double bond is also important. Thus, *cis*- and *trans*-5-dodecen-1-yl acetate (XXI) were found to be less attractive to cabbage loopers than *cis*-7-dodecen-1-yl acetate (XXII) (Warthen and Jacobson, 1968) (Fig. 7).

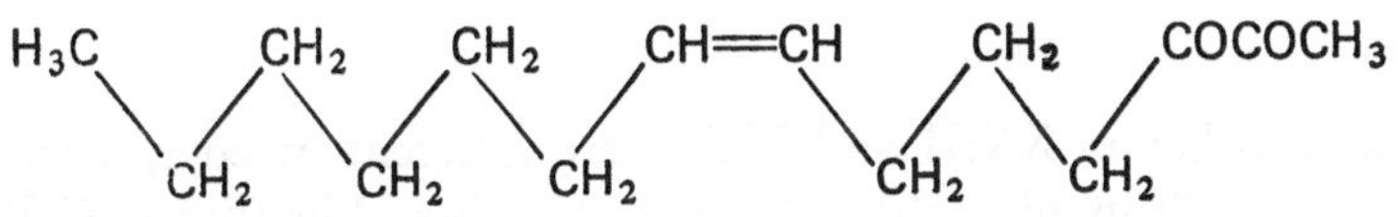

XXI

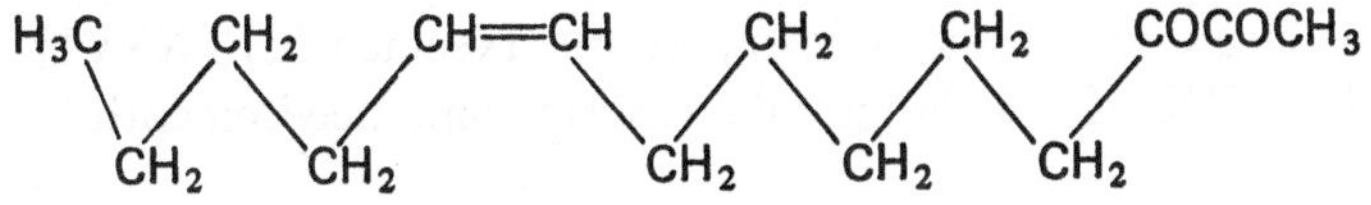

XXII

FIG. 7

The problem of geometric isomerism and its relationship to chemoreception must become more interesting and important in substances having more than one double bond. The realization of this fact will undoubtedly be the cause of more intensive investigations in the future.

Geometric isomerism: rings

Geometric isomerism also results from the different spatial arrangement of substituents in cyclic compounds. The cyclohexyl compounds are perhaps the most common and have been studied in considerable detail. They are particularly important in investigations of natural products as a large number of terpenoids are either cyclohexyl in nature or can easily be converted to cyclohexane derivatives. Furthermore, ring isomerism is of special interest in the study of the relationship of structure to odor. Of the numerous examples described in the literature, the differences in the olfactory characteristics of menthols seem to have attracted the most attention (Fig. 8).

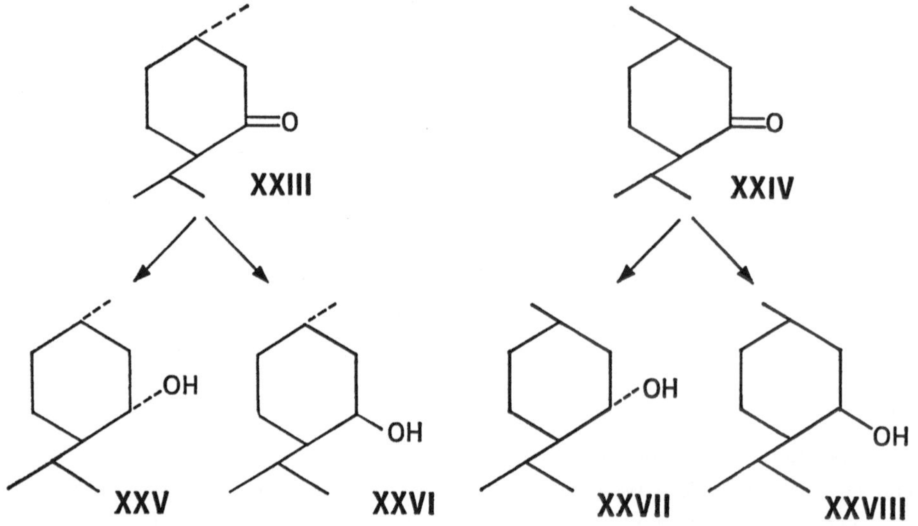

FIG. 8

The two menthones, XXIII and XXIV, are different in odor, although it takes a trained nose to distinguish them from each other. Menthone XXIII is minty, whereas isomenthone XXIV has a sweet, green-minty character. Surprisingly, the alcohols differ from each other much more markedly. For instance, only menthol (XXV) has the clean, minty odor of peppermint. Neomenthol (XXVI) has a musty odor; isomenthol (XXVII) is disagreeably musty, and neo-isomenthol (XXVIII) is

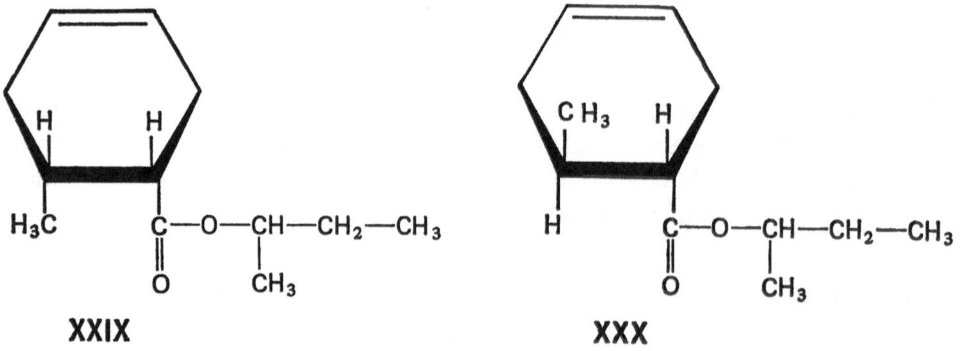

FIG. 9

rather unpleasant. The experienced perfumer can easily detect the presence of any of the other isomers in the proportion of several parts to a thousand of menthol (Bliss and Glass, 1940).

It has been found that insects exhibit distinct differences in their choice of cyclohexanic isomers. In a study of sex attractants for the Mediterranean fruit fly (*Ceratitis capitata*), several thousand compounds were screened and a compound was discovered which proved to be particularly useful even though it bears no relation to naturally occurring pheromones (Beroza et al., 1961). It was also found that the *cis* form (XXIX) of this compound had little value as a sex attractant whereas the *trans* form (XXX) was a powerful attractant (Fig. 9).

The importance of identifying the ring isomer for olfactory and chemoreception studies is thus quite apparent.

Conformational isomerism

The effect of the conformational form on the odor of compounds has not been sufficiently investigated, but there is evidence that the conformational state of a molecule may affect its odor.

It is well known that cyclohexanic compounds may exist in both boat (XXXI) and chair (XXXII, XXXIII) forms (Fig. 10). In the case of cyclohexane, it is pos-

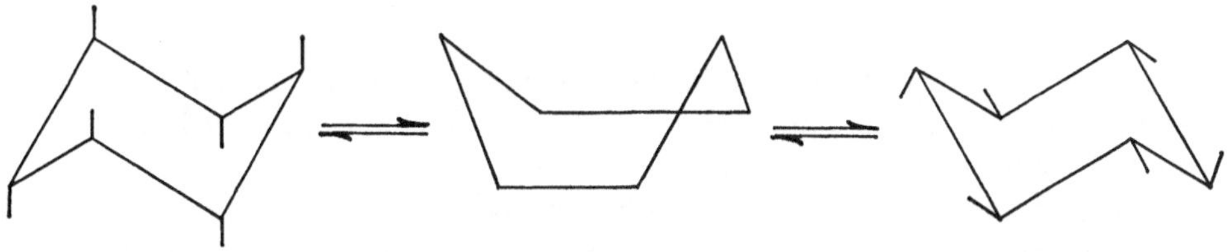

XXXII   XXXI   XXXIII

FIG. 10

sible to detect the interconversion by means of NMR spectral studies at different temperatures. If the activation energy for a molecule to change from one form to another is great enough, the isomers can be separated.

Neomenthol (XXVI) and neo-isomenthol (XXVIII) are known to exist in isomeric conformational forms (Fig. 11). Esterification studies also indicate that neo-isomenthol reacts with the equatorially situated hydroxyl group to the extent of 12 to 20 percent. Neomenthol, on the other hand, reacts with an equatorially situated hydroxyl group to a rather insignificant degree (Winstein and Holnes, 1955).

It would be interesting to choose odorous compounds which have olfactory properties as well as sufficient energy differences in order to isolate the various forms and investigate their odor characteristics.

It would also be very desirable in the future to place intensified emphasis on investigations of the relationship of conformation and odor.

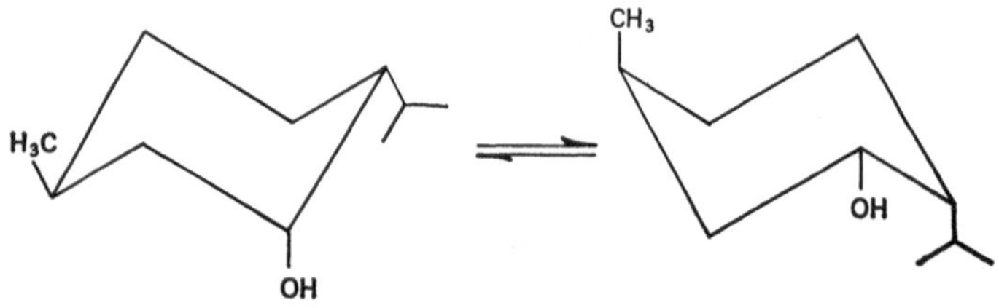

XXVI

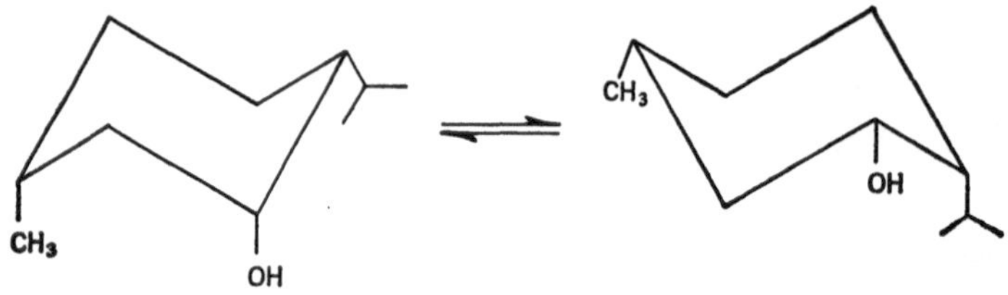

XXVIII

FIG. 11

Isomerism due to asymmetric carbon atom

Olfactory and other biologic differences between the enantiomers of a compound are particularly interesting, and there has been considerable work as well as a certain amount of controversy on this subject.

Although some question exists regarding the odor differences between the *dextro* and *levo* isomers, no instances have been reported where one isomer had an odor and the other had none.

One of the problems in determining olfactory differences between enantiomers is the necessity for carefully purifying the two isomers. Trace impurities affect the end result, often making it impossible to arrive at reliable conclusions. This difficulty is ever present in chemoreception studies of enantiomers.

An early report (Werner and Conrad, 1899) that *dextro*-dimethyl-*trans*-hexahydrophthalate had a strong odor while the *levo* isomer was almost odorless has since been disproved and the difference found to be due to an impurity in the *levo* isomer (Posvic, 1953). The carefully purified enantiomers were olfactorily indistinguishable.

It appeared that *dextro*-2,6-dimethyl octanol had a fresher, stronger odor than its racemic mixture (von Braun and Kaiser, 1923) but further studies revealed that these compounds were not in pure form (Naves, 1946; Naves and Ardizio, 1948) and that the previous results could not be considered conclusive.

It has been noted (Naves, 1957; Langenau, 1968) that no differences in odor between the enantiomers of linalool (XXXIV), citronellal (XXXV), and citronellol (XXXVI) can be detected (Fig. 12).

In preparing naturally occurring terpene alcohols, it is necessary to carry purification processes to the extreme. It is common knowledge among perfumers that the

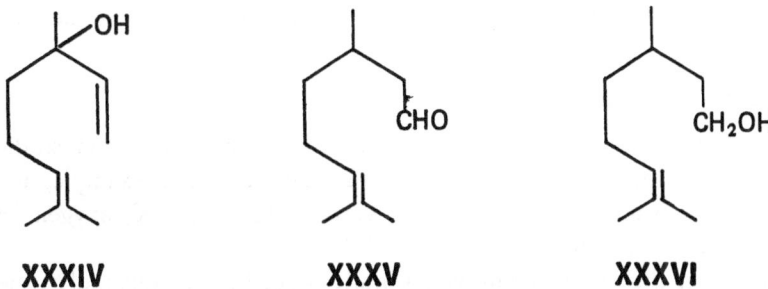

XXXIV   **XXXV**   **XXXVI**

FIG. 12

source of linalool or geraniol is easily determined by merely smelling the compound. Trace impurities which are found in the original source betray the origin of the product. One method of testing the reliability of the purification method employed is to prepare the compounds from different sources and bring them to such a state where they cannot be distinguished from each other by smell. At that point, it would be rewarding to investigate the possibility that some member of the animal kingdom may exhibit selective preferences.

In the cases mentioned above, there is no olfactory difference between the enantiomers, but in other instances, the differences are quite apparent. Among the most intensely studied compounds is menthol. *Levo*-menthol is about three times stronger in odor than *dextro*-menthol (Dell and Bournet, 1948). It is interesting, however, that at their minimal perceptible concentration (20 μg/liter), to twenty times that amount, the odors of the two menthol enantiomers were indistinguishable, whereas beyond that point, the *levo* became stronger. It is also notable that the racemic mixture exhibited an odor strength which was the average of the two enantiomers.

An outstanding example of pronounced differences in odor between enantiomers is carvone. Pure *levo*-carvone has a typical spearmint odor whereas the *dextro* isomer smells of caraway (Langenau, 1968). There are other cases where such differences are found, e.g., in carveols (Johnson and Read, 1934).

It is known that often one isomer is physiologically active whereas the other is not or is much less so. We can therefore expect differences between enantiomers in chemoreception studies. Nevertheless, this may not always be the case since some enantiomers possess identical odors while others are different.

In the case of the gypsy moth, it was found that the *dextro* and racemic sex attractants were equally potent (Jacobson and Read, 1934).

Keto-enol equilibrium

Certain compounds, notably those possessing an active alpha hydrogen adjacent to a carbonyl, tend to undergo enolization. The classic example of this type of tautomerism is the enolization of ethyl acetoacetate.

The olfactory properties of the keto and enol forms of a compound have not been reported to my knowledge and an investigation of the differences in the olfactory and other biologic properties of the two isomers would indeed be interesting.

Other types of isomerism

Other types of isomers which await olfactory or chemoreception studies occur as a result of the presence of isotopes in the molecule. For example, in nature, deuterium occurs in about 1 part in 7,000; carbon ^{13}C, 1 in 100; oxygen ^{18}O, 1 in 500, and nitrogen ^{16}N, 1 in 250.

It is apparent that all compounds, naturally occurring or prepared in the laboratory, contain measurable amounts of isotopes. An asymmetric carbon atom has four different groups, two of which may be oxygen ^{16}O and ^{18}O. The enantiomers may well have olfactory differences. Investigations of olfactory differences of isotopic stereoisomers offer a fruitful field of research. Studies of the biologic activity of such compounds would be of value.

Differences in the olfactory or chemoreception properties of nuclear spin isomers or gauche-anti-conformational forms are as yet beyond the grasp of the researcher.

CHEMICAL FACTORS IN CHEMORECEPTION STUDIES

In reviewing some of the olfactory investigations carried out in the past, one is frequently amazed to note a total disregard of the chemical reactions that may be taking place during the course of the experiments.

When chemicals are evaporated in air so that there is an intimate physical molecular contact with the components of the air, the possibility of a chemical reaction taking place becomes very real.

The reactions under the heading of oxidation may involve a number of processes, sometimes resulting in the creation of a new odorant or perhaps in the destruction of one.

The organic chemist is skeptical of olfactory research carried out with a substance such as benzaldehyde. This aldehyde is prone to such rapid oxidation that if a spilled sample is wiped with a cloth, the heat of oxidation may cause the cloth to catch fire. When this type of substance is evaporated in air, most of it will oxidize to odorless benzoic acid within a very short time. Experimental procedures cannot be meaningful unless this factor is taken into consideration.

In some cases where an unsaturated compound is used for olfactory research, there is the risk that new compounds will be formed because of autoxidation processes. It is well known that in fats peroxide formation at the double bond produces powerful odorants. The presence of mere traces of unsaturated aldehydes thus formed makes a food unacceptable for human consumption, even though the amounts may be so small that they cannot be detected analytically.

The role of moisture in the air requires more study, but there is sufficient evidence to indicate that it does play a part in the acuity of perception of odorants. A compound which undergoes hydration or some related reaction may give unpredictable results if the relative humidity is changed.

Certain chemical compounds are known to react with odorants in a manner that is as yet not clearly understood but apparently involves chemical reactions at surfaces (Kulka, 1965). Air fresheners often act on this principle.

The reactivity of organic compounds at high dilution is another area that needs investigation.

The influence of light should not be ignored when considering the properties of highly diluted compounds. Light-induced reactions encompass a wide field, including cyclization, isomerization, and polymerization.

Physical Factors in Chemoreception Studies

Some physical phenomena are often entirely overlooked in chemoreception studies involving chemicals in high dilution.

The surface of the container possibly poses the most serious problem in high dilution studies. If a picogram of odorant (10^{-12} g) is dissolved in a liter of air, how much of it is actually absorbed or adsorbed by the walls of the container? It is true that some substances adsorb much less than others, but the total effect may be relatively enormous when dealing with very small quantities of materials. Measurement of the relative absorption capacity of such diverse materials as metals, glass, china, and others, have been subjected to preliminary studies (Zwaardemaker, 1926).

The effect of formation of azeotropic mixtures should also not be overlooked. Water forms azeotropic mixtures with a large number of compounds; consequently, the relative humidity on different days may have a marked bearing on experimental results. The formation of binary azeotropic mixtures between two odorous materials may give unpredictable results.

Measurement of Quantities in Chemoreception Studies

When dealing with chemoreception in animals or with human olfaction, the quantities involved in the studies are often so small that none of the available methods of measurement can be applied. The problem can be solved only by mathematical calculations or extrapolations.

In carrying out laboratory experiments in olfaction, it becomes necessary to present the odorant material diluted with air to the subject. Determination of threshold values are particularly susceptible to error; for this reason, much of the research reported in the past is suspect.

In a study by Bach (1937), a known quantity of a substance was evaporated in a measured volume of air, and further dilutions were obtained by mixing a determined volume of this with known volumes of air. The process was repeated until the odor of the substance was just perceptible—that dilution value being the threshold value. The concentrations were expressed as milligrams per cubic meter (about

1,250 g) of air. According to the report, the value for vanillin was 2×10^{-7} mg/m³ air. The meaning of the data is described below:

The molecular weight of vanillin is 152. There are 6×10^{23} molecules per mole (Avogadro's number). Therefore the weight per molecule is:

$$\frac{152 \text{ g/mole}}{6 \times 10^{23} \text{ molecules/mole}} \times 10^3 \text{ mg/g}$$

or about 2×10^{-19} mg/molecule.

There are thus $\frac{2 \times 10^{-7}}{2 \times 10^{-19}}$ or 10^{12} molecules in 2×10^{-7} mg vanillin.

We may assume that for olfaction we need 20 ml per sniff. The number of molecules of vanillin in this 20 ml sniff (at 2×10^{-7} mg/m³ air) is

$$\frac{20 \text{ cm}^3/\text{sniff} \times 10^{12} \text{ molecules/m}^3}{10^6 \text{ cm}^3/\text{m}^3}$$

or 2×10^7 molecules per sniff.

There are 6×23 molecules of air per 22.4 liters.

Therefore, in 20 ml there are 6×10^{20} molecules of air.

In other words, we are smelling 2×10^7 molecules of vanillin in 6×10^{20} molecules of air, or about 1 molecule of vanillin per 6×10^{13} molecules of air.

The method used in determining the threshold value of vanillin can be criticized on the following points:

What were the sources of error in the process of diluting the vanillin?

What about factors such as absorption and adsorption? It is well known that glass adsorbs odorous materials. This may not be a serious consideration when dealing with relatively high concentrations, but at very high dilutions, adsorption may remove most of the material from the air so that the amount of vanillin entering the olfactory organs could be far below the calculated amount.

Other materials can be cited which behave in a similar fashion. For instance, skatole has approximately the same threshold value as vanillin (4×10^{-7} mg/m³ air). Yet it was quite recently reported (Benyon and Saunders, 1960) that zone-refined skatole is free from its characteristic objectionable smell. This claim is disputed by others who have examined zone-refined skatole (Somerville, 1968). The answer may not be simple: in many cases where highly purified products are odorless, an odor develops on standing, probably because of the formation of trace products through the interaction of skatole with oxygen in the air. One is tempted to theorize that in the case of freshly zone-refined skatole, the product is odorless but, on standing, gradually develops the characteristic odor. If this is the case, we may be dealing with a sensitivity far greater than indicated in the threshold value given.

This controlled experiment in olfaction, though just one example, serves to illustrate the type of problems confronting the worker in olfaction or chemoreception.

The gypsy moth (*Porthetria dispar*) presents us with some interesting problems. The female of the species does not fly, and the male is attracted to the female by scent. Marked males have managed to find the female at a distance of over two miles.

As early as 1913, the abdominal sections of two insects were extracted and the extract was shown to attract the male moth. Beroza and co-workers (1960) extracted

500,000 abdominal segments from virgin female gypsy moths and obtained 3.4 mg of a substance containing only 25 percent active material. This material was (+)-acetoxy-1-hydroxy-*cis*-hexadecane; boiling point 160°C at 0.2 torr, molecular weight 300. The above workers state that 10^{-7} μg or 10^{-13} g of this compound was an active attractant.

It is seen that one molecule of this compound weighs:

$$\frac{300 \text{ g/mole}}{6 \times 10^{23} \text{ molecules/mole}}, \text{ or } 5 \times 10^{-22} \text{ g/molecule.}$$

A sphere of one meter radius has a volume of:

$$1 \times \frac{4}{3}\pi r^3 \text{ or } 4 \times (1000)^3 \text{ mm}^3 \text{ or } 4 \times 10^9 \text{ mm}^3.$$

Let us assume that the above perceptible quantity of 10^{-13} g of pheromone is diffused in a sphere of 1 m radius. If an insect contacts 1 mm³ air in that sphere, he will come in contact with $\dfrac{10^{-13} \text{ g/m}^3}{4 \times 10^9 \text{ mm}^3/\text{m}^3}$ or 3×10^{23} g of the pheromone, which is less than a molecule.

Olfactory perception of pheromones at a distance of several kilometers therefore requires exceedingly antientropic dissemination pathways if sufficient concentrations of molecules are to be available to serve as stimuli.

Problems of the same kind and degree must be dealt with when investigating the alarm reaction in fish, migration or homing of fish, and other chemosensory phenomena.

REFERENCES

Bach, H. 1937. Von der Geruchsmessung. *Gesundheits-Ingenieur*, 60:222.

Barnes, R. B., Eyring, H., and Webb, T. J. 1948. Symposium on Purity and Identity of Organic Compound. *Anal. Chem.* (Feb.) p. 96.

Bedoukian, P. 1963. Leaf alcohol. *Amer. Perfumer and Cosmetics*, 78 (12):31.

——— 1966. Unpublished data.

Benyon, J. H., and Saunders, R. A. 1960. Purification of organic materials by zone refining. *Brit. J. Appl. Physics*, 11:128.

Bercza, M., et al. 1960. New attractants for the Mediterranean fruit fly. *J. Agr. Food Chem.*, 9:361.

Bliss, A. R., and Glass, H. B. 1940. A chemical and pharmacological comparison of the menthols. *J. Amer. Pharm. Ass.*, 29:171.

Buchi, G. and Yang, N. C. 1955. Cis-a-Jonon. *Helv. Chim. Acta*, 38:1338.

Carpenter, M. 1954. Industrie Parfumerie, p. 128.

——— 1957. Phenyl ethyl alcohol. Givaudanian (March).

Crombie, L., and Harper, S. H. 1950. Leaf alcohol and the stereochemistry of the cis- and the trans-n-hex-3-en-1-ols. *J. Chem. Soc.* [Org.], p. 874.

Dell, W., and Bournet, K. 1948. Uber den Geruch optischer Antipoden; Ber. von Variochem VVB Schimmel, p. 159. Miltitz bz. Leipzig.

Gildemeister, E., and F. Hoffmann. 1960. Die Atherischen Ole, 4th ed. Berlin, Akademie-Verlag, Vol. IIIa, p. 539 and 568.

Hopf, P. 1968. Personal communication.

Houlihan, W. J. 1961. The synthesis and odour of the 2,6-dimethyl octanols. *Perfumery Essential Oil Record* [London], 52:782.

Jacobson, M. and Jones, W. A. 1962. The synthesis of highly potent gypsy moth sex attractant and some related compounds. *J. Org. Chem.*, 27:2523.

————— Beroza, M., and Jones, W. A. 1960. *Science*, 132:1011.

Johnson, R. G., and Read, J. 1934. Researches in the carvone series. *J. Chem. Soc.* [Org.], p. 235.

Kulka, H. 1965. Novel air fresheners. *Specialties*, 1(14):3.

Langenau, E. E. 1968. Correlation of objective-subjective methods as applied to the perfumery & cosmetic industries. *ASTM Special Technical Publication* (*Philadelphia*), (440):71.

Mann, C. 1967. The role of trace ingredients in perfumes. *The American Society of Perfumers, Proceedings of the Thirteenth Annual Symposium* (April 13), New York.

Naves, Y. R. 1946. Sur l'identification de citronellol en presence de geraniol et de nerol. *Helv. Chim. Acta*, 29:1453.

————— 1957. The Relationship between the Stereochemistry and Odorous Properties of Organic Substances *S.C.I. Monograph No. 1*, London.

————— and Ardizio, P. 1948. Absorption dans l'ultra-violet moyen de derives de terpenes et de sesquiterpenes aliphatiques. *Helv. Chim. Acta*, 31:1246.

Posvic, H. 1953. The odors of optical isomers. *Science*, 118:358.

Somerville, W. 1968. Personal communication.

Stoll, M., Hulstkamp, J., and Rouve, A. 1948. Synthese de la civettone naturelle. *Helv. Chim. Acta*, 31:543.

Von Braun, J., and Kaiser, W. 1923. Geruch und molekulare Asymmetrie; *Ber. Deutschen Chem. Gesellschaft*, 56:2268.

Warthen, D., and Jacobson, M. 1968. 5-Dodecen-1-ol acetates, analogs of the cabbage looper sex attractant. *J. Med. Chem.*, 11 (2):373.

Werner, A., and Conrad, H. E. 1899. Ueber die optisch-activen Transhexahydrophthalsauren; *Ber. Deutschen Chem. Gesellschaft*, 32:3048.

Winstein, S., and Holnes, N. J. 1955. t-Butylcyclohexyl derivatives-quantitative conformation analysis. *J. Amer. Chem. Soc.*, 77:5562.

Zwaardemaker, H. 1926. International Critical Tables 1, 359, Utrecht, Holland.

Chemical Communication in Insects:
Behavioral and Ecologic Aspects

COLIN G. BUTLER

Rothamsted Experimental Station
Harpenden, Herts.
England

INTRODUCTION 35
TRAIL-MARKING PHEROMONES 36
OLFACTORY SEX ATTRACTANTS 37
APHRODISIACS 42
TRAIL-MARKING PHEROMONES OTHER THAN SEX ATTRACTANTS 44
OTHER OLFACTORY MARKERS 47
GUIDANCE MECHANISMS WHEN FOLLOWING AIR-BORNE ODOR TRAILS 50
PHEROMONES AND AGGREGATION 52
ALERTING PHEROMONES 55
CONTROL OF DEVELOPMENT OF SEXUAL MATURITY 57
CONTROL OF QUEEN-REARING IN SOCIAL INSECTS 61
SOME GENERAL CONCLUSIONS 65
ACKNOWLEDGMENT 65
REFERENCES 65

INTRODUCTION

Most insects live solitary lives and communicate with other insects only when seeking mates, courting, and mating.* However, an interesting minority of insects are gregarious for at least part of their lives (e.g., locusts, some ladybirds) and others exhibit various degrees of sociality (e.g., earwigs, ants, bees, wasps, termites), and among some of these communication has developed to a remarkable extent. Indeed, colony life as it exists in many species of ants and termites and in the honeybees, depends on intricate systems of communication by visual, auditory, tactile, and olfac-

---

* The use of chemicals by some insects to repel would-be predators and parasites is not considered as "communication" in the sense adopted in this chapter.

tory and gustatory signals, and combinations of them. The olfactory and gustatory signals—i.e., the chemical signals—alone are considered in this chapter, except when a chemical signal is known to be combined with visual or tactile ones.

The chemical substances involved in insect communication are, of course, pheromones which by definition (see Karlson and Butenandt, 1959; Karlson and Lüscher, 1959) are

> secreted to the outside by an individual and received by a second individual *of the same species*, in which they release a specific reaction, for example, a definite behavior or a developmental process.

Insect pheromones have been classified in several ways by different workers. For example, Karlson (1960) divided them into *olfactorily acting* pheromones, in which he included sex attractants, marking scents, and alerting (alarm) substances, and *orally acting* pheromones, such as the queen substance of the honeybee and those pheromones of termites controlling the production of supplementary reproductives, whereas Wilson (1963a) classified them as *releaser* substances and *primer* substances. Releaser pheromones, which include the olfactory sex attractants, produce an immediate and reversible change in the behavior of the recipient, on whose central nervous system the chemical substance apparently acts more or less directly. Primer pheromones, such as the substances produced by some members of a community that accelerate or retard sexual maturation in other members of the community, trigger off a chain of physiologic changes in the recipient. These important physiologic changes take place without causing immediate changes in behavior. Instead, they equip the insect with a new behavioral repertory, which can henceforth be evoked by appropriate stimuli.

For the purpose of this chapter, I have adopted the simple plan of classifying the insect pheromones according, as far as is known, to their biologic function, but even this is less simple than it might seem, because some pheromones have different functions under different conditions.

TRAIL-MARKING PHEROMONES

By definition (see Karlson and Butenandt, 1959) only those chemical trails that are laid by one individual and perceived by another individual of the same species whose behavior is thereby influenced can be ascribed to pheromones. Host plant and similar odors are thus eliminated and have nothing directly to do with communication, although as Pitman et al. (1968) showed with the scolytid beetles, *Dendroctonus brevicornis* Lec., *D. ponderosae* Hopk., and *D. frontalis* Zimm., swarms of which aggregate on suitable trees, they sometimes seem to act together with and, perhaps, reinforce the attraction of olfactory pheromones produced by those female insects who first discover suitable new host trees. Only trails marked out with olfactory, and possibly sometimes gustatory, pheromones will therefore be considered.

For convenience, chemical trails can be divided into two fairly distinct categories: aerial trails and terrestrial trails.

Aerial trails

This category consists of those trails in which the marking pheromones are windborne and may be perceived some meters downwind of their source. Such trails require continuous renewal and are produced either by flying insects (e.g., a virgin queen honeybee on her nuptial flight) or by stationary or slowly moving (crawling) insects (e.g., a nubile female silkmoth). They include most, perhaps all, olfactory sex attractants (see below) but are not necessarily connected with sexual activity (e.g., scents emitted by a mated queen honeybee when swarming, which enable her workers to maintain contact with her).

Terrestrial trails

This category consists of trails in which the marking pheromones are placed either on the ground or on some other substrate on which the insects are running, such as a limb of a tree or a wall, and extend more or less continuously along the trail (e.g., many ant and termite trails: for ants, see Carthy, 1951; Moser and Blum, 1963; Wilson, 1962; for termites, see Lüscher, 1961a; Stuart, 1961, 1963a), or they are placed on discrete objects, such as stones, posts, twigs or other herbage along the route (e.g., males of some bumblebees and solitary bees: Frank, 1941; Haas, 1946, 1949, 1960; and foragers of some stingless bees, Meliponinae: Kerr, 1960; Lindauer and Kerr, 1958, 1960; Stejskal, 1962). (See also Trail Marking Pheromones Other than Sex Attractants, below.)

OLFACTORY SEX ATTRACTANTS

Some kinds of virgin female moths, mainly members of the families Lasiocampidae, Saturniidae, and Bombycidae, have long been known to release scents that attract males over distances. This phenomenon is certainly much more widespread than has hitherto been realized, especially in species with small population densities, in many of which auditory or visual signals are unlikely to play much part in bringing the sexes together for reproduction. Indeed its frequency is already becoming evident from intensive research engendered by the need to determine the incidence and distribution of insect pests in inaccessible places such as forests, and to augment insecticidal and cultural methods of controlling harmful insects.

The wide distribution of olfactory sex attractants among the various orders of insects is shown in the very abbreviated list in Table 1.

TABLE 1
Some insects in which olfactory sex attractants have been found

Dictyoptera	*Blattella orientalis* (L.)	Roth and Willis (1952)
	B. germanica (L.)	Roth and Willis (1952)
	Nauphoeta cinerea (Olivier)	Roth and Dateo (1966)
	Periplaneta americana (L.)	Wharton, Miller, and Wharton (1954)
Hemiptera homoptera	*Matsucoccus resinosae* (Bean and Godwin)	Doane (1966)
Hemiptera heteroptera	*Belostoma indica* (Vitalis)*† *Dicranocephalus* spp.	Butenandt and Tam (1957) Lansbury (1965)

TABLE 1 (cont.)

Some insects in which olfactory sex attractants have been found

Neuroptera	*Agulla* spp.	Acker (1966)
Mecoptera	*Bittacus* spp.†	Bornemissza (1966)
	Harpobittacus australis (Klug)	Bornemissza (1964)
	H. nigriceps (Selys)	Bornemissza (1964)
Lepidoptera‡	*Alabama argillacea* (Hübner)	Berger (1968)
	Bombyx mori (L.)	Butenandt, Beckman, and Hecker (1961)
	Carpocapsa pomonella (L.)	Barnes, Peterson, and O'Connor (1966)
	Harrisina brillans (B. & McD.)	Barnes, Robinson, and Forbes (1954)
	Heliothis virescens (F.)	Gentry, Lawson, and Hoffman (1964)
	Lobesia botrana (Schiff.)	Chaboussou and Carles (1962)
	Pectinophera gossypiella (Saund.)	Graham and Martin (1963)
	Porthetria dispar (L.)	Collins and Potts (1932)
	Prionoxystus robiniae (Peck)	Solomon and Morris (1966)
	Rhyacionia frustrana (Comstock)	Wray and Farrier (1963)
	Sitotroga cerealella (Olivier)	Keys and Mills (1968)
	Trichoplusia ni (Hübner)	Shorey (1964)
	Zeadiatrea grandiosella (Dyar)	Davis and Henderson (1967)
Diptera nematocera	*Culiseta inornata* (Williston)	Kliever, Muira, Husbands, and Hurst (1966)
Diptera cyclorrhapha	*Drosophila malerkotlina*	Narda (1966)
	Musca domestica (L.)†	Rogoff, Beltz, Johnsen, and Plapp (1964)
Hymenoptera symphyta	*Diprion similis* (Htg.)	Coppel, Casida, and Dautermann (1960)
	Neodiprion pratti pratti (Dyar)	Bobb (1964)
Hymenoptera aculeata§	*Apis cerana* (F.)	Butler, Calam, and Callow (1966)
	A. florea (F.)	Butler, Calam, and Callow (1966)
	A. mellifera (L.)	Gary (1962)
	Eulaema tropica (L.)	Lopez (1963)
	Acanthomyops claviger (Roger)†	Law, Wilson, and McCloskey (1965)
	Lasius neoniger (Emery)†	Law, Wilson, and McCloskey (1965)
	L. alienus (Förster)†	Law, Wilson, and McCloskey (1965)
Coleoptera	*Anthonomus grandis* (Boheman)	Keller, Mitchell, McKibben, and Davich (1964)
	Attagenus megatoma (F.)	Silverstein, Rodin, Burkholder, and Gorman (1967)
	Diabrotica balteata (Le Conte)	Cuthbert and Reid (1964)
	Ips confusus (Le Conte)	Kliefoth, Vite, and Pitman (1963)
	I. ponderosae (Swaine)	Kliefoth, Vite, and Pitman (1963)
	Limonius californicus (Mann.)	Jacobson and Harding (1968)
	Rhopaea spp.	Soo Hoo and Roberts (1965)
	Scolytus spp.	Norris (1965)
	Tenebrio molitor (L.)	Valentine (1931)
	Trogoderma granarium (Everts)	Finger, Stanic, and Shulov (1965)

* Waterborne rather than airborne.
† Olfactory sex attractant probable but not certain.
‡ For various other moths, see: Allen, Kinard, and Jacobson, 1962; Gaston and Shorey, 1964; Hecker, 1959; Jacobson, 1965; Richards, 1927.
§ For bumblebees and various solitary bees, see: Haas, 1946, 1949, 1952, 1960; Kullenberg, 1956; Stein, 1963.

Sex-attractant scents or lures are usually produced by females and attract males exclusively. However, there are a few species in which the usual roles are reversed and the males produce scents that attract the females, e.g., boll weevil, *Anthonomus grandis* (Keller et al., 1964); scorpion flies, *Harpobittacus australis* and *H. nigriceps* (Bornemissza, 1964); swift moth, *Hepialus hectus* L. (Richards, 1927); bark beetle, *Ips confusus* (Wood and Bushing, 1963); cockroach, *Leucophaea maderae* (Engelmann, 1965); ants, *Lasius neoniger*, *L. alienus* and *Acanthomyops claviger* (Law et al., 1965).

Some insects are attracted to, and aggregate on, suitable host plants where they mate, by one or other of the sexes emitting an olfactory pheromone that attracts both sexes. For example, Wood and Bushing (1963) found that males of the wood-boring scolytid beetle *Ips confusus* produce, probably in the hind gut, and excrete, with the frass in their tunnels, an olfactory pheromone that attracts both sexes but especially the females; also Pitman et al. (1968) showed that newly emerged females of the bark beetles *Dendroctonus brevicornis* Lec., *D. frontalis* Zimm., and *D. ponderosae* Hopk., in recently infested trees produce an olfactory pheromone (probably *trans*-verbenol) that acts synergistically with volatile material produced by the host to attract both sexes. A somewhat similar phenomenon occurs in the solitary bee *Andrena flavipes* Panzer, of which not only males patrolling in search of nubile females, but also females seeking nest sites, are attracted by the odor from nest burrows, part of which comes from the females occupying them (Butler, 1965a). Eisner and Kafatos (1962) showed that a scent produced by males of the lycid beetle *Lycus loripes* Chev., attracts both sexes, and Hodek (1960) confirmed the conclusion of Yakhontov (1938) that olfactory pheromones are important in causing males and females of the coccinellid beetle *Semiadalia undecimnotata* L., to aggregate. The attraction of females by the scent of other females of the same species is shown not only by the solitary bee *Andrena flavipes* (Butler, 1965a), but also by gravid females of the desert locust, *Schistocerca gregaria* Forskål, which are attracted to an oviposition site by the scent of other females already settled there (Norris, 1963b). Similarly, gravid females of the wood-boring beetles *Scolytus multistriatus* Marsh., and *S. quadrispinosus* Say, are attracted by the odor of females in galleries in the trunks of trees (Norris, 1965).

Olfactory sex attractants serve to bring male and female insects close enough to see or touch one another, which may need to be very close. For example, the drone honeybee, *Apis mellifera*, has to get within one meter of a queen before he can see her, whereas the queen's olfactory sex attractant lures drones from distances of as much as 60 meters downwind (Butler and Fairey, 1964). The action of olfactory sex attractants is comparable with those of the auditory sex attractants (stridulation) of many Orthoptera and with the visual ones produced by the photogenic organs of fireflies and glowworms (Coleoptera:Lampyridae).

Although many—probably most—olfactory sex attractants are species-specific, there are enough exceptions known to show that such specificity is not essential to prevent indiscriminate cross-mating. Even with closely related species occupying the same general habitat, interspecific mating is unusual in nature (although less so in the artificial conditions of the laboratory), probably because of slight but important differences in courtship behavior. However, even relatively large differences in the shape of the male intromittent organ between one species and another often seem not to prevent successful copulation. The following are examples of lack of specificity of olfactory sex attractant: the sex attractant of the female gypsy moth, *Porthetria dispar* attracts males of *P. dispar* and of the nun moth, *P. monacha* L., and vice versa according to Schneider (1962); but Beroza and Jacobson (1963) state that although the sex attractant of a female *P. dispar* attracts males of both species, that of *P. monacha* attracts only males of its own species. The olfactory sex attractant of the female tobacco moth, *Ephestia elutella* Hübner, not only excites males of her own species, but also those of the Mediterranean flour moth, *E. kühniella* Zeller, and may even make them try to copulate (Beroza and Jacobson, 1963). Similarly, males of the silk moth, *Bombyx mori*, respond slightly to the odors of the female gypsy moth, *Porthetria dispar*, and those of several female saturniids (Schneider, 1963). According to Collins and Weast (1961) intergeneric copulation between silk moths is known but seldom leads to viable offspring. Again, with honeybees, the odors of queens of *Apis cerana* and *A. florea* attract free-flying drones of *A. mellifera*, suggesting that these three species either share the same sex attractant [known to be 9-oxodec-*trans*-2-enoic acid in *A. mellifera* (Butler and Fairey, 1964] or similar ones (Butler, Calam, and Callow, 1967). Nevertheless, interspecific matings are unlikely in regions where two or more species of honeybees coexist, because the drones differ so much in size and anatomy. Interspecific mating is rare between moths whose males respond to the olfactory sex attractants of females of other species living in the same district, and is probably prevented by such factors as: the different species being active at different times of the night, or perhaps flying at different altitudes or in different ecologic sites, or because, even when the male of one species has found a nubile female of another species, he fails to excite her to mate because of unsuitable courtship behavior or lack of the correct aphrodisiac (see Aphrodisiacs below).

Although the olfactory sex attractant of the females of one moth species may stimulate males of their own and of other species, they seem not to affect females of their own or other species (see Schneider, 1963). By contrast, the sex attractant of a queen honeybee, although not stimulating other queens, does stimulate other females—i.e., workers—when presented under the right conditions, such as during swarming (Butler and Simpson, 1967).

Some female insects release their olfactory sex attractants only at specific times of day, which may be a device to prevent interspecific mating between closely related species. Examples of insects known to restrict the release of sex attractant are: the grape vine moth, *Lobesia botrana* to the evening, the moth *Clysiana ambiguella* Hübner to between 2 and 6 a.m., the moth *Sparganothis pilleriana* Schiff. to between about 11 a.m. and 4 p.m. (Beroza and Jacobson, 1963). The same phenomenon is probably true for some African doryline ants whose males fly at different times during the night (Haddow et al., 1966).

It is often several days after female insects emerge that they produce enough olfactory sex attractant to lure males—for example, 10 days for the banded cucumber beetle, *Diabrotica balteata* (Cuthbert and Reid, 1964), three days for cockchafers, *Rhopaea* spp. (Soo Hoo and Roberts, 1965), five to seven days for the queen honeybee (Butler, unpubl.). Also, many female insects stop releasing, and probably producing, the sex attractant either during or soon after mating—e.g., the scarabid beetles *Rhopaea magnicornis* Blackburn, *R. morbillosa* Blackburn, and *R. verreauxi* Blanchard (Soo Hoo and Roberts, 1965) and the Virginia-pine sawfly, *Neodiprion pratti pratti* (Bobb, 1964)—and this is probably what usually happens. However, the queen honeybee, *Apis mellifera*, continues to produce and release the sex attractant throughout the several years of her life (Butler, 1960b) but for purposes other than sexual attraction once mating has occurred—for example, to prevent the production of queens and also of egg-laying by the workers (Butler, Callow, and Johnston, 1961).

Several conditions have probably to be satisfied for male insects to respond to the olfactory sex attractants of their females. For example, honeybee drones are attracted by a queen only when she is flying at a height that depends on the wind velocity (Butler and Fairey, 1964). Drones and nubile queens fly almost exclusively in the afternoons of warm, sunny days, and drones are not attracted by a queen when in their hives or on their alighting-boards, nor, indeed, near the hive entrances. Similarly, males of the solitary bee *Andrena flavipes*, which recognize the female by her orange-colored hind legs, seldom alight beside a female and attempt to copulate with her unless she is within the area of the communal nest site with its associated odor (Butler, 1965a).

Olfactory sex attractants are produced in various glands, mostly abdominal ones. In Lepidoptera these glands are near the tip of the female abdomen (segments 8 and 9) and are derived from membranous intersegmental folds, according to Hecker (1959). They may consist of eversible 'scent rings' (*Cucullia verbasci* L. and *C. argentea* Hufn.), movable 'scent folds' (*Phalera bucephala* L., *Porthetria similis* Foussl., *Plodia interpunctella*) or intersegmental 'scent sacs' (*Aglia tau* L., *Bombyx mori*, *Saturnia pavonia* L.) which are sometimes eversible (*Bombyx mori*). (For details see: Götz, 1951; Hecker, 1959; Jacobson, 1965; Richards, 1927). The male scorpion flies *Harpobittacus nigriceps* and *H. australis* produce their olfactory sex attractants in two pairs of eversible, reddish-brown vesicles between abdominal segments 6–7 and 7–8. When exposed, these vesicles expand and contract slowly and rhythmically, dispersing the musty-smelling scent that attracts the females. The vesicles have no reservoirs and consist, like those of Lepidoptera, of specialized intertergal membrane, the cells of which presumably discharge the secretion directly on to the membrane surface (Bornemissza, 1964).

Blattella germanica, *B. orientalis*, *Periplaneta americana*, and other species of cockroaches have three types of scent glands—anal, tergal and sternal—one or more of which can occur in the same species (Roth and Willis, 1952). According to Stürckow and Bodenstein (1966) the sex odor of *P. americana* is produced by differentiated areas of the integument located on various parts of the body. Females of the scolytid beetle *Dendroctonus pseudotsugae* apparently produce their attractive odor in their Malpighian tubules (Zethner-Møller and Rudinksy, 1967). The sawfly *Diprion similis* produces its sex attractant in glands in the female's abdomen (Coppel et al., 1960), as do the beetles *Diabrotica balteata* (Cuthbert and Reid, 1964), *Rhopaea* spp. (Soo Hoo and Roberts, 1965), and *Tenebrio molitor* (Valentine, 1931). By contrast the honeybee produces it in the paired mandibular glands of the queen (Gary, 1962), which are pear shaped with large, central reservoirs, and lie in the head with their ducts opening at the bases of the mandibles. The queen spreads the mandibular gland secretion over her body, probably when grooming (Butler, 1961).

According to Barth (1961), production of female sex pheromone in the Cuban cockroach, *Byrsotria fumigata*, is controlled by a hormone from the corpora allata, which thus regulates mating behavior. Such hormonal control of the production of sex attractant has not yet been demonstrated in other insects.

In 1927 Richards pointed out that the males of insects that assemble to the female usually have much more complex antennae than the female and these are almost certainly responsible for their scent perception. Later evidence supports this idea. For instance, electrophysiologic methods have shown this to be so in the silk moths *Bombyx mori* (Schneider, 1957), *Antheraea pernyi* Guerin-Meneville, (Schneider et al., 1964), in the honeybee, in which the plate organs (sensilla placodea) of the drone are the actual receptors (Lacher and Schneider, 1963), and in the gypsy moth, *Porthetria dispar* (Schneider, 1963). In the male cockroach *Pycnoscelus surinamensis* L., the antennal segments from number nine onwards bear the sensillae (Roth, 1965). Similarly, in females of the cockroaches *Nauphoeta cinerea* Olivier, *Byrsotria fumigata*, and *Leucophaea maderae*, which respond to the sex attractant odors of the males from some distance away, the antennae are the principal receptor organs concerned, although sensillae on the terminal segments of the maxillary and labial palps also respond slightly (Roth, 1965).

APHRODISIACS

Aphrodisiacs are substances produced by either sex, usually by the male and often as only part of a complex pattern of courtship behavior, that prepare the opposite sex for copulation after the pair have been brought together. Aphrodisiacs usually stimulate the recipient insect through its olfactory organs, as with many butterflies (see Freiling, 1909; Richards, 1927), but can also stimulate its gustatory organs, as with the wood cricket, *Nemobius sylvestris* Bosc. (Gabbutt, 1954) and some cockroaches (Roth and Willis, 1952). They need distinguishing from sex attractants, although some of these also act as aphrodisiacs in some insects. Bringing male and female insects together without the sex-attractant-aphrodisiac being released, can impair or prevent mating. Thus, for example, when a flying drone honeybee is induced to approach a model of a queen honeybee suspended in air free from the olfactory sex attractant, 9-oxo-*trans*-2-decenoic acid, he does not attempt to seize the model even when its sting chamber is open, but with 9-oxodecenoic acid exposed on or near such a model with open sting chamber he will often seize and try to copulate with it (Butler, 1967). Or again, as Riddiford and Williams (1967) showed, the female polyphemus moth, *Antheraea polyphemus*, only releases her sex pheromone in the laboratory when she smells an odor (shown to be that of *trans*-2-hexenal by Riddiford, 1967) that is produced by oak leaves but not by the leaves of several other kinds of trees on which the larvae will feed satisfactorily. Until this pheromone is released the male does not become sexually aroused even in the presence of young females, and males without antennae fail to mate even when they contact receptive females.

Many species of butterflies produce olfactory sex attractants in addition to distinct aphrodisiac scents; the nubile virgin females produce sex attractants whereas males of the same species have aphrodisiac scent glands consisting of special glandular scales or 'androconia.' One of the best known examples is the grayling butterfly, *Hipparchia semele* L.; during courtship the male clasps the antennae of the female between his wings, so bringing them in contact with the scent scales on these. The scent glands of some male butterflies are associated with special brushes used to disperse the scented androconia around the female and stimulate her to copulate (see Freiling, 1909; Richards, 1927). Glands, thought to have a similar function, have been described in Neuroptera, Trichoptera, Diptera and Hymenoptera (see Richards, 1927).

Use of the males' so-called 'hairpencils' by *Danaus* butterflies seem a prerequisite to mating. The males evert their hairpencils and insert them into their wingpockets for a few seconds several times a day. Both their hairpencils and their wingpockets produce scents that stimulate the female (Brower and Jones, 1966). Meinwald and Meinwald (1966) have identified three major components. 2, 3-dihydro-7-methyl-*1H*-pyrrolizin-1-one, cetyl (*n*-hexadecyl) acetate and *cis*-vaccenyl acetate, in the hairpencil secretion of the male danaid butterfly *Lycorea ceres ceres* Cramer, but an odorous trace substance remains unidentified (Meinwald et al., 1966). Courtship behavior of the night-flying moth *Phlogophora meticulosa* L. has recently been studied by Aplin and Birch (1968) who confirmed that when flying males approach a young female in response to her olfactory sex attractant, which she releases only about dawn, they evert their abdominal scent brushes, thereby releasing a strong almond odor around the female with whom they then try to mate. On no occasion was copulation attempted without the brushes first being everted and the scent dispersed; as with the scents of at least two other male noctuid moths (*Leucania impura* Huebner and L. *conigera* Schiff), the major component is benzaldehyde. As all three species are common grass-feeeders and active at the same time of year, crossbreeding is probably prevented by their flying during different periods of the night.

There is a little evidence that species-specific aphrodisiacs of the males help to prevent cross-mating between some species of *Drosophila*. Mayr (1946) found that whereas *Drosophila pseudoobscura* and D. *persimilis* usually will not mate together, they do after removing the females' antennae, which bear the principal receptors of olfactory stimuli. The evidence is far from conclusive, however, because when courtship begins the most important stimuli of female receptiveness are the wing vibrations of the male (Ewing and Manning, 1963), and Mayr (1950) himself produced evidence suggesting that the female's antennae bear the receptors concerned. Indeed, Petit (1959) showed that removing the distal segments of a female's antennae, and thus most of her olfactory sensillae, has a much smaller effect than amputating both antennae completely.

Substances thought to be olfactory and probably also gustatory aphrodisiacs are produced in the dorsal abdominal glands of some male cockroaches—for instance *Blattella germanica* and *Supella supellectilium* Serville (Roth, 1952), *Blatta orien-*

talis (Roth and Willis, 1952), *Nauphoeta cinerea* (Roth, 1962), *Byrsotria fumigata* (Roth and Barth, 1964), *Leucophaea maderae* (Roth, 1965). When a male of one of these cockroaches approaches a female to whom he has been attracted, at least in part, by her olfactory sex attractant, he behaves in a characteristic way, raising his wings and tegmina and so exposing his aphrodisiac gland. The pheromone from this gland stimulates the female who mounts the male and 'feeds' on this glandular secretion just before copulating. Male crickets of the genus *Oecanthus* have metanotal glands said to serve a similar function (Imms, 1957), and in certain American species of crickets of the genus *Nemobius* the proximal internal spine on the hind tibia of the male attracts the female, as does the right tegmen of the male of *Nemobius sylvestris* Bosc. (Gabbutt, 1954). Gabbutt (1954) describes how the female *N. sylvestris* 'feeds' by running her labial palps over the male's tegmen and does not use her mandibles, suggesting that she 'tastes' the secretion rather than 'feeds' on it. Gurney (1947) suggested that, when the female has mounted the male, these glands attract her attention away from the spermatophore so that she does not eat it before the sperm mass has passed into her spermatheca. However, Gabbutt (1954) points out that this seems not to be so with *N. sylvestris* in which the remains of the spermatophore are eaten by the female 30 to 60 minutes after its transference to her genitalia, whether or not she 'fed' on the tegmen of the male.

The complex mating behavior of the male hen flea, *Certatophyllus gallinae* Schrank, is triggered when he accidentally collides with a female and chemoreceptors on his maxillary palps are stimulated by a substance on her abdomen (Humphries, 1967).

Lansbury (1965) described a pair of organs in the males of several species of *Dicranocephalus* (Hemiptera: Heteroptera), similar to an organ in a South African Pentatomid of the genus *Boercias,* which he considers are almost certainly associated with mating. Perhaps they produce aphrodisiacs or olfactory sex attractants.

TRAIL-MARKING PHEROMONES OTHER THAN SEX ATTRACTANTS

Although this group of pheromones includes a few that are used to mark aerial trails—for example, the scent emitted by a queen honeybee when flying with a swarm (Butler and Simpson, 1967)—most are used to mark terrestrial trails and are placed on the ground or other substrate over which the insects are crawling. These trails, which have been most thoroughly investigated with ants, sometimes exhibit polarity based on the shapes of the discrete drops of material of which they are composed: each drop is a narrowing streak pointing in the direction the trail-laying insect is moving (MacGregor, 1948; Carthy, 1951). However, such polarity seems fortuitous and merely reflects the way the drop of pheromone leaves the tip of the trail-laying insect's abdomen when the substrate is momentarily touched with it; at all events the ant *Lasius* (*Dendrolasius*) *fuliginosus* Latreille, which produces well-polarized trails that can be made visible by dusting with lycopodium powder, does not make use of their polarity when orienting (Carthy, 1951). Kerr et al. (1963) suggested that the odor trails of some stingless bees (Meliponidae) are polarized, their odor spots being more intense near the food source, and that foragers use this polarity when orienting, but the evidence is not conclusive.

In general, ant odor trails are made only by successful foragers who lay them during the return journey from food source to nest. An odor trail left by an individual, such as one who has discovered a profitable source of food, is often, indeed usually, overlaid and reinforced by odor trails left by other individuals who later follow the trail from nest to source of food. Thus any polarity of the trail from the shapes of the individual drops of pheromone of which it is composed soon becomes lost. Indeed, a well-used odor trail can quickly become as much as six ants wide, and a wandering ant finding an empty trail and attempting to follow it cannot tell in which direction to turn to reach the goal, but must, of course, on the average take the correct direction on half of such occasions (Carthy, 1952). When a wandering ant finds a trail along which ants are running in both directions, she can probably tell in which direction to turn to reach either the nest or the food because most home-returning ants will be carrying food whereas those running towards the food will have none. An ant leaving the nest cannot go wrong as the trail leads directly from nest to food.

There seems no evidence to support Forel's (1886) suggestion that ants use their antennae as contact-odor calipers to determine the orientation of scent shapes and so find their way along a polarized trail. Most ants seem to use an odor trail simply as a general guide; examining the trail from time to time with one or other of the antennae enables them to follow it. Occasionally, however, an ant will follow a trail very closely, turning her head from side to side and examing the trail first with one antenna and then with the other as she runs along (Carthy, 1951; Vowles, 1955). Indeed, an ant behaves in much the same way as a dog following a scent trail, and whenever the trail is broken she will cast around until she either succeeds in finding the rest of the trail or, failing to do so, abandons her efforts.

The life of an odor trail laid by one individual is usually brief, often remaining perceptible for only a few minutes, e.g., 5 or 6 minutes with the ant *Lasius fuliginosus* (Carthy, 1951), 2 minutes with the fire ant *Solenopsis saevissima* Fr. Smith (Wilson, 1962, 1963a), and 9 to 14 minutes with stingless bees, *Trigona* spp. (Kerr et al., 1963); indeed, their efficiency would obviously be greatly diminished if they persisted after the food supply was exhausted. Because odor trails are usually laid by successful foragers, the intensity of the odor of a trail, and therefore its attraction to foragers, depends on the amount and quality of the food supply; as a food supply and the odor intensity of a trail to it diminish, fewer foragers are attracted. However, well-trodden odor trails can become so well established as to persist for several days, or exceptionally for weeks, and may even withstand rain without further reinforcement, e.g., army ant, *Eciton* spp. (Blum and Portocarrero, 1964).

Regularly used trails often become so well trodden as to be visually distinguishable from their surroundings, e.g., the main foraging tracks radiating from nests of the wood ant, *Formica rufa* L., or may become demarcated with fecal pellets, as by the termite *Nasutitermes corniger* Motschulsky (Stuart, 1961), though an olfactory pheromone from the sternal gland still plays the predominant role (Stuart, 1963a). A crawling insect, such as an ant or termite, probably uses several different senses, separately or together, when following trails, e.g., scent, sight, touch, sun-compass reaction, and orientation by polarized light from a blue sky—the actual sense or senses used depending both on the species and the conditions at the time.

Crawling worker honeybees also lay odor trails, which, in contrast to those of ants, are more or less continuous from their inception; the unidentified pheromone is deposited by the feet although present on most parts of a worker's body (Butler, Fletcher, and Watler, 1969). A trail is laid, for example, when a hive is turned so that a side wall faces in the direction previously faced by the entrance. Home-returning foragers tend to alight on this side wall and run about in all directions until they find the entrance. Even when they have done so, many of them on returning from further foraging expeditions alight on the side wall and run around the corner to the entrance, partly guided by the odor previously left along the route (Lecomte, 1956). A similar odor trail is also laid between syrup dish and hive when honeybees find food in a darkened arena in which they cannot fly (Butler, unpubl.).

The odor streaks of a trail laid by an individual ant or a termite are only slightly separated from one another, and usually do not long remain discrete because other individuals reinforce the trail by adding their own streaks. In contrast, those of such flying insects as the males of bumblebees and of a few solitary bees (Frank, 1941; Haas, 1946, 1949, 1960), and of foragers of some stingless bees, Meliponidae (Kerr, 1960; Lindauer and Kerr, 1958, 1960), remain separated by several centimeters or even meters, as the odor spots are, apparently deliberately, placed on definite objects such as stones, posts, twigs, leaves, and so on, at irregular intervals along the flight trail. The scent-trail pheromones of ants and termites come from the abdomen, e.g., from the gut in the ants *Eciton hamatum* F. (Blum and Portocarrero, 1964), *Myrmelachista ramulorum* Wheeler and *Paratrechina longicornis* Latreille (Blum and Wilson, 1964), and probably also *Lasius fuliginosus* (Carthy, 1951); from the venom gland in the ants *Atta texana* Buckley (Moser and Blum, 1963) and *Tetramorium guineense* F. (Blum and Ross, 1965); from Dufour's gland in the ant *Solenopsis saevissima* (Wilson, 1959); from ventral abdominal glands in several dolichoderine ants (Wilson and Pavan, 1959) and in some termites (Lüscher, 1961a; Stuart, 1961). In contrast those of the male bumblebees, scent-trail pheromones of solitary bees and Meliponidae mentioned above are all produced in the mandibular glands (Haas, 1960; Lindauer and Kerr, 1958, 1960). The substances with which these flying insects mark objects along their trails, as those used by many ants and termites, all seem to be species-specific, in contrast to their alerting (alarm) substances. Indeed, they may be colony specific, as McGregor (1948) found that the foraging ants of one colony of *Myrmica ruginodis* Nyl. pay no attention to the scent trails of others, probably because the trail-marking pheromone of this species comes from the anus and consists of excrement with, or without, added glandular material. This might be expected to be colony-specific, as it is well established that the composite odors of colonies of ants, honeybees, and other social insects are highly specific, enabling them to distinguish between members of their own colonies and intruders from other colonies of the same species. Scent-marking substances from special glands are less likely to be colony-specific—for example, the Nassanoff gland scent of the honeybee, which is equally attractive to bees from other colonies and to those from the scenting bee's own colony (Renner, 1960). Stein (1963) is almost certainly mistaken in supposing that the trail-marking substances of male bumblebees are not species-specific, and his identification of them as farnesol is very doubtful. Although it is true that different bumblebee species tend to lay their odor trails at slightly different heights above the ground and in slightly different places, it seems certain that the trails of the different species would lead to confusion were the trail-marking pheromones not distinct.

The odor trails of Meliponidae, as of ants and termites, are laid only by successful foragers and connect their nests with sources of food.

The males of bumblebees and some solitary bees mark places (objects) along routes that are regularly followed by many individuals of the same species, often from several colonies. These trails are probably of sexual significance, as it has often been suggested that virgin females are attracted to the scent-marked places males are visiting, thereby increasing the chance of nubile males and females meeting, but this does not seem to have been demonstrated.

The odor trails laid by Meliponidae, male bumblebees, and other flying insects by marking stationary objects do not persist for long without reinforcement and need renewing at least once a day. Those of *Trigona* spp. last for an average of 9 to 14 minutes (Kerr et al., 1963).

There are a few records of worker ants regularly using the odor trails of other species (Forel, 1898; Wheeler, 1921; Wilson, 1965b).

OTHER OLFACTORY MARKERS

In addition to such olfactory pheromones as alerting substances, sex attractants, aphrodisiacs, and pheromones encouraging aggregation, there is a less well-defined group of scents used by social insects which attract other individuals either to themselves or to particular places such as sources of food or the entrances to their nests. The marking substances belonging to this group are, by definition, species-specific and sometimes also colony-specific. They are best known in ants and honeybees, and include what Wilson (1965a) calls the 'surface pheromones'— substances that are either absorbed in waxes and other fatty substances on the body surface and released slowly into the atmosphere, or are so nonvolatile as to be perceptible only by contact chemoreception. These are, according to Wilson (1965a), the colony odors, which include the species odors, the caste-recognition scents, the releasers of grooming behavior, and, in some social insects at least, the secretions that stimulate food exchange between individuals. Each of these odors may be composite and consist not only of the products of one or more exocrine glands of the insects concerned but also of other odoriferous substances acquired in various ways. For example, a worker honeybee's body odor—that is, the colony odor shared by all members of her colony (Butler and Free, 1952; Kalmus and Ribbands, 1952)—is probably composed not only of substances from one or more of her own exocrine glands but also of other scents absorbed during contact with other members of the colony, and with combs, larvae, food, and the nest atmosphere (see Butler, 1954a; Buttel-Reepen, 1900; Renner, 1955, 1960). Although there is no doubt that Kalmus and Ribbands (1952) were correct in stating that differences between the odors of different colonies of honeybees largely reflect differences in the proportions of the food these colonies obtain from different kinds of flowers, they were mistaken, as Renner (1955, 1960) showed, in supposing that this is because a colony's odor depends on the aromatic waste products of its common food supply being excreted as a composite, communal scent via the Nassanoff glands of its members. On the contrary, the composite odors of individual colonies of a species probably differ because of differences in the absorbed (acquired) substances, including flower scents, rather than because of differences in the glandular secretions they contain.

These 'surface pheromones,' which are important in colony defense and other aspects of social life, are difficult to study, partly because of difficulties in devising sufficiently sensitive methods of bioassay and partly because of the chemical problems in dealing with very small quantities of mixtures of substances. Although their chemistry is unknown, the biology of a few has been explored. One substance that is left on dishes and other objects crawled over by honeybees occurs on most parts of a worker's body, but especially her thorax, and is deposited by her feet (Butler, 1966a). It is very attractive to foraging honeybees, possibly even more so than the Nassanoff gland scent of the worker, and persists for 10 minutes or more when exposed in the open air during summer (Butler, 1966a). It also seems to be the trail substance of workers and serves the important function of marking the entrances of their hives (Butler, Fletcher, and Watler, 1968). The social wasp *Vespula vulgaris* L. has a nest entrance-marking pheromone chemically distinct from that of the honeybee (Butler et al., 1968), and many, if not all, social insects probably use them. Another surface pheromone, present on the body surface of a virgin queen honeybee, enables workers to distinguish her from a mated, laying queen and stimulates the workers of a colony headed by a mated queen to kill any strange virgin queen introduced into their nest. This pheromone can be wiped from a virgin queen with filter-paper, which will be attacked when offered to the workers of a colony headed by a mated, laying queen. It can also contaminate the body of a worker bee who touches a virgin queen, and such contaminated workers are liable to be killed by other members of the colony (Butler, 1954a). Similarly, distinctive scents probably enable other social insects to distinguish between mated and virgin queens. Perhaps the scents really depend on differences between the metabolism of queens with and without active ovaries, rather than on mating per se, because honeybees (Sakagami, 1958; Velthuis et al., 1965), bumblebees (Free, 1955) and *Polistes* wasps (Pardi, 1948) can similarly discriminate between workers with active and inactive ovaries. According to Velthuis (1967) differences in their surface pheromones also enable queen honeybees to recognize each other as rivals. Other surface pheromones probably enable worker honeybees to distinguish between drone and worker larvae of the same size when both are placed in worker cells, and to treat them accordingly.

One of the best known of the olfactory markers is that released by a worker honeybee when she exposes the surface of her Nassanoff scent gland on the dorsal surface of her abdomen. A worker of *Apis mellifera* (Frisch, 1923) or *A. cerana* ssp. *indica* (Butler, 1954a) often exposes her scent gland in this way when feeding at, or flying around, a dish containing abundant, concentrated, sucrose syrup, thus attracting other foragers flying nearby. However, there are only two records of *A. mellifera* doing so when visiting flowers (Frisch and Rösch, 1926; Free and Racey, 1967), but *A. cerana* workers often do so and *A. florea* F. workers sometimes do so (Butler, 1954a). Workers of *A. mellifera, A. cerana* and probably *A. florea* also expose their Nassanoff glands and, in addition, fan currents of air over them with their wings, thereby dispersing the scent and attracting workers from further away when some obstacle to their normal behavior is removed —for instance, when a searching forager finds the entrance to her hive which has been moved while she was out, or when a worker succeeds in finding the exit of a cage in which she has been imprisoned, or when a worker finds a missing queen (Butler, 1954a; Renner, 1960). Boch and Shearer (1962, 1963) stated that this Nassanoff gland odor was that of geraniol, but they made no biologic tests. Free (1962) found that geraniol was much less attractive to worker honeybees than the natural scent-gland odor. Later, Boch and Shearer (1964) found that Nassanoff gland secretion also contained nerolic and geranic acids and, after field trials, concluded that the odor of a mixture of these three substances was almost as attractive to honeybees as the odor of the natural secretion. However, Weaver et al. (1964) and Butler and Calam (1969) concluded that by far the most important ingredient of this scent is citral and showed that a mixture of synthetic citral (geranial) and geraniol is as attractive as the natural scent. The odors of nerolic and geranic acids seem to be of little or no consequence.

The dorsal abdominal glands of *Myrmica gulosa* and some other ants, which open on the intersegmental membrane between segments six and seven, may be analogous to the Nassanoff glands of honeybees (Cavill and Robertson, 1965). Perhaps their secretion serves a similar function.

The queens of colonies of at least some social insects—for example, honeybees (Butler and Simpson, 1965; Gary, 1961; Velthuis, 1967), and some ants (Stumper, 1956)—have on their bodies, and leave on objects they touch, marking scents that attract their workers; these scents retain their activity when extracted with various organic solvents. None has yet been identified, but Gary (1961) found that in the honeybee, *Apis mellifera*, the bulk of this scent originates in the mandibular glands of the queen. This scent, which attracts worker bees to their queen is distinct from, although it works together with, the odor, or perhaps taste, of another substance produced in a queen's mandibular glands, a fatty acid, 9-hydroxy-*trans*-2-decenoic acid, which stabilizes honeybee swarm clusters (Butler, Callow, and Chapman, 1964).

Renner and Baumann (1964) described other glands opening on the dorsal abdominal segments of queen honeybees and suggested that they may be scent producing. However, their function remains to be demonstrated.

GUIDANCE MECHANISMS WHEN
FOLLOWING AIR-BORNE ODOR TRAILS

The idea that an insect merely has to fly up an odor gradient to reach its source is untenable for various reasons but seems to die hard. Such odor gradients must be steepest in still air, but in still air a flying insect cannot orientate itself towards the source of an attractive odor. When the wind drops while insects are flying towards such a source, they become disorientated and are no longer able to direct their course towards it (for examples see Kellog et al., 1962; Wood and Bushing, 1963; Butler and Fairey, 1964; Soo Hoo and Roberts, 1965). In a wind, even under the most favorable field conditions, a uniform odor gradient cannot exist for more than a few centimeters from the source, and the odor trail soon becomes very irregular with a complex pattern of streamers of different concentrations that frequently change shape and break up (Sutton, 1947, 1949; Wright, 1964b). Also the average concentration in the cone-shaped odor trail downwind of the source changes so slowly that an insect flying up its horizontal axis would have to fly a long way—possibly several hundred meters—before the average scent intensity would change enough to be detected by the insect (Schwinck, 1956; Bossert and Wilson, 1963; Hocking, 1963; Wright, 1964b).

Wright (1958) thought that the irregularity of the scent distribution might be a necessary guiding factor. However, when adult fruit flies, *Drosophila melanogaster* Meigen, were released at the leeward end of a specially constructed wind tunnel (for details of the tunnel see Kellog and Wright, 1962) down which a wind bearing an attractive food odor was moving, the flies flew straight to the source of the odor when the wind was uniformly permeated with scent, but when a trickle of pure air was introduced to the uniformly scented wind, thus creating irregularities in the scent trail, the flies followed very devious courses (Kellog et al., 1962).

Earlier, Schwinck (1954, 1955), working with moths, concluded that perception of an attractive odor causes an insect that is flying about at random with frequent changes of direction to turn into the wind and fly steadily upwind (positive anemotaxis) as long as it continues to receive the olfactory stimuli. When it loses the odor the insect casts about, making frequent turns, until it finds it again when it once more turns and flies upwind. Schwinck further concluded that such a guiding system can operate over distances of several kilometers and can explain reports of moths and other insects being attracted over long distances (for examples see Hecker, 1959). Similarly, Kellog et al. (1962) found that fruitflies fly steadily upwind as long as they can smell an attractive food odor, but within about one-fifth of a second of losing it, turn and make a series of casts across the wind stream. On regaining the scent they promptly turn upwind once more and the the process is continued.

Electrophysiologic studies by Schneider (1963) on olfactory perception by male silk moths, *Bombyx mori,* strongly support the idea that insects reach odor sources simply by flying upwind. Schneider found that the reaction of male silk moths to their females' sex attractant 'bombykol' (isolated and identified by Butenandt, Beckman, and Hecker, 1961) remains almost constant at low concentrations, but as greater concentrations are reached, the intensity of the response increases slowly at first, then very rapidly. It seems, therefore, that although intensity discrimination is excellent at the greater concentrations, at the smaller ones the insect cannot distinguish between different molecular densities. This means that when a male some distance downwind of a female receives just enough scent molecules to excite him, he turns upwind and flies towards the female; he will then continue to be very slightly stimulated and keep flying upwind, until, rather suddenly, when he gets within a few centimeters of the female, the stimulation greatly increases and indicates she is near. Because he cannot discriminate between differences in concentration when amounts are small, local differences in scent concentration caused by turbulence in the airstream are unlikely to deflect him from his upwind course.

Seeking the source of an attractive airborne odor poses less of a problem for a terrestrial insect than for a flying insect. (An aquatic insect seeking the source of a waterborne odor is an much the same situation as a flying insect). For a terrestrial or flying insect to reach the source it must be able to find out from what direction the wind is blowing and so in which direction to proceed. A terrestrial insect has only to turn until it 'feels the wind on its face,' but a flying insect cannot do this because the wind it feels is the result of its own movement. If a flying insect is to direct its flight upwind it must have some way of perceiving its own drift in the wind. Mosquitoes (Kennedy, 1939) and *Drosophila* (Kellog and Wright, 1962) can face directly into the wind only when they can see the ground beneath them. Probably even nocturnal flying insects can see the ground because their eyes are well-adapted to vision in dim light (see Wigglesworth, 1965). Many insects can orientate successfully when flying over water but, at least for honeybees (Heran and Lindauer, 1963), the surface of the water must be disturbed.

Male insects are stated by some workers to have been attracted for long distances by the scent of females of the same species; for example, the silk moth, *Actias silene* Hubner, 11,000 m (Mell, 1922); gypsy moth, *Porthetria dispar*, 3,800 m (Collins and Potts, 1932); ailanthus moth, *Philosamia cynthia* Du., 2,400 m (Hecker, 1959). However, more recent work with other insects indicates much shorter distances, e.g., pine sawfly, *Diprion similis*, 30-60 m (Casida et al., 1963); vine tortricid moth, *Lobesia botrana*, 50 m (Chaboussou and Carles, 1962); banded cucumber beetle, *Diabrotica balteata*, 15 m (Cuthbert and Reid, 1964); pasture cockchafer, *Rhopaea magnicornis*, 27 m (Soo Hoo and Roberts, 1965); honeybee, 60 m (Butler and Fairey, 1964). Although marked male moths have found females several kilometers away from the place where the males were released, probably this was not because they perceived and were guided by the females' scent trails throughout these distances, but rather they fortuitously reached points perhaps 100 m or less downwind of a female where the concentration of her scent molecules reached the threshold value to elicit a response.

The threshold value of an olfactory sex attractant can be very small. For example, male cockroaches, *Periplaneta americana*, respond to 10^{-20}g, or about 30 molecules (based on average molecular weight of 200) of their females' sex attractant (Jacobson and Beroza, 1963); the sex attractant of the gypsy moth, *Porthetria dispar,* is effective at about 10^{-18}g (Beroza and Jacobson, 1963) and that of *Bombyx mori* at about 10^{-18}g (Schwink, 1956). Butler and Fairey (1964) found that drone honeybees were attracted from a distance of no more than 60 m even when a surface area of about 50 cm² carrying several milligrams of sex attractant was exposed. Indeed, within the range 0.25 to 50.0 mg the numbers of drones attracted in unit time differed little and it is doubtful whether they were being attracted from further afield by the larger than by the smaller amounts (Butler unpubl.).

When a drone honeybee reaches a hidden source of sex attractant not associated with a queen or anything he can confuse with one, he overshoots the source of the scent by about 9 m, in a wind of about 5m/sec, before he turns across the wind [much as the fruitflies of Kellog et al. (1962) did on losing the scent of their food] and casts around with many rapid changes of direction until he finds the scent and turns upwind again, or loses it altogether (Butler and Fairey, 1964). It seems likely that, on smelling a queen, a drone flies upwind for about 9 m looking for her (i.e., the distance he overshoots a hidden source of scent) and should he fail to see her, or to receive at least one further scent stimulus, he stops flying upwind and casts around for her scent. Because of his poor visual acuity a drone has to get within 1 m of an object the size of a queen before he can see it. A drone's problem in finding a queen honeybee somewhere in the air greatly exceeds that of male insects whose females secrete sex attractants and remain almost stationary for long periods.

A queen honeybee always starts and finishes her nuptial flight at the same point, her hive or nest. Therefore, whatever she does during her flight, the sum of the distances she moves in various directions is zero relative to the ground. This means that her upwind air times are longer than her downwind ones, so that she leaves longer scent trails when travelling upwind than when travelling downwind. Therefore, when a drone finds a scent trail the queen is more likely to be upwind than downwind of him and he is more likely to get nearer to the queen by turning upwind than by flying at random. This does not mean that the queen is equally likely to be found whatever she does. She is much more likely to be found if she hovers in one place. Therefore, although we know nothing about the flight paths of nubile queens, it seems probable that she will be found to fly around in a circumscribed space as this would make it much easier for the drones to find her.

Pheromones and Aggregation

Many species of insects belonging to several orders form aggregations for various purposes including mutual protection and mating: temporary aggregations (hibernating Coccinellidae, mating swarms of Ephemeroptera, and so on) or persistent aggregations (colonies of social insects). Such aggregations may be composed of insects of either or both sexes, and of one species or, occasionally, of several. Many biologic and physical factors of the environment are concerned both in promoting and in maintaining aggregations. With some species, one or more pheromones are clearly involved, whereas with others none has been demonstrated, and it seems unlikely that any will be.

Temporary aggregations

Many Diptera, especially Nematocera, form temporary aerial swarms, the so-called 'dancing' or 'mating' swarms (Cecidomyiidae, Chiang, 1961; Ceratopogonidae, Downes, 1955; Chironomidae, Gibson, 1945; Culicidae, Corbet, 1964), as do many Ephemeroptera (Speith, 1940) and Formicidae (Wilson, 1963b). Olfactory pheromones probably play a part not only in the genesis of the swarms of some species but also in maintaining their cohesion.

Such swarms often consist only of males, and the females are thought to be attracted by the sight, sound, or scent of a swarm, and on approaching it to be chased and seized by one of the males who mates with her. Although Nielson and Haeger (1960), who watched swarms of male mosquitoes without seeing any females attracted to them, questioned this hypothesis, it is clear that dancing swarms sometimes have a sexual function, either directly, by attracting the opposite sex as in Chironomidae (Gibson, 1945), Ephemeroptera (Speith, 1940), and Tabanidae (Bickle, 1959), or indirectly, by increasing the sexual excitement or otherwise affecting the reproductive physiology of the participants, perhaps by raising their body temperatures as Corbet and Haddow (1962) suggested in some Tabanidae.

Pheromones are also sometimes involved in the formation and maintenance of hibernating and aestivating aggregations of insects. In the Coccinellid beetle *Semiadalia undecimnotata*, for example, an olfactory pheromone produced by the adults, even long-dead ones, plays an important part in causing individuals to congregate in hibernacula where they pair before dispersing (Hodek, 1960). Similarly, an olfactory pheromone produced by males of the lycid beetle *Lycus loxipes* attracts both sexes and causes them to form clusters where they mate (Eisner and Kafatos, 1962). Aggregating pheromones are also produced by females of the bark beetles, *Dendroctonus frontalis*, *D. brevicornis*, and *D. ponderosae*, and by males of *Ips confusus* (Vité and Pitman, 1968). In these *Dendroctonus* spp. the pheromone is probably *trans*-verbenol and it acts synergistically with volatile material from the tissues of the host tree (Pitman et al., 1968). The odors of host degradation products and of a pheromone produced by ovipositing females together cause the hickory bark beetle *Scolytus quadrispinosus* Say to aggregate (Goeden and Norris, 1965). The desert-living tenebrionid beetle *Blaps sulcata* Castlenau is stimulated to aggregate under flat stones by day by an olfactory pheromone from the anal gland; when dilute it is an attractant and when concentrated a defensive secretion (Kaufmann, 1966), thereby having biologic functions resembling those of the so-called 'alarm' (or, preferably in my opinion, 'alerting') pheromones (see Alerting Pheromones, p. 55).

Norris (1963) found that a pheromone, probably one on the body surface, assists female desert locusts, *Schistocerca gregaria*, to aggregate at an oviposition site, and I concluded that one or more pheromones help the solitary bee *Andrena flavipes* to aggregate at a nesting site (Butler, 1965a). Batra (1966) reached a similar conclusion about the primitive social bee *Lasioglossum zephyrum*.

Pheromones are also probably important in maintaining the brief protective relationships between some female subsocial insects and their offspring, for example, the common earwig, *Forficula auricularia* L., and the pentatomid *Elasmosthethus interstinctus* L. Perhaps they also help to maintain the cohesion of 'colonies' of larvae of such insects as the lackey moth, *Malacosoma neustria* L., the processionary moth, *Thaumetopoea processionea* L., and the sawfly *Diprion pini*.

Although a pheromone is clearly sometimes an important, even the most important, factor in promoting and maintaining aggregations of insects for protection, hibernation, aestivation, mating, and oviposition, equally clearly it is usually only one of several factors involved.

Persistent aggregations

Only social ants, bees, wasps, and termites form persistent aggregations or colonies in which the well-being of each individual depends on its campanions. Such colony cohesion is always maintained by several factors, which often include one or more pheromones. These pheromones may be produced by the queen or queens, the workers, brood, and sometimes, perhaps, the males.

Colonies of those insects whose social lives are the most highly developed—honeybees and some ants and termites—all exhibit the phenomenon of extensive, direct, food-sharing between adults (ants, Wilson and Eisner, 1957; honeybees, Delvert-Salleron, 1963; Nixon and Ribbands, 1952; termites, Gösswald and Kloft, 1958). The same phenomenon probably occurs to a lesser extent in social wasps (*Vespula* spp., Montagner and Courtois, 1963; *Polistes* sp., Morimoto, 1960), in stingless, Meliponine bees (*Trigona iridipennis* Fr. Smith, Butler, unpubl.), but apparently not in bumblebees (Free and Butler, 1959), except perhaps through the food stores (Lecomte, 1963). Such a mechanism is ideal for spreading pheromones rapidly through a colony; with honeybees (Butler, 1964) and probably other social insects, it is important in helping to maintain colony cohesion, and allows small quantities of pheromones, obtained by a few individuals from their queen, to be distributed to all the workers.

The maintenance of colony cohesion has been more thoroughly studied with honeybees than other social insects. In a normal honeybee (*Apis mellifera*) colony unidentified olfactory pheromones from the queen's mandibular glands (Butler, 1960a; Gary, 1961; Pain, 1961) and Koschewnikow glands (Butler and Simpson, 1965) are, together perhaps with other pheromones from her mandibular and other glands, responsible for maintaining colony cohesion. Colonies that have lost their queens soon seem to obtain a pheromone, which promotes cohesion and prevents excessive queen rearing, from the larvae the workers are rearing as queens (Butler, 1954b). The cohesion of colonies of ants (Carr, 1962; Stumper, 1956), wasps (Ishay, 1964) and termites (Verron, 1963) also seems to be largely caused by substances obtained from their queens. Stimuli from the larvae also probably have a part to play, as workers will not readily desert them; also an olfactory pheromone from the adult workers themselves, possibly from their Nassanoff glands, is important in stabilizing queenless groups of workers (Free and Butler, 1955).

Simpson (1963) demonstrated that the cohesion of a honeybee (*Apis mellifera*) swarm depends on a pheromone from the head of its queen, later shown to be 9-hydroxy-*trans*-2-decenoic acid (Butler, Callow, and Chapman, 1964), one of more than 20 substances, many of which may be pheromones, present in the mandibular gland secretion of the queen honeybee (Callow et al., 1964).

ALERTING PHEROMONES

When alarmed, many insects release caustic, nauseous, sticky, and other kinds of protective secretions that incommode enemies, such as parasitic or predatory animals. Most of these defensive or warning substances are not pheromones because they do not affect the behavior of other members of the same species. However, some defensive secretions—such as the formic acid, which workers of the ant *Formica rufa* project at enemies—also serve to alert to danger other members of the alarmed individual's own species and are, therefore, pheromones. These pheromones have been called 'alarm' substances (Maschwitz, 1964a, 1964b; Wilson, 1963a, 1965a), but it is perhaps better to call them 'alerting' substances to distinguish them from purely defensive substances.

Alerting pheromones occur in many social insects: they have been demonstrated in at least 21 species of ants, in the honeybee, in the wasps *Vespula germanica* F. and *V. vulgaris* L. (Maschwitz, 1964b), and in some termites (Moore, 1964; Stuart, 1963b); but Maschwitz (1964a) could not find them in the wasp *Polistes dubia* Kohl, in the bumblebees *Bombus lucorum, B. hortorum* L., and *B. hypnorum* L., or in the ants *Ponera coactata* and *Myrmecina graminicola*, all of which have small colonies with relatively few workers compared with those social insects that have alerting pheromones. Therefore, it probably became biologically desirable to supplement visual and contact alerting signals by chemical or other means when large colonies evolved and difficulties of communication between widely separated nestmates increased. Alerting pheromones will also probably be found to play a part in the defensive behavior of those gregarious insects that form large groups, such as the nymphs and adults of locusts.

In all the social insects in which they have been studied, the alerting substances are produced by the workers and, except for the queen honeybee, *Apis mellifera*, also by the female reproductives, never by the males (Maschwitz, 1964b). Like trail substances, they are produced in various glands in different insects— for example, in the mandibular glands of the ants *Pogonomyrmex badius* Latreille (Wilson, 1958) and *Lasius fuliginosus* (Pavan, 1956), *Acanthomyops claviger* Roger (Chadra et al., 1962), *Atta sexdens rubropilosa* Forel (Butenandt, Linzen, and Lindauer, 1959), and of the Meliponine bees, *Trigona tataria* (Kerr and Cruz, 1961) and *Lestrimelitta limao* Fr. Smith (Blum, 1966); in the anal glands of the ants *Iridiomyrmex pruinosus* Mayr, *Monacis bispinosa* Olivier and *Tapinoma sessile* Say (Wilson and Pavan, 1959); in the sting glands (unicellular glands near the base of the sting) of the honeybee, *Apis mellifera* (Ghent and Gary, 1962; Boch et al., 1962); and possibly in the frontal glands of nasute soldiers of some termites (Moore, 1964). It has also been suggested that a scented substance, which accumulates in the mandibular glands of foraging worker honeybees (Simpson, 1960) and was subsequently identified as 2-hepatanone (Shearer and Boch, 1965), acts both as a weak attractant and as an alerting pheromone for other honeybees (Maschwitz, 1964a 1964b; Shearer and Boch, 1965). However, I was unable to confirm either of these suggestions, using either the natural product or synthetic 2-heptanone in a wide range of concentrations, but instead obtained clear evidence supporting Simpson's (1966) conclusion that it repels worker honeybees (Butler, 1966b). Its biologic function therefore remains a matter for conjecture; possibly it is used by bees that have entered the nest of another colony and are robbing a honeycomb to deter the colony's defenders from attacking them; also, perhaps, as Simpson (1966) suggested, to discourage nurse bees from feeding larvae that have just been fed.

Although I think it very doubtful that the 2-heptanone in the mandibular gland secretion of honeybee foragers serves the dual purpose of attracting and alerting workers, some odor trail pheromones of other insects seem to do so, sometimes serving, at small concentration, to attract workers to the site of an attack, instead of the more usual purpose of attracting them to a source of food or to a nest site, and at great concentration to elicit characteristic defensive behavior. This happens both in the fire ant, *Solenopsis saevissima* (Wilson, 1963a) and in the termite *Zootermopsis nevadensis* Hagen (Stuart, 1963b) and possibly in the stingless bee *Trigona tataria* (Kerr and Cruz, 1961). In another stingless bee, *Lestrimelitta limao* Fr. Smith, the mandibular gland secretion contains citral and is released and serves as a recruiting (alerting) pheromone when *L. limao* workers attack the nests of other stingless bees and may perhaps also function as an offensive substance (Blum, 1966). Citral also occurs in the mandibular gland secretion of the myrmecine ant *Atta sexdens rubropilosa* and functions as an alerting substance (see Butenandt, Linzen, and Lindauer, 1959) and in the Nassanoff gland of the worker honeybee, *Apis mellifera*, in which species it serves to attract other workers (see Weaver et al., 1964; Butler and Calam, 1969). A ketone, 4-methyl-3-heptanone, has been identified in the mandibular gland secretion of six species of harvester ants and for one, *Pogonomyrmex barbatus*, it acts as an alerting and recruiting pheromone. It probably has similar functions in the other species (McGurk et al., 1966). Gunnison (1966) obtained gas chromatograms of the liquid fraction of honeybee venom showing at least 13 peaks, one of which was probably isopentyl acetate. Although both isopentyl acetate and the liquid fraction of bee venom greatly excite honeybees, a further stimulus is necessary to elicit stinging.

More alerting pheromones have been isolated and identified than any other class of pheromone. Most are terpenes, with molecular weights smaller than 200; they are very volatile and short lived once they have been discharged, thus being, as Bossert and Wilson (1963) demonstrated, admirably suited to their purpose. For example, Wilson (1963a) points out that when a worker of the harvester ant *Pogonomyrmex badius* Latreille is alarmed, she discharges a puff of alerting pheromone from her mandibular glands. In still air this puff of volatile material expands so rapidly that within 15 seconds it occupies a sphere of air about 6 cm in diameter. The small concentration of pheromone at the periphery of the sphere attracts other ants towards its source, the point of disturbance, where the concentration is enough to elicit defensive behavior often able to repel or overwhelm an adversary. The concentration around its point of discharge soon diminishes and fails to evoke a response of any kind after 35 seconds. If the recruited ants themselves become very alarmed, they, in their turn, discharge more alerting pheromone, thus recruiting further defenders. This mechanism ensures that, although each of the frequent local disturbances to which an ant nest is exposed elicits the necessary response at the appropriate place, the number of ants responding is always determined by the requirements of the individual incident; further, the state of alertness ends as soon as it is no longer necessary. Honeybees, termites, and wasps operate a similar system. However, olfactory alerting pheromones form only a part of the releaser of defensive behavior; color, movement, and possibly other signals are also important, as Free (1961) showed with the honeybee.

The following olfactory alerting pheromones have been identified: citral (ant, *Atta sexdens rubropilosa*, Butenandt, Linzen, and Lindauer, 1959); citronellal (ant, *Acanthomyops claviger*, Chadra et al., 1962); a furan, dendrolasin (ant, *Lasius fuliginosus*, Pavan, 1956); 2-heptanone (ants, *Conomyrma pyramica* Roger, Blum, and Warter, 1966; *Iridomyrmex pruinosus*, Blum et al., 1966); propyl isobutyl ketone (ant, *Tapinoma* sp., Wilson and Pavan, 1959); isopentyl acetate (honeybee, *Apis mellifera*, Boch et al., 1962); possibly α-pinene or another monoterpenoid hydrocarbon (termites, *Nasutitermes exitiosus* Hill, *N. walkeri* Hill, and *N. graveolus* Hill, Moore, 1964).

Alerting pheromones seem less species-specific than sex attractants and other pheromones, perhaps because many were probably defensive or offensive secretions before acquirinng this second function. For instance, the volatile alerting substances produced in the anal glands of any one of the ants *Iridiomyrmex pruinosus, Monacis bispinosa, Liometopum occidentale* Emery, and *Tapinoma sessile* alert the others too (Wilson and Pavan, 1959).

CONTROL OF DEVELOPMENT OF SEXUAL MATURITY

Sexual maturity of some individuals of some gregarious and many social insects seems to be influenced by pheromones produced by other members of their communities. Maturation of some, such as worker honeybees, is retarded or prevented, whereas of others, such as desert locusts, it is accelerated. Sometimes the same species has both retarding and accelerating pheromones.

When an adult male desert locust, *Schistocerca gregaria*, becomes sexually mature, he changes color from brownish-pink to bright yellow (Norris, 1952), and the presence of such males ready to mate in a crowd accelerates maturation in any immature adult male and female (Norris, 1954). Mature females stimulate maturation in others less strongly than do mature males, and the stimulus from the male is probably a scent which is effective over only a short distance. Norris also found that, whereas mature males and females stimulate sexual development in imature specimens, females younger than 8 days old retard male maturation. She concluded, therefore, that this species has both retarding and accelerating pheromones, and crowding has both accelerating and synchronizing effects. Synchronization works both ways, accelerating development of the laggards when most of the individuals in a crowd are approaching sexual maturity, and retarding the potentially precocious when most are still juvenile.

In 1958, Loher showed that the accelerating pheromone of mature desert locusts is a volatile substance covering their bodies. It is perceived over short distances by immature adults of both sexes, by antennal olfactory organs; it causes them to move their antennae, then their maxillary and labial palps, and, finally, to vibrate their hind femora. This behavior was used to measure the activity of extracts of the substance, which was found to be produced in epidermal glands and to be soluble in fat solvents (Loher, 1961). One five-thousandth part of the average amount of this substance obtainable from one mature yellow male at a given time elicited the characteristic response in an immature adult locust. Loher further showed that β-carotene, the yellow pigment of the mature male, is distinct from this pheromone whose identity is still unknown, though 2-choloroethanol elicits the same responses (Carlisle and Ellis, 1964).

A similar pheromone stimulating sexual maturation has not yet been demonstrated in mature males of the African migratory locust, Locusta migratoria R. and F., although it probably exists because the members of a crowd of this species also all reach maturity at about the same time (Norris, 1954).

Removing the corpora allata from immature adult male desert locusts prevents them attaining sexual maturity and producing the accelerating pheromone (Loher, 1961). Indeed, Norris and Pener (1965) showed that allatectomized males and females not only fail to accelerate sexual maturation in normal males but retard it as do very young adults. Allatectomy seems to prevent maturation and to prolong the period while the maturation of others can be inhibited.

Pheromones that help to synchronize sexual maturation can be expected to occur in those insects, such as some ladybirds (Coccinellidae), which sometimes aggregate as immature adults, become sexually mature, and mate before dispersing. Perhaps a single pheromone aids both aggregation and sexual maturation of some species.

Crickets (Acheta domesticus L.) reared in groups grow faster than when alone (Chauvin, 1958), and a pheromone seems responsible for the stimulation (McFarlane, 1966).

Colonies of social Hymenoptera consist essentially of many female insects of which one or, with some species, a few individuals dominate the others and produce most, if not all, of their colony's eggs. The degree of sexual superiority of such dominant, egg-laying individuals (queens) over the non-egg-laying individuals (workers) differs not only between species but even at different times in a given colony of any one species. The females of a colony of the halictine bees Augochlorella persimilis Viereck and A. striata Prov., which are morphologically indistinguishable and all about the same size, often are all sexually mature and several may have mated, but in summer one dominant individual produces her colony's eggs (Ordway,

1965). At the other extreme, the queens of all ants (except a few parasitic forms) and the honeybees *Apis cerana, A. florea,* and *A. mellifera* are morphologically distinct from the workers who cannot mate. There are many intermediates between these two extremes; in some, although the females are morphologically very similar and able to mate, only one, or a few, in each colony do, the rest behaving as workers—for example, the halictine bees (*Lasioglossum imitatum* (Michener and Wille, 1961) and *L. rhytidophorum* Moure (Michener and Lange, 1958).

In some social insects, such as *Polistes* wasps and bumblebees (see Pardi, 1951; Free and Butler, 1959), dominance of one individual over her colony companions seems to be associated with greater development of her ovaries. By confining a queen bumblebee with some workers, Free (1955) showed that the dominant position she attained in the dark interior of a nest is first established by physical violence towards the other potential egg-laying females (workers), who later learn to avoid her when they smell her. This pheromone is apparently associated with the activity of the queen's ovaries.

Similar olfactory/gustatory pheromones have probably evolved in most social Hymenoptera whose colonies always contain a sexually mature, mated queen (or queens), who inhibits sexual maturity developing in the workers. Leaving a colony queenless, or making it difficult for the workers to *touch* the queen causes the workers' ovaries to develop and they lay eggs. Examples of inhibition of ovary development in workers by a queen have been described in the ants, *Formica rufa pratensis* Retz (Bier, 1956), *Dolichoderus quadripunctatus.* L. (Torossian, 1967), *Leptothorax tuberum unifasciatus* F. (Bier, 1954); in the bees, *Apis cerana* (Sakagami, 1954), *A. mellifera* (see reviews by Butler, 1959a; Pain, 1961; Verheijen-Voogd, 1959), *Bombus* spp. (Cumber, 1949), *Lassioglossum* spp. (see Ordway, 1965), *Trigona frieremaiai* Fr. Smith (Sakagami et al., 1963); in the wasps, *Polistes* spp. (Deleurance, 1946), *Vespula* spp. (Imms, 1957). By contrast, morphologically distinct workers with well-developed ovaries are common in colonies with mature, mated queens of some species, such as the ant *Myrmica rubra* (Brian, 1953), and eggs laid by workers seem to be required for feeding brood. They are also common in colonies of several kinds of meliponine bees (Sakagami et al., 1963), of the Cape variety of *Apis mellifera* (Anderson, 1963), and in *Apis cerana* ssp. *japonica* F. (Sakagami, 1954).

Workers' ovaries develop slightly during preparations for swarming by honeybee colonies headed by sexually mature queens (Perepelova, 1928), also when the queen is old and the workers are rearing another queen to supersede her (Butler, unpubl.). They develop even further in *A. mellifera* colonies headed by immature queens, and very readily indeed during preparations both for swarming and for supersedure in colonies of *Apis cerena* ssp. *indica* with sexually mature queens (Butler, unpubl.). Ovaries often develop in workers of *A. cerana* to the stage where they lay eggs.

Unfertilized eggs laid by workers usually produce males, which are often fully functional sexually, but can also produce females—both workers and queens. For instance, eggs laid by the workers of several species of *Camponotus* produce further workers ants (Skaife, 1961). Eggs laid by workers of the Cape variety of *Apis mellifera* also produce workers (Anderson, 1963). Indeed, although the vast majority of eggs laid by workers of most varieties of *A. mellifera* give rise to males, a few produce females (Mackensen, 1943).

Only from the honeybee has a pheromone that inhibits development of workers' ovaries been isolated and identified. Pheromones produced by mature queens of some other social insects may be related to it, for I found that extracts of mature queens of the honeybees, *Apis cerana* and *A. florea,* of a termite, *Odontotermes* sp., and of an ant, *Formica fusca,* contained substances able to inhibit oogenesis in queenless workers of *Apis mellifera* in a way similar to an extract of a mature queen of this species (Butler, 1965). By contrast, extracts of mature queens of the ant *Myrmica rubra,* the social wasp *Vespula germanica,* and the bumblebees *Bombus terrestris* and *B. pratorum* had no such effect.

Oogenesis in worker honeybees, *Apis mellifera,* is usually inhibted by one or more pheromones (Butler, 1957a; Pain, 1961; Voogd, 1955) produced in their queens' mandibular glands (Butler and Simpson, 1958; Butler, 1959b). Secretion from these glands is spread by the queen over her body when grooming, so becoming available to her workers, some of whom lick it from her and share it in regurgitated food with other members of their colony (Butler, 1956). Two substances in the queen's mandibular gland secretion seem responsible—a keto acid, 9-oxo-*trans*-2-decenoic acid, and an unidentified scented substance (Butler, Callow, and Johnston, 1961; Butler and Fairey, 1963), the two together constituting the "queen substance" originally postulated by Butler (1954a, 1954b). My colleagues and I have obtained considerable, but incomplete, inhibition with a synthetic preparation of 9-oxodecenoic acid alone (Butler, Callow, and Johnston, 1961; Butler and Fairey, 1963), which Pain (1961) was unable to do, claiming that both the acid (her pheromone 1) and a scent (her pheromone 2) must be presented together to produce any effect. The two pheromones—the 9-oxodecenoic acid and the scented substance, which may well be 9-hydroxy-*trans*-2-decenoic acid (see Butler and Callow, 1968)— do not act synergistically and, even when presented together, are less effective than a live queen (Butler and Fairey, 1963). Possibly another factor, such as another kind of signal or better distribution of the inhibitory materials by a moving queen, is also involved.

How pheromones from the queen inhibit oogenesis in workers is uncertain. To be effective worker honeybees and ants must touch the queen and, apparently, lick her, not merely smell her (see Bier, 1954; Butler, 1954b). Experiments in which worker ants (Bier, 1954) and worker honeybees (Butler, 1954b) were taken from groups of workers with mature queens and given to groups of workers without queens, show that inhibition can be carried by individuals who have recently been with a queen. Also, inhibitory 'queen substance' recovered from the crops of worker honeybees after they have licked a queen, was biologically active when fed to queenless workers (Butler, 1956).

Queen substance ingested by a worker directly may inhibit development of her ovaries, or it may stimulate chemoreceptors and act through the brain, exerting control by means of an endocrine secretion. Injection experiments with 9-oxodecenoic acid have failed to give a definite answer (Butler and Fairey, 1963).

In a queenless honeybee colony, the ovaries of many, but not all, the workers develop (Butler, 1957a; Voogd, 1955), whatever their ages (Butler, 1959b). Those workers whose ovaries develop first probably inhibit ovaries of the others from developing. However, Pain and Barbier (1960), Pain, Barbier and Roger (1967), and I (unpubl.) failed to demonstrate that sexually mature (laying) workers produce an inhibitory pheromone. Velthuis et al., (1965) claim to have done so, but we (Butler and Chapman, unpubl.) were neither able to confirm the biologic results of their experiments nor to find any trace of 9-oxodecenoic acid in workers from queenless colonies caught directly after they had laid eggs and some of whom were apparently being treated by other workers as workers treat mature queens. However, workers from colonies with mature queens contained 9-oxodecenoic acid, presumably obtained from their queens.

CONTROL OF QUEEN-REARING IN SOCIAL INSECTS

The control of queen rearing in colonies of social insects has received much attention during the last few years, stimulated perhaps by the concept of control by pheromones.

Queen rearing in colonies of social insects takes two forms: first, the production at a given time of several, often many, queens for colony reproduction, and, second, the production of one, or a few, queens per colony to replace those that have been lost or have become ineffective. In the first type of reproduction, colony reproduction, many males are usually produced at the same time as the queens, or shortly before. Because at any moment the colonies of a given species differ in their degree of development, males are often being produced in some colonies when queens are being produced in others, so lessening the chances of inbreeding.

Colonies of all social insects, whether they last for only one season (e.g., bumblebees and wasps) or are more or less perennial (e.g., ants, honeybees, meliponine bees, and termites), practice colony reproduction; but only those whose colonies are perennial replace lost or failing queens, by rearing replacement queens (honeybees), or 'supplementary reproductives' (termites), or by accepting into their nests (some ants) newly mated queens, reared for colony reproduction, either by their own or by strange colonies. This last method of acquiring a new queen to replace a lost one is also common in annual colonies. For example, colonies of *Bombus* and *Vespula* that lose their queens early in the season, may adopt overwintered mated queens who try to enter their nests. The entrance of a queen is vigorously resisted by the workers of any colony with a mated queen.

Loss of a queen by an annual colony of *Bombus* or *Vespula* has been said to expedite the production of males and queens (Imms, 1957). From discussion with other observers and from my own experience, I think this is very doubtful. If it were true, it would presumably be because loss of the queen and the eggs she lays soon diminishes the number of larvae requiring food and allows more food to be given to some of the remainder, thus producing royal forms. However, even if queen rearing is accelerated in this way in some annual colonies, the queens produced do not replace the lost ones and head their colonies, in direct contrast to what happens in the perennial colonies of honeybees.

When a honeybee colony loses its queen the workers soon modify one or more worker cells containing young female larvae or, more rarely, eggs into emergency queen cells. They start to rear a new queen while they still have the means (female larvae) to do so. Why a few larvae, out of many of the same age and sex, are chosen in preference to others is not known but may be because the chosen cells contain more of a pheromone from the queen (Butler, unpubl.). Experiments in which honeybee queens were separated from their workers to various extents strongly suggested that the workers of a colony obtain something from their queen which can inhibit them from rearing queens (Butler, 1954b). Later, I showed that this inhibitory 'queen substance' is present on all parts of a queen's body and that the workers obtain it by licking her and share it widely with other workers in regurgitated food (Butler, 1956). Extracts of queen substance can be obtained from queens by extraction in ethanol or other organic solvents and will inhibit queen rearing when given in suspension in water and in other ways (Butler and Gibbons, 1958; Butler et al., 1961).* The inhibitory material is scarce in newly emerged queens (Pain, Hugel, and Barbier, 1960), but increases rapidly during the first week of life (Butler and Paton, 1962), and usually remains plentiful until the queens are old (Butler and Paton, 1962; Pain, Barbier, Bogdanovsky, and Lederer, 1962). It is produced in the queen's mandibular glands (Butler and Simpson, 1958), the most abundant component of the secretion of which is 9-oxo-*trans*-2-decenoic acid which was synthesized and shown to be largely responsible for inhibiting queen rearing (about 0.13 μg/bee is required) (Butler et al., 1961). The rest of the inhibition seems to be caused by an unidentified, volatile 'inhibitory scent' which acts synergistically with the fatty acid (Butler, Callow, and Johnston, 1961; Butler, 1961). This 'inhibitory scent' is produced not only in the queen's mandibular glands, but in other glands, perhaps in many dermal ones. It has recently been identified as 9-hydroxy-*trans*-2-decenoic acid (Butler and Callow, 1968). Continued production of this pheromone (possibly even compensatory production of it), which is found all over a queen's body, may explain the discovery by Gary and Morse (1962) that queens whose mandibular glands have been removed retain some inhibitory power.

Identification of 9-oxo-*trans*-2-decenoic acid as the principal pheromone inhibiting queen rearing has been confirmed by several workers, including Barbier and Lederer (1960) and Barbier et al. (1960), and the *cis* isomer has also been shown to be active (Barbier and Hügel, 1961).

Sometimes, even though it has a mated, laying queen, a honeybee colony begins rearing new queens. The rearing of such queens may either be preparatory to superseding the old queen whose output of 'queen substance' (9-oxodecenoic acid + 9-hydroxydecenoic acid) has become deficient, or preparatory to colony reproduction (swarming). In the latter case, the queen may either be producing insufficient 'queen

---

* Pain (1961) is incorrect in saying that I found it necessary in my bioassays of queen substance to add sucrose syrup to make it attractive to workers and effective in inhibiting queen rearing by them.

substance' to inhibit queen rearing, as in a case of supersedure, or, more frequently, although she is producing enough 'queen substance,' it is not reaching the workers in the overcrowded hive; perhaps because jostling occurs and prevents workers taking 'queen substance' from the queen and sharing it with other workers (Butler, 1960b). A queen whose production of 'queen substance' is deficient and who swarms with part of her old colony and founds a new one elsewhere is soon afterwards superseded by a new queen that is reared because of her failure to produce enough inhibitory material. Why reproductive swarming is sometimes associated with queen supersedure and other times is not, remains to be discovered. Queen cells can also appear—but are seldom tolerated for long—when the queen is widely separated from groups of bees in other parts of the hive and movement of bees from the group near the queen and peripheral groups is impeded, perhaps by a 'queen-excluder.' In summer some colonies with apparently normal queens will even tolerate, for a few days, occupied queen cells placed experimentally in the middle of their brood nests. In 1959 Simpson and I (unpubl.), arguing that those colonies that accepted queen cells were deficient in queen substance, attempted, without success, to use this behavior to measure the queen substance deficit, if any, of each of 100 colonies once a week throughout the summer. It seems more difficult to induce worker honeybees to destroy occupied queen cells than to prevent workers starting to rear queens.

The way in which 9-oxodecenoic acid inhiibts worker honeybees from rearing queens is uncertain. The fatty acid is probably appreciated by a bee's chemoreceptors and works through the brain, rather than directly on the recipient's metabolism, because injection with this substance has no apparent effect (Butler and Fairey, 1963). Johnston et al. (1965) showed that this fatty acid is metabolized quickly enough in a worker's gut to account for the manifestation of queenless behavior by workers separated from a queen for a short time. They also proposed the interesting, but unproven, hypothesis of a pheromone cycle in honeybees—that the workers receive 9-oxodecenoic acid from the queen and convert it into an inactive form with the same carbon chain. They supposed that the inactive form may then be transferred to the workers' mandibular glands and passed back to the queen in food. The queen could then convert it back into the active form by a simple enzymic reaction. Such a scheme would save the large amount of energy required for the complete synthesis of the fatty acid chain.

Termite colonies, in contrast to those of other social insects, contain a royal pair (king and queen), the primary reproductives. When these die, or are removed, they are replaced in a few days by secondary, or supplementary, reproductives, which develop directly from the larvae or nymphs, in which both sexes are represented in approximately equal numbers. The drywood termite, *Kalotermes flavicollis* Fabr., which has been considerably studied, has no fixed worker caste, and the full-grown larvae function as workers, which are therefore called pseudoworkers or pseudergates. Pseudergates usually remain such for a long time, but can, when the primary reproductives die, develop directly into supplementry reproductives.

Lüscher (1956) showed that provided a male of *Kalotermes flavicollis* is present, the pseudergates of a stable colony have only to receive the excrement of a queen for all to remain inhibited from developing into supplementary reproductives. The pheromone concerned is contained in the queen's excrement, which is collected by, and shared among, the pseudergates. An individual male primary reproductive exerts no measurable inhibitory effect; two produce a significant effect; but an individual female primary reproductive does inhibit the production of female supplementaries to some extent. Male and female reproductives must be present together for complete inhibition. These results suggest that each sex of a reproductive pair produces a sex-specific substance that inhibits the development into supplementary reproductives of pseudergates of the same sex. Further, Lüscher (1956) suggested that the primary reproductive of each sex stimulates the production of inhibitory pheromone by the opposite sex. Lüscher also showed that the male primary reproductive produces another pheromone which, in the absence of a female reproductive, stimulates production of female supplementaries. None of these pheromones has yet been isolated and identified. But it is interesting to note that Hrdý, et al. 1960) showed that extracts of queen honeybees (queen substance) largely inhibited the development of supplementary reproductives when fed to queenless pseudergates of *K. flavicollis*. Similarly, I have found that extracts of a mature queen of a termite, *Odontotermes* sp., inhibited queen rearing by queenless workers of *Apis mellifera* (Butler, 1965b). Possibly, therefore, pheromones of somewhat similar composition are involved in termites and honeybees.

In 1944, Light, working with termites of the genus *Zootermopsis*, produced similar evidence to that obtained by Lüscher with *Kalotermes flavicollis* that a pheromone inhibiting the development of female supplementary reproductives is present in extracts of the heads and thoraces, but not of other parts, of primary female reproductives. He was disappointed, however, that he could not obtain complete inhibition; but here, as in *Kalotermes*, it now seems likely that a pheromone from the male of the royal pair is also necessary for complete inhibition.

Lüscher (1960) produced evidence that these pheromones controlling the development of supplementary reproductives in termites work by interfering with the endocrine system of the pseudergates.

When too many supplementary reproductives are produced, other sex-specific pheromones seem to ensure their destruction by pseudergates; male reproductives induce elimination of males, females of females. These pheromones may come from dermal glands scattered over the body and probably stimulate chemoreceptors on the antennae of the pseudergates (Lüscher, 1961b).

There is also evidence that pheromones control the production of other castes, such as soldiers, in termite colonies (Light, 1943).

Evidence of the control of queen rearing in the colonies of other social insects is much less than in honeybees and termites. The possibility does not seem to have been explored in meliponine bees, but some data have been obtained for ants. One or more pheromones from a queen probably inhibits queen rearing in colonies of *Eciton* spp. (Schneirla and Rosenblatt, 1961). Bier (1958) demonstrated that inhibitory material from the queen is distributed among the workers of colonies of *Formica rufa pratensis*, and Brian (1963) showed the same thing in *Myrmica rubra*. Clearly much more work remains to be done before we shall know the importance of pheromones in controlling queen production, and perhaps male production, in ants and other social insects.

SOME GENERAL CONCLUSIONS

It is becoming increasingly clear that pheromones play a major part in the lives of very many species of insects by helping to bring the sexes together preparatory to mating (sex attractants) and in stimulating them to mate when they have found one another (aphrodisiacs). They are also important in promoting aggregation in such essentially solitary insects as ladybirds (Coccinellidae) for hibernation, aestivation, and sometimes mating, but, as would be expected, are more important in the lives of gregarious insects and still more in those of social insects.

The olfactory sex attractants of some closely related species of moths (e.g., silk moths) consist of the same or very similar substances. Similarly, the olfactory sex attractants of several species of honeybees (*Apis* spp.) probably all contain 9-oxodecenoic acid, and such alerting pheromones as citral occur not only among related species but also in different orders. The behavioral and ecologic systems that prevent interspecific mating and similar 'mistakes' raise problems of great interest.

Biologic economy is also shown in the way that within a single species a particular substance (pheromone), either alone or in association with others, serves several distinct purposes. For example, the 9-oxodecenoic acid produced in the mandibular glands of the queen honeybee, *Apis mellifera*, serves on its own as an olfactory sex attractant and as an aphrodisiac; together with 9-hydroxydecenoic acid it is an inhibitor of queen rearing by workers and of oogenesis in workers, and during swarming it enables the flying workers to find the queen. Or again, a given pheromone can convey a different message according to its concentration; thus, substances that when concentrated act as alerting pheromones for ants are aggregating ones when dilute.

In order that a pheromone may exert its influence it must be presented in the right context. For example, the olfactory sex attractant of a nubile virgin queen honeybee, *Apis mellifera*, does not attract drones or stimulate them in any way either within the hive or near its entrance, but does so only when she is flying at an altitude that depends on the wind speed at the time.

Clearly, in addition to the many chemical problems of isolating and identifying pheromones, there are many intriguing biologic ones awaiting investigation.

ACKNOWLEDGMENT

I am very grateful to the Cambridge Philosophical Society for permission to use my review "Insect Pheromones," which was published in *Biological Reviews*, 42:42-87, 1967, as the basis of this chapter.

REFERENCES

Acker, T. S. 1966. Courtship and mating behavior in *Agulla* species (Neuroptera: Raphidiidae). *Ann. Entom. Soc. Amer.*, 59:1-6.

Acree, F., Jr., Beroza, M., Holbrook, R. F., and Haller, H. L. 1959. The stability of hydrogenated gypsy moth sex attractant. *J. Econ. Entom.*, 52:82-85.

Allen, N., Kinard, W. S., and Jacobson, M. 1962. Procedure used to recover a sex attractant for the male tobacco hornworm. *J. Econ. Entom.*, 55:347-351.

Anderson, R. H. 1963. The laying worker in the Cape honeybee, *Apis mellifera capensis. J. Apicult. Res.*, 2:85-92.

Aplin, R. T., and Birch, M. C. 1968. Pheromones from the abdominal brushes of male noctuid lepidoptera. *Nature (London)*, 217:1167-1168.

Barbier, M., and Hügel, M. F. 1961. Synthèse de l'acide céto-9-décène-2-*cis*-oïque, isomère *cis* de la 'substance royale'. *Bull. Soc. Chim. France*, 202:1324-1326.

—— and Lederer, E. 1960. Structure chimique de la 'substance royale' de la reine d'abeille (*Apis mellifica* L.). *C. R. Acad. Sci. (Paris)*, 250:4467-4469.

—— Lederer, E., and Nomura, I. 1960. Synthèse de l'acide céto-9-décène-2-*trans*-oïque ('substance royale') et de l'acide céto-8-nonène-2-*trans*-oïque. *C. R. Acad. Sci. (Paris)*, 251:1133-1135.

Barnes, M. D., Peterson, M., and O'Connor, J. J. 1966. Sex pheromone gland in the female codling moth, *Carpocapsa pomonella* (Lepidoptera: Olethreutidae). *Ann. Entom. Soc. Amer.*, 59:732-734.

Barnes, M. M., Robinson, D. W., and Forbes, A. G. 1954. Attractants for moths of the western grape leaf skeletonizer. *J. Econ. Entom.*, 47:58-63.

Barth, R. H., Jr. 1961. Hormonal control of sex attractant production in the Cuban cockroach. *Science (N.Y.)*, 133:1598-1599.

Batra, S. W. T. 1966. The life cycle and behavior of the primitively social bee, *Lasioglossum zephyrum* (Halictidae). *Univ. Kansas Sci. Bull.*, 46:359-423.

Berger, R. S. 1968. Sex pheromone of the cotton leafworm. *J. Econ. Entom.*, 61:326-327.

Beroza, M., and Jacobson, M. 1963. Chemical insect attractants. *World Rev. Pest Control*, 2:36-48.

Bickle, R. H. 1959. Observations on the hovering and mating of *Tabanus bishoppi* Stone (Diptera, Tabanidae). *Ann. Ent. Soc. Amer.*, 41:403-412.

Bier, K. 1954. Über den Einfluss der Königin auf die Arbeiterinnen-Fertilität im Ameisenstaat. *Insectes Sociaux*, 1:7-19.

—— 1956. Arbeiterinnfertilität und Aufzucht von Geschlechsteiren als Regulationsleistung des Ameisenstaates. *Insectes Sociaux*, 3:177-184.

—— 1958. Die Regulation der Sexualität in den Insektenstaaten. *Ergebn. Biol.*, 20:97-126.

Blum, M. S. 1966. Chemical releasers of social behavior. VIII. Citral in the mandibular gland secretion of *Lestrimelitta limao* (Hymenoptera:Apoidea:Melittidae). *Ann. Entom. Soc. Amer.*, 59:962-964.

—— and Portocarrero, C. A. 1964. Chemical releases of social behavior. IV. The hindgut as the source of the odor trail pheromone in the Neotropical Army Ant genus *Eciton. Ann. Entom. Soc. Amer.*, 57:793-794.

—— and Ross, G. N. 1965. Chemical releasers of social behavior. V. Source, specificity, and properties of the odour trail pheromone of *Tetramorium guineense* (F.) (Formicidae:Myrmicinae). *J. Insect Physiol.*, 11:857-868.

—— and Warter, S. L. 1966. Chemical releasers of social behavior. VII. The isolation of 2-heptanone from *Conomyrma pyramica* (Hymenoptera:Formicidae: Dolichoderinae) and its modus operandi as a releaser of alarm and digging behavior. *Ann. Entom. Soc. Amer.*, 59:744-749.

———Warter, S. L., and Traynham, J. G. 1966. Chemical releasers of social behaviour. VI. The relation of structure to activity of ketones as releasers of alarm for *Iridomyrmex pruinosus* (Roger). *J. Insect Physiol.*, 12:419-427.

——— and Wilson, E. O. 1964. The anatomical source of trail substances in formicine ants. *Psyche (Cambridge)*, 71:28-31.

Bobb, M. L. 1964. Apparent loss of sex attractiveness by the female of the Virginia-Pine sawfly, *Neodiprion pratti pratti*. *J. Econ. Entom.*, 57:829-830.

Boch, R. and Shearer, D. A. 1962. Identification of geraniol as the active component in the Nassanoff pheromone of the honey bee. *Nature, Lond.*, 194:704-706.

——— and Shearer, D. A. 1963. Production of geraniol by honey bees of various ages. *J. Insect Physiol.*, 9:431-434.

——— and Shearer, D. A. 1964. Identification of nerolic and geranic acids in the Nassanoff pheromone of the honey bee. *Nature, Lond.*, 202:320-321.

——— Shearer, D. A., and Stone, B. C. 1962. Identification of isoamyl acetate as an active component in the sting pheromone of the honeybee. *Nature (London)* 195:1018-1020.

Bornemissza, G. F. 1964. Sex attractant of male scorpion flies. *Nature (London)*, 203:786-787.

——— 1966. Specificity of male sex attractants in some Australian scorpion flies. *Nature (London)*, 209:732-733.

Bossert, W. H., and Wilson, E. O. 1963. An analysis of olfactory communication among animals. *J. Theor. Biol.*, 5:443-469.

Brian, M. V. 1953. Oviposition by workers of the ant *Myrmica*. *Physiol. Comp. Oecol.*, 3:25-36.

——— 1963. Studies of caste differentiation in *Myrmica rubra* L. 6. Factors influencing the course of female development in the early third instar. *Insectes Sociaux*, 10: 91-102.

Brower, L. P., and Jones, M. A. 1966. Precourtship interaction of wing and abdominal sex glands in male *Danaus* butterflies. *Proc. Roy. Entom. Soc.*, [A], 40:147-151.

Butenandt, A., Beckmann, R., and Hecker, E. 1961. Über den Sexuallockstoff des Seidenspinners. I. Der biologische Test und die Isolierung des reinen Sexuallockstoffes Bombykol. *Hoppe-Seyler's Z. Physiol. Chem.*, 324:71-83.

——— Linzen, B., and Lindauer, M. 1959. Über einen Duftstoff aus der Mandibulardrüse der Blattschneiderameise *Atta sexdens rubropilosa* Forel. *Arch. Anat. Micr. Morph. Exp.*, 48:13-19.

——— and Tam, N. D. 1957. Über einen geschlechtsspezifischen Duftstoff der Wasserwanze *Belostoma indica* Vitalis. *Hoppe-Seyler's Z. Physiol. Chem.*, 308:277-283.

Butler, C. G. 1954a. The World of the Honeybee. London, Wm. Collins.

——— 1954b. The method and importance of the recognition by a colony of honeybees (*A. mellifera*) of the presence of its queen. *Trans. Roy. Entom. Soc.*, 105:11-29.

——— 1956. Some further observations on the nature of 'queen substance' and of its role in the organisation of a honey-bee (*Apis mellifera*) community. *Proc. Roy. Entom. Soc.* [A], 31:12-16.

———1957a. The control of ovary development in worker honeybees (*Apis mellifera*). *Experientia*, 13:256-257.

——— 1957b. The process of queen supersedure in colonies of honey-bees (*Apis mellifera* L.). *Insectes Sociaux*, 4:211-223.

——— 1959a. Queen substance. Bee World 40:269-275.

——— 1959b. The source of the substance produced by a queen honey-bee (*Apis *mellifera* L.) which inhibits development of the ovaries of the workers of her colony. *Proc. Entom. Soc.* [A], 34:137-138.

———— 1960a. Queen recognition by worker honeybees (*Apis mellifera* L.) *Experientia*, 16:424-426.

———— 1960b. The significance of queen substance in swarming and supersedure in honey-bee (*Apis mellifera* L.) colonies. *Proc. Roy. Entom. Soc.* [A], 35:129-132.

————1961. The scent of queen honeybees (*A. mellifera* L.) that causes partial inhibition of queen rearing. *J. Insect Physiol.*, 7:258-264.

————1964. Control of behaviour in the honeybee colony. *Proc. Roy. Instit. Great Brit.*, 40:82-91.

———— 1965a. Sex attraction in *Andrena flavipes* Panzer (Hymenoptera:Apidae), with some observations on nest-site restriction. *Proc. Roy. Entom. Soc.* [A], 40:77-80.

———— 1965b. Die Wirkung von Königinnen-Extrakten verschiedener sozialer Insekten auf die Aufzucht von Königinnen und die Entwicklung der Ovarien von Arbeiterinnen der Honigbiene. *Z. Bienenforsch.*, 8:143-147.

———— 1966a. Pheromones of worker honeybees. *Rep. Rothamsted Experimental Station for 1965*, pp. 197-198.

———— 1966b. Mandibular gland pheromone of worker honeybees. *Nature (London)*, 212:530.

———— 1967. A sex attractant acting as an aphrodisiac in the honeybee (*Apis mellifera* L.) *Proc. Roy. Entom. Soc.* [A], 42:71-76.

———— and Calam, D. H. 1969. Pheromones of the honey bee—The secretion of the Nassanoff gland of the worker. *J. Insect Physiol.*, 15:237-244.

———— Calam, D. H., and Callow, R. K. 1967. Attraction of *Apis mellifera* drones by the odours of the queens of two other species of honeybees. *Nature (London)*, 213:423-424.

———— and Callow, R. K. 1968. Pheromones of the honeybee (*A. mellifera* L.), The 'inhibitory scent' of the queen. *Proc. Roy. Entom. Soc* [A], 43:(in Press).

———— Callow, R. K., and Chapman, J. R. 1964. 9-hydroxydec-*trans*-2-enoic acid, a pheromone stabilizing honeybee swarms. *Nature (London)*, 201:733.

———— Callow, R. K., and Johnston, N. C. 1961. The isolation and synthesis of queen substance, 9-oxodec-*trans*-2-enoic acid, a honeybee pheromone. *Proc. Roy. Soc.* [*Biol*], 155:417-432.

———— and Fairey, E. M. 1963. The role of the queen in preventing oogenesis in worker honeybees (*A. mellifera* L.). *J. Apicult. Res.*, 2:14-18.

———— and Fairey, E. M. 1964. Pheromones of the honeybee: biological studies of the mandibular gland secretion of the queen. *J. Apicult. Res.*, 3:65-76.

———— Fletcher, D. J. C., and Watler, D. 1969. Nest-entrance marking with pheromones by the honeybee, *Apis mellifera* L., and by a wasp, *Vespula vulgaris* L. *Anim. Behav.*, 17:142-147.

———— and Free, J. B. 1952. The behavior of worker honeybees at the hive entrance. *Behaviour*, 4:263-292.

———— and Gibbons, D. A. 1958. The inhibition of queen rearing by feeding queenless worker honeybees. (*A. mellifera*) with an extract of 'queen substance'. *J. Insect Physiol.*, 2:61-64.

———— and Paton, P. N. 1962. Inhibition of queen rearing by queen honeybees (*Apis mellifera* L.) of different ages. *Proc. Roy. Entom. Soc.* [A], 37:114-116.

———— and Simpson, J. 1958. The source of the queen substance of the honey-bee (*Apis mellifera* L.). *Proc. Roy. Entom. Soc.* [A], 33:120-122.

———— and Simpson, J. 1965. Pheromones of the honeybee (*Apis mellifera* L.). An olfactory pheromone from the Koshchewnikow gland of the queen. Scientific Studies, Univ. Libcice, Czechoslovakia 4:33-36.

—— and Simpson, J. 1967. Pheromones of the queen honeybee (*Apis mellifera* L.) which enable her workers to follow her when swarming. *Proc. Roy. Entom. Soc.* [*A*], 42:149-154.

Buttel-Reepen, H. von. 1900. Sind die Bienen Reflexmaschinen? *Biol. Zbl.*, 20:1-82.

Callow, R. K., Chapman, J. R., and Paton, P. N. 1964. Pheromones of the honeybee: chemical studies of the mandibular gland secretion of the queen. *J. Apicult. Res.*, 3:77-89.

Carlisle, D. B., and Ellis, P. E. 1964. Effect of a 2-chloro-ethanol on maturation of locusts. *J. Endocr.*, 30:153-154.

Carr, C. A. H. 1962. Further studies on the influence of the queen in ants of the genus *Myrmica*. *Insectes Sociaux*, 9:197-211.

Carthy, J. D. 1951. The orientation of two allied species of British ant. *Behaviour*, 3:275-318.

—— 1952. The return of ants to their nest. *Trans. IXth Int. Congr. Ent.* vol. 1:365-369.

Casida, J. E., Coppel, H. C., and Watanabe, T. 1963. Purification and potency of the sex attractant from the introduced pine sawfly, *Diprion similis*. *J. Econ. Entom.*, 56:18-24.

Cavill, G. W. K., and Robertson, P. L. 1965. Ant venoms, attractants and repellents. *Science (New York)*, 149:1337-1345.

Chaboussou, F., and Charles, J. P. 1962. Observations sur le piégeage sexuel des mâles d'endémis (*Lobesia botrana* Schiff.). *Rev. Zool. agric. Appl.*, 61:81-98.

Chadra, M. S., Eisner, T., Munro, A., and Meinwald, J. 1962. Defense mechanisms of arthropods. VII. Citronellal and citral in the mandibular gland secretion of the ant *Acanthomyops claviger* (Roger). *J. Insect Physiol.*, 8:175-179.

Chauvin, R. 1958. L'action de groupement sur la croissance des grillons (*Gryllulus domesticus*). *J. Insect Physiol.*, 2:235-248.

Chiang, H. C. 1961. Ecology of insect swarms. I. Experimental studies of the behaviour of *Anarete near felti* Pritchard in artificially induced swarms. *Anim. Behav.*, 9:213-219.

Collins, C. M., and Potts, S. F. 1932. Attractants for the flying gypsy moth as an aid in location of new infestations. *Techn. Bull. U.S. Dep. Agric.*, No. 336.

Collins, M. M., and Weast, R. D. 1961. Wild silk moths of the United States. Cedar Rapids, Iowa, Collins Radio Co.

Coppel, N. C., Casida, J. E., and Dauterman, W. C. 1960. Evidence for a potent sex attractant in the introduced pine Sawfly *Diprion similis* (Hymenoptera: Diprionidae). *Ann. entom. Soc. Amer.*, 53:510-512.

Corbet, P. S. 1964. Observations on the swarming and mating of mosquitoes in Uganda. *Proc. Roy. Entom. Soc.* [*A*], 39:15-22.

—— and Haddow, A. J. 1962. Diptera swarming high above the forest canopy in Uganda, with special reference to Tabanidae. *Trans. Roy. Entom. Soc.*, 114:267-284.

Cumber, R. A. 1949. The biology of bumble-bees with special reference to the production of the worker caste. *Trans. Roy. Entom. Soc.*, 100:1-45.

Cuthbert, J. P., Jr., and Reid, W. J., Jr. 1964. Studies of sex attractant of banded cucumber beetle. *J. Econ. Entom.*, 57:247-250.

Davis, F. M., and Henderson, C. A. 1967. Attractiveness of virgin female moths of the southwestern corn borer. *J. Econ. Entom.*, 60:279-280.

Deleurance, E. 1946. Une régulation à base sensorielle périphérique. L'inhibition de la ponte des ouvrières par la présence de la fondatrice des *Polistes* (Hymenoptera-Vespidae). *C. R. Acad. Sci. (Paris)*, 223:871-872.

Delvert-Salleron, F. 1963. Étude, Au moyen de radio-isotopes, des échanges de nourritures entre reines, mâles et ouvrières d'*Apis mellifica* L. *Ann. Abeille*, 6:201-227.

Doane, C. C. 1966. Evidence for a sex attractant in females of the red pine scale. *J. Econ. Entom.*, 59:1539-1540.

Downes, J. A. 1955. Observations on the swarming flight and mating of *Culicoides*. *Trans. Roy. Entomol. Soc. Lond.*, 106:213-236.

Eisner, T., and Kafatos, F. C. 1962. Defense mechanisms of arthropods. X. A pheromone promoting aggregation in an aposematic distasteful insect. *Psyche Cambridge*, 69:53-61.

Engelmann, F. 1965. Pheromones and mating behaviour. *Proc. XII Int. Congr. Ent., London (1964)*, p. 289.

Ewing, A. W., and Manning, A. 1963. The effect of exogenous scent on the mating of *Drosophila melanogaster*. *Anim. Behav.*, 11:596-598.

Finger (Bar-Ilan), A., Stanic, V., and Shulov, A. 1965. Attracting substance (pheromone) produced by virgin females of *Trogoderma granarium* Everts (Col. Dermestidae). *Riv. Parassit.*, 26:27-29.

Forel, A. 1886. Études myrmecologiques. *Ann. Soc. Entom. Belg.*, 30:215-313.

———— 1898. La paraboise chez les fourmis. *Bull. Soc. Vaud. Sci. Nat.*, 34:380-384.

Frank, A. 1941. Eigenartige Flugbahnen bei Hummelmännchen. *Z. vergl. Physiol.*, 28:467-484.

Free, J. B. 1955. The behaviour of egg-laying workers of bumblebee colonies. *Brit. J. Anim. Behav.*, 3:147-153.

———— 1961. The stimuli releasing the stinging response of honeybees. *Anim. Behav.*, 9:193-196.

———— 1962. The attractiveness of geraniol to foraging honeybees. *J. Apicult. Res.*, 1:52-54.

———— and Butler, C. G. 1955. An analysis of the factors involved in the formation of a cluster of honeybees. *Behavior*, 7:304-316.

———— and Butler, C. G. 1959. Bumblebees. London, Wm. Collins.

———— and Racey, P. A. 1967. Pollination in glasshouses. *Rep. Rothamsted Experimental Station for 1966*, p. 213.

Freiling, H. 1909. Duftorgane der weiblichen Schmetterlingen nebst Beiträgen zur Kenntnis der Sinnesorgane auf dem Schmetterlingsflügel und der Duftpinsel der Männchen von *Danais* und *Euploea*. *Z. Wiss. Zool.*, 92:210-290.

Gabbutt, P. D. 1954. Notes on the mating behaviour of *Nemobius sylvestris* (Bosc.) (Orth., Gryllidae). *Brit. J. Anim. Behav.*, 2:84-88.

Gary, N. E. 1961. Queen honey bee attractiveness as related to mandibular gland secretion. *Science (N.Y.)*, 133:1479-1480.

———— 1962. Chemical mating attractants in the queen honey bee. *Science (N.Y.)*, 136:773-774.

———— and Morse, R. A. 1962. Queen cell construction in honey bee (*Apis mellifera* L.) colonies headed by queens without mandibular glands. *Proc. Roy. Entom. Soc. [A]*, 37:76-78.

Gaston, L. K., and Shorey, H. H. 1964. Sex pheromones of Noctuid moths. IV. An apparatus for bioassaying the pheromones of six species. *Ann. Entom. Soc. Amer.*, 57:779-780.

Gentry, C. R., Lawson, F. R., and Hoffman, J. D. 1964. A sex attractant in the tobacco budworm. *J. Econ. Entrom.*, 57:819-821.

Ghent, R. L., and Gary, N. E. 1962. A chemical alarm releaser in honey bee stings (*Apis mellifera* L.). *Psyche* (*Cambridge*), 69:1-6.

Gibson, N. H. E. 1945. On the mating swarms of certain Chironomidae (Diptera). *Trans. Roy. Entom. Soc.* [*London*], 95:263-294.

Goeden, R. D., and Norris, D. M., Jr. 1965. Some biological and ecological aspects of ovipositional attack in *Carya* spp. by *Scolytus quadrispinosus* (Coleoptera: Scolytidae). *Ann. Entom. Soc. Amer.*, 58:771-777.

Gösswald, K., and Kloft, W. 1958. Radioaktive Isotope zur Erforschung des Staatenlebens der Insekten. *Umschau*, 58:743-745.

Götz, B. 1951. Die Sexualduftstoffe an Lepidopteren. *Experientia*, 7:406-418.

Graham, H. M., and Martin, D. F. 1963. Use of cyanide in pink bollworm sex-lure traps. *J. Econ. Entom.*, 56:901-902.

Gunnison, A. F. 1966. An improved method for collecting the liquid fraction of bee venom. *J. Apicul. Res.*, 5:33-36.

Gurney, A. B. 1947. A new species of *Pristoceuthopilus* from Oregon and remarks on certain special glands of Orthoptera. *J. Wash. Acad. Sci.*, 37:430-435.

Haas, A. 1946. Neue Beobachtungen zum Problem der Flugbahnen bei Hummelmännchen. *Z. Naturforsch*, 11:596-600.

———— 1949. Gesetzmässiges Flugverhalten der Männchen von *Psithyrus sylvestris* Lep. und einiger solitärer Apiden. *Z. vergl. Physiol.*, 31:671-683.

———— 1952. Die Mandibeldrüse als Duftorgan bei einigen Hymenopteren. *Naturwissenschaften*, 39:484.

———— 1960. Vergleichende Verhaltensstudien zum Paarungsschwarm solitären Apiden. *Z. Tierpsychol.*, 17:402-416.

Haddow, A. J., Yarrow, I. H. H., Lancaster, G. A. and Corbet, P. S. 1966. Nocturnal flight cycle in the males of African doryline ants (Hymenoptera:Formicidae). *Proc. Roy Entom. Soc.* [*A*], 41:103-106.

Hecker, E. 1959. Sexuallockstoffe-hockwirksame Parfüms der Schmetterlinge. *Umschau*, 59:465-467.

Heran, H., and Lindauer, M. 1963. Windkompensation und Seitenwindkorrektur der Bienen beim Flug über Wasser. *Z. vergl. Physiol.*, 47:39-55.

Hocking, B. 1963. The use of attractants and repellents in vector control. *Bull. WHO*, 29 (Suppl.):121-126.

Hodek, J. 1960. Hibernation-bionomics in Coccinellidae. *Časopis Československé společnosti entomologické, Praha.*, 57:1-20.

Hrdý, I., Novák, V. J. A., Škrobal, D. 1960. Influence of the queen inhibitory substance of honeybees on the development of supplementary reproductives in the termite, *Kalotermes flavicollis*. *In* The ontogeny of Insects. Acta Symposie de Evolutione insectorum, Prague, 1959, pp. 172-174.

Humphries, D. A. 1967. The mating behaviour of the hen flea *Ceratophyllus gallinae* (Schrank) (Siphonaptera:Insecta). *Anim. Behav.*, 15:82-90.

Imms, A. D. 1957. A general textbook of entomology, 9th ed. (revised by O. W. Richards and R. G. Davies), London, Methuen.

Ishay, J. 1964. Observations sur la biologie de la Gûepe orientale *Vespa orientalis* F. *Insectes Sociaux*, 11:193-206.

Jacobson, M. 1965. Insect sex attractants. New York, Interscience Publishers.

———— and Beroza, M. 1963. Sex attractant of the American cockroach. *Science* (N.Y.), 142:1258-1261.

────── and Harding, C. 1968. Sex attractant of the wireworm. *Limonius californicus*. *Science* (N.Y.), 159:208.

Johnston, N.C., Law, J. H., and Weaver, N. 1965. Metabolism of 9-ketodec-2-enoic acid by worker honeybees (*Apis mellifera* L.) *Biochemistry*, 4:1615-1621.

Kalmus, H., and Ribbands, C. R. 1952. The origin of the odours by which honeybees distinguish their companions. *Proc. Roy. Soc.* [B], 140:50-59.

Karlson, P. 1960. Pheromones. *Ergebn. Biol.*, 22:212-225.

────── and Butenandt, A. 1959. Pheromones (ectohormones) in insects. *Ann. Rev. Entom.*, 4:39-58.

────── and Lüscher, M. 1959. 'Pheromones': a new term for a class of biologically active substances. *Nature* (*London*), 183:55-56.

Kaufmann, T. 1966. Observations on some factors which influence aggregation by *Blaps sulcata* (Coleoptera:Tenebrionidae) in Israel. *Ann. Entom. Soc. Amer.*, 59:660-664.

Keller, J. C., Mitchell, E. B., McKibben, G., and Davich, T. B. 1964. A sex attractant for female boll weevils from males. *J. Econ. Entom.*, 57:609-610.

Kellog, F. E., Frizel, D. E., and Wright, R. H. 1962. The olfactory guidance of flying insects. IV. *Drosophila*. *Canad. Entom.*, 94:884-888.

────── and Wright, R. H. 1962. The olfactory guidance of flying insects. III. A technique for observing and recording flight paths. *Canad. Entom.*, 94:486-493.

Kennedy, J. S. 1939. The visual responses of flying mosquitoes. *Proc. Zool. Soc. Lond.* [A], 109:221-242.

Kerr, W. E. 1960. Evolution of communication in bees and its role in speciation. *Evolution*, 14:386-387.

────── and Cruz, C. da C. 1961. Funções diferentes tomadas pela glândula mandibular na evolucão das abelhas em geral e em 'Trigona (*Oxytrigona*) *tataira*' em especial. *Rev. Bras. Biol.*, 21:1-16.

────── Ferreira, A., and Simões de Mattos, N. 1963. Communication among stingless bees—additional data (Hymenoptera:Apidae). *J. N.Y. Entom. Soc.*, 71:80-90.

Keys, R. E., and Mills, R. B. 1968. Demonstration and extraction of a sex attractant from female Angoumois grain moths. *J. Econ. Entom.*, 61:46-47.

Kliefoth, R. A., Vité, J. P., and Pitman, G. B. 1963. A laboratory technique for testing bark beetle attractants. *Contrib. Boyce Thompson Inst.*, 22:283-290.

Kliever, J. W., Muira, T., Husbands, R. C., and Hurst, C. H. 1966. Sex pheromones and mating behavior of *Culisetta inornata* (Diptera:Culicidae). *Ann. Entom. Soc. Amer.*, 59:530-533.

Kullenberg, B. 1956. Field experiments with chemical sexual attractants on Aculeate Hymenoptera males, I. *Zoologiska bidrag från Uppsala.*, 3:253-354.

Locher, V., and Schneider, D. 1963. Elektrophysiologischer Nachweis der Reichfunktion von Porenplatten (sensilla placodea) auf den Antennen der Drohne und der Arbeitsbiene (*Apis mellifica* L.) *Z. vergl. Physiol.*, 47:274-278.

Lansbury, I. 1965. New organ in Stenocephalidae (Hemiptera-Heteroptera). *Nature* (*London*), 205:106.

Law, J. H., Wilson, E. O., and McCloskey, J. A. 1965. Biochemical polymorphism in ants. *Science* (N.Y.), 149:544-545.

Lecomte, J. 1956. Uber die Bildung von 'Strassen' durch Sammelbienen, deren Stock um 180° gedrecht wurde. *Z. Bienenforsch.*, 3:128-133.

────── 1963 Étude des échanges de nourriture de la colonie de Bourdons au moyen de radioisotopes. *C. R. Acad. Sci.* [D] (*Paris*), 257:3664-3665.

Light, S. F. 1943. The determination of the castes of social insects. *Quart. Rev. Biol.*, 18:46-63.

———— 1944. Experimental studies on ectohormonal control of the development of supplementary reproductives in the termite genus *Zootermopsis* (formerly *Termopsis*). *Univ. Calif. Publ. Zool.*, 43:413-454.

Lindauer, M., and Kerr, W. E. 1958. Die gegenseitige Verständigung bei den stachellosen Bienen. *Z. vergl. Physiol.*, 41:405-434.

————and Kerr, W. E. 1960. Communication between the workers of stingless bees. *Bee World*, 41:29-41, 65-71.

Loher, W. 1958. An olfactory response of immature adults of the desert locust. *Nature (London)*, 181:1280.

———— 1961. The chemical acceleration of the maturation process and its hormonal control in the male of the desert locust. *Proc. Roy. Soc. [Biol.]* 153:380-397.

Lopez, D. F. 1963. Two attractants for *Eulaema tropica* L. *J. Econ. Entom.*, 56:540.

Lüscher, M. 1956. Hemmende und fördernde Faktoren bei der Entstehung der Ersatzgeschlechtstiere bei der Termite *Kalotermes flavicollis* Fabr. *Rev. Suisse Zool.*, 63:261-267.

———— 1960. Hormonal control of caste differentiation in termites. *Ann. N.Y. Acad. Sci.*, 89:549-563.

———— 1961a. Demonstration of a trail pheromone in termites. *Sympos. Genet.*, 11:189-192.

———— 1961b. Social control of polymorphism in termites. *In* Kennedy, J. S., ed. Insect Polymorphism Symposium No. 1. *Roy. Entom. Soc. (London)*, pp. 57-67.

MacGregor, E. G. 1948. Odour as a basis for orientated movement in ants. *Behaviour*, 1:267-296.

Mackensen, O. 1943. The occurrence of parthenogenetic females in some strains of honeybees. *J. Econ. Entom.*, 36:465-467.

McFarlane, J. E. 1966. Studies on group effects in crickets. I. Effect of methyl linolenate, methyl linolenate, and vitamin E. *J. Insect Physiol.*, 12:179-188.

McGurk, D. J., Frost, J., Eisenbraun, E. J., Vick, K., Drew, W. A., and Young, J. 1966. Volatile compounds in ants: identification of 4-methyl-3-heptanone from *Pogonomyrmex* ants. *J. Insect Physiol.*, 12:1435-1441.

Maschwitz, U. W. 1964a. Alarm substances and alarm behaviour in social hymenoptera. *Nature (London)*, 204:324-327.

———— 1964b. Gefahrenalarmstoffe und Gefahrenalarmierung bei sozialen Hymenopteren. *Z. vergl. Physiol.*, 47:596-655.

Mayr, E. 1946. Experiments on sexual isolation. VII. The nature of isolating mechanisms between *Drosophila pseudoobscura* and *Drosophila persimilis*. *Proc. Nat. Acad. Sci. U.S.A.*, 32:128-137.

———— 1950. The role of the antennae in the mating behaviour of female *Drosophila*. *Evolution*, 4:149-154.

Mell, R. 1922. Biologie und Systematik der Chinesischen Sphingiden. Berlin.

Meinwald, J., and Meinwald, Y. C. 1966. Structure and synthesis of the major components in the hairpencil secretion of a male butterfly, *Lycorea ceres ceres* (Cramer.). *J. Amer. Chem. Soc.*, 88:1305-1310.

———— Meinwald, Y. C., Wheeler, J. W., Eisner, T., and Brower, L. P. 1966. Major components in the exocrine secretion of a male butterfly (*Lycorea*). *Science (N.Y.)*, 151:583-585.

Michener, C. D., and Lange, R. B. 1958. Observations on the behaviour of Brasilian halictid bees. V. *Chloralictus. Insectes Sociaux*, 5:379-407.

———— and Wille, A. 1961. The bionomics of a primitively social bee, *Lasioglossum inconspicuum*. *Kansas Univ. Sci. Bull.*, 42:1123-1202.

Montagner, H., and Courtois, G. 1963. Données nouvelles sur le comportement alimentaire et les échanges trophallactiques chez les Gûepes sociales. *C. R. Acad. Sci. (Paris)*, 256:4092-4094.

Moore, B. P. 1964. Volatile terpenes from *Nasutitermes* soldiers (Isoptera, Termitidae). *J. Insect Physiol.*, 10:371-375.

Morimoto, R. 1960. On the social co-operation in *Polistes chinensis antennalis* Perez. IX. Studies on the social Hymenoptera of Japan. *Kontyû*, 28:198-208.

Moser, J. C., and Blum, M. S. 1963. Trail marking substance of the Texas leaf-cutting ant: source and potency. *Science (N.Y.)*, 140:1228.

Narda, R. D. 1966. Analysis of the stimuli involved in courtship and mating in *D. malerkotlina* (Sophophora, Drosophila). *Anim. Behav.*, 14:378-383.

Nielsen, E. T., and Haeger, J. S. 1960. Swarming and mating in mosquitoes. *Misc. Publs. Entom. Soc. Amer.*, 1:71-95.

Nixon, H. L., and Ribbands, C. R. 1952. Food transmission within the honeybee community. *Proc. Roy. Soc. [Biol.]*, 140:43-50.

Norris, D. M. 1965. In-flight dispersal and orientation of two *Scolytus* species (Coleoptera) to their host plants for ovipositional purposes. *Proc. XII. Int. Congr. Entom. London (1964)*, p. 293.

Norris, M. J. 1952. Reproduction in the desert locust (*Schistocerca gregaria* Forsk.) in relation to density and phase. *Anti-Locust Bull.*, No. 13.

———— 1954. Sexual maturation in the desert locust (*Schistocerca gregaria* Forsk.) with special reference to the effect of grouping. *Anti-Locust Bull.* No. 18.

———— 1963a. Laboratory experiments on gregarious behaviour in ovipositing females of the desert locust (*Schistocerca gregaria* Forsk.) *Entomol. Exp. Appl.*, 6:279-303.

———— 1963b. Laboratory experiments on gregarious oviposition in the desert locust (*Schistocerca gregaria* Forsk.). *Anim. Behav.*, 11:408-409.

———— and Pener, M. P. 1965. An inhibitory effect of allatectomized males and females on the sexual maturation of young male adults of *Schistocerca gregaria* (Forsk.) (Orthoptera:Acrididae). *Nature (London)*, 208:1122.

Ordway, E. 1965. Caste differentiation in *Augochlorella* (Hymenoptera, Halictidae). *Insectes Sociaux*, 12:291-308.

Pain, J. 1961. Sur la phérormone des reines d'abeilles et ses effects physiologiques. *Ann. Abeille*, 4:73-152.

———— and Barbier, M. 1960. Mise en évidence d'une substance attractive extraite du corps ouvrières d'Abeilles non orphelines (*Apis mellifica* L.). *C.R. Acad. Sci. [D] (Paris)*, 250:1126-1127.

———— Barbier, M., Bogdanovsky, D., and Lederer, E. 1962. Chemistry and biological activity of the secretions of queen and worker honeybees (*Apis mellifica* L.). *Comp. Biochem. Physiol.*, 6:233-241.

———— Barbier, M., and Roger, B. 1967. Dosages individuels des acides céto-9-décène-2-oïque et hydroxy-10-décène-2-oïque dans les têtes des reines et des ouvrières d'abeilles. *Ann. Abeille*, 10:45-52.

———— Hügel, M. F., and Barbier, M. 1960. Sur les constituants du mélange attractive des glandes mandibulaires des reines d'Abeilles (*Apis mellifica* L.) à differents stades de leur vie. *C.R. Acad. Sci. [D] (Paris)*, 251:1046-1048.

Pardi, L. 1948. Dominance order in *Polistes* wasps. *Physiol. Zool.*, 21:1-13.

———— 1951. Studio della attività e della divisione di lavoro in una società de *Polistes gallicus* (L.) dopo la comparsa delle operaie. *Arch. Zool. Ital.*, 36:363-431.

Pavan, M. 1956. Studi sui Formicidae. II. Sull'origine, significato biologico e isolamento della dendrolasina. *Ric. Sci.*, 26:144-150.

Perepelova, L. I. 1928. Laying workers, the egg-laying activity of the queen, and swarming. *Bee World.* 10:69-71.

Petit, C. 1959. De la nature des stimulations responsibles de la sélection sexuelle chez *Drosophila melanogaster*. *C. R. hebd. Séanc. Acad. Sci., Paris.* 248:3484-3485.

Pitman, G. B., Vité, J. P., Kinzer, G. W., and Fentiman, A. F. (Jun.) 1968. Bark beetle attractants: *trans*-verbenol isolated from *Dendroctonus*. *Nature (London)*, 218:168-169.

Renner, M. 1955. Neue Untersuchungen über die physiologische Wirkung des Duftorganes der Honigbiene. *Naturwissenschaften*, 42:589.

——— 1960. Das Duftorgan der Honigbiene und die physiologischt Bedeutung ihres Lockstoffes. *Z. Vergl. Physiol.*, 43:411-468.

——— and Baumann, M. 1964. Über komplexe von subepidermalen Drüzenzellen (Duftdrüsen?) der Bienenköniginnen. *Naturwissenschaften*, 51:68-69.

Richards, O. W. 1927. Sexual selection and allied problems in insects. *Biol. Rev.*, 2:298-360.

Riddiford, L. M. 1967. *Trans*-2-hexenal: mating stimulant for polyphemus moths. *Science (N.Y.)*, 158:139-141.

——— and Williams, C. M. 1967. Volatile principle from oak leaves: role in sex life of polyphemus moth. *Science (N.Y.)*, 155:589-590.

Rogoff, W. M., Beltz, A. D., Johnsen, J. O., and Plapp, F. W., Jr. 1964. A sex pheromone in the housefly, *Musca domestica* L. *J. Insect Physiol.*, 10:239-246.

Roth, L. M. 1952. The tergal gland of the male cockroach, *Supella supellectilium*. *J. Morph.*, 91:469-477.

——— 1962. Hypersexual activity induced in females of the cockroach *Nauphoeta cinerea*. *Science (N.Y)*, 138:1267-1269.

——— 1965. Loci of sensory end-organs used by cockroaches in mating behavior. *Proc. XII. Int. Congr. Entom. London (1964)*, p. 305.

——— and Barth, R. H., Jr. 1964. The control of sexual receptivity in female cockroaches. *J. Insect Physiol.*, 10:965-975.

——— and Dateo, G. P. 1966. A sex pheromone produced by males of the cockroach *Nauphoeta cinerea*. *J. Insect Physiol.*, 12:255-265.

——— and Wills, E. R. 1952. A study of cockroach behavior. *Amer. Midl. Nat.*, 47:66-129.

Sakagami, S. F. 1954. Occurrence of an aggressive behaviour in queenless hives, with considerations on the social organisation of honeybee. *Insectes Sociaux*, 1:331-343.

——— 1958. The false queen: fourth adjustive response in dequeened honeybee colonies. *Behaviour*, 13:280-296.

——— Beig, D., Zucchi, R., and Akahira, Y. 1963. Occurrence of ovary-developed workers in queenright colonies of stingless bees. *Rev. Bras. Biol.*, 23:115-129.

Schneider, D. 1957. Elektrophysiologische Untersuchungen von Chemo- und Mechanorezeptoren der Antenne des Seidenspinners *Bombyx mori* L. *Z. vergl. Physiol.*, 40:8-41.

——— 1962. Electrophysiological investigation of the olfactory specificity of sexual attracting substances in different species of moths. *J. Insect Physiol.*, 8:15-30.

——— 1963. Electrophysiological investigation of insect olfaction. *Proc. 1st Int. Sympos. on Olfaction and Taste*, London, pp. 85-103.

——— Lacher, V., and Kaissling, K-E. 1964. Die Reaktionsweise und das Reaktionsspektrum von Riechzellen bei *Antheraea pernyi* (Lepidoptera, Saturniidae). *Z. vergl. Physiol.*, 48:632-662.

Schneirla, T. C., and Rosenblatt, J. S. 1961. Behavioral organisation and genesis of the social bond in insects and mammals. *Amer. J. Orthopsychiat.*, 31:223-253.

Schwinck, I. 1954. Experimentelle Untersuchungen über Geruchsinn und Strömungswahrnehunung in der Orientierung bei Nachtschmetterlingen. *Z. vergl. Physiol.*, 37:19-56.

———— 1955. Wietere Untersuchungen zur Frage der Geruchsorientierung der Nachtschmetterlinge. Partielle Fühleramputation bei Spinnermännchen, insbesondere am Seidenspinner *Bombyx mori*. *Z. vergl. Physiol.*, 37:439-458.

———— 1956. A study of olfactory stimuli in the orientation of moths. *Proc. Xth Int. Congr. Entom.*, 2:577-582.

Shearer, D. A., and Boch, R. 1965. 2-Heptanone in the mandibular gland secretion of the honey bee. *Nature (London)*, 206:530.

Shorey, H. H. 1964. Sex pheromones of noctuid moths. II. Mating behavior of *Trichoplusia ni* (Lepidoptera:Noctuidae) with special reference to the role of the sex pheromone. *Ann. entom. Soc. Amer.*, 57:371-377.

Silverstein, R. M., Rodin, J. O., Burkholder, W. E., and Gorman, J. E. 1967. Sex attractant of the black carpet beetle. *Science (N.Y.)*, 157:85-87.

Simpson, J. 1960. Functions of salivary glands of adult honeybees. *Ann. Rep. Rothamsted Exp. Stn. 1960*, pp. 172-173.

———— 1963. Queen perception by honey bee swarms. *Nature (London)*, 199:94-95.

———— 1966. Repellency of the mandibular gland scent of worker honey bees. *Nature (London)*, 209:531-532.

Skaife, S. H. 1961. The study of ants. London, Longmans.

Solomon, J. D., and Morris, R. C. 1966. Sex attraction of the carpenterworm moth. *J. Econ. Entom.*, 59:1534-1535.

Soo Hoo, C. F., and Roberts, R. J. 1965. Sex attraction in *Rhopaea* (Coleoptera: Scarabaeidae). *Nature (London)*, 205:724-725.

Speith, H. T. 1940. Studies on the biology of the Ephemeroptera. II. The nuptial flight. *J. N.Y. Entom. Soc.*, 48:379-390.

Stein, G. 1962. Über den Feinbau der Mandibeldrüse von Hummelmännchen. *Z. Zellforsch. Mikrosk. Anat.*, 57:719-736.

———— 1963. Über den Sexuallockstoff von Hummelmännchen. *Naturwissenschaften*, 50:305.

Stejskal, M. 1962. Duft als 'Sprache' der tropischen Bienen. *Südwestdtsch. Imker*, 14:271.

Stuart, A. M. 1961. Mechanism of trail-laying in two species of termites. *Nature (London)*, 189:419.

———— 1963a. Origin of the trail in the termites *Nasutitermes corniger* (Motschulsky) and *Zootermopsis nevadensis* (Hagen), Isoptera. *Physiol. Zoöl.*, 36:69-84.

———— 1963b. Studies on the communication of alarm in the termite *Zootermopsis nevadensis* (Hagen), Isoptera. *Physiol. Zoöl.*, 36:85-96.

Stürckow, B., and Bodenstein, W. G. 1966. Location of the sex pheromone in the American cockroach, *Periplaneta americana* (L.). *Experientia*, 22:851-853.

Stumper, R. 1956. Sur les sécrétions attractives des Fourmis femelles. *C. R. hebd. Séanc. Acad. Sci., Paris*, 242:2487-2489.

Sutton, O. G. 1947. The problem of diffusion in the lower atmosphere. *Quarterly Journal of the Royal Meteorological Society, London*, 73:257-280.

———— 1949. The application of micrometeorology to the theory of turbulent flow over rough surfaces. *Quarterly Journal of the Royal Meteorological Society, London,* 75:335-350.

Torossian, C. 1967. Recherches sur la biologie et éthologie de *Dolichoderus quadripunctatus* (L.) (Hym. Form. Dolichoderidae). IV. Étude des possibilités évolutives des colonies avec reine, et des femelles isolées désailées. *Insectes Sociaux,* 14:259-280.

Valentine, J. M. 1931. The olfactory sense of the adult mealworm beetle *Tenebrio molitor* (Linn.). *J. Exp. Zool.,* 58:165-227.

Velthuis, H. H. W. 1967. On abdominal pheromones in the queen honeybee. *Internat. Apic. Congress XXI, Univ. Maryland, Abstract 11.*

———— Verheijen, F. J., and Gottenbos, A. J. 1965. Laying worker honey bee: similarities to the queen. *Nature (London),* 207:1314.

Verheijen-Voogd, C. 1959. How worker bees perceive the presence of their queen. *Z. vergl. Physiol.,* 41:527-582.

Verron, H. 1963. Rôle des stimuli chimiques dans l'attraction sociale chez *Calotermes flavicollis* (Fabr.). *Insectes Sociaux,* 10:167-301.

Vitè, J. P., and Pitman, G. B. 1968. Bark beetle aggregation: effects of feeding on the release of pheromones in *Dendroctonus* and *Ips. Nature (London),* 218:169-170.

von Frisch, K. 1923. Über die 'Sprache' der Bienen. *Zool. Jahrb. Abt. Allg. Zool. Physiol.* Tiere, 40:1-186.

———— and Rösch, G. A. 1926. Neue Versuche über die Bedeutung von Duftorgan und Pollenduft für die Veständigung im Bienenvolk. *Z. vergl. Physiol.,* 4:1-21.

Voogd, S. 1955. Inhibition of ovary development in worker bees by extraction fluid of the queen. *Experientia,* 11:181-182.

Vowles, D. M. 1955. The foraging of ants. *Brit. J. Anim. Behav.,* 3:1-13.

Weaver, N., Weaver, E. C., and Law, J. H. 1964. The attractiveness of citral to foraging honeybees. *Progr. Rep. Tex. Agric. Exp. Stn.,* no. 2324, pp. 1-7.

Wharton, D. R. A., Miller, G. L., and Wharton, M. L. 1954. The odorous attractant of the American cockroach, *Periplaneta americana* (L.) *J. Gen. Physiol.,* 37:461-469.

Wheeler, W. M. 1921. A new case of parabiosis and the 'ant gardens' of British Guiana. *Ecology,* 2:89-103.

Wigglesworth, V. B. 1965. The Principles of Insect Physiology, 6th ed. London, Methuen.

Wilson, E. O. 1958. A chemical releaser of alarm and digging behavior in the ant *Pogonomyrmex badius* (Latreille). *Psyche (Camb.),* 65:41-51.

———— 1959. Source and possible nature of the odor trail of the fire ant *Solenopsis saevissima* (Fr. Smith). *Science (N.Y.),* 129:643-644.

———— 1962. Chemical communication among workers of the fire ant, *Solenopsis saevissima* (Fr. Smith). 1. The organisation of mass foraging. 2. An information analysis of the odour trail. 3. The experimental induction of social responses. *Anim. Behav.,* 10:134-164.

———— 1963a. Pheromones. *Sci. Amer.,* 208:100-114.

———— 1963b. The social biology of ants. *Ann. Rev. Entom.* 8:345-368.

———— 1965a. Chemical communication in the social insects. *Science (N.Y.),* 149:1064-1071.

———— 1965b. Trail sharing in ants. *Psyche (Camb.),* 72:2-7.

———— and Eisner, T. 1957. Quantitative studies of liquid food transmission in ants. *Insectes Sociaux.* 4:157-166.

———— and Pavan, M. 1959. Glandular sources and specificity of some chemical releasers of social behavior of dolichoderine ants. *Psyche (Camb.)*, 66:70-78.

Wood, D. L., and Bushing, R. W. 1963. The olfactory response of *Ips confusus* (Le Conte) (Coleoptera:Scolytidae) to the secondary attraction in the laboratory. *Canad. Entom.*, 95:1066-1078.

Wray, C., and Farrier, M. H. 1963. Response of the Nantucket pine tip moth to attractants. *J. Econ. Entom.*, 56:714-715.

Wright, R. H. 1958. The olfactory guidance of flying insects. *Canad. Entom.* 90:81-89.

———— 1964a. Insect control by nontoxic means. *Science (N.Y.)*, 144:487.

———— 1964b. The science of smell. London, Allen and Unwin.

Yakhontov, V. V. 1938. Practical results of an experiment. A biological method of controlling pests of lucerne and cotton. *Rev. Appl. Entom.* 26:238-239.

Zethner-Møller, O., and Rudinsky, J. A. 1967. Studies on the site of sex pheromone production in *Dendroctonus pseudotsugae* (Coleoptera:Scolytidae). *Ann. entom. Soc. Amer.*, 60:575-582.

5

The Role of Chemicals in
Termite Communication

ALASTAIR M. STUART
University of North Carolina
North Carolina State University at Raleigh
Raleigh, N.C. 27607

INTRODUCTION 79
PHEROMONES 81
CHEMICALS OTHER THAN PHEROMONES 98
CONCLUSION 101
ACKNOWLEDGMENTS 102
REFERENCES 103

INTRODUCTION

In the social insects, communication by chemicals is perhaps the commonest method used to transmit information from one member of a colony to another, though other means of communication are also employed. Work on the chemicals involved in communication has been widespread in the social Hymenoptera, especially in the bees (see Butler, 1967), and in the ants (Wilson, 1963); but similar work on the only order of insects which is wholly social, the Isoptera, has been less extensive. There are several reasons for this, but the principal one is probably that it is exceedingly difficult to experiment with these "cryptic" insects in a meaningful way. Another reason could be that although chemical communication is obvious when looked for, it is usually not so spectacular in termites as in the Hymenoptera. In this latter order, it is difficult not to notice the spectacular alarm behavior initiated by volatile chemicals. As will be discussed later, it is possible that the cryptic habit of most termites is not as well suited for communication by highly volatile chemicals as is the free-ranging habit which is characteristic of many of the ants and bees.

Nonetheless, today work is progressing in the analysis of the behavior of several species of termites where chemicals are important (Lüscher and Müller, 1960;

Stuart, 1960, 1963a, 1963b, 1967) while chemicals involved in the sense of smell have been implicated in termite behavior for many years. Bugnion (1914) believed that the termites followed trails using a chemical sense, while Beaumont (1889-1890), the Superintendent of Motive Power of the Panama Railroad, considered that in *Nasutitermes* smell was important in the recognition of comrades and enemies. It is now known that chemicals are present and important in many aspects of termite behavior—from caste recognition to alarm.

There are many ways in which chemicals used in communication can be classified and each method has both advantages and disadvantages. Some of the more common methods by which the chemicals have been classified are according to: (1) their chemical formulas: this is the preferred method of the organic chemist (e.g., Jacobson, 1966); (2) the behavioral response they evoke in the animal: attractants, repellents, aphrodisiacs (Butler, 1967; Wilson, 1965); (3) their place of origin in the animal: "surface pheromones" (Wilson, 1965); (4) the category of behavior in which they function: sex pheromones, alarm pheromones, orientation pheromones, trail pheromones (Butler, 1967; Karlson and Butenandt, 1959); (5) whether they have a "releaser" effect, or a "primer" effect resulting in a physiologic alteration in the endocrine or reproductive system (Wilson and Bossert, 1963; Wilson, 1965).

These categories are by no means mutually exclusive and many workers employ more than one depending on the situation (e.g., Wilson, 1965; Butler, 1967). Other workers have proposed other terms and subdivided some of the existing categories; e.g., Karlson and Butenandt (1959) suggested that substances which acted on an insect through its sense organs might be termed "telomones" while the term "pheromone" (Karlson and Lüscher, 1959) might be restricted to substances which act in a more biochemical manner. In essence a "telomone would have much in common with a "releaser" pheromone (Wilson and Bossert, 1963), while the "pheromone" of Karlson and Butenandt (1959) has something in common with a "primer" pheromone" (Wilson and Bossert, 1963). Recently Brown (1968) has suggested that a new term "allomone" be used for situations in which there is communication between different species: he defines an allomone as:

> a chemical substance produced or acquired by an organism which, when it contacts an individual of *another species* in the natural context, evokes in the receiver a behavioral or developmental reaction adaptively favorable to the transmitter.

Besides encompassing any substances given off by the various inquilines of ants and termites, which prevent them from being attacked by their hosts, such a definition could also be applied to defense secretions used by insects. These surely evoke in the animal receiving the secretion a response (which may lead to death) adaptively beneficial to the insect discharging the secretion.

There have been other attempts (Novák, 1966) to bring order to the various chemical substances involved in development, regulation, and behavior, but at this time they perhaps tend more to confuse than to elucidate. Karlson and Butenandt (1959) felt that it was premature to try to categorize pheromones and they pleaded that it was more important that the mechanism of operation of known pheromones be investigated and that they be obtained in pure forms and identified. As far as termites are concerned, this plea still seems eminently logical. Therefore, the classification used in this chapter is not intended to be a general one but merely a convenient arrangement in which chemical communication in termites can be discussed. The format used separates the chemicals used in communication into: (1) those which are pheromones (in the general sense) and (2) those which are involved in the transmission of information, yet do not seem to fit the original definition of a pheromone. Brown (1968) might call some of the chemicals in this category "allomones." The pheromones are further subdivided according to whether the behavioral effect is a relatively immediate one or whether it is of long duration. In some ways these subdivisions correspond to the "releasers" and "primers." It was difficult, though, to categorize the slow-acting surface pheromones of termites as "releasers" (Tinbergen, 1948) yet neither do they seem to meet the criteria for a "primer" pheromone, such as the caste determination ones. However, they can fit into the second of the subdivisions used here.

The behavior of an insect or any animal is the result of many stimuli (even though at any one time one or more may be predominant). An attempt is made to relate the use of chemical communication to other methods of communication occurring concurrently or sequentially in certain known patterns of behavior. These other types of communication will be commented on here only when they are associated in a definite manner with chemical communication, or when there is a difference of opinion as to whether a certain piece of behavior is indeed elicited by chemicals rather than by other modalities.

Finally, it is my belief that after any piece of behavior has been carefully analyzed, an attempt should be made to reconstruct and synthesize the total behavior of the animal. This has always been the goal of the zoologist studying animal behavior (Darling, 1937; Russell, 1930) but it is only recently that the necessary tools, techniques, and information have become available (Clark, 1967). Therefore, wherever possible an attempt is made to re-integrate the parts into the whole.

PHEROMONES

Constantly produced pheromones having a long-term response

COLONY AGGREGATION AND COHESION. Anyone who has observed a termite colony, either in the field or in the laboratory, must have been impressed by the fact that the individuals of the various castes are not homogeneously mixed together. In the field it has often been noticed that the young and eggs are found in a specific region of the nest and that the reproductives are usually found together, even in those species (e.g., Zootermopsis) where a special gallery for the queen and king is absent. Nymphs about to molt, pre-alates, and unflown alates are also found grouped together and segregated from the rest of the colony. There are many ways in which this grouping could come about but the most likely are by: (1) mutual attraction between the same caste, (also between males and females); (2) mutual or unilateral repulsion of one caste by another; (3) low activity causing an aggregation due to

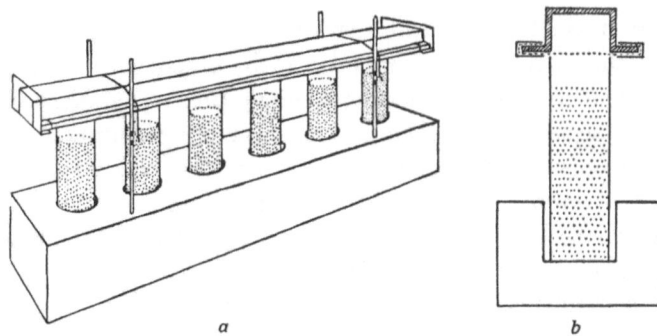

FIG. 1. Diagram of the olfactometer used by Verron (1963). (a) Shows the complete apparatus made of Plexiglass, 30.0 cm long and 5.0 cm wide. It is marked off in centimeters. Six tubes, 9.0 cm high and of a diameter of 2.5 cm, are placed equidistantly on the underside of the wire gauze forming the running surface of the apparatus. The tubes are filled with moist sand up to 1.5 cm from the top. The termites used as the stimulus are always placed in the third tube. Usual precautions to avoid contamination are taken. (b) Transverse section through the apparatus. (*From* Verron. 1963. *Insectes Sociaux,* 10:187.)

an orthokinetic effect (cf. Fraenkel and Gunn, 1940); (4) a response to various environmental factors such as temperature, humidity, odor, and so on; (5) reactions to various combinations of the above categories.

It is most likely that pheromones could be involved in (1) and (2), to some extent (4), and (5), and there is now some evidence that chemicals are involved in some aggregations of termites (Verron, 1963).

Verron (1963) set out principally to examine the olfactory interaction between the various castes of *Kalotermes flavicollis* (F.) To do this he constructed an olfactometer (Fig. 1) in which the termites being tested were prevented from coming into direct contact with the immobile or "stimulant" group. Basically he conducted two sets of experiments: the first recorded the degree of "reactivity" exhibited by various castes to other castes, while the second examined the relative ability of the castes to stimulate the other castes ("stimulation"). The second set is, of course, the corollary of the first. Since the measured value was not an absolute, it is necessary to summarize Verron's techniques. The termites used as the "stimulant" group, usually numbering 1, 5, 10, 15, or 20, were always placed in the third tube (Fig. 1): by having six tubes the apparatus could be divided into six regions. The termite to be tested was introduced into the chamber and its position was recorded every three minutes during one hour. Twenty readings were thus obtained and the number of times the termite was observed in the experimental zone (3) was recorded and totalled. The experiment was repeated 20 times and the results were examined for statistical significance using the *t*-test. A summation of 20 replicates was used as the measure of "reactivity."

In considering Verron's work, it must be realized that single individuals were being tested so that social interaction between "reactors" was eliminated. The results obtained were extensive and a complete discussion would not be warranted here. Nevertheless, Table 1 is an attempt to summarize the main results obtained by Verron, and both the experiments on stimulation and reactivity have been com-

TABLE 1.

Attraction between the various castes of *Kalotermes flavicollis*

"REACTIVITY" GROUP	"STIMULANT" GROUP							
		NYMPHS		ALATES		REPRODUCTIVES		
	LARVAE	EARLY STAGE	LATE STAGE	BEFORE FLIGHT	AFTER FLIGHT	PRI-MARY	SEC-ONDARY	SOLDIERS
Larvae	5	5	4	2	5	2	2	2
Alates	1	1	1	1	3	1	1	1
Primary reproductive	—	—	—	—	2	—	—	—
Secondary reproductive	5	4	—	—	—	—	3	—
Soldier	2	2	2	—	—	2	—	2

Numerical values derived from measurements of heights of "reactivity" histograms figured in Verron (1963): they represent the height of the appropriate histogram to the nearest cm. Values of 1 and 2 represent no significant attraction or reactivity. Values 4 and 5 represent strong attraction.

bined. The greater the numerical value in the table the greater the attraction. The following points are evident from Verron's results: (a) the nymphs (both apterous and brachypterous) apparently show a strong attraction towards other nymphs, and towards alates after (but not before) they have flown; (b) the nymphs show some reactivity to other castes, but not a great deal; and (c) the secondary reproductives are the only caste, other than the nymphs, to show any great reactiviy. They seem to be attracted strongly to the nymphs but less so to other secondaries. This is strange, as the primary reproductives are quite unreactive to other castes; the point has often been made by workers in insect sociobiology that secondary neoteinic reproductives should not be considered a separate caste. From Verron's experiment it would seem that, at least to this extent, polyethism is present between the imagos and the secondary reproductives. (d) The reactivity of the alates to each other after the flight is considered to be related to sex (Verron, 1963). Verron further studied the origin of the attractive substance or substances, and found that poor feeding decreases the attractive power of a group of individuals while starved termites have no ability whatsoever to attract. He concluded from this that the odoriferous substance has its origin in the digestion of the wood by the termite, and he states that the fresh gut exerts a characteristic attractive effect. Verron has also isolated a compound, 3-hexen-1-ol, which he considers to be the substance responsible for the attraction. This substance belongs to a group of naturally occurring attractants formed in plants. However, as termites apparently use the substance in communication, it seems reasonable to consider the substance a pheromone as originally defined.

Verron concluded from his experiments that the attraction was correlated with trophallaxy (Wheeler, 1928) and that as sexual differentiation increases, reactivity decreases (Table 1). Fielde (1905) suggested this for ants, but her results have not been substantiated in that group of insects (Sudd, 1967). Verron did not comment extensively on the bearing of his results to colony aggregation and cohesion, but rather emphasized the trophallactic aspects of behavior. Still, his results can be used to shed some light on the factors involved in the aggregation of the various castes.

Alates. It appears (Table 1) that there is little or no attraction between alates in *Kalotermes*. The aggregation in a colony would not seem to depend on any volatile chemical (Verron's experimental set-up could only test volatile substances). The evidence from these experiments would tend to favor an aggregation mechanism such as (c) or (d) above, though a contact chemical could have an enhancing effect. In *Zootermopsis* a definite rhythm is present in addition to any other factor (personal observation).

Soldiers. The reactivity of the soldiers in *Kalotermes* to each other and the other castes seems rather low. This is difficult to understand since, in *Zootermopsis* at least, they are usually associated with the reproductives (primary or secondary). Furthermore, their food is principally stomodeal and proctodeal, so they might have been expected to show a strong reactivity towards the larvae and nymphs on whom they are dependent for food. Again, it is possible that contact pheromones are present in addition to the volatile 3-hexen-1-ol found by Verron. More work is needed before any definite statements can be made on the importance of chemicals in keeping the soldiers together: obviously other factors must be involved.

Secondary reproductives. The other surprising fact evident from Verron's studies is that the secondary reproductives are more strongly attracted to the nymphs than to themselves. Verron explains these findings as due to the importance of food exchange. The secondary reproductives are, however, found together in nature and not scattered randomly throughout the colony as are the nymphs. It seems that another mechanism producing aggregation in the secondary reproductives must be operating that negates the chemical attraction of 3-hexen-1-ol at least to some extent.

Larvae. As Stuart (1968) has pointed out, the reactivity of the larvae or nymphs to newly flown alates may be a reflection of the antagonistic behavior of the nymphs to imagos that attempt to return to the colony after having flown. This behavior, while not strictly aggregation, may be usefully treated here as it can lead to a type of grouping of individuals. Such a phenomenon was observed in *Reticulitermes lucifugus* by Buchli (1961), and has also been noted in *Zootermopsis* (Stuart, 1968). Very few of the returning alates survive. If alates are prevented from leaving the colony (e.g., in a laboratory colony) then they too are ultimately dispatched by the nymphs.

Queen. No pheromones have yet been isolated from queens. However, it is obvious from the many observations on the higher termites that the physogastric queen is producing a substance which attracts the workers grooming her, and the soldiers surrounding her. The most classic instance of this is seen in *Termes bellicosus* Smeathman and reported by Escherich (1908).

Verron's studies provide a start in the investigation of the chemical factors involved in colony cohesion and aggregation, but much more work is needed to clarify the several points that have been discussed above. Since the chemical involved in Verron's studies belongs to a group of well-known attractants (some occurring in leaves: Moore, 1968b; Riddiford, 1968) it would be of value to determine whether the 3-hexen-1-ol is exuded through the cuticle in general, whether it is an anal secretion, or whether it is voided in the feces and adsorbed onto the body. If the latter occurs, Verron's findings would reflect either a differential adsorption according to sex or else some behavioral mechanism which ensures that certain castes have a greater contact with the substance than others. It should be noted that Verron and Barbier (1962) have reported that 3-hexen-1-ol is present in the galleries of termites (*Microcerotermes edentatus* Wasmann) and in the wood on which they were found.

An important sidelight of Verron's work is the method by which *Kalotermes* responds to the attractive substance. After ablating antennae, and labial and maxillary palpi, Verron found that a response could still be obtained. However, in conditions of high oxygen tension, normal individuals gave no response and Verron suggests that the tracheal system is involved in the perception of this odor. If this is indeed so, many previous experiments on sensory perception in termites that involved ablations of the known sense organs might require a reexamination.

CASTE RECOGNITION AND DETERMINATION. Caste recognition has been separated from colony cohesion because the recognition of one caste of termites by another does not necessarily imply that attraction is involved (by chemical or other means), but only that information is transmitted from one termite to another that it is of a certain caste. As will be seen, the termite receiving the information may react in various ways depending on other conditions: it may exhibit cannibalism, grooming, alarm, attraction, and so on. While vision has been implicated in certain aspects of caste recognition, e.g., that between newly de-alated imagos (Fuller, 1915), experiments and observations have shown that in many instances chemicals are involved. The work on "reactivity" and "stimulation" carried out by Verron (1963) and reviewed above, showed that odors do exist by which the various castes of *Kalotermes* could recognize each other. Verron emphasized this point: his criterion was, of course, attraction and to that extent his conclusions are indirect. Lüscher (1961) states that the actual presence or absence of a caste (soldiers or reproductives) is noted by individuals in the colony and is evident from their changed behavior in experimental situations. Using the classic technique of Huber (1810), he separated the sexual pair of a colony of *Kalotermes* from the rest of the colony by means of a wire gauze barrier; the colony produced secondary reproductives, but these were eliminated by the workers as soon as they were formed. In a similar experiment a double wire gauze was used, thus stopping antennal contact. Supplementaries were again produced but this time they survived. Lüscher concluded that a pheromone perceived by antennal contact indicates the presence of the original reproductives and this information causes the workers to eliminate the new secondaries. It should once again be noted that a volatile chemical cannot be involved since in the double-barrier situation the information would have been transmitted. Lüscher (1961) suggested that contact chemoreceptors on the antennae of the pseudergates of *Kalotermes* are involved. In the original experiment of Lüscher (1952a) the reproductives involved were primaries, and it is possible that they produce a specific identifying pheromone which is different from that produced by the supplementaries. In later papers the use of the term "functional reproductives" for the original primary reproductives is confusing as this could refer just as well to functional secondary neoteinics in colonies lacking primaries. It would also seem that this experiment has been further modified (Lüscher, 1961) by leaving only *one* sex of the functional (primary?) reproductives in one half of the divided colony. Under these conditions, Lüscher claims that there is selective elimination by sex. The only supplementaries to be eliminated are those of the same sex of the reproductive retained. It would seem that this may only be true for *Kalotermes*, as Buchli (in Lüscher, 1952b) has stated that in *Reticulitermes* neoteinics can develop in the presence of the royal imaginal couple. Buchli further states that elimination occurs indiscriminately in that genus (the old primary reproductives or the new neoteinics being equally liable to elimination by cannibalism), but that elimination is reduced if more living space is given to the colony. In *Zootermopsis* there would seem to be no selection involved in elimination, when it occurs, and each supplementary formed seems to have the same chance as any other of being killed and eaten, (unpublished observations). The reconciliation of the results obtained from different genera awaits further rigorous experimentation and the isolation of any chemicals involved.

On reflection, however, it would seem that the mechanism for caste elimination proposed by Lüscher (1952a, 1961) may prove to be too simple in some respects, while too complicated in others. If the hypothesis put forward and the experiments are indeed correct then:

(a) The original reproductives in a colony (primary or secondary) must indicate that they are: (1) functional, (2) the original reproductives, (3) male or female.

(b) The newly formed reproductives must indicate they are: (1) functional, (2) "new," (3) male or female.

It is presumed that this information is conveyed chemically. In both cases it could therefore be presumed that chemicals indicating sex and maturity (1 and 3) are the same for both. For (2) it could be that there are either separate pheromones for new and old reproductives or that one or other of the reproductives lacks a pheromone. This could become even more complicated if differences occur between primary and secondary reproductives.

Before eliminating a reproductive, a worker or nymph must make a decision based on these variables. Such a system implies a complicated evaluation of information by a nymph. For example, it would only kill a newly formed reproductive if an old reproductive of the same sex is present. It is more likely that other factors are involved or that selection may not be as strict as Lüscher's experiment would seem to indicate.

It should perhaps be noted here that in termites there is normally no fighting between the reproductives themselves as may happen in ants. Elimination is by the nymphs or pseudergates. The only reported case of fighting between reproductives was by Emerson (1933a) when he reported that one queen killed another. The situation was not at all clear and Emerson believed that a form of "conditioning" was occurring: the evidence for this was not good.

From behavior seen in other species, we may infer that other types of caste recognition occur in termites. There is no doubt that the physogastric queen of

African and Central American termites is recognized by the workers. This must almost certainly be due to a pheromone and much work remains to be done in elucidating the chemicals involved in such behavior.

Caste determination. Lüscher (1961, 1967) has summarized and reviewed his own work on caste determination in termites. Other reviews have also appeared quite recently (e.g., Highnam, 1964; Brian, 1965; Wilson, 1965; Weaver, 1966) which deal to a greater or lesser extent with this subject. To give another account at this time would seem superfluous and the reader is referred to Figure 2 where Lüscher's hypothesis is summarized and to the references just cited. Caution should be used in judging all factors shown in Figure 2 as proved since Lüscher (1961) himself points out that the interactions involved represent only a working hypothesis.

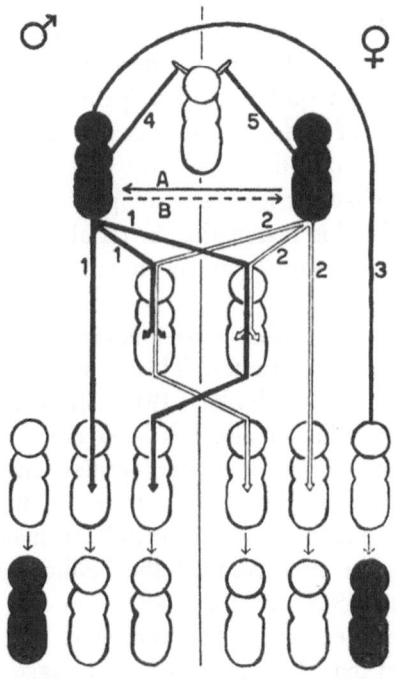

Fig. 2. Tentative representation of pheromone actions controlling the production and elimination of supplementary reproductives in *Kalotermes flavicollis*. Reproductives are shown in black; pseudergates in white. The thin arrows indicate molts. Males are represented on the left, females on the right of the diagram. The numbers 1 through 5 indicate the postulated different pheromones; the letters A and B, stimuli of an unknown nature. The female reproductive stimulates the male by factor (A) to give off its sex-specific inhibitory pheromone (1). The female reproductive also gives off a sex-specific inhibitory pheromone (2), but the stimulation by the male (B) is less evident. The inhibitory pheromones are absorbed by the pseudergates of the appropriate sex and prevent their development into supplementary reproductives. Without the appropriate inhibitory pheromone the competent pseudergates change into supplementaries, more females doing so in the presence of the stimulatory pheromone (3) which is given off by the male reproductives. This pheromone probably increases competence in female pseudergates. The pheromones (4) and (5), which are postulated to cause elimination of excess reproductives, are also produced by the sexuals and are probably equally sex-specific. (*From* Lüscher. 1961. *Symposium No. 1 Roy. Entom. Soc. (London).*)

Pheromones eliciting a more or less
immediate response

SEX PHEROMONES.   Sex pheromones are ubiquitous in insects, and they may be of
several types. It seems useful to separate the types into three categories as failure
to do so has obviously produced misunderstandings in the literature (see Vité,
1967; Wood et al., 1967). Wilson and Bossert (1963) have pointed out that sex
pheromones release an attraction response, a sexual response [the aphrodisiacs of
Butler (1967)] or both. A third type of pheromone that might be added (see below)
is a generalized excitatory one which only secondarily produces sexual behavior
other than copulation. Because such a pheromone causes excitation, it would have
similarities with alarm pheromones.

Sexual behavior in termites is quite different from that in the social Hymenoptera. Briefly, imagos of both sexes are formed and leave the nest together in a dispersal rather than mating flight. They fly clumsily, settle on the ground and then
break off their wings. In many species tandem pairs of a male and a female are
formed and the termites run off and excavate a cell in the soil or in wood. Depending on the species, one or both of the sexes may take part in the excavation. It is
only later, when the initial chamber has been completed and sealed off, that copulation occurs. It is probably for this reason that in termites no spectacular response to
a pheromone has yet been discovered similar to those of ants or indeed of other
nonsocial insects.

Nevertheless, there is evidence that a pheromone may be involved in the "calling" behavior in some species of termites. This behavior has been excellently described in South African termites by Fuller (1915). It occurs immediately after the
female loses her wings. The behavior also has been noted and studied many times
in American and European species of *Reticulitermes*. In that genus the female
raises her abdomen through 90 degrees while remaining still: when a male touches
her she drops her abdomen and begins to run and the male follows in tandem. If
contact is lost the female once more resumes the "calling attitude." This has been
well described by Emerson (1929, 1933b, 1949) who also conducted experiments
showing the importance of the contact between male and female in tandem behavior. It has been supposed that the female, when she takes up the calling attitude,
releases a scent which attracts the male. This idea was put forward by Goetsch
(1936), but Grassé (1942) studying *Kalotermes flavicollis* felt unable to support
this as he was unable to find a gland with the appropriate functional qualifications
for producing a scent. The sternal gland (see Stuart, 1964) was dismissed, as no
ducts could be observed opening from it to the exterior, and he felt glands such as
this probably had more of an endocrine function.

Buchli (1960) reinvestigated the post-flight behavior of *Reticulitermes* and he
supported the contention of Goetsch that the female releases a pheromone which
the male recognizes. By presenting various dummies to males that had been separated from females, he was able to show that only a glass dummy, previously
touched to the abdomen of a calling female, elicited a response in the males. Buchli
also noted that when the sexes were segregated it was possible to induce males to

follow males by exposing them to sunlight. The females in a similar situation never formed tandem pairs. In less extensive experiments with *Kalotermes*, Buchli found that the presence of a female was required to produce tandems between males placed in a box; exposure to sunlight was insufficient in this genus. Stuart (1968) in an independent study on *Reticulitermes flavipes* (Kollar) obtained, in general, results similar to those of Buchli. It was found that, in the absence of the females, males could be induced at will to form tandem chains of up to eight individuals by the relatively simple stimulus of tapping the petri dish containing them. This response could still be obtained with males after 12 days of confinement at 23°C and approximately 75 percent relative humidity. Any other stimulus that disturbed the termites also produced the same response.

Buchli concluded from his results that (a) the female produces a scent which attracts the male "chemotropically"; and (b) when touched by the male's antennae, a calling female will begin to run, but if contact is lost the female will again assume the calling attitude: it is these actions which produce tandem behavior. The additional study of Stuart (1968) raises some questions regarding the type of chemical that might be produced. Is it a simple attractant or is it a sex recognition pheromone? It cannot be considered strictly as an aphrodisiac as it does not induce copulation. Stuart (1968) has suggested that if a pheromone is indeed produced then it will probably have a general excitatory effect on the animal and it will be sex specific. In other words, the substance would have the same effect as any stimulus (sunlight, jarring, air motion) that produces accelerated motion in the animals. In this regard it may prove to be somewhat similar to an aphrodisiac pheromone or an alarm pheromone except that it acts at a time when the males will follow any moving object roughly their own size: instances have been noted of males following workers as well as other males (Buchli, 1960). Buchli suggests that the accessory glands of the female, which open on the anterior border of the ninth sternite near the genital aperture, may produce the postulated pheromone.

Tandem behavior and the associated "calling" behavior of the female do not occur in all termites and may even be absent or present within a single genus, e.g., *Cryptotermes*. In *C. havilandi* (Sjöst) it is absent (Wilkinson, 1962), while in *C. cynocephalus* Light it is present (Kalshoven, 1960). In *Zootermopsis* calling behavior and tandem behavior seem to be absent. Some following does occur but it is very sporadic and not at all efficient. Preliminary studies (unpublished) have shown that in the newly de-alated reproductives of that genus there is, however, a definite mutual sexual attraction between males and females. Ether extracts of whole females or males, and ether controls, tested against females and males indicated a mutual attraction between males and females, but little or no response was shown by males to extracts made from males. However, females do show some attraction to one another. Such an attraction would be useful biologically to ensure that most males and females do end up in the same excavated cavity.

When it is realized that termites do not usually fly far from their nest and that copulation does not occur until some time after the pairing,* it is not surprising

---

* In *Zootermopsis* copulation takes place sometime within the first two weeks after the beginning of the excavation, and the first eggs are laid about 16 days after pairing. This time varies in various genera and species. Mr. Frank J. Gay has kindly provided the following information

that highly volatile and spectacular sex pheromones are absent. Again, since the two sexual individuals (or a relatively small number, when secondary reproductives are present) are confined in very close proximity to each other in the nest no highly specific pheromone (e.g., aphrodisiacs such as are found in the related cockroaches) is needed to ensure copulation. The animals must, of course, be able to recognize one another as reproductives. In *Zootermopsis* copulation is quite unspectacular (Stuart, 1968) and seems to happen almost by accident. The act is similar in the primary reproductives and in the secondary reproductives, including the fertile soldiers which are found in this genus. Indeed, among the secondary reproductives one cannot use the act of copulation of an egg-laying reproductive with another to be able to state that the other partner is a male. By marking and dissection, it has been found that often two females are found *in copula* (Stuart, 1968).

It can be seen then that chemical communication is present in sexual behavior in termites. Much more work is needed to evaluate properly the role and importance of any pheromones. It would seem they are less important than in some other orders, but caution must be exercised as the reports may merely indicate a lack of extensive study.

ALERTING OR ALARM PHEROMONES. Alarm pheromones are difficult to categorize as "alarm" is a subjective concept. It can mean an increase in locomotion with fleeing, in some cases immobility, or sometimes lowering of an animal's attack threshold. Some authors (e.g., Butler, 1967) prefer to use the term "altering" pheromone to separate the communicatory substances from chemical defense secretions, but this word again has anthropomorphic overtones. Perhaps it is better to retain the term "alarm reaction" and to be sure that the behavior being described fits a stated definition of this term. Such a definition of alarm in social insects partly based on Maschwitz (1967) could be: the ability of an individual to inform nest mates quickly of an enemy attack, and of violent physical perturbations of the environment such as flooding, the breaking of a nest, and so on. Alarm pheromones would then be pheromones which play some part in these reactions. In this chapter this is what is meant by the term "alarm pheromone."

In termites, while there is at least one pheromone that is known to be involved in an alarm reaction, in itself it does not fall under the category of a substance which excites other nest mates. Rather it is connected with orientation of the termites and will therefore be treated under orientation.

Recently among the termites some substances have been implicated in alarm which are considered similar in action to those found in the Hymenoptera. There is some doubt as to whether they ought to be considered alarm pheromones (Maschwitz, 1967) and, therefore, it may be useful to discuss their function here.

---

regarding the appearance of the eggs in some Australian termites which gives some idea of the variability present in the Order. Termopsidae: *Porotermes adamsoni* (Froggatt), earliest eggs at 24 days, commonly present at 28 days. Rhinotermitidae: *Coptotermes acinaciformis* (Froggatt) earliest eggs at 7 days, commonly present at 9 days; *C. frenchi* Hill, earliest eggs at 4 days, commonly present at 5 days; *C. lacteus* (Froggatt), earliest eggs at 8 days, commonly present at 11 days; *Schedorhinotermes intermedius intermedius* (Brauer), earliest egg at 6 days commonly present at 7 to 8 days. Termitidae: *Nasutitermes exitiosus* (Hill) earliest eggs at 4 days; *Amitermes neogermanus* Hill earliest eggs at 21 days, commonly present at 24 to 28 days.

Soldiers of the genus *Nasutitermes* have heads that are drawn out anteriorly in a nose or nasute at the tip of which there is the opening of the cephalic or frontal gland. In such termites, this gland takes up a large amount of space in the head. When the nasute soldiers are disturbed by the presence of a moving object, by their being touched, or when a foreign odor reaches them, the secretion produced by the gland is expelled violently. If the object causing this alarm is an ant or a termite from another colony, the intruder is immobilized somewhat by the secretion which rapidly becomes sticky on exposure to air. This has long been known as a defense mechanism (Beaumont, 1889-1890). In addition to this function, the secretion has also been claimed to have an excitatory effect on other termites and to cause "alarm" (Ernst, 1959). Moore (1964, 1968a) supports the views of Ernst and in addition has shown that, at least in the Australian nasutes, the secretion is composed mainly of pinene. When pinene is presented to other soldiers in a manner similar to that used by Ernst, it produces the same effect of stimulating these soldiers to fire further shots. Maschwitz (1967) does not regard the nasute secretion as a true alarm substance, and in experiments conducted by himself on *Nasutitermes ephratae* (Holmgren), he concluded that the secretion has no specific alarm effect. No details are yet available on Maschwitz's investigation. In *Nasutitermes corniger* (Motschulsky), the defense secretion produced by the soldiers is not necessary in itself for the transmission of alarm. If a *Nasutitermes* nest is broken into, then reactions quite similar to that which would occur in *Zootermopsis* in the same situation take place (Stuart, 1967). In the Panamanian nasutes excitation is transmitted by mechanical means. There is again recruitment of other termites to the site of alarm by a chemical trail. No secretion is ejected by the soldiers unless a direct mechanical stimulus or odor stimulus is presented near them. It should also be noted that the defense actions of all termites vary with the degree of the original stimulus and the caste involved (Stuart, 1967). It is suggested here that the pinenes which would be ejected in situations where attack is actively in progress (e.g., a moving object near a break in a nasute nest) cause an increase in the level of excitation in the other termites. They would thus have a temporary synergistic effect at the alarm site, but alarm would still be communicated by the mechanical bumping of an alarmed termite. It can be seen, then, that the role of the cephalic gland secretion of *Nasutitermes* in alarm transmission is not yet clear and more work is needed to resolve the different views. The definitive experiment would show whether or not a secretion can excite other termites under conditions where other factors that could have a communicatory function (contact, sound, visual) have been eliminated. Such experiments were conducted by Stuart (1963b) on *Zootermopsis* and the apparatus used (Fig. 3) could be adapted for other genera of termites. In one experiment, a colony of 50 nymphs of *Zootermopsis* was placed equally in two halves of a Wilson nest which had been divided by double-wire gauze. The nest was placed on a sponge rubber pad to prevent jarring, and two soldiers were added to one half and one to the other. When one half of the nest was maximally alarmed, no transmission of the alarm to the other half (15 mm away) was noted (Fig. 3a). In a second experiment no alarm could be evoked when air was actively drawn over from an alarmed colony to its other half (Fig. 3b). In either case, when contact was permitted between the halves, alarm was transmitted. Thus, it

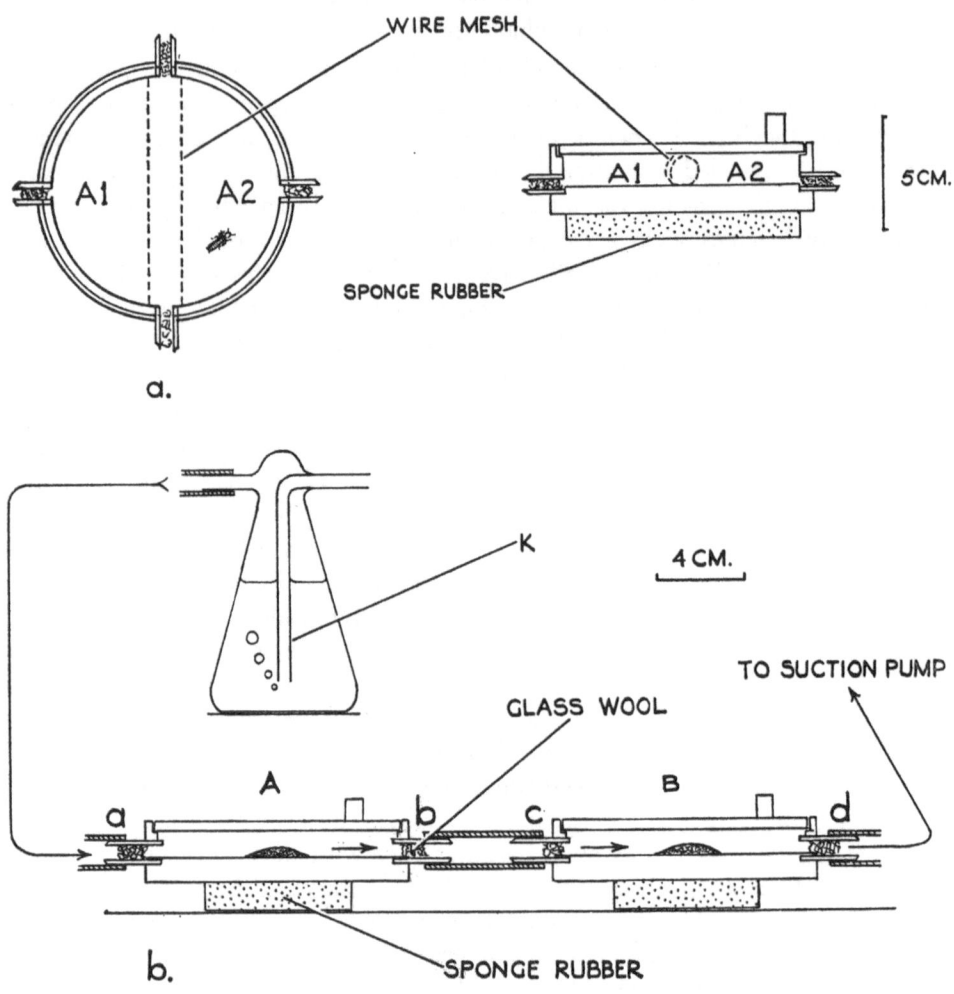

FIG. 3. (a) Apparatus used to see whether alarm could be transmitted from one-half of a colony to another without actual contact. Sound and odor could easily travel the short distance 15.0 mm between the two halves. (b) Apparatus used to see whether alarm could be transmitted from one-half of a colony to another by a volatile chemical. The effect of sound was eliminated in this set-up. (Drawn to scale except for gas-washing flask.) In both instances the termites were left undisturbed by the experimenter for from 3 to 6 days. See text for explanation. (*From* Stuart. 1963b. *Physiol. Zool.*, 36:85-96.)

was shown that in *Zootermopsis* alarm is not transmitted by chemicals (at least not by volatile ones) but is propagated by mechanical contact. On observation this contact can be seen to consist of a vigorous zig-zag bumping movement (Stuart 1963b).

However, chemicals have been implicated in the alarm behavior of at least two other species of termites. Moore (1968a, 1968b) reports that limonene has been identified as the alarm pheromone of *Drepanotermes rubriceps* (Froggatt) while he presumes terpinolene plays a similar role in *Amitermes herbertensis* Mjöb.

The criterion for alarm in these species is reflected in the bioassay used by

Moore for the substance produced in *Drepanotermes*: a piece of filter paper with the substance was presented to a soldier termite at a distance of a few millimeters, a similar untreated piece of filter paper served as the control. If the soldier gave a short-lived snapping and biting frenzy, the assay was judged positive. While limonene causes such attack behavior it is debatable whether it can definitely be called an alarm substance until it has been shown that it causes general excitation and the alerting of the colony rather than attack. In this regard, Maschwitz (1967) studying alarm in *Apis* showed that the alarm secretion only attracts and excites worker bees, whereas attack is triggered by other stimuli. Though different mechanisms are involved, attack also seems triggered by many stimuli after recruitment in the primitive termite *Zootermopsis*. When soldiers of this genus are placed in a situation like that used for *Drepanotermes* and odoriferous substances presented to them on filter paper (cf. Moore); then depending on the chemical, the termites will exhibit different degrees of snapping and alarm. The same result could, of course, again be evoked by mechanical stimulation. In any situation where transmission of alarm is being investigated, a whole colony must be observed to see what the effect of the alarming of one termite or a group of termites does to the behavior of the rest of the colony in experimental situations. As the reactions of social insects to simple stimuli are not necessarily equally simple, great caution must be used in interpreting experiments and designing bioassays. This will be emphasized again in relation to trail laying, but in general an attempt should be made to experiment on the insects in as natural a situation as possible (Stuart, 1968).

If the substance (limonene) under the conditions just stated proves truly to be an alarm substance in *Drepanotermes*, then it is possible that the termites might have followed an evolution in the production of a chemical alarm system similar to that of the *Hymenoptera*. In that order the more primitive members do not have alarm pheromones whilst the advanced groups do (Maschwitz, 1967). It is interesting that in *Polistes* an excitation, reminiscent of an alarm reaction, can be produced by foreign scents (Maschwitz, 1967). This also is true of *Zootermopsis* (Stuart, 1963b, 1968). It is also possible that in a higher termite such as *Drepanotermes* an alarm produced by a chemical would be more useful biologically than it would in a termite such as *Zootermopsis*. Thus, one might expect a harvesting termite to have lines of communication which are much extended in comparison with those of a primitive termite. For the harvester, a rapidly diffusing volatile chemical might transmit alarm faster than the mechanical method found in *Zootermopsis* and thus be advantageous. *Zootermopsis* lives entirely in wood where nest and foraging area are the same, and in the galleries and tunnels of such a termite a volatile chemical (even one with a short fade-out time) would almost certainly excite more termites than necessary to deal with any one situation. This will be discussed further under orientation behavior which is a component of alarm behavior in *Zootermopsis*. Moore (1968a, 1968b) considers that in both *Amitermes herbertensis* and *Drepanotermes* the original defensive role of the secretion has been lost, though in *Amitermes vitiosus* Hill a sticky terpene secretion is still produced.

If this proves correct there would seem to be a sequence of defense/alarm in *Nasutitermes* and some *Amitermes* to a purely alarm function as in *Drepanotermes* and *A. herbertensis*.

ORIENTATION. Insects orient in many different ways; e.g., to water flow, light, visual stimuli, and so on. Many use chemicals to aid in orientation: these may be (a) general attractants that produce a gradient, (b) substances that stimulate the insect to orient using some other directional factors: e.g., the anemotaxis of the male *Bombyx* (see Butler, 1967) when stimulated by the scent of the female, or (c) trail pheromones. Apart from the previously discussed sex attractants, which can be considered as a special sort of orientation pheromone, only pheromones concerned with trail laying have been investigated in the termites (Lüscher and Müller, 1960; Stuart, 1960, 1961, 1963a). If alarm is extended to include the "Futteralarm" of Goetsch (1936), then in termites the trail is always interrelated with alarm behavior (Stuart, 1963b, 1967, 1968). In other words, the finding of food excites termites in a manner similar to that when they discover some foreign substance or are attacked. In the latter case, both the excitatory stimulus and the reaction are of a higher order.

The use of trails was first discovered in foraging behavior and some early accounts are given by Beaumont (1889-1890) and Grassi and Sandias (1896-1897). One of the most spectacular trail-laying genera is that of *Nasutitermes*. Many members of this genus construct carton nests on the limbs of trees and forage on the forest floor. They maintain a connection with their aboreal nest by means of the well-known covered runways or arcades. It is now known that these runways are preceded by an open trail (Grassé and Noirot, 1951; Pasteels, 1965, 1967; Stuart, 1960, 1961, 1963a, 1967, 1968). If a portion of carton is taken into the laboratory and placed on a clean glass plate at a distance of about 200 cm from a piece of wood, in a very short time a trail is established between the piece of nest and the wood. The trail is marked by fecal droppings, the pavé of Grassé and Noirot (1951), but it is not the origin of the trail as was thought by Grassé and Noirot. It is now known that a trail pheromone is produced by the sternal gland of *Nasutitermes* (Stuart, 1961, 1963a). A trail substance is also produced by the sternal glands of other genera including *Zootermopsis* (Lüscher and Müller, 1960; Stuart, 1960, 1961, 1963a) and *Reticulitermes* (Stuart, 1968). The sternal gland is found quite universally in termites and it is probable that it has a similar function in all genera. The action of the gland has been studied mainly in *Zootermopsis* (Stuart, 1964; Stuart and Satir, 1968), and as a knowledge of its structure and function is required to understand the behavior of the insect, a short account will be given here.

Basically, the gland is modified epidermis and is situated in the fifth abdominal sternite. There are no ducts opening to the exterior, but electron microscopy (Stuart and Satir, 1968) has indicated the presence of what appears to be a terpenoid secretion in spaces in the cuticle also seen in a layer of secretory cells. A diagram of the gland, based on electron micrographs, is shown in Figure 4. In addition to having the three layers shown in the figure the gland is studded with approximately 200 campaniform sensilla. These sensilla are postulated to provide sensory information about the degree of compression of the sternites (Stuart, 1964). Compression occurs when a termite actively laying a trail presses its abdomen on the ground (see Fig. 5). The postulated feedback mechanism is diagrammed in Figure 6. When the sternites are compressed the secretion is expelled from the external reservoir formed by the overlapping of segment 5 by

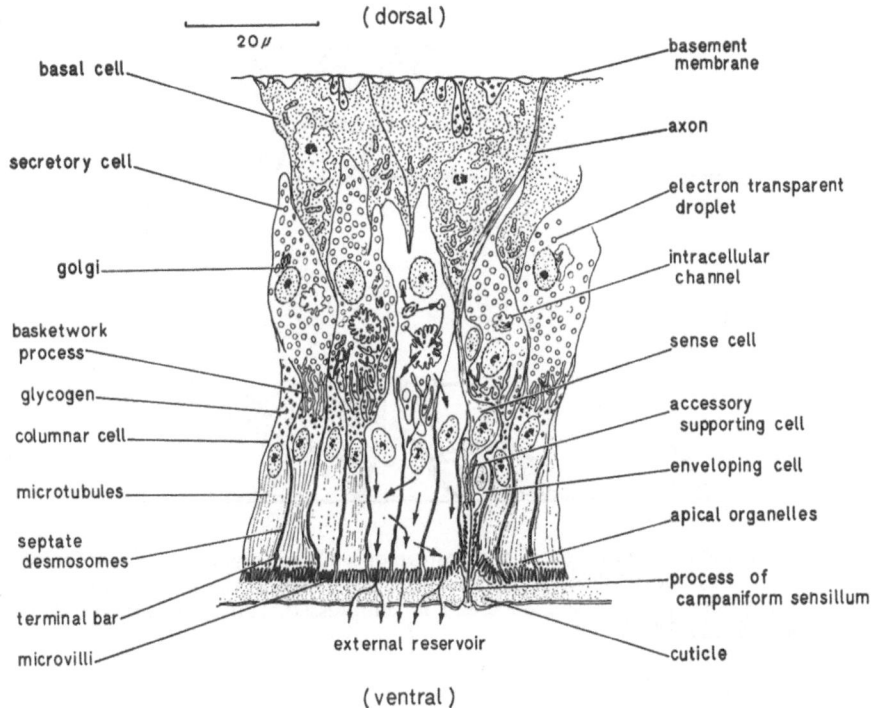

(dorsal)

20 μ

basal cell

secretory cell

golgi

basketwork process

glycogen

columnar cell

microtubules

septate desmosomes

terminal bar

microvilli

basement membrane

axon

electron transparent droplet

intracellular channel

sense cell

accessory supporting cell

enveloping cell

apical organelles

process of campaniform sensillum

external reservoir

cuticle

(ventral)

Fig. 4.   Diagrammatic representation of a portion of the *Zootermopsis* sternal gland showing prominent cellular features. In the central region of the diagram, details of cell structure are simplified to indicate (arrows) a probable pathway of pheromone transport from the secretory cell to the gland reservoir. The pheromone is presumably manufactured and packaged in the electron-lucent, smooth membrane-bounded droplets, originally derived from the Golgi region. These droplets fuse to form the microvillous channel. Molecules of pheromone move from the channel through the regions of tight junction in the basketwork processes into the columnar cells. Transport through the columnar cells may be free, vesicular, or as a component of the microtubule wall. At the columnar cell border, pheromone is liberated into the lipid micelle meshwork of the cuticle, and it passes through the cuticle to the gland's reservoir (outside the animal) formed by the overlapping of sternite V by sternite IV. (*After* Stuart and Satir. 1968. *J. Cell Biol.*, 36:527-549.)

segment 4. It can be seen that although the pheromone is secreted continuously it is only applied to the substrate while the termite drags its abdomen. The termite, thus, could have control over the amount of pheromone laid down as this would be a function of the compression of the sternites (Fig. 6).

At present the identification of the pheromone or pheromones remains to be done* Moore (1966) has found a substance extracted from whole termites of *Nasutitermes exitiosus* (Hill) that will produce trail following when a proper assay is used. The chemical has been characterized as an unsaturated diterpenoid hydrocarbon of molecular weight 273 and empirical formula $C_{20}H_{32}$. It is assumed that in this species the secretion originates in the sternal gland as in *Nasutitermes*

* Since this article was completed, n-caproic acid (n-hexanoic acid) has been implicated as the trail pheromone of *Zootermopsis nevadensis* (Hagen) by Hummel and Karlson (1968) and Karlson et al. (1968).

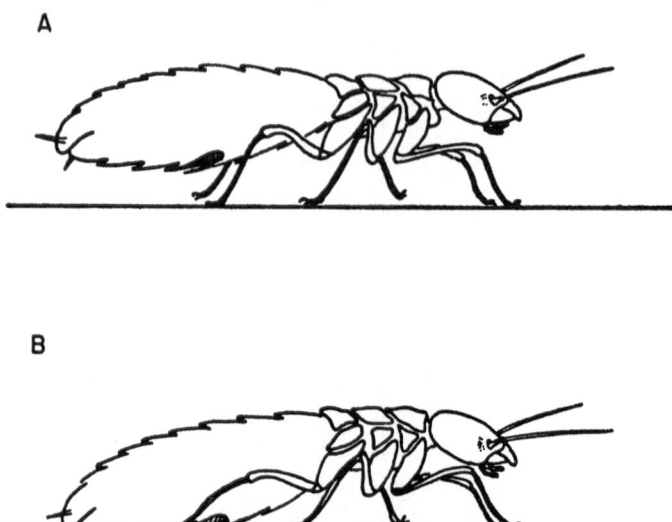

Fig. 5. Schematic representation of the posture of a 12 mm nymph of *Zootermopsis* during normal activity (A) and while laying a trail (B). Note that in the latter position sternite IV overlying the sternal gland (hatched area) is compressed. (*After* Stuart and Satir. 1968. *J. Cell Biol.*, 36:527-549.)

corniger Motschulsky (Stuart, 1960, 1961). However, caution again must be exercised in unequivocally regarding the substance as originating in the gland. Trail laying mimics are now being found, and Becker (1966) states that the ink from a ballpoint pen contains a substance which causes trail following in 15 species of termites belonging to the families Rhinotermitidae and Termitidae. Recently Matsumura, Coppel, and Tai (1968) have implicated *cis*-3-*cis*-6-*trans*-8-dodecatriene-l-ol as the trail pheromone of *Reticulitermes flavipes*. It would seem (though details regarding numbers of termites used and experimental conditions were not given) that their bioassay is very much like that of Smythe et al. (1967). This assay as Stuart (1968) has pointed out cannot be considered valid. Under the experimental

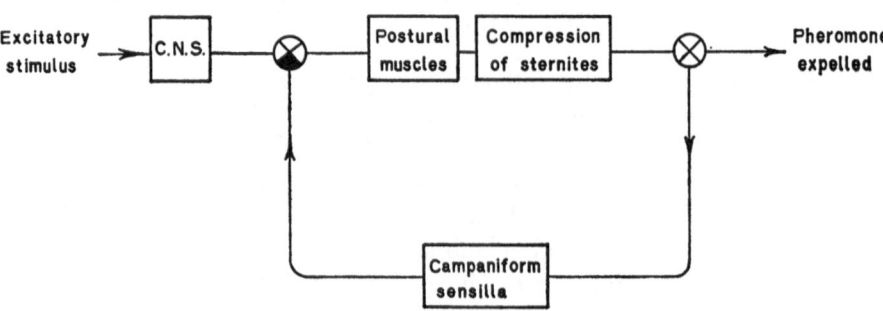

Fig. 6. Control diagram showing the suggested propioceptive feedback loop from campaniform sensilla during trail-laying in *Zootermopsis*. The excitatory stimulus may be of several origins (see Stuart, 1967). Only the pathways leading to trail-laying are indicated. A usual notation for control systems is use, the shaded quadrant indicating subtraction. (*From* Stuart. 1968. *Biology of Termites*. Courtesy of Academic Press, Inc.)

conditions used by these workers in their assay, various attractants when drawn out in a trail can give a positive response. Until the substance is tested in the manner of Stuart (1963a) or in the similar procedure of Moore (1968b) its identification as the trail-following pheromone must await confirmation. Once more it must be emphasised that whole termites were again used for the extraction and (especially as it is stated that a similar substance can be extracted from wood (aspen) infected with the fungus *Lenzites trabea* Pers. ex. Fr.) it is always possible that gut contents of the termite are the source of the substance.

In the original studies of Stuart (1960, 1961, 1963a), bioassays were employed in which the whole colony or at least a large number of termites from a colony were used in the tests. Furthermore, the test utilized squashes and extracts of the extirpated gland. Only Pasteels (1965), among the subsequent investigators, has made extracts from the gland itself, and only he and Moore (1968b) have used meaningful assays.

In addition to finding the origin and composition of a pheromone, it is equally important to find its biologic function. More studies should be concerned with the latter aspect. In the Nasutitermitinae and other termites that forage at great distances from their nest, the trail obviously serves to maintain contact between the nest and the food source. When a food source has been found, a trail laid on the way out by the scout is reinforced on the way back and the excitation of finding food (low-level alarm) is transmitted by contact (Stuart, 1960, 1968) when one termite meets another. In *Zootermopsis*, a termite which lives in damp decaying logs in its native California, such extensive foraging does not occur. However, Lüscher (1960) has stated that the function of the trail in *Zootermopsis* is in foraging. He confined six to ten *Zootermopsis* nymphs in an artificial 1.5 by 7.0 cm nest for two days. A running surface of 7.0 by 9.5 cms was available to the termites when the nest was opened, and a similar nest to the first was filled with moist sawdust and placed at the other end of the running surface. Eventually, a nymph found the sawdust and returned carrying a particle of food to the first nest. The termites then migrated to the second nest.

Stuart (1960) carried out a similar experiment with *Zootermopsis* using larger numbers in a more natural situation, but was unable to induce the termites to forage. The termites followed the usual behavior of *Zootermopsis* and deposited fecal material at the exit thus sealing themselves off. In many instances the termites starved within a short distance of food. It would seem that in natural conditions *Zootermopsis* would not forage extensively if at all. The most important use of the trail in *Zootermopsis* has been shown to be in the recruitment of workers and soldiers to the site of alarm. (Stuart, 1963b, 1967, 1968). When *Reticulitermes* (unpublished) are placed in a situation similar to that of Lüscher (1960), foraging does occur and a trail is established between the nest portion and food. Similar behavior occurs in the nasutes (Pasteels, 1967; Stuart, 1960, 1965). Although no food is carried back to the nest mates by the first termite to reach the food, alarm is transmitted by the well-known bumping movement (Stuart, 1963b). Also, a trail is laid which causes the termites to migrate to the paper used as the food source.

It is noteworthy that the trail pheromone in *Zootermopsis* seems remarkably stable (Stuart, 1960, 1963a; Lüscher and Müller, 1960). In *Nasutitermes* if the diterpenoid is indeed the trail substance—and this seems likely because the secretion seen in the sternal gland of the *Zootermopsis* at least has appearance of a terpene (Stuart and Satir, 1968)—then it would not be highly volatile with a molecular weight of 273 (Moore, 1966). The olefinic primary alcohol of Matsumura et al. (1968) would probably still be only moderately volatile. One of the criteria of Wilson and Bossert (1963) regarding trail substances is that they should have a short fade-out time. They contend that without this the insects would become trapped on trails and the advantage of speedy recruitment would be lost. In termites, it would seem that the trails are more permanent even when covered runways are absent and the question arises as to why a termite follows any one trail rather than another. Three possibilities, which are not mutually exclusive, are: (a) that there is a concentration effect, (b) that a reinforcement could occur, and (c) that the excitation which is transmitted mechanically regulates recruitment. The latter possibility has been demonstrated (Stuart, 1963b): greater excitation caused by a mechanical prodding produced greater recruitment. A disturbed or excited termite will tend to follow a trail (Stuart, 1963b, 1968). The disadvantage of a highly volatile alarm pheromone being produced in a confined space has previously been pointed out.

CHEMICALS OTHER THAN PHEROMONES

Odors wholly or partly derived from the environment

COLONY ODOR.   It is difficult to classify the odors that seem specific to a colony as they may be derived partly from the environment, partly from the exudations of the various castes, and partly from exudations of the individual insect itself. As Butler (1967) points out, such odors are difficult to study because of difficulties in devising sensitive methods of bioassay. In termites, colony odor is known mainly in connection with colony defense and the recognition of intruders. There is no strong evidence that the odor is derived from a pheromone. Since the bioassay for a colony odor pheromone would be the negative one of preventing the attack or alarm response in another termite, it would be difficult to test. There is also the possibility that the odor is learned and again if this were the case it is debatable whether any chemical substance isolated could be called a pheromone, though nothing in the original definition precludes this. In view of these difficulties, colony odor in termites will be treated separately from pheromones.

Colony odor is probably one of the most primitive types of chemical communication possible. In termites it is quite noticeable, and it has been known for many years that when two colonies of the same species are brought in contact with each other (usually under experimental conditions) the individuals of one colony will mutually repel those of the other one, and in a confined space fighting will occur. Though similar behavior had been observed in ants (Lubbock, 1882; McCook, 1887; Fielde, 1905) and odor implicated, Beaumont (1889-90) and

Grassi and Sandias (1896-1897) were probably the first to show that odor had played a part in the similar behavior in termites. Beaumont (1889-1890) conducted some elementary experiments. These showed that if a termite was washed and replaced in its own colony, it was investigated vigorously by its nest mates almost as if it were an alien. Conversely, when placed in an alien colony a washed termite was not attacked to the same extent as an unwashed comrade. Andrews (1911) conducted much more critical experiments concerning the effect of washing a termite on antagonism between colonies and concluded that although the subject was complex, the principal factor involved was certainly colony odor ("nest aura"). He was cautious to point out that the behavior of the termite could have been changed in other ways by the immersion. Andrews also noted that it was possible to mix colonies of *Nasutitermes ripperti* (Ramber) if large numbers are used, and he also noted that the degree of antagonism varied with the individual colony. Dropkin (1946) was able to combine colonies by first chilling them to immobilization, mixing them, and then allowing them to become active once more. The physiologic effect of cooling is unknown, but Stuart (1968) has suggested that the condensation formed on the termite when it is returned to a warmer temperature may be masking an odor just as immersion in water does. There always remains the possibility that the cooling has some effect on the nervous system of the insects and perhaps affects their memory for their original colony odor. Dropkin (using his technique) was able to mix colonies of *Kalotermes jouteli* Banks and *K. schwarzi* Banks with colonies of other species of *Kalotermes* and also with colonies of *Zootermopsis* and *Neotermes*.

There is certainly no doubt that odor is the principal factor if not the only one involved in recognition at the colony level. The origin of the odor (metabolic, pheromonal, environmental, or a mixture) is not known. In the bumble bees *Bombus agrorum* F. and *B. lucorum* L., Free (1958) showed that the scent of an alien colony could be adsorbed onto the body of an introduced bee, which would then be attacked when replaced in its own colony. Some preliminary experiments have been carried out on *Zootermopsis* to investigate the origin of the odor in that genus (Stuart, 1968). It was found that after three months, two halves of a divided colony of *Z. angusticollis* showed antagonism toward each other. Type of wood, and termite size and numbers, were controlled. In three months physiologic changes are known to occur so that the change in odor could be caused by changes in the mixture of existing pheromones. This could result when additional supplementary reproductives and soldiers develop. Such change is quite possible, and indeed Verron (1963) has shown that the different castes and instars show differential "reactivity" as a result of odors. Emerson (1929, 1939), too, seems to support the idea of an environmental origin of the odor.

It should perhaps be mentioned that eggs and young nymphs from different colonies of *Zootermopsis* are attacked in the same manner as older nymphs: there is no acceptance as has been reported for ants (Fielde, 1904; Plateaux, 1960). To sum up, the evidence at present suggests that colony odor in termites is derived from the environment (food, other members of the colony) and to some extent from a specific generic odor. The relative importance of these factors in colony odor and of colony odor itself (see Andrews, 1911) in caste recognition awaits further study.

The actual aggressive response deserves some comments as the same response will be given to a relatively nonodorous object which is in motion or static (Stuart, 1963b, 1967) and also to a drop of distilled water. Apparently, the termite is responding to anything that is different in odor from that of its own colony. Such a response raises interesting questions as to the neurophysiologic mechanisms involved.

Blood

Among other substances that can affect behavior in termites is the blood of a colony member. This can hardly be called a pheromone yet when a termite is injured, it is immediately investigated by other members of its colony and the blood seems to cause further feeding which leads to cannibalism. Cannibalism is, of course, a factor in caste elimination, q.v. It is difficult to determine whether odor or taste is involved in this reaction without critical experimentation. In insects less distinction is made between gustatory and olfactory responses. The reaction of termites to the blood could, of course, be merely the same antagonistic behavior produced by many stimuli including odor different from that of the colony (see above). That the injured termite is eaten rather than buried is probably a secondary reaction (Stuart, 1967).

Fungal and other attractants

Many environmental attractants are found in nature and, as they cannot be thought of as communication except in a very broad sense, little will be said of them here. In the case of the attractancy of rotting wood, some comment is needed: Esenther and Coppel (1964) have inferred that a substance from *Lenzites trabea* Pers. ex Fr. acts like a trail pheromone. Stuart (1968) has discussed this work and has pointed out that the bioassays used are not good ones for indicating the presence or absence of trail following. It is difficult to believe that the termites find the *Lenzites* by following an odor gradient; or, as Esenther and Coppel (1964) have suggested, by turning in the direction of the source and then exhibiting a preference to move in a straight line. No previous example of the latter situation has been recorded (cf. Fraenkel and Gunn, 1940).

Smythe et al. (1967)* themselves seem to state that when a substance they have extracted from *Lenzites* is drawn out as a trail, it only causes a termite to follow if the termite actually crosses it. This suggests that a substance of relatively low volatility exists and it would seem to negate the idea that such a substance could produce a significant odor gradient in nature. *Reticulitermes* has been observed while foraging (unpublished observations) and when this termite finds an attractive source of food it will recruit other members to the food site in a similar manner to that of *Nasutitermes* (Stuart, 1960, 1967, 1968). That is, it lays a

---

* It should be again noted that while the authors of this paper refer to the extract tested as "sternal gland attractant," the extraction was actually made from 370g (approximately 235,000 termites) of whole termites, *R. virginicus*. They assume that the substance has its origin in the sternal gland.

pheromone trail and communicates excitation, presumably caused by the finding of food, by contact. These observations seem borne out by photographs of the tests made by Becker (1965) concerning the extent to which *Reticulitermes* and *Kalotermes* are attracted by woods infected with several brown rot fungi. Becker's illustrations show that relatively straight constructions are made to those blocks of wood which are rated attractive: fewer or no constructions are built to the other blocks. Therefore, it is suggested that the more likely mechanism operating in *Reticulitermes* is that foraging is relatively random, but when an attractive food source is found the termite is excited to a greater or lesser extent and will return to the nest, laying or reinforcing a trail and communicating the excitation. It would seem that wood invaded by *Lenzites* contains a substance which is more attractive to *Reticultitermes* than any substance in sound wood. In any choice situation more termites would thus tend to be recruited to *Lenzites* wood than to the sound wood. Such a mechanism would accord wih the results of Becker (1965) and a similar mechanism (without trail-laying, of course) is known in the nest selection of swarming bees (Lindauer, 1961).

Conclusion

An attempt has been made to show how chemicals play a part in the various activities of termites. It was pointed out in the introduction that at present any classification of chemicals that produce behavior cannot be too precise. Classification itself is an attempt by man to order the study of chemical behavior so that discrepancies can be noted and investigated. To the animal, chemicals and odors may be extremely important factors in its environment, but they are not the only ones operating. Several pheromones as well as other chemicals may be present at the same time, or they may be presented in different orders to produce different patterns of behavior. A beginning has been made in the analysis of at least one major piece of termite behavior, that of nest construction (Stuart, 1967), and the relationship of this behavior to other blocks of behavior such as alarm and defense. By referring to Figure 7 it can be seen that chemicals play an important role in construction behavior and interact with other means of communication such as physical contact. Pheromones are involved as are colony odors, and the whole social integrating of a colony will doubtless be shown to contain many other systems which operate at the organismic and colony level.

Work on the chemical identification of substances thought to be involved in communication is advancing rapidly with the recruitment of natural product chemists to the field and with the increasing availability of analytic instruments such as gas-liquid chromatographs and mass spectrometers. It cannot be too strongly emphasized, however, that without proper knowledge of the biology of the insects and without relevant and repeatable bioassays, even the most careful chemical analysis is worthless. With the relevant knowledge of the behavior and physiology of the insects and meaningful test procedures, the work of the chemist is invaluable. For this reason, work on pheromones in termites should proceed in the following order: (1) the observation of the phenomenon (e.g., trail-following); (2) the

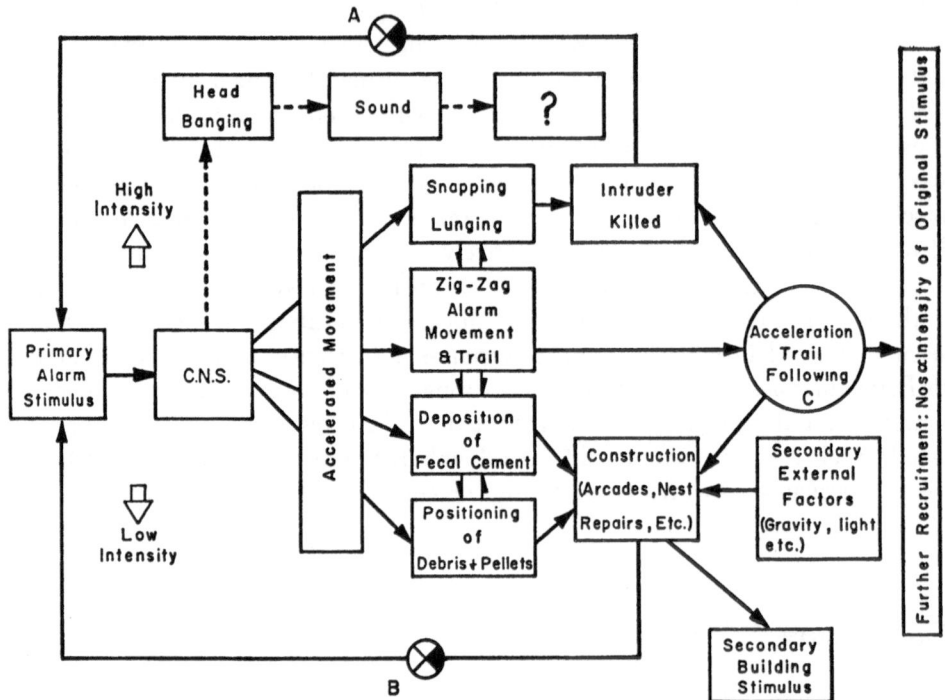

FIG. 7. Diagram of some of the reactions of termites to alarm, showing two feedbacks involved in homeostasis. Probable relationship of "head-banging" behavior—mechanisms indicated by broken lines. A and B are negative feedback pathways. C represents a second termite. (*After* Stuart. 1967. *Science*, 156:1123-1125. Copyright 1967 by the American Association for the Advancement of Science.)

implication and extraction of a pheromone; involving the designing of a meaningful bioassay; (3) the location of the source of pheromone in the animal by dissection and extractions of various organs and tissues; (4) the purification, identification, and synthesis of the pheromone; (5) the mode of synthesis and secretion in the animal (histology, histochemistry, physiology, etc.); and (6) the relating of the use of the pheromone to the total behavior of the animal. To some extent the last three steps can be done concurrently (see also Wilson, 1962).

In insects which are interesting in their own right and at the same time of great economic importance, future work on the pheromones of termites should proceed rapidly now that a foundation has been laid.

ACKNOWLEDGMENTS

Thanks are due to Dr. R. J. Bartell, Dr. L. Barton Browne, Mr. F. J. Gay, Dr. B. P. Moore, and Dr. Elizabeth S. Stuart for critically reading the manuscript.

Much of the author's work reported here was carried out with the aid of grant No. GB 5051 from the U.S. National Science Foundation, and the writing was completed during the tenure of a U.S. Public Health Service Special Research

Fellowship (No. 1-F3-GM-37, 610-01) with the Division of Entomology, CSIRO, Canberra, Australia.

REFERENCES

Andrews, E. A. 1911. Observations on termites in Jamaica. *J. Anim. Behav.*, 1:193-228.

Beaumont, J. 1889-1890. Observation on the termites or white ants of the Isthmus of Panama. *Trans. N.Y. Acad. Sci.*, 8:85-114; 9:157-180.

Becker, G. 1965. Versuche über den einfluss von braunfaulepilzen auf wahl und ausnutzung der holznahrung durch termiten. *Mater. Organismen* (*Berlin*), 1:95-156.

———— 1966. Spurfolge-reaktion von termiten auf glycolverbindingen. *Z. Angew. Zool.*, 53:495-498.

Brian, M. V. 1965. Social Insect Populations. New York, Academic Press, Inc.

Brown, W. L. 1968. An hypothesis concerning the function of the metapleural gland in ants. *Amer. Natur.*, 102:188-191.

Buchli, H. 1960. Les tropismes lors de la pariade des imagos de *Reticulitermes lucifugus* R. *Vie Milieu*, 11:308-315.

———— 1961. Les relations entre la colonie naternelle et les jeunes imagos ailés de *Reticulitermes lucifugus*. *Vie Milieu*, 12:627-632.

Bugnion, E. 1914. La biologie des termites de Ceylan. *Bull. Mus. Nat. Hist. Natur. Paris*, 20:170-204.

Butler, C. 1967. Insect pheromones. *Biol. Rev.*, 42:42-87.

Clark, R. B. 1967. Zoology: The Study of Animals. Inaug. Address Univ. of New-castle-on-Tyne, England.

Darling, F. F. 1937. A Herd of Red Deer. London, Oxford University Press.

Dropkin, V. H. 1946. The use of mixed colonies of termites in the study of host-symbiont relations. *J. Parasit.*, 32:247-251.

Emerson, A. E. 1929. Communication among termites. *Trans. 4th Int. Congr. Ent. Ithaca.* 2:722-727.

———— 1933a. Conditioned behavior among termites (Isoptera). *Psyche* 40:125-129.

———— 1933b. The mechanism of tandem behavior following the colonizing flight of termites. *Anat. Rec.*, 57:61 (Abstr.).

———— 1939. Social coordination and the superorganism. *Amer. Midl. Nat.*, 21:182-209.

———— 1949. The organization of insect societies. *In* Allee, W. C. et al., eds. Principles of Animal Ecology. Philadelphia, W. B. Saunders Company, pp. 419-435.

———— 1961. Vestigial characters of termites and processes of regressive evolution. *Evolution*, 25:115-131.

———— and Simpson, R. C. 1929. Apparatus for the detection of substratum communication among termites. *Science*, 69:648-649.

Ernst, E. 1959. Beobachtungen beim spritzakt, der *Nasutitermes* Soldaten. *Revue Suisse Zool.*, 66:289-295.

Escherich, K. 1908. Aus dem Leben der Termiten oder "Weissen Ameisen." *Leipziger Illustr. Zeit.*, 131:513-518.

Esenther, G. R., and Coppel, H. C. 1964. Current research on termite attractants. *Pest Control*, 32:34-46.

Fielde, A. M. 1904. Observations on ants in their relation to temperature and to submergence. *Biol. Bull.*, 7:170-174.

—— 1905. The progressive odor of ants. *Biol. Bull.*, 10:1-16.

Fraenkel, G., and Gunn, D. L. 1940. The Orientation of Animals: kineses, taxes and compass reactions. New York, Oxford University Press, Inc.

Free, J. B. 1958. The defence of bumble bee colonies. *Behaviour*, 12:233-242.

Fuller, C. 1915. Observations on some South African termites. *Ann. Natal Mus.*, 3:329-505.

Goetsch, W. 1936. Beiträge zur biologie des termitenstaates. *Z. Morph. Ökol. Tiere*, 31:490-560.

Grassé, P. P. 1942. L'essaimage du termites. Essai d'analyse causale d'un complexe instinctif. *Bull. Biol. Fr. Belg.*, 76:347-382.

—— and Noirot, C. 1951. Orientation et routts chez les termites. Le "balisage" des pistes. *Année Psychol.*, 50:273-280.

Grassi, B., and Sandias, A. 1896-1897. The constitution and development of the society of termites: Observations on their habits: with appendices on the parasitic protozoa of Termitidae, and on the Embiidae. *Quart. J. Microsc. Sci.*, 39:245-322; 40:1-82.

Higham, K. C. 1964. Hormones and behavior in insects. *In* Carthy, J. D., and Duddington, C. L., eds., *Viewpoints Biol.* 3:119-136.

Howse, P. E. 1964. The significance of the sound produced by the termite *Zootermopsis angusticollis* (Hagen). *Anim. Behav.*, 12:284-300.

Huber, P. 1810. Recherches sur les Moeurs des Fourmis Indigènes. Pashoud, Paris.

Hummel, H. and Karlson, P. (1968) Hexansäure als Bestandteil des Spurpheromons der Termite *Zootermopsis nevadensis* (Hagen). *Hoppe-Seyler's Z. physiol. Chem.* 349:725-727.

Jacobson, M. 1966. Chemical insect attractants and repellents. *Ann. Rev. Entom.*, 11:403-422.

Kalshoven, L. G. E. 1960. Biological notes on the *Cryptotermes* species of Indonesia. *Acta Trop.*, 17:263-272.

Karlson, P., and Butenandt, A. 1959. Pheromones (Ectohormones) in insects. *Ann. Rev. Entom.*, 4:39-58.

—— and Lüscher, M. 1959. "Pheromones": a new term for a class of biologically active substances. *Nature (London)*, 183:55-56.

—— Lüscher, M. and Hummel, H. (1968) Extraktion und Biologische Auswertung des Spurpheromons der Termite *Zootermopsis nevadensis*. *J. Insect. Physiol.* 14:1763-1771.

Lindauer, M. 1961. Communication among Social Bees. Cambridge, Harvard University Press.

Lubbock, Sir J. 1882. Ants Bees and Wasps (Reprint 1929). London, Kegan Paul Trench.

Lüscher, M. 1952a. Die produktion und elimination von ersatizgeschlechtstieren bei der termite *Kalotermes flavicollis* Febr. *Z. Vergl. Physiol.*, 34:123-141.

—— 1952b. New evidence for an ectohormonal control of caste determination. *Trans. 9th Int. Congr. Ent. Amsterdam*, 1:289-294.

—— 1960. Socialwirkstoffe bei termiten. *Proc. 11th Int. Congr. Ent. Vienna.*, pp. 579-582.

—— 1961. Social control of polymorphism in termites. *In* Kennedy, J. S., ed. Insect polymorphism. Symposium No. 1 *Roy. Entom. Soc. (London)*, pp. 57-67.

—— 1967. *In* Beament, J., and Treherne, J., eds. Insects and Physiology. Edinburgh. Oliver and Boyd.

—— and Müller, B. 1960. Ein spurbildendes sekret bei termiten. *Naturwissenschaften,* 27:503.

Maschwitz, U. W. 1967. Alarm substances and alarm behaviour in social insects. *In* Harris, Loraine, and Wool, eds. Vitamins and Hormones. London, Academic Press, Inc.

Matsumura, F., Coppel, H. C., and Tai, A. 1968. Isolation and identification of termite trail-following pheromone. *Nature (London),* 219:963-964.

McCook, H. 1887. The mound making ants of the Alleghennies. *Trans. Amer. Entom. Soc.,* 6:253.

Moore, B. P. 1964. Volatile terpenes from *Nasutitermes* soldiers (Isoptera, Termitidae). *J. Insect Physiol.,* 10:371-375.

—— 1966. Isolation of the scent-trail pheromone of an Australian termite. *Nature (London),* 211:746-747.

—— 1968a. Studies on the chemical composition and function of the cephalic gland secretion in Australian termites. *J. Insect Physiol.,* 14:33-39.

—— 1968b. Biochemical studies in termites. *In* Krishna, K., and Lechtleitner, F., eds. Biology of Termites. New York, Academic Press, Inc., vol. 1.

Novák, V. J. 1966. Insect Hormones. London, Methuen.

Pasteels, J. 1965. Polyéthisme chez les ouvriers de *Nasutitermes lujae* (Termitidae, Isoptères). *Biol. Gabonica,* 1:191-205.

—— 1967. Polyéthisme chez les ouvriers de *Nasutitermes lujae* lors de l'éstablishhemént d'une piste de récolte (Isoptères, Termitidae). *C. R. du Ve Congres de l'Union Internationale pour l'Etude des Insects Sociaux.* Toulouse, 1965.

Plateaux, L. 1960. Adoptions expérimentales de larves entre des fourmis de genres différents: I. *Leptothorax nylanderi* Förster et *Solenopsis fugax* Latreille. II. *Myrmica laevinodis* Nylander et *Anergates atratulus* Schenck. III. *Anergates atratulus* Schenck et *Solenopsis fugax* Latreille. IV. *Leptothorax nylanderi* Förster et *Tetramorium caespitum* L. *Insectes Soc.,* 7:163-170, 221-226, 345-348.

Riddiford, L. M. 1968. Artificial diet for Cecropia and other Saturniid silkworms. *Science,* 160:1461-1462.

Russell, E. S. 1930. The Interpretation of Development and Heredity. A study in biological method. New York, Oxford University Press, Inc.

Smythe, R. V., Coppel, H. C., Lipton, S. H., and Strong, F. M. 1967. Chemical studies of attractants associated with *Reticulitermes flavipes* and *Reticulitermes virginicus. J. Econ. Entom.,* 60:228-233.

Stuart, A. M. 1960. Experimental studies on communication in termites. Ph.D. Thesis, Harvard University, Boston.

—— 1961. Mechanisms of trail-laying in two species of termites. *Nature (London),* 189:419.

—— 1963a. The origin of the trail in the termites *Nasutitermes corniger* (Motschulsky) and *Zootermopsis nevadensis* (Hagen), Isoptera. *Physiol. Zoöl.,* 36:69-84.

—— 1963b. Studies on the communication of alarm in the termite *Zootermopsis nevadensis* (Hagen), Isoptera. *Physiol. Zoöl.,* 36:85-96.

—— 1964. The structure and function of the sternal gland in *Zootermopsis nevadensis* (Hagen), Isoptera. *Proc. Zool. Soc. (London)* 143:43-52.

—— 1967. Alarm, defense and construction behavior relationships in termites (Isoptera). *Science,* 156:1123-1125.

—— 1968. Social behavior and communication. *In* Krishna, K., and Lechtleitner, F., eds. Biology of Termites. New York, Academic Press, Inc.

—— and Satir, P. 1968. Morphological and functional aspects of an insect epidermal gland. *J. Cell Biol.*, 36:527-549.

Sudd, J. H. 1967. An introduction to the behaviour of ants. London, Arnold.

Tinbergen, N. 1948. Social releasers and the experimental method required for their study. *Wilson Bull.*, 60:6-51.

Verron, H. 1963. Rôle des stimuli chimiques dans l'attraction sociale chez *Calotermes flavicollis* (Fabr.). *Insectes Soc.*, 10:167-336.

—— and Barbier, M. 1962. L'hexène-3-ol-1 substance attractive des termites: *Calotermes flavicollis* Fab. et *Microcerotermes edentatus C.R. Acad. Sci.* [D], (*Paris*), 254:4089-4091.

Vité, J. P. 1967. Sex attractants in frass from bark beetles. *Science*, 156:105.

Weaver, N. 1966. Physiology of caste determination. *Ann. Rev. Entom.* 11:79-102.

Wheeler, W. M. 1928. The Social Insects. New York, Harcourt, Brace & Co.

Wilkinson, W. 1962. Dispersal of alates and establishment of new colonies in *Cryptotermes havilandi* (Sjöstedt) (Isoptera, Kalotermitidae). *Bull. Entom. Res.*, 53:265-286.

Wilson, E. O. 1962. Chemical communication among workers of the fire ant *Solenopsis saevissima* (Fr. Smith) 3. The experimental induction of social responses. *Anim. Behav.*, 10:159-164.

—— 1963. The social biology of ants. *Ann. Rev. Entom.*, 8:345-368.

—— 1965. Chemical communication in the social insects. *Science*, 149:1064-1071.

—— and Bossert, W. H. 1963. Chemical communication in animals. *Recent Prog. Horm. Res.*, 19:673-716.

Wood, D. L., Silverstein, R. M., and Rodin, J. O. 1967. Sex attractants in frass from bark beetles. *Science*, 156:105.

6

Responses of Olfactory and Gustatory Receptor Cells in Insects

BRUNHILD STÜRCKOW

Department of Biology
Northeastern University
Boston, Mass. 02115

INTRODUCTION 107
HISTORY OF THE RESEARCH ON CHEMORECEPTIVE SYSTEMS 108
GOALS OF THE STUDIES ON CHEMORECEPTIVE SYSTEMS 110
PERCEPTION OF INSECTS COMPARED WITH THAT OF MAN 111
STRUCTURE OF CHEMORECEPTIVE SENSILLA 119
NEUROPHYSIOLOGY OF CHEMORECEPTIVE SENSILLA 126
COMPARATIVE STUDIES BETWEEN NEURAL AND BEHAVIORAL RESPONSES 139
WORKING HYPOTHESES OF RECEPTIVE MECHANISMS 143
METHODOLOGY: THE FLOW CAPILLARY USED IN STUDIES ON GUSTATION OF INSECTS 145
REFERENCES 148
DISCUSSION 158

INTRODUCTION

Responses to stimulation by odorous vapors and flavored solutions have been studied in three ways: (1) as behavioral responses, which allow conclusions about the attractiveness, repellency, or other nature of the stimulus, (2) as neural responses recorded mainly from the sensillum, and (3) as comparative studies between neural and behavioral responses. A sensillum is a receptive unit in the integument, composed of one to several sensory neurons and generally two accessory cells. These secondary cells have altered the integument during molting to an outgrowth (such as a hair, a cone, or a peg), or to an ingrowth (such as a cavity, a pit, or a plate organ), beneath which the sensory cell bodies are located. Later, the accessory cells seem to continue to assist in the function of the sensillum.

As in other sensory modalities, the site and nature of the initial receptive mechanisms and the decoding of their message to the central nervous system are the main points of interest. This paper, guided by recent research, extends the survey of some of the numerous aspects of chemoreception dealt with in preceding reviews.

HISTORY OF THE RESEARCH ON CHEMORECEPTIVE SYSTEMS

During the time of the early microscopic studies, from about 1880 to 1930, as much attention was paid to chemoreceptive sensilla as to visual and mechano-receptive systems; however, statements about the function of the sensillum studied were missing because of the lack of adequate proof. The location of a sensillum on antennae or mouthparts and the supply through more than one sensory neuron were the only information regarding a probable chemoreceptive function (one sense cell was considered to belong in all cases to a mechanoreceptive sensillum). This guess, sometimes still used today, was probably mostly correct; however, it offered no challenge for more intensive studies.

Other motives for the lack of functional morphology were the great variety of forms of chemoreceptive sensilla and the minuteness of dendritic fibers, which can be traced only with the electron microscope. A missing unequivocal nomenclature added to the disorder. Terms such as peg, cone, rod, sensillum basiconicum, sensillum stylonicum, "Riechkegel," and so forth have been used for the same sensillum; similar examples are abundant. Furthermore, these names did not give a comprehensive description of a sensillum; even subdivisions into, for example, thin- and thick-walled sensilla basiconica did not describe definitively all forms of sensilla basiconica (DuBose, 1967). An appropriate nomenclature remains an unsolved problem.

Strictly speaking, Barrows (1907), von Frisch (1919), and Minnich (1921) founded the investigation of insect chemoreception and stimulated subsequent studies. (This work is covered in general reviews by Dethier and Chadwick, 1948a, Dethier, 1947, 1956, 1963, and Hoffmann, 1961; in reviews by Dethier, 1954, on olfaction, by Frings and Frings, 1949, on the loci of taste receptor cells; and in general textbook articles by Wigglesworth, 1965, Jahn and Wulff, 1950, von Buddenbrock, 1952, Dethier, 1953, and Prosser, 1961.) Barrows, von Frisch, and Minnich determined, by using behavioral responses, the operating range and discrimination properties of some chemoreceptive systems and the site of their sensilla. Previous attempts without convincing results had been made by Hauser (1880), Kraepelin (1883), Will (1885), Schenk (1903), Forel (1908), McIndoo (1914), and others.

Barrows (1907) experimented with the pomace fly, *Drosophila ampilophila*, in a simple olfactometer—the first used with insects—and found that various concentrations of attractive odors evoke different degrees of response. He also determined, through partial amputation, the site of an olfactory pit on the terminal antennal segment.

Von Frisch (1919) trained bees to respond to odors by offering the odor together with a sugar solution in a box with an entrance hole. After the bees were trained to the fragrant food, he removed the sugar solution and observed whether the bees would search in the box with the odor used for training or in control boxes provided with other odors or none. Many details of olfaction in the bee became known through these preference tests with trained specimens. Von Frisch (1921) also showed that the location of olfactory sensilla in the bee is restricted to the eight distal segments of the antenna and that the two proximal segments and all other parts of the body are devoid of odor receptor cells (compare Lefebvre's similar results with the wasp, 1838). He also tried to prove that one of the densely located sensilla, the sensillum placodeum, is *the* site of olfactory reception, by dissecting all chemoreceptive segments of both antennae except for a part of one segment which was supplied with three sensilla placodea and some hairs. He considered these hairs to be mechanoreceptors; however, Vogel (1923) showed that several of them were supplied with five to eight neurons and, therefore, they were probably also chemoreceptive sensilla. Incontestable proof of the function of the sensilla placodea was first possible through neurophysiologic work (Lacher and Schneider, 1963; Lacher, 1964).

Minnich touched parts of butterflies (1921), certain muscid flies (1926a, 1929; compare Barrow's similar results with *Drosophila*, 1907) and the bee (1932) with a brush wetted with taste solutions. As soon as he touched the tarsi, or the mouthparts (in the bee, also the tip of the antennae) with a sugar solution, the specimens responded with the extension of the proboscis. Minnich (1926b) obtained this distinct response also by stimulating a single sensillum: a taste hair of the proboscis of the blowfly. Many details of gustation became known through his experiments and similar ones carried out later by other workers.

A turning point in the study of chemoreceptive systems was the introduction of neurophysiologic methods used first, less successfully, on entire antennae by Dethier (1941), Chapman and Craig (1953), and Boistel and Coraboeuf (1953), then used with some success by Boistel (1953) and Roys (1954), and applied with more success by Schneider (1955, 1957b) to record impulses of a single sensillum, by Schneider (1957a, 1957b) to record an overall response of an antenna (the electroantennogram), and by Morita and Yamashita (1961) and Boeckh (1962) to record generator potentials of a single olfactory sensillum. Hodgson et al. (1955) first recorded impulses of a single gustatory sensillum, and Morita and Yamashita (1959) first recorded generator potentials of a taste sensillum.

The histologic aspects of chemoreception remained reserved until Slifer et al. (1957) first investigated with the electron microscope an olfactory (or gustatory) sensillum on the antenna of a grasshopper, and Adams (1961) studied the fine structure of the taste hair of the stable fly.

The understanding of chemoreception in insects has experienced a boom since the application of neurophysiologic and electron-microscopic methods (reviewed by Hodgson, 1955, 1958, 1964, 1965; Hoffmann, 1961; Dethier, 1963; Horridge, 1965; Marler and Hamilton, 1966; for olfaction by Schneider, 1961b, 1963a, 1963b, 1965, 1967; for gustation by Dethier, 1962; Oakley and Benjamin, 1966; and with regard to host plant selection by Schoonhoven, 1968); however, our knowledge of chemoreceptive systems is still far behind the understanding of vision and to a lesser degree of mechanoreception. Although there is abundant literature, detailed morphologic and functional analyses of chemosensitive sensilla are rare compared with their great efficiency and variety of forms.

GOALS OF THE STUDIES ON
CHEMORECEPTIVE SYSTEMS

Responses of chemoreceptor cells have been studied behaviorally and neurophysiologically with the same or overlapping goals: (1) to locate the site of reception of the stimulus on entire appendages, definite segments of appendages, or certain types of sensilla, (2) to determine adequate stimuli, their threshold intensity, effective concentration range, and attractiveness, repellency, or other nature, and (3) to study the relation between configuration and physicochemical characteristics of a molecule and its stimulating effect. In principle, the methods used have been the same in olfaction and gustation, except for the stimulation of a single taste sensillum in behavioral experiments (Minnich, 1926b; Dethier, 1955; and Stürckow, 1967). Neurophysiologic recordings of single sensilla and neurons have been made in olfaction and gustation, although by different techniques, which will be discussed later.

Behavioral studies, which at present have no broad parallel in neurophysiologic experiments, are conditioning tests with larvae. Thorpe and Jones (1938), Thorpe (1939), Cushing (1941), and reviewed by Thorpe (1963) showed that olfactory stimuli applied during larval life can become attractants in adult life. Saxena (1967) and Schoonhoven (1967a) found a loss of host-plant specificity after feeding an artificial diet during several instars. However, Evans (1961) found that certain sugars mixed into the diet of larvae significantly decreased the sensitivity of the adults not only to the same sugars but also to other sugars which apparently react on the same site of the chemoreceptive membrane. Conditioning experiments with insecticides will be discussed in the section on NEUROPHYSIOLOGY OF CHEMORECEPTIVE SENSILLA (p. 126). Comparative studies of neural and behavioral responses in specimens that have been conditioned should indicate whether the development of the chemoreceptive membrane, or the state of the central nervous system, or both are influenced—a question, which touches evolutionary problems. Schoonhoven (1967a) made such a preliminary comparison and found that the change in behavior after conditioning was at least partly due to a change in the sensitivity of the receptor cells.

Within a certain concentration range of the stimulus, all studies on chemoreceptive systems have shown a dependence on the intensity of the stimulus, expressed in molecules per unit stimulating air or solution; this principle governs the "receptive response" as well as the "perceptual response." The afferent message of a chemoreceptor cell in reply to a stimulus is understood as the receptive response. The perceptual response can be measured only indirectly through a behavioral criterion, which must be chosen critically to eliminate the possibility that perception and behavioral response occur at significantly different intensities of the stimulus (compare Dethier's considerations, 1954).

Both responses, the neural and the behavioral, result in a stimulus-percent response curve, which is synonymous with a dose-percent effect curve. A nearly linear relation can be derived when the percent of the response is expressed in probability units (probits) and the intensity of the stimulus is plotted in logarithms (Bliss, 1935). Although already Dethier and Chadwick (1948b) have thoroughly applied statistic approaches to the evaluation of chemoreceptive responses in insects, the evaluation of stimulus-percent response lines has been mainly developed through pharmacologic and toxicologic studies. Probably, the most convenient and, at the same time, most reliable method for the evaluation of dose-percent effect data is offered by Litchfield and Wilcoxon (1949). It allows a quick judgment of the degree of homogeneity between data and line; gives confidence limits of the slope, the median effective dose (ED_{50}), and any other dose; provides means for the use of 0 and 100 percent responses; and gives a rapid test for parallelism of two lines with an easy computation of their relative potency.

Insects and man perceive chemical stimuli as agents without significance, attractants, or repellents. An agent without significance can become at higher concentrations first an attractant and then a repellent, or only one of both. An example in olfaction where a change in odor quality occurs with different intensities is isovaleraldehyde, which has been tested with the housefly (Dethier, Hackley, and Wagner-Jauregg, 1952). It evokes a maximal attraction at 1×10^{-5} M and a maximal repellency at 6×10^{-5} M, if applied with an olfactometer. Between these concentrations it is attractive or repellent (further examples are reviewed by Dethier, 1947). An example in gustation for a change in quality is given by alkaloid glycosides, which at low concentrations are sensed by the Colorado potato beetle, *Leptinotarsa*, as agents without significance or weak attractants, but which become repellents at higher intensities (Stürckow and Löw, 1961).

The perception of each quality of a stimulus will be reflected in an individual stimulus-percent response line. The transitional stage from one quality to another are points of special interest for comparative studies of neural and behavioral responses. They have not as yet been carried out except for a preliminary trial with *Leptinotarsa*. The impulse pattern (steady-state level) of a single taste sensillum of the beetle in response to alkaloid glycosides changed from a low frequency of single impulses to irregularly occurring volleys of impulses of a high frequency (compare Fig. 10, *right*). In neurophysiologic tests, the threshold for the change to volleys of impulses was found at concentrations (Stürckow, 1959, 1961b) at which reduced food intake commenced in behavioral tests (Stürckow and Löw, 1961). This suggests that volleys of impulses signify the response to a substance with a repellent taste.

PERCEPTION OF INSECTS COMPARED WITH THAT OF MAN

When von Frisch (1919) made his experiments with trained bees, statistical methods for the evaluation of tests were not as well known as they are today. He compensated for this with logical thinking. Von Frisch showed that the

TABLE 1

The perception of 12 odors by the human and the bee

ODORANT PAIR		COMPARATIVE EFFECTS OF ODORS			
		MAN	BEE (average number visiting each odor)		
A	B	A : B	A : B		C (control)
methyl anthralinate (NH_2, $COO-CH_3$ on benzene)	β-naphthyl methyl ether (OCH_3 on naphthalene)	same (n = 1)	138 "similar" 58		4 (n = 1,949)
isobutyl benzoate ($COO-CH_2-C-H$, CH_3, CH_3)	isoamyl salicylate (OH, $COO-CH_2-CH_2-CH$, CH_3, CH_3)	same (n = 4)	106 "similar" 16		8 (n = 3,418)
nitrobenzene (NO_2)	benzaldehyde (CHO)	similar (n = 1)	26 "same" 15		1 (n = 628)
p-cresyl methyl ether (CH_3, OCH_3)	m-cresyl methyl ether (CH_3, OCH_3)	similar (n = 1)	90 "similar" 12		1 (n = 645)
β-bromostyrene ($CH:CHBr$)	phenyl acetaldehyde (CH_2-CHO)	similar (n = 1)	200 "similar" 6		2 (n = 648)
amyl acetate ($CH_3-COO-CH_2-CH_2-CH$, CH_3, CH_3)	6-methyl-5-hepten-2-one (CH_3, CH_3, $C=CH-CH_2-CH_2-\overset{O}{C}-CH_3$)	similar (n = 2)	152 "similar" 6		1 (n = 1,158)

The odors of each pair evoke in man either almost the same or similar (but clearly different) sensations. The perception in the bee is likewise quoted as almost the "same" or "similar" (but clearly different); the sensations of the bee are indirectly judged by the number of bees (trained to odor A of each pair) that visited in the experiment odor A, odor B, or the control C (another or no odor). All means of bees within each pair (A:B, B:C, A:C) are significantly different from each other ($t <<$ or $\sim 2\%$), except for nitrobenzene benzaldehyde, where $t >30\%$ for A:B, but $t \sim 1\%$ for B:C. (Adapted from von Frisch. 1919. Zool. Jahrb. Physiol., 37:168-192.)

112

threshold for the acceptance of a sugar solution did not depend upon the intensity or quality of the odor used for training and repeated his experiments until he felt certain about the results. In addition, he followed the rules for control tests and used pure chemicals or noted when a change in a chemical had occurred. I evaluated statistically those of von Frisch's data that were collected for a comparison between the sensations of the human and the probable sensations of the bee, and found no misinterpretation. In Table 1, his results are combined with inserted values for t, which have been obtained with the Student's t-test or by following the outlines of Snedecor and Cochran (1967, section 4.14) in the case of unequal variances. The bees were trained to feed on a sugar solution accompanied by the first odor of each pair and were observed for whether they would confuse the first with the other odor when the odors were offered in boxes without food. Von Frisch tested the bee very carefully but neglected the study of sensations in the human. His statement referring to the perception of the human are mainly based on his own observations only.

The bees confused one pair of odors, which we probably would have differentiated, and differentiated the odors of two pairs, which we might have confused. The odors of the last three pairs were differentiated by the bees, as we would have done. Thus, von Frisch found similarities and differences between the sensations of man and the bee. However, if he had performed the experiments with the human also as preference tests with trained and rewarded individuals, the apparently more sensitive discrimination properties of the bee might have been found to be untrue. Engen and Pfaffmann (1959) and Engen (1960, and Engen, Chap. 12) found in man an effect of practice on the judgments of odor itensity and quality.

Ribbands (1955) doubted the correctness of von Frisch's results. Ribbands thought that he had proved that the bee possesses an ever-present body odor in addition to the scent-producing organ, which causes the bees to follow one another. I evaluated Ribbands' data statistically but found no significant difference between the number of bees visiting the vial that was used throughout the experiment and the number of bees visiting the control vials that were replaced nine times by new ones (test vial versus control vial with the high number of visiting bees: $t < 70$ percent). If Ribbands' statement had been correct, the data of von Frisch should have shown an increasing number of bees at boxes that had been visited frequently, but this was not the case. Thus, a body odor, which might be inherent to the bee, did not play a significant role in the experiments with trained bees.

Ribbands (1955) also compared the thresholds, which he had obtained with two pure chemicals for the human and the bee, with those found by von Frisch (1919, pp. 118-139) and found a significant difference; but he did not consider that he and von Frisch had used different diluents (glycol and paraffin oil), which, together with a different experimental arrangement, could have caused the different results. Ribbands might have had a point for discussion, but without proof of comparable experimental conditions a comparison of the results is not valid.

Several years later, the odor of the scent-producing organ of the bee was overestimated a second time by Johnson (1967) and Wenner (1967). They thought from their evidence that the information on a new feeding place given by a forager-bee to the younger field bees in the hive is mainly based on two scents: the fragrancy of the forager-bee caused by the blossoms it had visited, and the secretion of its scent-producing organ at the feeding place. Von Frisch (1967, 1968), however, interprets their experimental data differently: at distances between 100 to 4,000 m from the hive, these two scents play a significant role in still air only in final, close-by determinations. In windy conditions, they are of more importance, but only then when the new feeding place is approached upwind.

Schwarz (1955) and Fischer (1957) indirectly supported von Frisch's results by determining and comparing the thresholds for different odors in the human and the bee. Schwarz determined these thresholds for 15 pure odors and found them for 11 odors in the same logarithmic unit of molecules per cm^3 of stimulating air and for four odors within two logarithmic units—partly higher, partly lower for the bee. The thresholds of the human for these four odors were unfortunately taken from another work—a report of von Skramlik (1948) that is based on studies of different authors and sometimes indicates a considerable variation (von Skramlik, 1926), or dated back to Ohma, for whom the literature citation was omitted.

Schwarz used an olfactometer with constantly flowing air, whereas Fischer offered the test substance in still air, as did von Frisch. A comparison of the data of both authors is possible only with nerol, the only substance used by both of them. Schwarz found a reciprocal value of the thresholds: human/bee of 1.8 and Fischer of 2.5, which is in good agreement. Fischer tested mainly flower perfumes and found also a close agreement between the thresholds of the human and the bee, except for the fragrance of the scent-producing organ of the bee, to which the bee was more sensitive than the human. He could not determine the extent of the difference between these two sensitivities, since the amount of extracted substance was unknown.

Although a threshold does not furnish information about the corresponding sensation, it was shown that the worker bee perceived these in a total of 29 stimuli at thresholds similar to those of man, and that none of these odors, more or less pleasant to the human, affected the bee as a strong repellent. Fatty acids made the training difficult but did not repel the bees, who found and sipped the sugar solution (Schwarz, 1955).

Schwarz (1955) cites similar comparative studies between olfactory thresholds of other insects and the human. His Table 6 needed some completion and correction (besides errors in calculation, the values of Dethier, 1941, 1943; and Dethier and Yost, 1952, were taken as too high, since these authors expressed their data as "median threshold," which represents the median effective dose: ED_{50}). After completion, Schwarz's Table 6 shows thresholds several times higher for insects than for man: 10^4 to 10^6 more molecules per cm^3 of stimulating air for skatole, which was tested with three beetles, and 10 to 10^3 times more molecules per cm^3 of stimulating air for benzaldehyde, benzene, and ethanol, which were tested with two caterpillars, a parasitic wasp, and two flies.

Wirth (1928), who studied the parasitic wasp, *Habrobracon*, and Warnke (1931), who experimented with the dung beetle, *Geotrupes*, defined the slightest vibrations of the antennae as the threshold criterion; Abbott (1937) investigated the senses of the sexton beetle, *Necrophorus*, and took the attraction of the greatest odorous dilution over 2 to 4 m in air moving towards the beetle or 50 to 70 cm in still air as the threshold. These authors were partly in doubt whether the chosen criteria represented the "true perceptual threshold" and thus could be compared with the threshold determined for the human, or whether the observed responses occurred at concentrations higher than the minimum dose necessary for sensation.

In experiments with flies, two screen windows of a dark testing-chamber were connected to an olfactometer (Wieting and Hoskins, 1939; Dethier and Yost, 1952). Slowly streaming air of 41° C or illumination from outside attracted the flies to these windows. An attractive odor applied through the test window caused the flies to aggregate at this opening, while a repellent odor caused them to move to the control window. In this way, determinations of stimulus-percent response curves were possible, and the thresholds were found to be roughly comparable with those obtained with trained bees—at least with respect to attractive odors. The studies of Ritter (1936) with the water beetle, *Hydrous*, seem also to be roughly comparable with the data found for the bee. She had conditioned the beetle to food blended with an odor, and rewarded eating movements in response to this odor by feeding. The differences between her tests and those with the bee were the testing millieus—water and air—and the use of preference tests in the bee in contrast with Ritter, who offered one stimulus at a time. In addition, the steady search for food, which is characteristic of the worker bee, is not characteristic of the water beetle.

Dethier (1941, 1943), who was aware of the difficulty connected with the determination of a threshold in the conventional sense (minimum effective dose), used the median effective dose, the ED_{50}, and called it misleadingly the "median attractant or repellent threshold," which could mean the average minimum dose; however, determinations of the ED_{50} and the complete stimulus-percent response line have also been shown to be difficult; apparently not so much for afferent neural responses as for perceptual responses, which can be measured only behaviorally. The difficulty lies in finding a criterion for the stimulus to be tested to which almost 100 percent of the specimens respond. It was successfully found with the search for food in the worker bee, which steadily gathers food for the colony. In adult insects, the search for food is subject to extreme variations (caused by additional interests): from none at all, as in some lepidoptera, to the eagerness of worker insects of social groups. If the control stimulus is not responded to by 95 to 100 percent of the test specimens, the determination of a stimulus-percent response line starts already with a mistake and will result in too flat a slope or range at too high intensities. An example are behavioral tests with adult potato beetles of different ages: 1- to 10-day-old beetles were completely tuned to food intake (Stürckow and Löw, 1961), while older beetles never responded with 95 or 100 percent to the control (Stürckow, 1959). Feeding tests with older beetles, which are primarily interested in mating or diapause (depending on light and temperature), intelligibly resulted in less sensitive stimulus-percent response lines. To find the developmental stage of specimens, which is tuned to the stimulus to be tested, is a skillful operation. Generally, it requires a comprehensive series of behavioral tests.

Thus, a comparison between the sensations of insects and the human on the base of thresholds is a dangerous one. Different forms of application of the stimulus are another often overlooked pitfall for comparisons (see the criticism of Johnston, 1967). If, therefore, we consider these thresholds determined for the human and some insects as subjected to an error of $\pm$ 2 logarithmic units of molecules per cm^3 of stimulating air, before we compare them, we should not make a major mistake. Then, the beetles, *Geotrupes, Necrophorus,* and *Hydrous,* are less sensitive to skatole than man, the bee has sensitivities similar to the human for 29 odors, and, likewise, a parasitic wasp, two caterpillars, and two flies have similar sensitivities to the human for one or two odors (Schwarz, 1955; Fischer, 1957).

Kay et al. (1967) have compared neurophysiologically determined thresholds of odors in the fly, *Lucilia,* with perceptual thresholds of the human. According to the tip diameter of their electrode (ϕ 2 to 3 μ), they probably measured the sum phenomenon of the generator potentials of a few sensilla in an olfactory pit of the antenna [compare an area of 2 to 3 μ^2 in the blowfly, which is supplied with seven to twelve sensilla (Dethier, Larsen, and Adams, 1963)]. Their threshold values for the human and the fly were also within the same or two logarithmic units of molecules per cm^3 of stimulating air with an apparently greater sensitivity for the fly. This greater sensitivity might have been caused by the short rise time of the stimulus, since the fly was stimulated by a flow system in a test chamber of 0.3 ml. Although the presentation of the odors to the human was not described, man cannot inhale odors to produce a comparable rapid onset of stimulation.

The often-mentioned assumption that the number of sensory neurons reflects the receptive capacity of a sense organ is probably true as to the width of the qualitative spectrum; however, the sensitivity to a certain odor or taste in specimens with a limited number of receptor cells is not necessarily surpassed by specimens with a great number of sensory neurons. The human louse, *Pediculus,* with 160 odor receptor cells (Table 2) might be more sensitive to a specific odor—for example, that of the human—than an insect with several hundred thousand sensory neurons or mammals with 10 to 200 million odor receptor cells (Moulton and Beidler, 1967). The estimates in Table 2 concern primarily pure odor receptor cells, but likewise include carbon dioxide, humidity, and cold-receptor cells (Lacher, 1964; Loftus, 1966; Schoonhoven, 1966) and possibly include some gustatory receptor cells (in some insects, taste hairs have been found on the antennae—reviewed by Frings and Frings, 1949, 1956, 1959; and Schoonhoven, 1968). The main site of olfaction in the insects cited in Table 2 is known to be on the antennae. Additional olfactory sensilla, which might occur on palpi or other parts of the body should be a trivial number and may be omitted for the estimates.

Similar comparisons between perceptual thresholds of man and insects have been made in gustation (reviewed by Dethier and Chadwick, 1948a). The same criticisms made of olfactory thresholds should be applied to gustatory data. These concern comparable applications of the stimulus, correct behavioral criteria, effects of practice, mistakes in calculation, and comparable test specimens. Since Minnich (1922a, 1922b), differences have been observed between thresholds of hungry and

TABLE 2

Estimated numbers of chemoreceptor cells on both antennae of nine species

ORDER	SPECIES (FAMILY)	FEMALE	MALE	AUTHOR
Hymenoptera	*Apis mellifica* (Apidae)	160,000 for workers, probably less for queens	500,000	after Vogel, 1923, Dostal, 1958, and Lacher, 1964
Orthoptera	*Tenodera angustipennis* (Mantidae)	100,000	400,000	after Slifer, 1968
	Carausius morosus (Phasmidae)	50,000	—	after Weide, 1960, and Slifer, 1966
	Melanoplus differentialis (Acrididae)	190,000	190,000	after Slifer et al., 1959
Lepidoptera	*Telea polymorphis* (Saturniidae)	70,000	300,000	from Boeckh et al., 1960
	Bombyx mori (Bombydidae)	70,000	70,000	after Schneider and Kaissling, 1957, and Schneider, 1965
Coleoptera	*Trypodendron lineatum* (Scolytidae)	2,700	2,700	after Moeck, 1968
	Ips confusus (Scolytidae)	3,400	3,400	after Borden, 1968
Anoplura	*Pediculus humanus corporis* (Pediculidae)	200	200	after Wigglesworth, 1941*

* The human louse bears, according to Wigglesworth, about 160 olfactory cells on both antennae. Here 200 chemoreceptor cells are cited because of the few hairs on the antennae provided with a single neuron, which also might be chemoreceptive instead of tactile sensilla.

saturated specimens, and between thresholds of tarsal, oral, or antennal gustatory sensilla.

The high threshold of the bee to sucrose is striking; it is 10- to 100-fold higher compared with the thresholds of six other insects (von Frisch, 1934; Hassett et al., 1950). Judged by Evans' results (1961) with larvae and adults of the blowfly reared on sugar-containing media, the higher threshold of the bee might be due to the high content of sucrose in honey, which is used as the last larval food. Similar conditioning experiments seem to be of primary significance in a further understanding of chemoreceptive systems.

Viewed from afar, there should be no great difference between olfactory and gustatory perceptions of humans and insects that occupy the same environmental space and have similar interests in plants and their fruits for nourishment. Exceptions have been found in substances that are of special significance for one but not the other species. Schneider (1961b) reported that we cannot perceive the odor of the sex pheromone of the silk moth, *Bombyx*, and Nedel (1960) stated that we do not sense the mandibular gland secretion of the honey bee queen, which is composed of several essential pheromones for the bee. Stürckow and Löw (1961) found that the alkaloid glycoside, leptine, is a food repellent for the potato beetle but is tasteless for us, whereas other alkaloid glycosides, which are less repellent for the beetle, have a bitter taste for us.

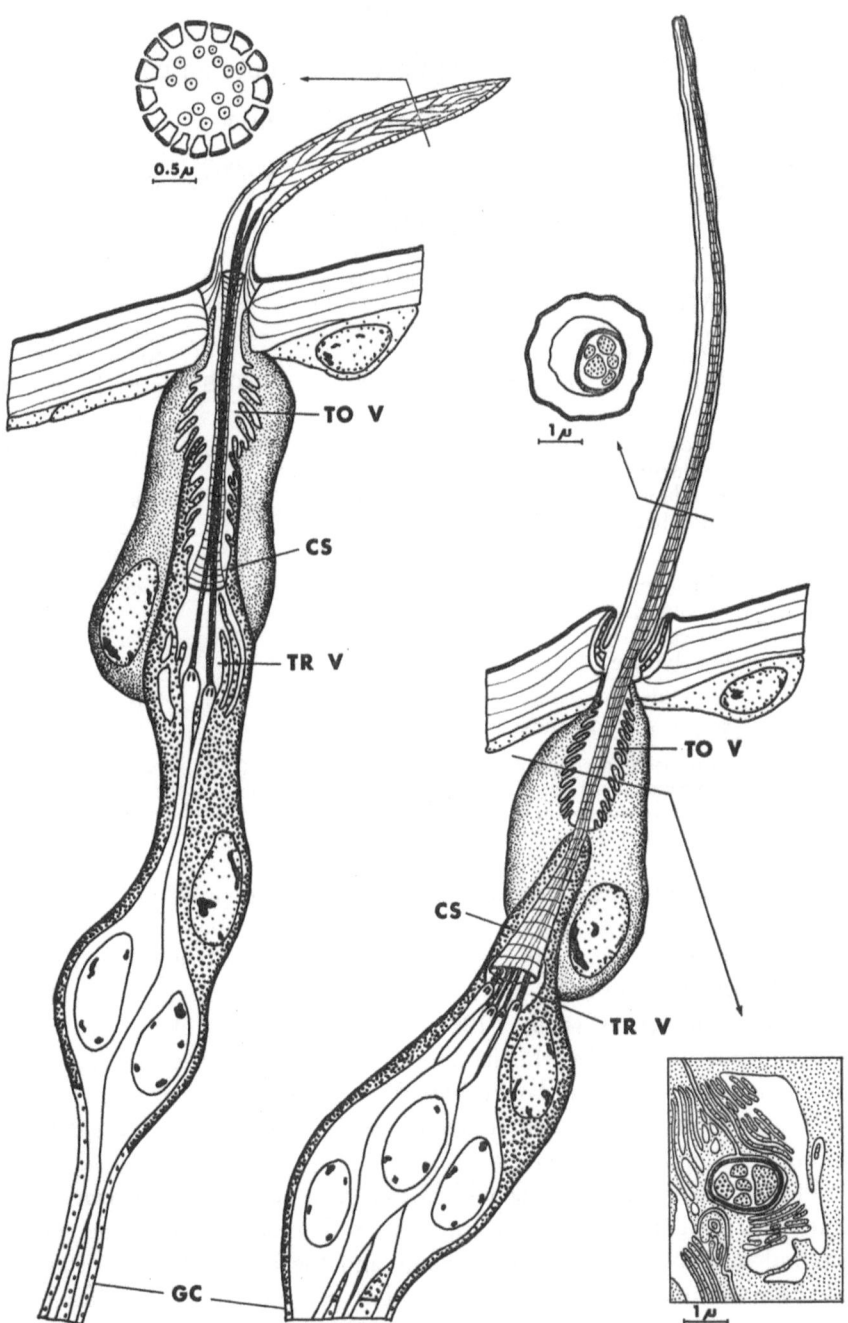

FIG. 1. Representatives of olfactory (*left*) and gustatory (*right*) sensilla. *Left*: schematic diagram of a sensillum basiconicum of the antenna of *Trypodendron lineatum*; *right*: schematic diagram of a sensillum trichodeum, type III, of the antenna of the same beetle.

The dendrites of the sensory neurons undergo a constriction of ciliary structure before they enter the cuticular sheath (CS), which is thin and ends at the base of the hair in the olfactory sensillum and is thick and continues to the tip of the hair in the gustatory sensillum. Cross sections through the extensions show the dendritic fibes free in the lumen (*left*) or enclosed in a channel (*right*). The cuticle of the olfactory hair exhibits

STRUCTURE OF CHEMORECEPTIVE SENSILLA

Recent reviews on the structure and forms of chemoreceptor organs in insects have been written on olfactory systems only (Slifer, 1961, 1967; Schneider, 1964). Our interests in connection with the function of chemoreceptive sensilla will be focused on the configuration of the chemoreceptive site, its supply of dendritic fibers, and the insulation of the neurons.

In olfactory systems each sensillum consists, in addition to two accessory cells, of one (Boeckh, 1962; Dethier et al., 1963; Lacher, 1964; Schneider et al., 1964) to several neurons [in the extreme about 60 (Slifer et al., 1959; Slifer and Sekhon, 1963; Slifer, 1966)], and in gustatory systems of generally four or five sensory neurons [in the extreme three to eight neurons (Lewis, 1954a; Dethier, 1955)]. As in other sensilla, the chemoreceptive neurons are descendants of one epidermal cell, which has divided into two mother cells before molting. One cell has yielded the accessory cells, and the other, in successive divisions, the sensory neurons (Peters and Richter, 1963; Peters, 1965).

The dendrites of sensory neurons grow centrifugally to the site of stimulus reception, whereas the axons develop centripetally and form the nerve (Henke and Rönsch, 1951; Wigglesworth, 1953). Fusion of axons, which has been suspected in a few instances, does not occur, judging by electron-microscopic studies (Schneider, 1965; Stürckow et al., 1967a; Moeck, 1968). Insect sensory neurons are true primary neurons (Snodgrass, 1935; Weber, 1933, 1954), the axons of which contact the corresponding central ganglion without the aid of internuncials. A steady replacement of taste receptor cells through new neurons, as found by Beidler and Smallman (1965) in the rat, is therefore improbable.

Figure 1 shows representatives of olfactory and gustatory sensilla. Although the variation of chemoreceptive systems is immense, the perforated area of cuticle for the reception of stimuli always seems to be large and generously spread over the extension of the sensillum in olfactory systems, and seems always to be restricted to a small area at the tip of the sensillum in gustatory systems. The difficulty in distinguishing between olfactory and gustatory receptor cells is well known (Dethier and Chadwick, 1948a). This structural criterion may turn out to be valid, although some overlapping may occur.

FIG. 1. (*cont.*)
many pores, while the gustatory hair has one pore at the tip. The axons of the sensory neurons are enveloped by glial cells (*GC*). The trichogen cell (*coarse punctation*) embraces the sensory cell bodies and beginning dendrites and in turn is enveloped by the tormogen cell (*fine punctation*). Both accessory cells form extracellular vacuoles (*TR V* trichogen vacuole, *TO V* tormogen vacuole), which might be in complete fluid exchange with each other and support the extracellular vacuole in the extension of the sensillum. A cross section through the tormogen cell at the level of its extracellular vacuole (*lower right*) shows the neuroglia type of cell, which is a feature of both accessory cells. The borderline between the trichogen and tormogen cell is not as clear as shown in the figure (*see text*). Considerable interdigitation might occur in distad direction beyond the borderlines given for both trichogen cells. (*After* Moeck. 1968. *Canad. J. Zool.* 46(3):521-556. By permission of the National Research Council of Canada.)

The dendrites of probably all chemoreceptive sensilla undergo, as in other sensilla, a constriction of ciliary structure soon after they leave the cell body. At present, this ciliary structure has been found only in olfactory sensilla (Krause, 1960; DuBose, 1967; Moeck, 1968; and reviewed by Slifer, 1967) and a probable gustatory sensillum (Fig. 1, *right*); however, through systematic study it should likewise be found in all gustatory sensilla. The ciliary region divides each dendrite into an inner and an outer segment. In olfactory hairs, the outer segment branches more or less extensively after it has entered the distal portion of the sensillum (the dendrites of the sensillum placodeum of the bee seem to be an exception, Slifer and Sekhon, 1961). In gustatory sensilla, the branching of dendrites is unusual and occurs only as an irregularity (Schoonhoven and Dethier, 1966, see sensillum stylonicum).

The envelopment of dendrites and sensory cell bodies has not been well understood (as discussed in detail by Larsen, 1962). The morphogenetic studies of Peters (1965) and the electron-microscopic investigations of Moeck (1968) are the only studies which have shed light on this problem. Moeck obtained ultrathin sections of desired areas by dissecting thick sections with a microscalpel and re-embedding and reslicing them. According to his results, which are in agreement with evidence from electron micrographs of other authors, the accessory cells behave like neuroglia because of their lamellate and villiform way of enveloping the sensory cell bodies and dendrites. Contrary to the unknown origin of neuroglia cells, the epidermal descent of the accessory cells is known.

Moeck (1968) found (in contradiction to Zacharuk, 1962, and Peters, 1965, who both used the light microscope) that the trichogen cell completely envelops the sensory cell bodies and the proximal dendrites and joins the neuroglia interdigitally in the region of the beginning axon. This interpretation cannot be read clearly from Moeck's figures (1968, Figs. 53 and 54), but he pointed out (personal communication) that the glial cell (*GC* in Fig. 54), which is visible on both sides of the beginning axon, borders the trichogen cell a short way distal from the lettering *GC*. The trichogen cell embraces the cell bodies of both neurons in Figure 54 and is, at its proximal end, rich in mitochondria (*MI*). In sensilla with numerous neurons, several pairs of accessory cells each belong to a small group of neurons (Slifer, Sekhon, and Lees, 1964).

The extension of the sensillum is formed entirely by the trichogen cell (Snodgrass, 1935; Weber, 1933, 1954), which grows out together with the dendrites and, at the same time, in the taste hair of the blowfly, forms a tube around the dendrities (Peters, 1965). After the formation of cuticular details, a vacuole of the trichogen cell fills the lumen of the hair. Peters (1965) was not able to determine with the light microscope whether this vacuole is extracellular or intracellular. Electron-microscopic sections of different hairs, cones, and pegs show no cell membrane in the lumen of these extensions. The dendrites are embedded only in a fluid, either within a separate cuticular sheath or free in the lumen of the extension (Fig. 1). Therefore, the trichogen cell has probably completely retracted its cytoplasm from the extension of the sensillum and has left an extracellular vacuole. It also forms an extracellular vacuole at the level of the proximal end of the cuticular sheath around the dendrites (Fig. 1). However, some cross sections of this region show the dendrites in the sheath apparently enveloped by plasma (Slifer and Sekhon, 1954a; Slifer, 1967). This cytoplasm might belong to microvilli of the trichogen cell that reach into the proximal opening of the cuticular tube. Further distad, the dendrites have always been found surrounded by a fluid.

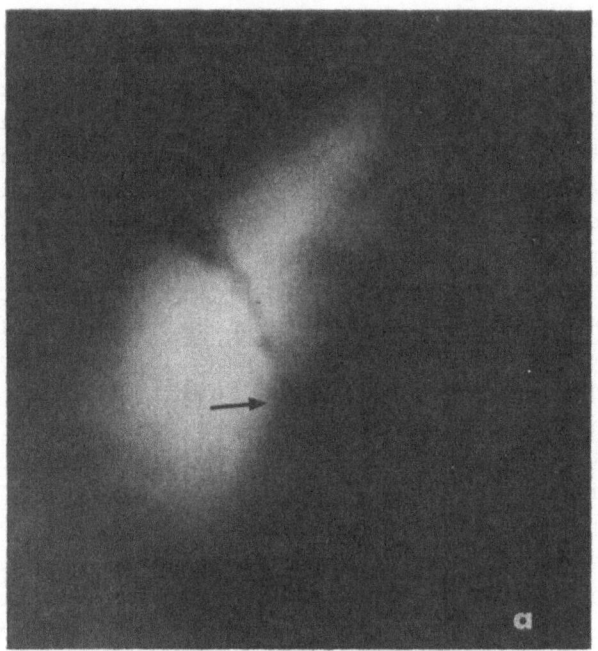

FIG. 2. View distad into the socket of a labellar taste hair of the blowfly, *Calliphora erythrocephala*.

a, methylene, blue, vital staining, teased preparation; × 2,100 (× 1,200 light microscope). b, explaining drawing. The *coarse arrow* (lower right) indicates the distal course of the cuticular sheath containing the dendrites. One dendrite leaves the sheath at the level of the *fine arrow* and ends at the socket. S, shadow of a group of sensory cells from a neighboring sensillum. (*After* Stürckow. 1961a. *Proc. XI Int. Congr. Entomol.*, 1:410-411.)

The tormogen cell in turn envelops the trichogen cell and forms the socket of the sensillum but does not reach further distad than to the base of the extension of the sensillum. According to Moeck (1968), it also exhibits an extracellular vacuole with many lamellae and microvilli wrapped like neuroglia around the cuticular sheath containing the dendrites (Fig. 1). In the olfactory sensillum investigated by Moeck (Fig. 1, *left*), the extracellular vacuoles of the two accessory cells seem to be separated only by the thin cuticular sheath (CS).

The sheath around the dendrites is a secretion of the trichogen cell (Peters, 1965). Its occurrence varies in that it may enclose the dendrites separately, in groups, or separately and in groups in the same sensillum, as well as it may end at the tip of the sensillum, at the base, or somewhere in the outer extension (DuBose, 1967; Moeck, 1968; reviewed by Slifer, 1967). The cuticular sheath may consist of a proximal and a distal portion as in the taste hair of the blowfly (Peters, 1965). Here, the proximal portion is nonsclerotisized and is of a different shape than the distal portion, which is sclerotisized. Thus far, all attempts to ascribe a function to the sheath, other than a mechanical protection of the dendrites, have failed. The sheath varies from being very thin to thick, or may not appear at all (Slifer, 1967). This variation suggests that the fluid within the sheath probably is of the same composition as that within the extracellular vacuoles. The attempts of Moeck (1968) to find border lines other than the cuticular sheath between the vacuoles of the two accessory cells were not successful. The distal trichogen and the extracellular tormogen vacuoles may have united, and it may be of no significance at which point the dendrites enter or contact the system of vacuoles. This interpretation is strengthened by the observation that dendrites penetrate the cuticular sheath at the base of the sensillum. Slifer and coworkers have reported this case for several olfactory sensilla and have shown a convincing electron micrograph of an olfactory sensillum of the grasshopper, *Melanoplus*, (Slifer and Sekhon, 1964b; Slifer, 1967). Another case of penetration of the cuticular sheath by a dendrite is the taste hair of the blowfly, *Calliphora* (Fig. 2). Peters and Richter (1963), using the electron microscope, found five dendrites in the proximal portion of the cuticular sheath and four in the distal portion, which is in agreement with Figure 2. One dendrite leaves the cuticular sheath within the socket of the taste hair, at least in some of the sensilla.

Figure 3 is a longitudinal section through the extension of an olfactory sensillum of the sexton beetle, *Necrophorus*. It shows pores in the cuticle which widen to a bay and narrow again. Similar pores have been found in several olfactory sensilla (reviewed by Slifer, 1967). The filaments, which seem to extend inwards from the pores are called "pore filaments" (Slifer, Prestage, and Beams, 1959). In two locations of Figure 3, these pore filaments are in contact with dendritic fibers, while others seem to end at a distance from the dendrites. These apparently blindly ending filaments are probably artifacts due to sectioning. In other insects, the pore filaments extend inward in a bouquet-like fashion (Slifer, Sekhon, and Lees, 1964; Slifer and Sekhon, 1964b; DuBose, 1967). Therefore, only such filaments can be found in contact with dendritic branches that make this contact within the plane of the section.

Contrary to earlier cautious statements (Slifer et al., 1959), Slifer and Sekhon

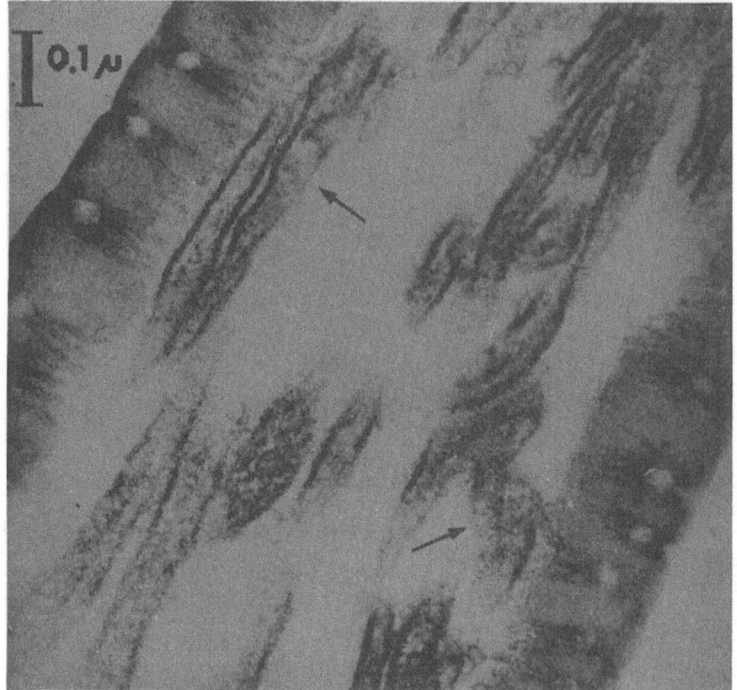

Fig. 3. Electron micrograph of a longitudinal section through the peg of an olfactory sensillum of the sexton beetle, *Necrophorus*.

The arrows point at locations (10 to 15 mm distant from the tip of the arrows) where some pore filaments contact the dendritic branches in the lumen of the peg; × 90,000. (*From* Schneider, Steinbrecht, and Ernst. 1966. *Naturwiss. Rundsch.*, 19(3):title page.)

(1964a) made the assertion without further evidence that the pore filaments are probably extensions of the dendrites, which reach into the pores. Richter (1962) had made the same statement, based on electron micrographs, which in one case apparently showed long strands of pore filaments. Since Hoffman (1961), the interpretation that dendritic filaments extend into the pores has appeared in most of the reviews without criticism.

Judged by observations on the taste hair of flies, these pore filaments do not represent direct extensions of the dendrites, although they seem to be in close contact with them. Because of their location and texture, their contents are probably homologous with a material called the "viscous substance" in the taste hair of the blowfly, *Calliphora* (Stürckow, 1967). Slifer et al. (1964) describe the appearance of the pore filaments as that of fine strands of woolen yarn, with or without a core of less density, and with irregular edges. In oblique sections through the wide portion of olfactory pores, densely packed vesicles or tubules are found (Fig. 4). All these features are in agreement with those of the viscous substance found in taste hairs of flies.

Figure 5 shows a longitudinal section through the tip of a taste hair of the stable fly, *Stomoxys*. A substance extrudes from a pore in the tip of the hair. The lining of the substance is irregular and granular or vesicular. Figure 6 shows

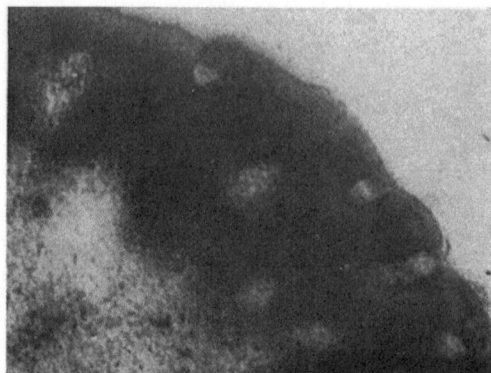

FIG. 4. Electron micrograph of an oblique cross section of an olfactory peg of the grasshopper *Melanoplus differentiales*.

In this section, the wide portion of the olfactory pores has been cross sectioned and is densely packed with vesicles or tubules; × 48,000. (*From* Slifer, Prestage, and Beams. 1959. *J. Morph.*, 105:189, Fig. 39.)

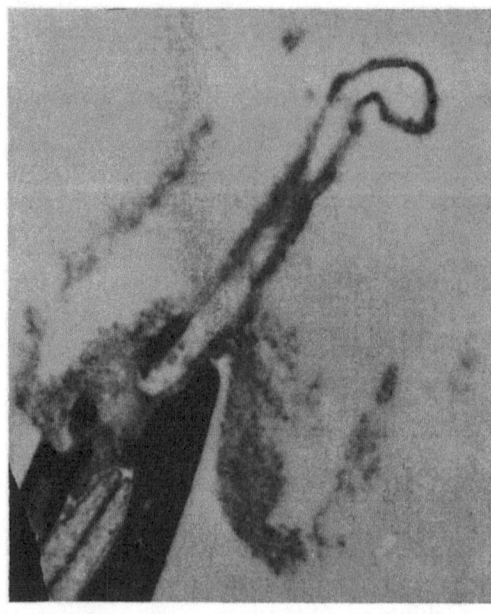

FIG. 5. Electron micrograph of a longitudinal section through the tip of a tarsal taste hair of the stable fly, *Stomoxys calcitrans*.

Two of the dendrites end beneath a pore in the cuticle, through which a substance with an irregular lining of granular or vesicular structure has been extruded. Fifty hr OsO_4 fumes, methacrylate; × 22,000. (*From* Adams. 1961. Doctoral dissertation. Courtesy of the author.)

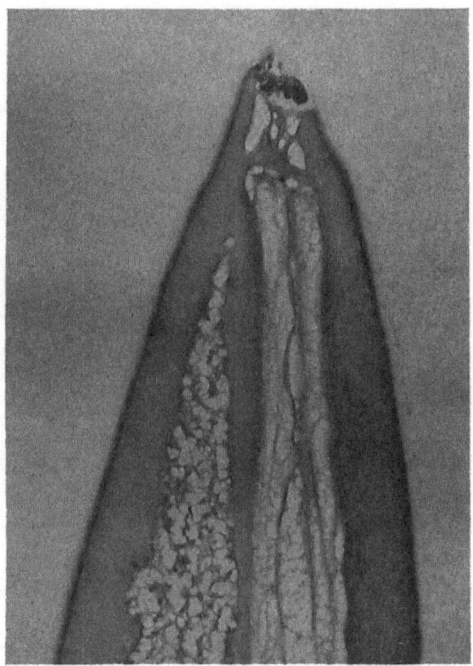

FIG. 6. Electron micrograph of a longitudinal section through the tip of a labellar taste hair of the blowfly, *Phormia regina*.

This is a typical section through the pore above the dendritic endings, which is connected through spongy cuticle with probably both channels of the hair. The dendrite-containing channel is to the right, the dendrite-free channel is to the left. In the outer portion of the pore, an electron-dense conglomerate is to be seen, which is clustered viscous substance originated probably from the dendrite-free channel. Glutaraldehyde, post-fixed with OsO_4 in Millonig's phosphate buffer, methacrylate; × 23,000.

a section through the tip of a taste hair of the blowfly, *Phormia*. Although this tip is differently formed than that of the taste hair of the stable fly (Adams, 1961; Stürckow et al., 1967b), it also has a pore distad from the endings of the dendrites, which is filled with congolomerates of viscous substance. Such conglomerates of electron-dense material at the tip are likewise known from other sections and from electron micrographs of intact hairs, which we made from hairs with freshly extruded viscous substance (Stürckow and Holbert, unpubl.). All fixatives tried caused this substance to collapse. Contrary to results obtained with the stable fly (Fig. 5), a 50-hour fixation in osmic acid fumes of the blowfly taste hair was unsuccessful.

Finally, the viscous substance was made visible in its living state through a replica, which was obtained by slightly pressing the tips of unfixed taste hairs into a jelly (Fig. 7). A vesicular texture is seen where the viscous substance is in focus. Although a replica showing such algae-like growth has been found only once in 47 replicas, there is little doubt that this extrusion is the viscous substance. It is known in this bouquet-like extension from electron micrographs of intact unfixed hairs and has been found in the collapsed form in three replicas (Stürckou and Holbert, unpublished).

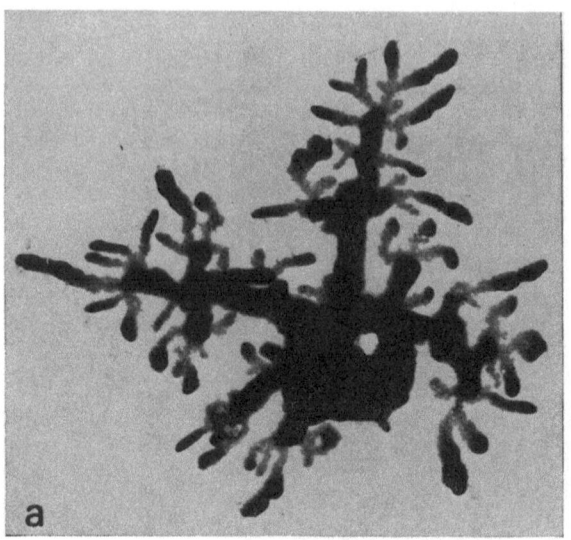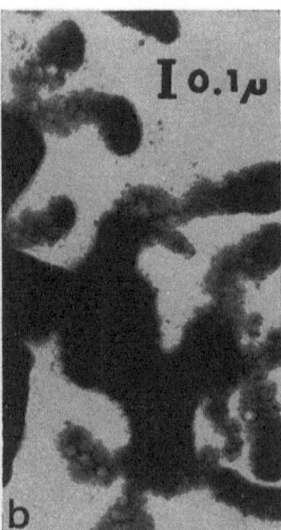

FIG. 7. Electron micrograph of a replica of the tip of a living labellar taste hair of the blowfly, *Phormia regina*.

a, the dark portion in the middle is the print of the hair tip showing clearly a pore or indentation. The algae-like growth around the tip is very probably the imprint of the viscous substance in its living unfolded state; × 16,000. **b,** a portion at higher magnification showing the vesicular texture of the viscous substance; × 45,000.

An extrusion of the degree shown in Figure 7 probably does not often occur. A small amount of viscous substance filling the pore seems to be normal and sufficient for the reception of a stimulus. The origin of this substance and of the content of the pore filaments in olfactory sensilla is still unknown and will be discussed in the section on WORKING HYPOTHESES OF RECEPTIVE MECHANISMS (p. 143).

In a crustacean decapod, a chemosensitive sensillum exhibits a thin, spongy cuticle without well-defined pores over a relatively large area; stain particles have constantly been found to accumulate on this cuticle but not on others (Ghiradella et al., 1968). These particles may be, at least partly, conglomerates of viscous substance.

NEUROPHYSIOLOGY OF CHEMORECEPTIVE SENSILLA

The function of chemoreceptor cells cannot be understood by studying neurophysiologically single receptor cells without reference to the entire system and without the indispensable guidance of behavioral studies (Hodgson, 1967). Behavioral experiments are of as basic importance to neurophysiology of sense organs as the light microscopy is to electron microscopy. On the other hand, neurophysiologic studies at different levels—entire appendages, groups of sensilla, single sensilla, and single receptor cells—have brought a new understanding of the function of chemoreceptive systems. Except for recordings made by Yamada (1968;

Yamada et al., 1968) from single units of the central nervous system, neurophysiologic data have been gained peripherally through recordings from the sensillum, and in one case from the nerve (Stürckow et al. 1967a).

Properly placed electrodes record two types of potentials from a receptor cell in response to an adequate stimulus: a slowly rising potential, referred to as the generator potential (the term "receptor potential" should be reserved for a potential at the receptive site of the stimulus, in agreement with Hodgson, 1964, 1965; and others), and a pattern of rapidly rising potentials, the impulses or spikes. The slow potential originates at the dendrite and is transmitted to the cell body. As soon as this potential has reached a certain height, fast-rising potentials of short duration, the impulses, are generated in the proximal region of the cell body, or in the beginning axon, or both. The height of the generator potential depends on the strength of the stimulus; the frequency of the impulses in turn depends on the height of the generator potential, thus, indirectly also on the intensity of the stimulus. Both the height of the generator potential and the frequency of the impulses can be used as a measurement of the reception of a stimulus.

Generator potentials

The recording of generator potentials of single neurons meets with difficulties because of summation effects. Generator potentials of neighboring neurons of the same sensillum (or neighboring sensilla, if a sense organ with a single neuron is selected) are recorded as a sum phenomenon. Since intracellular recordings of chemoreceptive neurons are not feasible (the cell bodies have an average diameter of 5μ, and penetration of the integument is required) extracellular recordings taken as close to the neuron as possible have been made.

The only examples in which undistorted generator potentials of single neurons have been recorded with high probability are those in Figure 8 recorded by Morita and Yamashita (1959) and Boeckh (1962). Not all of the generator potentials recorded by Morita and Yamashita in 1966 seem to be generated by single neurons. The relation of the frequency of the impulses to the height of the receptor potential together with the single type of impulses suggest that the recordings of Figure 8 probably originated from single neurons. Morita and Yamashita (1959) recorded with a microelectrode from the cracked shaft of a taste hair of the blowfly and obtained this single generator potential probably by the proper choice of stimulus and hair. A sugar solution generally excites two of the five neurons in the labellar taste hair of flies, the sugar- and the water-receptor cell; however, in some taste hairs, the water-receptor cell is missing (Wolbarsht, 1957; Stürckow, 1963), which must have been the case in the hair used by Morita and Yamashita (1959). Boeckh inserted a microelectrode into the vicinity of the dendrite-containing channel in the proximal portion of olfactory sensilla of the sexton beetle, Necrophorus. These sensilla were specifically sensitive to the odor of decaying meat and were mostly supplied with a single neuron (Boeckh, 1962). Carrion odor elicited an excitatory generator potential. Generator potentials of reverse polarity were obtained in response to propionic acid. Since

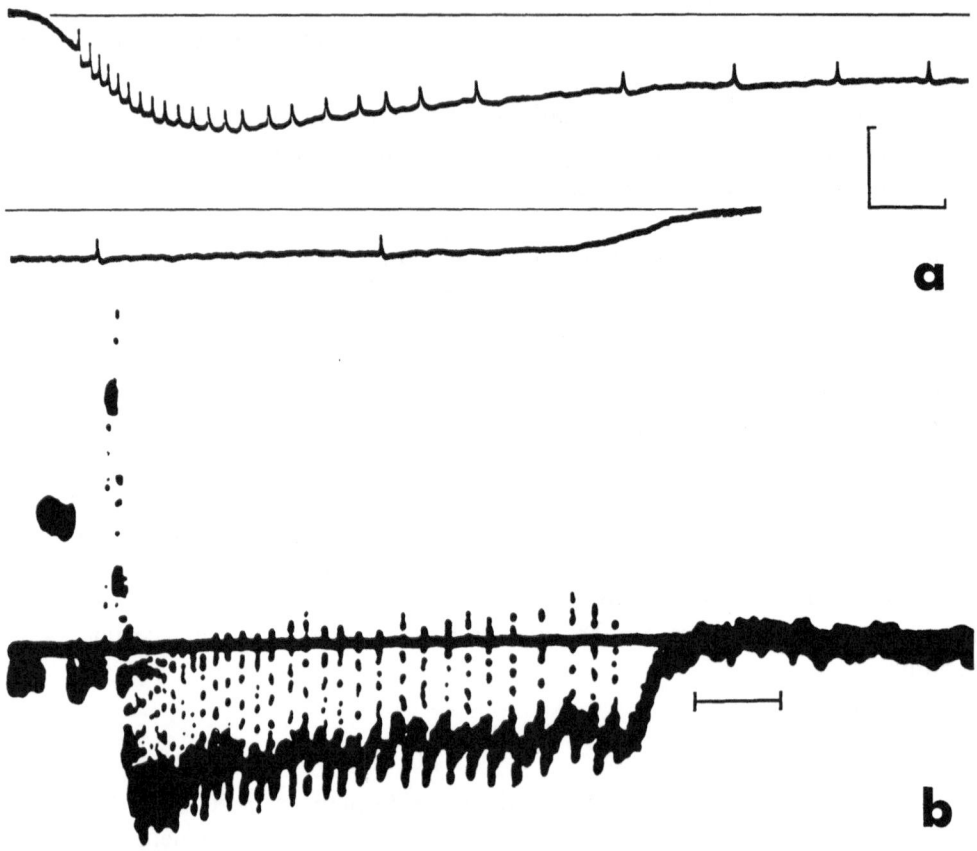

FIG. 8. a. Receptor potential and impulses of probably a single receptor cell of an olfactory sensillum of the sexton beetle, *Necrophorus vespillio*.

The stimulus was odor of decaying meat. The upper and the lower trace are interrupted for 300 msec; time scale 50 msec; vertical scale 2 mV. (*After* Boeckh.1962. Z. *Vergl. Physiol.*, 46:212-248. Courtesy of Springer-Verlag.)

b. Receptor potential of probably a single receptor cell of a gustatory sensillum of the blowfly, *Calliphora vomitoria*.

The stimulus was a sugar solution. Time scale 100 msec; rectangular calibration pulse of 1 mV before stimulation. (From Moritta and Yamashita. 1959. *Science*, 130(3380): 1922.)

they lacked the on-response known from single excitatory generator potentials [which excitatory potentials elicited by benzylmercaptan and heptanol also were missing (Boeckh, 1962, 1967)], they might be understood (contrary to Boeckh's interpretation) as a sum phenomenon of excitatory and inhibitory potentials of two or more receptor cells. Inhibitory slow potentials suppress impulses which have spontaneously occurred or have been generated in response to a previous excitatory stimulus. Boeckh (1967) probably is correct in suggesting that centrally, inhibitory potentials also become interpreted.

The remote chance of recording generator potentials from single neurons without interference from neighboring potentials (Wolbarsht and Hanson, 1967), makes this technique less valuable for the investigation of the reaction spectra of single neurons. The recording of impulses from single sensilla has been more successful. It was used on a broad base in gustation first by Hodgson and Roeder (1956) and Hodgson (1957), and in olfaction first by Schneider et al. (1964) and Lacher

(1964). Interfering impulses from neighboring neurons can be distinguished by deviating heights and frequencies. The recording of impulses is simpler than that of generator potentials, since AC amplification and any type of electrode can be used. This is in contrast to DC amplification with nonpolarizable electrodes which must be used for the proper recording of slow potentials.

Electroantennograms

Overall responses of entire appendages to a stimulus have been studied only with antennae of insects, and not with palpi or labellae bearing gustatory receptor cells. Such a recording, obtained by inserting electrodes at the tip and the base of an antenna, is called an "electroantennogram" (Schneider, 1957a) and is understood—similar to the electroretinogram—as a sum phenomenon of all slow potential changes in the antenna in response to a stimulus. The excitatory and inhibitory generator potentials of sensory neurons probably represent one of the main components of the antennogram.

Boeckh et al. (1965) and Schneider (1965, 1967) suggest a close agreement between the stimulus-percent response curves of antennograms and of single generator potentials; however, a significant conclusion cannot be expected from curves based on such unrepresentative samples submitted without statistical analysis. One of their curves was based on generator potentials of six single neurons and the other one on antennograms of 30 to 200 data per point of the curve. The antennograms are sum phenomena of several thousand neurons, which enhances the disproportion of their data. The results were obtained with the male of the silk moth, *Bombyx*, and synthetic bombykol as the stimulus, which is the sex attractant of the female moth.

According to Schneider (1965), each antenna of the male silk moth bears 15,000 to 20,000 olfactory neurons responding specifically to bombykol as well as to a few other compounds, plus 20,000 chemoreceptor cells responding to other stimuli. Thus, the olfactory capacity of the silk moth is probably not as one-sided as described by Marler and Hamilton (1966). The quantitative reaction spectra to bombykol of the special 15,000 receptor cells might vary widely (this cannot be judged from six trials). Part of the 20,000 less specific receptor cells may respond to bombykol with inhibition, which apparently has not been investigated and would be a factor in such an overall response as the antennogram.

Details of the antennogram have not been analyzed with the care and elegance accorded the electroretinogram; they have rarely been studied because of doubts about their correctness and utility. However, nowadays it is known that under constant conditions and within a certain intensity range of the stimulus this principle applies: the stronger the stimulus, the greater the magnitude of the antennogram, since more sensory neurons are responding and each develops a higher generator potential. This rule is also valid for inhibitory, inverse antennograms, which Schneider (1957a, 1957b) found, when he investigated the response of the silk moth to some narcotics and to cycloheptanone.

The first antennograms were recorded by Schneider and Hecker (1956) with the antenna of the silk moth. Later, Schneider used this method frequently (1957a, 1957b, 1961a, 1962, 1963a, 1963b; Schneider et al., 1967). Although shape

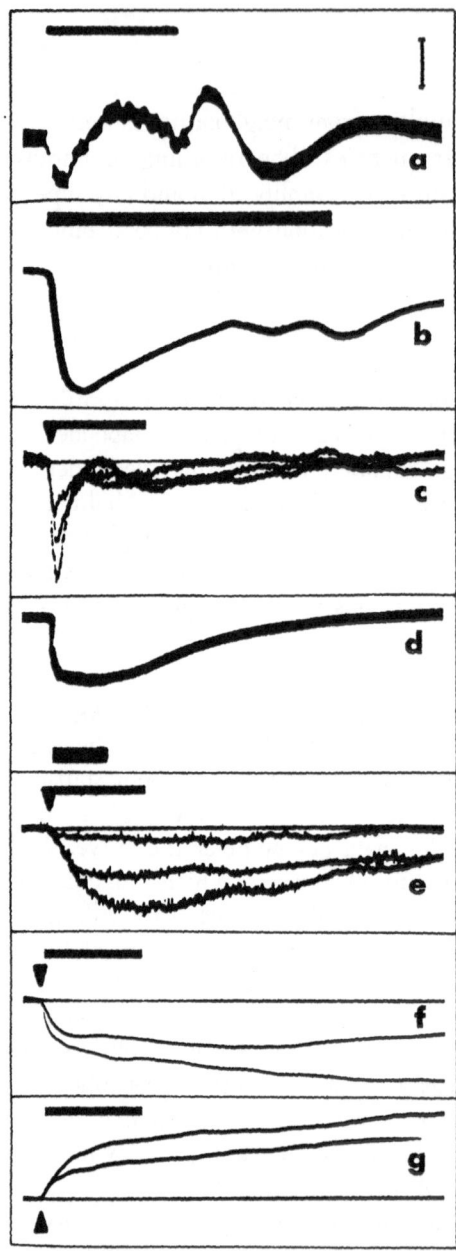

FIG. 9.   Combination of antennograms of four insects and an artificial antenna of cotton soaked in saline solution.

All antennograms were obtained with roughly comparable methods as to the application of the stimulus.

antennogram:	length of stimulus (sec):	height of vertical scale (mV):
a	0.9	0.4
b	2.8	0.4
c	0.5	1.0
d	0.7	0.8
e	0.5	1.0
f	0.5	1.0
g	0.5	1.0

Compare in b and c the influence of the duration of the stimulus.

and kinetics of the antennogram depend to a certain degree on the quality of the stimulus (Schneider, 1957a, 1957b, 1962; compare also summed responses of about ten sensilla, Kay et al., 1967), the antennogram cannot be used to identify an odor quality as indicated by Schneider (1957a, 1957b, 1961b), nor to judge its attractive, repellent, or other nature (as indicated by Abushama, 1966). Several odors evoke similar antennograms in the same specimen and also in different genera (Boeckh et al., 1963; Stürckow, 1965; Abushama, 1966). Figure 9 shows that shape and kinetics of the antennogram depend strongly on the kind of antenna used, just as the sum phenomenon of an electrical current depends on the layout of a circuit. *Bombyx* and *Porthetria* have combed antennae, about 5 and 7 mm long, which give similar antennograms to different odors. Zootermopsis nymphs have threadlike antennae of 2 mm, and Periplaneta of 5 cm in length; each shows its typical antennogram to different odors.

Biologic potentials must be distinguished from artificial potential changes, which can be obtained with an antenna made from a cotton thread wetted with Ringer's solution and attached to polarizable electrodes. Artificial cotton antennae of 5-cm length furnished longer-lasting responses than insect antennae (Fig. 9, compare f and g with a through e). Using the height of the first 0.3 sec for the determination of a stimulus-percent response line for amyl acetate, the slope of the line of the artificial antennae was the same as that of cockroach antennae except that the concentration range was 10^3 times higher for artificial antennae (Stürckow, unpubl.). When one of the electrodes was covered by a capillary, the potential nonbiologic change in response to the stimulus was abolished at this electrode; the same occurred, resulting in an inverse potential change, if the other electrode was covered with a capillary (Fig. 9, f and g). The nonbiologic potential can be abolished totally if both electrodes are threaded into a capillary or if unpolarizable Ag–AgCl electrodes are used. Therefore, the artificial "antennogram" is probably a summed contact potential of the wetted electrode surface, while the cotton thread

FIG. 9 (cont.)
a. Representative antennogram with marked on- and off-response of the nymph termite, *Zootermopsis angusticollis*, in response to three oils, and the odor of termite abdomens. (*From* Abushama. 1966. *Entomol. Exp. Appl.*, 9:343-348.)
b. Representative antennogram of the male of *Periplaneta americana* in response to amyl acetate and to the odor of the female cockroach. (*From* Boeckh, Priesner, Schneider, and Jacobson. 1963. *Science*, 141(3582):716.)
c. Representative antennogram of the male or female cockroach, *Periplaneta americana*, in response to three concentrations of *iso*-amyl acetate and amyl acetate; each stimulus is 100-times stronger than the preceding. The antennograms were recorded at 5-min intervals, and later photographically superimposed with the aid of orientation marks.
d. Representative antennogram of the male silk moth, *Bombyx mori*, in response to the odor of the female moth. (*From* Schneider and Hecker. 1956. *Z. Naturforsch.*, 11b(3):121-124.)
e. Representative antennogram of the male gypsy moth, *Porthetria dispar*, in response to amyl acetate, *iso*-amyl acetate, and the odor of the female moth; each stimulus is 100-times stronger than the preceding, superimposition as in c. (Cf. Stürckow. 1965. *J. Insect Physiol.*, 11:1573-1584.)
f and g. Nonbiologic potentials from an artificial antenna (see text). Each time, the first stimulus (response: low curve) and second stimulus (response: high curve) were given at an interval of 5 min; second stimulus with amyl acetate was 100-times stronger than the preceding; superimposition as in c.

serves as a conductor. Mozell (1961) reports a similar influence of polarizable electrodes on slow potentials. Antennograms recorded from dead antennae of the bark beetle, *Ips confusus* (Borden, 1968), can only be understood as nonbiologic potentials. Schneider (1956, 1957b) was not able to record antennograms with nonpolarizable electrodes from dead antennae of the silk moth.

Impulses

The most interesting results on the frontier of neurophysiology are based on recordings of impulses from single neurons or small groups of neurons.

REACTION SPECTRA.   In *gustation*, specialized receptor cells were first found in the taste hair of the blowfly. Each is activated primarily by salt, sugar (Hodgson et al., 1955), or water (Wolbarsht, 1957; Evans and Mellon, 1962a). The existence of a mechanoreceptor cell associated with taste hairs (Wolbarsht and Dethier, 1958) was questioned by Stürckow et al. (1967a), although in labellar taste hairs of the blowfly one dendrite ends at the base of the hair (Fig. 2); also, impulses of a mechanoreceptor cell were recorded from taste hairs of larvae and adults of *Vanessa* and *Pieris* (Takeda, 1961; Schoonhoven, 1967b). Similar specialized gustatory receptor cells were found, for example, in *Vanessa* (Morita et al., 1957; Takeda, 1961), the potato beetle, *Leptinotarsa* (Stürckow, 1959), the larva of the silk moth, *Bombyx* (Ishikawa, 1963), and some lepidopterous larvae, mainly *Protoparce* and *Pieris*, (Schoonhoven and Dethier, 1966; Schoonhoven, 1967b); a bitter-receptor cell was found in the larva of *Bombyx* (Ishikawa, 1966).

Attempts to understand the coding of receptor cells ceased to be fruitful as the search for specialized cells continued, since additional results began to complicate the picture. These were the findings of two different sugar receptor cells in one taste hair of the larva of *Bombyx* (Ishikawa, 1963); the indication of two salt receptor cells in the taste hair of *Leptinotarsa* (Stürckow, 1959), of the larva of *Bombyx* (Ishikawa, 1966), of *Calliphora* (Stürckow, 1963; den Otter and van der Poel, 1965; den Otter, 1967) of the lepidopterous larva, *Philosamia cynthia* (Schoonhoven and Dethier, 1966), and finally of *Phormia* itself (Steinhardt, 1966); further, one or two otherwise specialized cells respond with volleys of impulses to *Solanum* alkaloid glycosides in *Leptinotarsa* (Fig. 10, *right*) and *Calliphora* (Stürckow, 1960), and respond with a continued pattern of impulses to acid in *Leptinotarsa* (Fig. 10, *left*) and in the larva of *Protoparce* (Schoonhoven and Dethier, 1966).

Certain receptor cells occur regularly in some taste hairs but are missing in others (Stürckow, 1963; den Otter, 1967; den Otter and van der Starre, 1967). This distribution indicates functional differences in taste sensilla of similar structure. Schoonhoven and Dethier (1966) and Schoonhoven (1967b) obtained a similar result in maxillary taste sensilla of lepidopterous larvae. Each maxilla bears two morphologically equal sensilla, a lateral and a medial one; in *Protoparce* and *Pieris*, acceptable substances seemed to cause a higher total impulse activity in the medial sensillum, while unacceptable substances were more effective in the lateral sensillum, although exceptions were found. Whether this finding has a

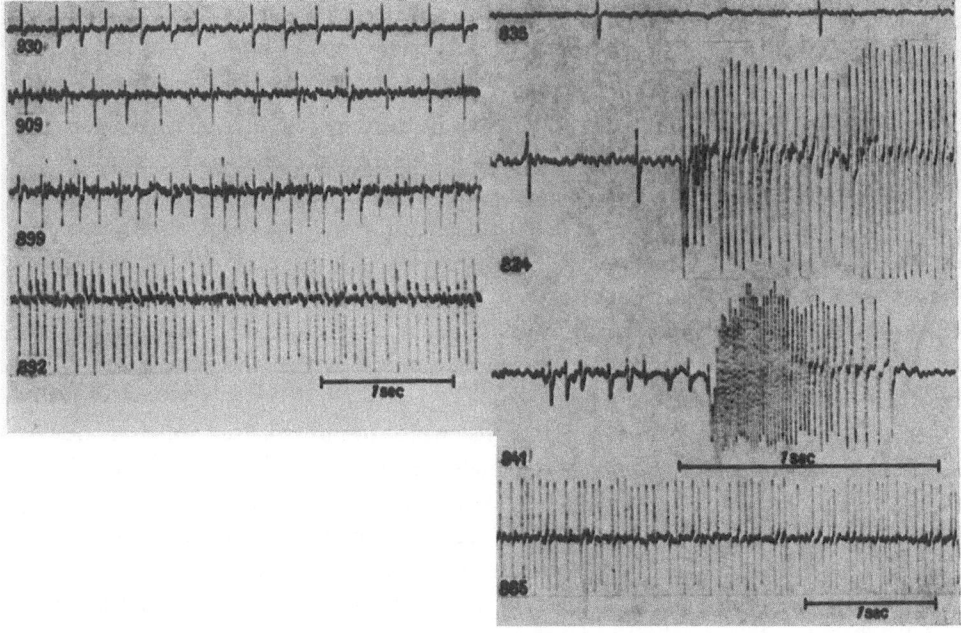

FIG. 10.   Impulse activity of eight taste hairs of *Leptinotarsa decemlineata* in response to solutions of different pH values.

Left: HCl of pH 6, 4, 3, 2 (upper to lower trace); *right*: tomatine 10^{-4} M of corresponding acidity, pH 6, 4, 3, 2. Steady-state response 30 to 105 sec after the onset of the stimulus. The salvo-like discharges in response to tomatine occurred at irregular intervals and were observed only between pH 3 and 5. Unbuffered solutions were prepared immediately before testing and adjusted with HCl. (*From* Stürckow. 1959. *Z. Vergl. Physiol.*, 42:255-302. Courtesy of Springer-Verlag.)

meaning for the central evaluation needs further investigation. A systematic study of functional differences in taste sensilla of the same species or specimen has not been made at present. The failure of a response of a taste hair has been neglected so far or interpreted as a lack of contact between the chemosensitive area and the stimulating solution, although a failure of impulse activity might be evaluated centrally by the insect.

Inhibition of the activity of specialized gustatory receptor cells by different compounds has been studied most thoroughly by Hodgson (1957), Steinhardt et al. (1966), Steinhardt (1966), Hodgson and Steinhardt (1967), and Morita (1967). The site of inhibition is not yet known. It might be either at the common receptive membrane or at the level of the receptor cells, where the cause of inhibition might be hyperpolarization of one or more sensory neurons.

Spontaneous activity is difficult to prove in gustatory sensilla, since water is already a stimulus. In blowflies, spontaneous activity was not found by Evans and Mellon (1962a) or by Stürckow et al. (1967a), but it was found by Hodgson and Steinhardt (1967). In a medial taste sensillum of the larva of *Pieris*, spontaneous activity was found by Schoonhoven (1967b). Since sensilla possessing a cold-receptor cell might occur among chemoreceptive sensilla, a spontaneously active receptor cell should also be stimulated by changes in temperature.

All receptor cells mentioned respond in the phasic-tonic manner with a steady-state activity, which can last for 10 minutes and longer (Stürckow, 1967). Exceptions to the phasic-tonic type are the response to alkaloid glycosides, which occurs in volleys of impulses at irregular and unpredictable intervals of seconds or minutes, and the spontaneously increased activity observed when flowing solutions are used for stimulation and apparently loosen some of the viscous substance (Stürckow, 1967 and unpubl.). The last exception indicates that the viscous substance in front of the dendritic endings is of significance for the reception of the stimulus and the direction of investigations will change from the search for *cells* with more or less specialized dendritic membrances to the search for *sensilla* with more or less specialized receptive membranes. How messages are transformed and differentiated from a common receptive membrane to a single dendrite will at the same time become a new field of investigation.

In *olfaction*, specialized receptor cells were found for sex pheromones in the male of the silk moth, *Bombyx*, the male of the saturnid silk moth, *Antherea*, and the drone of the bee (reviewed by Boeckh et al., 1965); for odors of decaying meat in the sexton beetle, *Necrophorus*, and the blowfly (reviewed by Boeckh et al., 1965); for odors of the queen in the worker bee (Beetsma and Schoonhoven, 1966); and for odors of grass in a locust (Boeckh, 1967b). These "odor specialists" respond in a similar manner to their special odor, but also to a few other odors. Systematic investigations on the quantitative reaction spectra of these odor specialists have not been performed on large assortments of insects. In contrast, each of the olfactory cells of vertebrates exhibits a qualitatively different reaction spectrum, although these cells are primary neurons as are the olfactory cells of insects (Gesteland et al., 1965; Gesteland, 1966; Matthews and Tucker, 1966).

Such "odor generalists" (Schneider, 1963b) have also been found abundantly in insects (reviewed by Boeckh et al., 1965; Lacher, 1967). They each show a different reaction spectrum to about 20 to 50 pure odors. An example of the impulse activity of two generalist receptor cells in response to four odors is given in Figure 11. The number of sensory neurons in this sensillum is four or five, of which only two respond. Excitation of the spontaneously active neuron (*high impulse*) is observed in response to the first two stimuli, whereas the last two stimuli elicit inhibition of the spikes of this neuron. Spontaneous activity is a widespread characteristic of olfactory sensilla. The other neuron (*low impulse*) responds with markedly less excitation to all four stimuli.

Since in no instance in olfaction and gustation (except for sensilla with one neuron) have the reaction spectra of *all* neurons of a sensillum been analyzed, it might be that one or more neurons of a sensillum respond with hyperpolarization to a wide range of stimuli and thus depress the impulse generation of neighboring neurons; however, inhibition of a firing cell could as well be due to hyperpolarization of the active neuron itself, as Boeckh (1967a) interprets the reverse generator potential of *Necrophorus* in response to propionic acid (compare p. 127, 128).

The classification into specialist and generalist receptor cells has met with difficulties, since Lacher (1967) was uncertain to which type some of the recep-

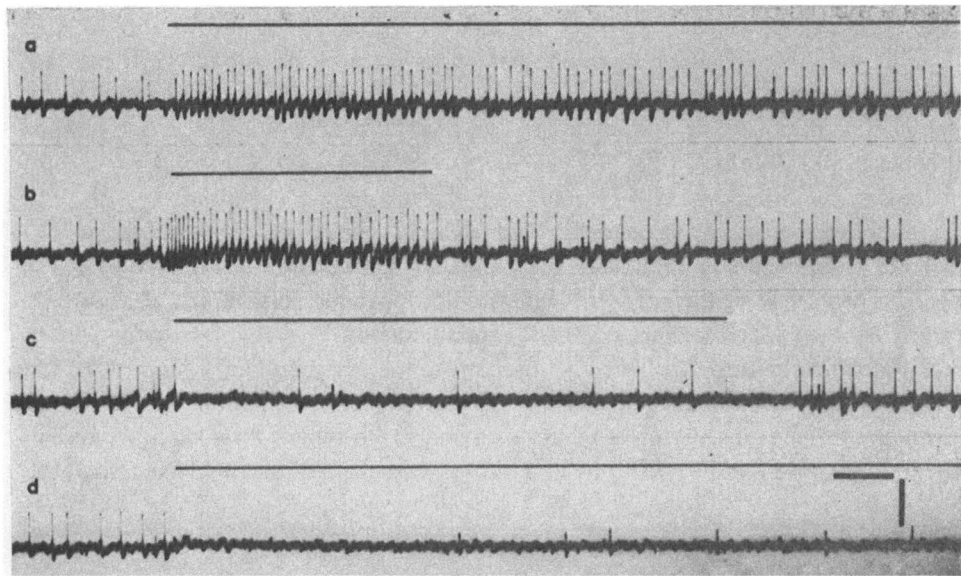

FIG. 11.   Response pattern of an olfactory sensillum of *Aedes aegypti*.
One electrode (ϕ 2μ) was inserted into the vicinity of the dendrite leading channel in the proximal portion of a sensillum of type A$_2$. The application of (a) terpineol, (b) citral, (c) eugenol, and (d) pentanoic acid is indicated by the line above each trace. Time scale 100 msec, vertical scale 0.5 mV. (*From* Lacher. 1967. *J. Insect Physiol.*, 13:1461-1470. Courtesy of Pergamon Press, Inc.)

tor cells of the mosquito, *Aëdes*, belong. As in the study of gustation, future work in olfaction might shift from the present search for specificity of *neurons* to the search for specificity of entire *sensilla*. Estimations of the numbers of molecules impinging at threshold intensities per sensory neuron (Boeckh et al., 1965; Schneider, 1965, 1967; Schneider et al., 1967) will then be irrelevant and would be replaced by the number of molecules per sensillum or per number of receptive sites of a sensillum. The neural response might reflect the response of the chemoreceptive membrane without distortion; however, at present this relationship is unknown.

The first step in the analysis of receptor sites per sensillum was made by Evans (1961, 1963), who showed for the taste hair of the black blowfly, *Phormia*, that glucose and fructose must combine with different sites of the chemoreceptive membrane. Both sites seem to give a message of their combination with a sugar molecule to the same neuron, the sugar-receptor cell; however, Evans required further investigation of this relationship with a clear criterion as to the type of receptor cell responding. Morita and Shiraishi (1968) recorded impulses of single taste hairs of the fleshfly, *Boettcherisca*; they did not find complete differentiation into two specific sites, one for glucose and one for fructose; however, they state that the sugar receptor site is composed of two subunits.

EFFECTS OF pH VALUES AND TEMPERATURE.   Experiments on the effect of the concentration of hydrogen ions on the activity of chemoreceptor cells have been made with taste hairs of two insects. In preliminary tests with *Phormia*, Evans and Mellon (1962b) measured the effect of acidity on the activity of the salt receptor cell. Gillary (1966a, 1966b) repeated these tests on a broader base with buffered and unbuffered solutions and found, in agreement with Evans and Mellon, no effect of the intensity of the hydrogen ions on the activity of the salt-receptor cell between pH 3 and 10 (Fig. 12).

Earlier, I (Stürckow, 1959) had carried out similar experiments with taste hairs of *Leptinotarsa* in response to the alkaloid glycoside, tomatine (Fig. 10). Responses with volleys of impulses were obtained between pH 5 and 3. Within this range, higher acidity facilitated the intensity of salvo-like discharges as shown in Figure 10. Since I had performed these experiments only with the threshold intensity of tomatine eliciting the salvo-like discharges, and had regularly found volley-like impulse patterns in response to higher concentrations of tomatine at pH 6, this dependence on the pH values can be considered to be typical only of the threshold concentration of tomatine (10^{-4} M). Thus, the faciliation of salvo-like discharges

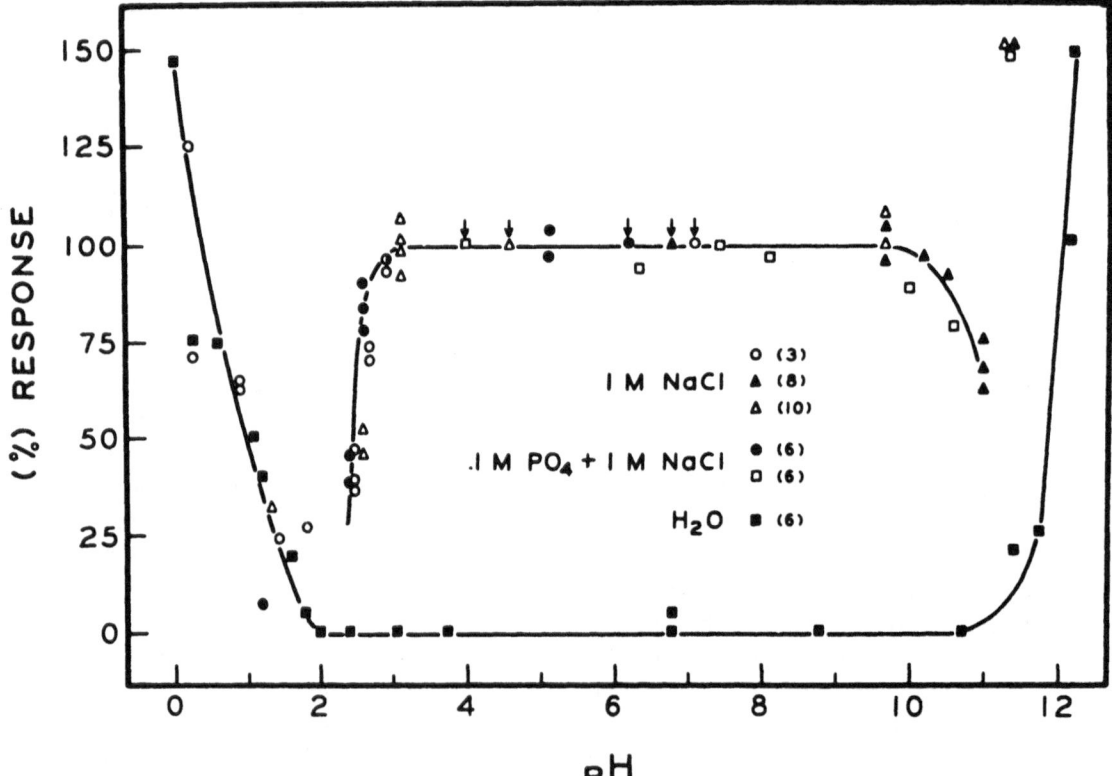

FIG. 12. Percent of response of the salt receptor cell of *Phormia regina* in dependence on the pH value of the NaCl solution.

Each point is the average of impulse frequencies measured 100 to 500 msec after the onset of the stimulus. Number of taste hairs tested in parentheses. The sequence was at random, and the responses of the plateau region were adjusted to 100 percent (arrows). Each experiment carried out on a different specimen has its own symbol. The test solutions were unbuffered or buffered with 0.1 M phosphate. Tests with water were the control; the pH value was adjusted with NaOH and HCl immediately preceding the test (*From* Gillary. 1966. *J. Gen. Physiol.*, 50(2):337-350.)

by higher acidity might not be found at higher intensities of tomatine. Evans and Mellon (1962b) also indicated for the blowfly a relationship (although a different one) between the critical concentration of hydrogen ions (pH 1 and 2) and the concentration of the salt stimulus. A continuation of such investigations might furnish insight into some fundamentals of the receptive mechanisms.

Gillary (1966a) mentioned an independence over a wide range of acidity for the sugar- and the water-receptor cells of the blowfly. Beneath pH 1.5 and above 11, he found injury to the receptive mechanisms if they were stimulated for a few minutes or repeatedly for shorter times. This result is in agreement with findings on chemoreceptor cells of the horseshoe crab, Limulus; Barber (1956) reported no damage between pH 1.7 and 7.4 for receptor cells, which respond to extracts of food, but higher acidity caused irreversible injury.

Gillary (1966c) found that low relative humidity decreased the activity of the salt receptor cell in Phormia; however, at the same time he found a significant rapid decrease in temperature at the tip of the stimulating capillary in low relative humidity, which caused him to be cautious with the interpretation of the results with regard to humidity in general. Contrary to Hodgson (1956), Dethier and Arab (1958), and Mellon (1961), Gillary observed a dependence on temperature for the salt-receptor cell when only the tip of the hair was exposed to solutions of different temperature (increased impulse frequency with higher temperature). His results were to be expected, since the activity of substances in solution changes with temperature, and the ionic activity is known to affect the response of the salt-receptor cell (Evans and Mellon, 1962a).

All authors found higher impulse activity if the entire preparation was warmed, or lower activity if it was cooled (see also Hodgson and Roeder, 1956); however, Hodgson (1956) at the same time observed in some taste hairs a decrease in activity when the temperature of the preparation was raised as little as 0.5° C. This result indicates the presence of a cold-receptor cell (compare Lacher, 1964; Loftus, 1966; Schoonhoven, 1966) and might be one reason for the contradictory results of temperature changes at the tip of the taste hair.

EFFECTS OF INSECTICIDES. Lewis (1954b) studied histologically the taste hairs of three flies to determine the site of uptake of contact insecticides. He found the most suitable place to be the unsclerotisized membrane, which connects the socket of the sensillum with its shaft.

Smyth and Roys (1955) studied neurophysiologically the effect of DDT vapor on afferent fibers of unknown function in a femur preparation of the housefly. Impulses, which occurred at uniform frequencies before treatment, occurred at higher frequency and in groups immediately after treatment. The effect was irreversible. In addition, they determined behaviorally the threshold to sugar of tarsal taste hairs of Musca and Phormia of untreated and DDT-poisoned flies and found it to be lowered about ninefold in poisoned specimens. Their explanation was the facilitation of impulse generation by DDT; however, the rejection thresholds for sodium chloride and ethanol were unchanged by poisoning. A resistant strain of Musca showed no decrease of the threshold to sucrose after application or injection of DDT.

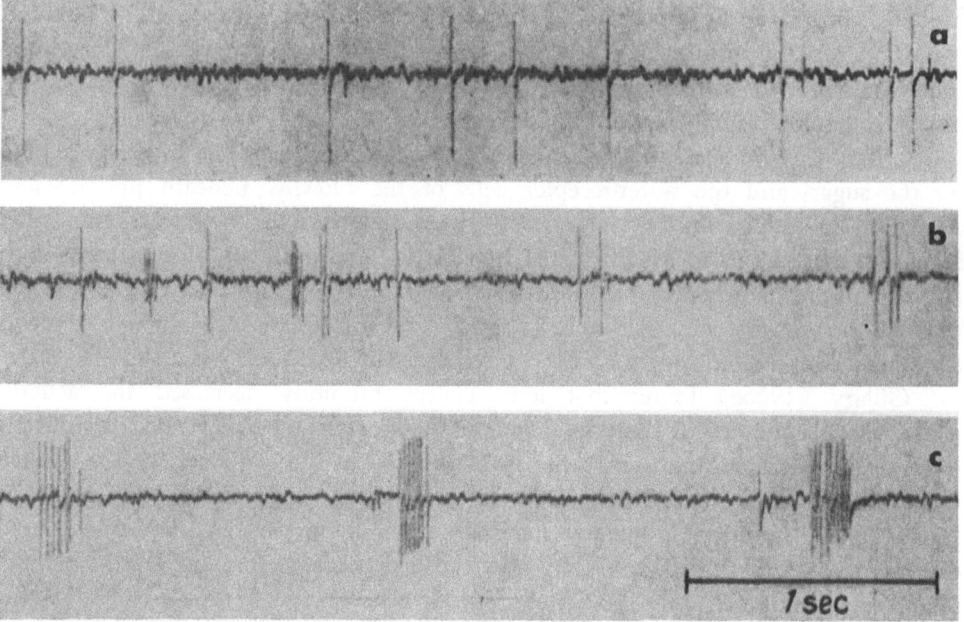

FIG. 13. Impulse activity of single taste hairs of *Leptinotarsa decemlineata* in response to NaCl 0.1 M.

a. representative steady-state response of an untreated potato beetle; b and c. representative steady-state responses of two taste hairs of a with Active Gesarol (containing DDT and benzene hexachloride) poisoned beetle. Two different types of receptor cells respond with grouping of impulses. (*From* Stürckow. 1959. Z. Vergl. Physiol., 42:255-302. Courtesy of Springer-Verlag.)

I (Stürckow, 1959) stimulated single taste hairs of *Leptinotarsa* with parathion forte (Bayer, Leverkusen) and Active Gesarol (containing DDT and benzene hexachloride, Geigy, Basel) and found in the impulse pattern no appreciable effect, even if stimulated for one hour with relatively high doses. However, the overflow of the stimulating solution over the socket of the taste hair and the poisoning of specimens with these insecticides immediately caused a change in the impulse patterns. Parathion resulted in a higher frequency and Active Gesarol elicited a grouping of impulses in agreement with the results of Smyth and Roys (1955). Figure 13 shows the normal response to a salt solution and the response of a poisoned beetle. The insecticides probably affected the neurons and not the chemoreceptive membranes, since stimulation of the taste hair tips with insecticides was ineffective.

Soliman and Cutcomp (1963) poisoned houseflies through applications of DDT and parathion and measured behaviorally their response to sucrose solutions. They found no noticeable change for parathion (even applied in a lethal dose$_{95}$), but for DDT—applied topically in a lethal dose—they found six times greater sensitivity to sucrose solutions within 1½ min. Doses lower than the lethal dose$_{12}$ produced no higher sensitivity to sucrose. Judged by these and Smyth and Roys' results, the grouping of impulses that I found in response to poisoning with Active Gesarol must have mainly been caused by the content of DDT. I had found in *Leptinotarsa* a grouping of impulses of a salt receptor cell and of another receptor cell (Fig. 13); but the rejective threshold of the blowfly for sodium chloride was found unchanged through poisoning by Smyth and Roys (1955). Such pretreated specimens offer many opportunities to investigate the receptive mechanisms and to decode the afferent impulse patterns through a comparison of neural and behavioral responses.

Specimens of a DDT-resistant strain of *Leptinotarsa*, that was reared on highly toxic potato leaves for several generations, were significantly less sensitive than normal potato beetles to the taste of a wild potato and the alkaloid glycoside, leptine, which makes this potato resistant to damage by the potato beetle (Stürckow and Löw, 1961). Whether these results are comparable with those discussed as conditioning experiments with larvae (see GOALS OF THE STUDIES ON CHEMORECEPTIVE SYSTEMS, page 110) is not known. Unfortunately, neurophysiologic responses of gustatory sensilla have not been investigated in this DDT-resistant strain.

COMPARATIVE STUDIES BETWEEN NEURAL AND BEHAVIORAL RESPONSES

Comparative investigations with quantitative determinations have been carried out infrequently. The experiments with *Leptinotarsa* and tomatine, mentioned in the section on GOALS OF THE STUDIES ON CHEMORECEPTIVE SYSTEMS (page 110) are a preliminary study, which showed a close agreement between neural and behavioral responses (Stürckow, 1959; Stürckow and Löw, 1961).

Lacher (1964, 1967) was the first to study this problem carefully. In neurophysiologic experiments, he recorded the impulse activity of a sensillum on the antenna of the bee, sensitive to changes of carbon dioxide and supplied with one neuron. He found a suprisingly close agreement between the stimulus-percent response curves derived by both methods, which shows that the reception of the stimulus is transmitted via the nervous system in a very efficient way. In the experiments of each (Lacher and Stürckow), the neural response of a single sensillum was compared with the behavior of the entire insect. The congruity between neural and behavioral criteria indicates that in insects—contrary to vertebrates—the excitation of a single sensillum is sufficient to trigger a behavioral response. [Compare also Minnich (1926b) and others, who elicited extension of the proboscis in the blowfly through stimulation of a single taste sensillum with sucrose].

Similar comparative studies with the human were made by Borg et al. (1967a, 1967b) who recorded, during operations on the middle ear, the impulse activity of the chorda tympani in response to gustatory stimuli. The same substances had been tested before the operation. Based on clear results, Borg et al. state a fundamental congruity between excitation of sensory neurons and perceptual intensity.

Other comparative studies used the electroantennogram as the neural response. Schneider et al. (1967) published final stimulus-percent response curves based on behavioral and neurophysiologic responses of *Bombyx* to stimulation by synthetic bombykol and its isomers (Fig. 14). For a comparison, my data (Stürckow, 1965)

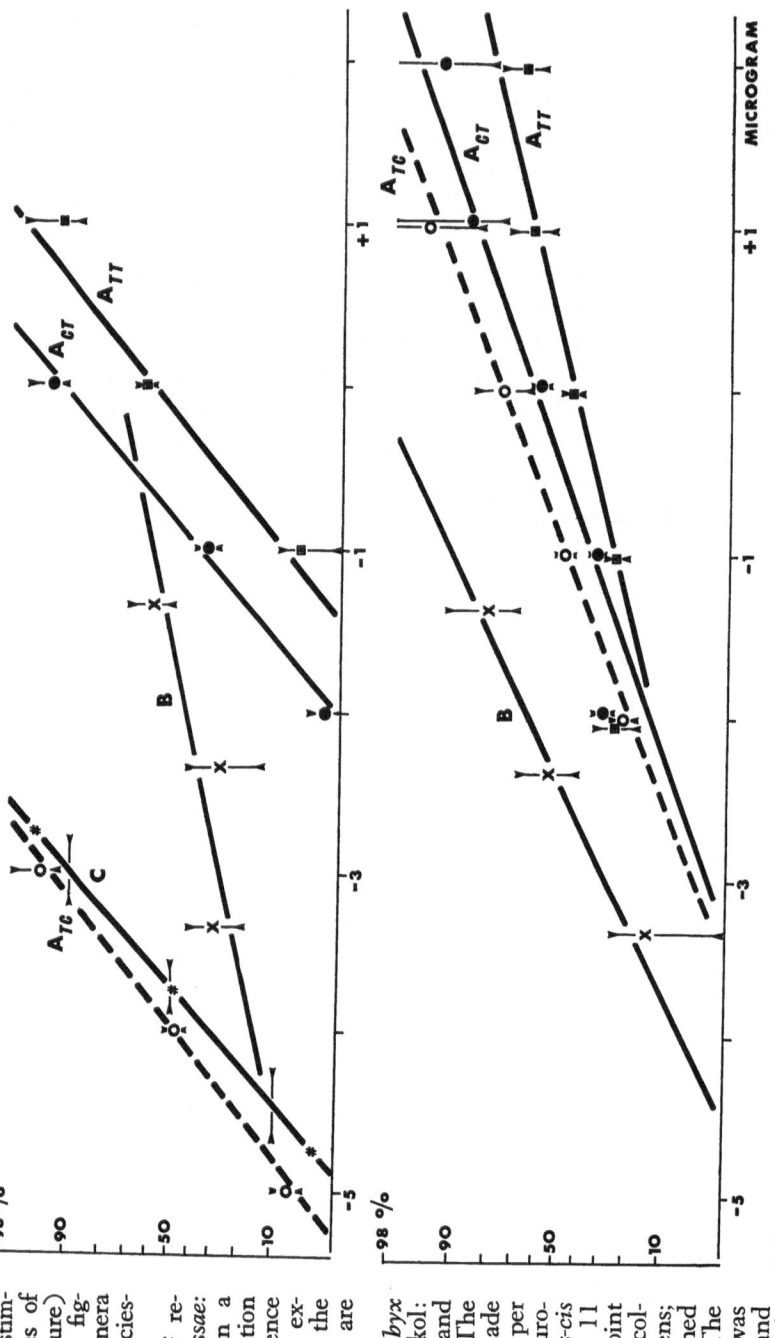

FIG. 14. Comparison of stimulus-percent response lines of behavioral tests (upper figure) and antennograms (lower figure) from three insect genera responding to their species-specific sex pheromone.

Ordinates: percent of response in probits, *abscissae*: stimulus in micrograms on a logarithmic scale, application see text or citations. Confidence limits 99.7% probability, except for line C, for which the limits of 95% probability are given.

A. tests with *Bombyx mori* and synthetic bombykol: TC *trans-cis*, CT *cis-trans*, and TT *trans-trans* isomers. The behavioral tests were made with 20 to 30 specimens per point of the line; the neurophysiological data for *trans-cis* bombykol are composed of 11 to 203 antennograms per point of the line, which were collected from over 40 specimens; less antennograms were gained for the other two isomers. The application of the stimulus was different in both methods and is described in the text. (*Adapted from* Schneider, Block, Boeckh, and Priesner. 1967. Z. Vergl. Physiol., 54:192-209. Courtesy of Springer-Verlag.)

B. tests with *Porthetria dispar* and an extract of abdominal tips of female moths, which was provided by Mr. M. Jacobson, Entomology Research Division, Agricultural Research Service, USDA, Beltsville, Maryland. The behavioral data are based on 31 specimens tested per point of the line; the neurophysiological data are composed of 5 antennograms per point of the line, which were made with 4 specimens. The application of the stimulus was similar in both methods. (*Adapted from* Stürckow. 1965. J. Insect Physiol., 11:1573-1583. Courtesy of Pergamon Press, Inc.)

C. tests with groups of ten males of *Periplaneta americana* and an extract of 32 females, 11 days old. A filter paper loaded with a measured amount of extract was held in the jar with males, and vibration of the wings was taken as the courtship criterion as for the lines A and B; 92 tests with 20 groups of homogeneously tuned males per point of the line; as to the calculation of the intensity see text.

were included; they concern similar studies with the gypsy moth, *Porthetria*, stimulated by an extract of abdominal tips of female moths. My behavioral tests cannot be considered as optimal results indicated by the flat slope of the line; however, my response lines agree in the concentration ranges with values for *Porthetria* published by Schneider (1963a), which apparently were obtained by a comparable application of the stimulus (not described by Schneider).

The neurophysiologic and behavioral methods resulted in response lines, which cover similar intensity ranges except for bombykol itself (*trans-cis*, Fig. 14, interrupted lines). But the calculations of Schneider et al. (1967), comparing the two application methods used in their behavioral and neurophysiologic tests, resulted in a 10^3-times weaker stimulation in the behavioral tests. For a comparison, this would mean that the data of the behavioral tests for bombykol and its isomers would range at concentrations 10^3-times weaker than shown in Figure 14, which equals about 10^{-6}-times weaker. On the other hand, their calculations [which have not been presented in detail (Boeckh et al., 1965; Schneider et al., 1967)] seem to be incorrect as judged by an estimate. They used the same cartridges in both tests filled with a filter paper soaked with testing substance. Air blown through these cartridges was applied in the following way:

	Antennograms	Behavioral tests
duration	1 sec	30 sec
distance	5 cm	50 cm
velocity	0.05 1/sec	40 1/sec

Since (a) the evaporation of the test substance in the cartridge is a constant and (b) the rise time of a stimulus is the crucial factor for its effect in behavioral as well as neurophysiologic tests (the height of the antennogram was evaluated), no striking indication is given for a weaker stimulation in the behavioral tests. The 10-times larger distance of the application of the stimulus might have been compensated for by the about 10^3-times greater velocity of the stimulus.

We cannot be certain whether neural activity exists beyond the sensitivity of the antennogram, since it is a weak reproduction of the real potential changes that occur in response to a stimulus. However, in this case, the behavioral response to bombykol (*trans-cis*) would be an exception to the other substances.

Schneider et al. (1967) pointed out correctly that a comparison of their behavioral stimulus-percent response line for bombykol with the data of Hecker (1960) and Butenandt et al. (1961) is not possible. These authors thought that they had determined the effective dose$_{50}$ for bombykol and its isomers; however, their values describe only the dilution of bombykol, into which they had dipped the glass rod for stimulation. The amount of bombykol carried out with this rod was not determined and was probably as large as the amount of bombykol used by Schneider et al. (1967) for the ED$_{50}$. The same criticism is true for the value given by Eiter et al. (1967).

One factor is striking in these comparative studies with antennograms: the lines based on neurophysiologic responses always cover more logarithmic units than the lines of the behavioral tests, which agree with determinations of Engen

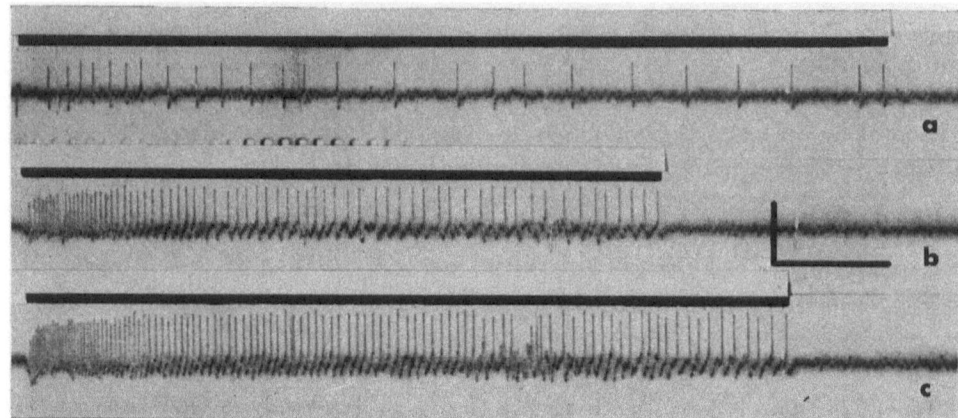

FIG. 15. Impulse activity of an olfactory sensillum on the antenna of *Locusta migratoria*.
The sensillum is supplied with 3 to 4 neurons, of which primarily one receptor cell responds, recognized by the uniform height and frequency of the impulses. The stimulus (line above each trace) is hexanoic acid, applied at 10-times higher concentration in b., and again 10-times higher intensity in c. Time scale: 250 msec, vertical scale: 300 μV. (*From* Boeckh. 1967. *Z. Vergl. Physiol.* 55:378-406.)

and Pfaffmann (1960) for the human in covering only two logarithmic units. The reason for the flat slope of the lines obtained neurophysiologically might be that additional neurons become active in response to higher concentrations, the signals of which are not evaluated centrally for the behavioral criterion studied. The quantitatively measured responses of two single receptor cells—the impulses of one are shown in Figure 15 and of the other are published by Lacher (1964, Fig. 17, p. 606)—show steep response lines, which favor this interpretation. However, if my explanation is correct, the stimulus-percent response lines of generator potentials of single neurons also should show a steep slope, contrary to the results of Boeckh et al. (1965), and Schneider (1967).

Stürckow, Bodenstein, and Lake (unpubl.) studied, as did Wharton et al. (1954a, 1954b), the slope of lines based on behavioral tests with groups of *Periplaneta americana* males in response to extracts of the cockroach pheromone. For a comparison, our control line, achieved with specimens that were homogeneously tuned to the pheromone, is inserted in Figure 14—based on the assumption that one mature female produces 1 μg of pheromone similar to the amount of pheromone per female of *Bombyx* (Steinbrecht, 1964). We carefully investigated the slope and concentration range of lines in connection with different problems (110 extracts from 3,300 cockroaches were offered in 15,600 tests to groups of ten males). Determinations of response lines made with specimens, that were heterogeneously tuned to the stimulus, resulted in less steep slopes than those determined with homogeneously sensitive specimens. Likewise, poor extracts tested with specimens tuned to the stimulus resulted in a flat slope of the response line, which could reach (depending on the degree of inadequacy) about the same threshold as a response line obtained with an extract rich in pheromone.

The different slopes of response lines based on behavioral tests and antennograms do not prevent a comparison between both methods. However, a comparison should be restricted to the two lower logarithmic units and should not anticipate a parallelism for both lines.

Working Hypotheses of Receptive Mechanisms

Although the type of receptive membranes and the coding of received stimuli probably differ between various organisms and between gustation and olfaction, the basic principles of the receptive mechanisms might be the same or similar for olfactory and gustatory systems in the human and in all animals that are supplied with sensory neurons connected to a central nervous system. The level at which the decoding of the afferent chemoreceptive message is studied is insignificant, since every level receives the meaning of the message; the level studied is important only for a comparison between studies with vertebrates and invertebrates. Afferent impulse patterns of insect chemoreceptor cells are directly comparable with impulse trains of olfactory receptor cells of vertebrates, but impulse patterns of gustatory nerves of vertebrates, which originated from peripheral internuncials, would be directly comparable only with impulse trains of the first central internuncials of insect chemoreceptive systems.

The theories available on chemoreceptive systems are of the type of working hypotheses on the reception, or coding of stimuli, or both. The array of theories signifies the growing interest in functions which we do not understand at present. Theories of olfactory systems are described by Moulton and Beidler (1967) except two later ones, those of Rosenberg et al. (1968) and Randebrok (1968). Because of the excellent and critical review of Moulton and Beidler (1967) on theories of chemoreception, they will not be discussed here. The theories on gustatory systems are described in several reviews, the latest one being that of Oakley and Benjamin (1966). In addition, a theory on the reception of glucose has been submitted by Evans (1963), which is in close agreement with the independently published theory of sweet tasting substances by Schallenberger and Acree (1967). These authors discussed the essential characteristics of molecules that have a sweet taste, while Evans gave evidence through alterations of substitutional groups of glucose and behavioral tests with the blowfly that the hydroxyl groups on C_3 and C_4 alone are responsible for stimulation (Fig. 16). "They are *trans*, but in the Cl conformation they are respectively inclined about 19° above and 19° below the plane of the ring" (Evans, 1963), which might be an additional necessity for stimulation. Another working hypothesis, in addition to those mentioned by Oakley and Benjamin (1966), is that of Erickson et al. (1965), who (a) outlined response functions for the gustatory system of rats, (b) for the first time tried to disclose stimulus dimensions, and (c) supported their approach with experimental evidence. The stimuli they used were seven salts, a base, an acid, a sugar, and a bitter-tasting substance.

The studies on chemoreception of insects, as described in the present review, are in favor of Rosenberg's et al. (1968) hypothesis on the nature and function of the receptive membrane, in favor of Evans' (1963) and Schallenberger and

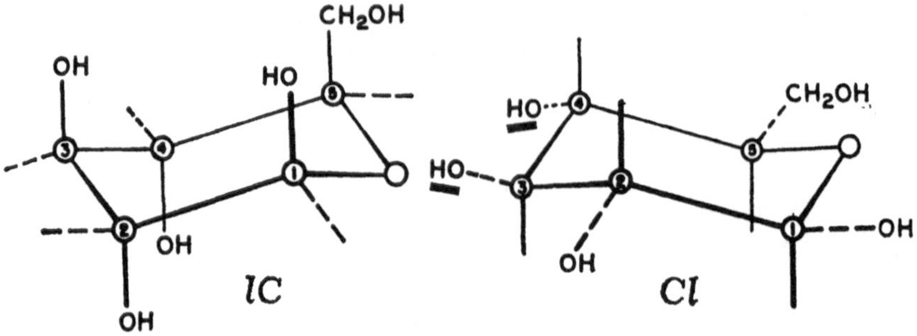

FIG. 16. β-D-glucopyranose in the 1C and Cl conformation.

The hydroxyl groups C_3 and C_4 (underlined) of glucose in the Cl form alone might be responsible for stimulation of the taste hair of *Phormia regina* judged by the extension of the proboscis. (*After* and in memory of the late Dr. D. R. Evans. 1963. *In* Zotterman, Y., ed., *Olfaction and Taste*, Vol. 1. Courtesy of Pergamon Press.)

Acree's (1967) outlines on the type of weak bond connections between molecules of sweet taste and the receptive membrane, and are in agreement with all hypotheses on coding, which are not too strongly based on narrow response spectra of single neurons. How differentiation of stimuli occurs, if reception takes place at a membrane common to several dendrites, is still unknown, although theoretically several suggestions could be made.

The linking structures between the environment and dendritic fibers of chemoreceptor neurons in insects are the viscous substance in the pore of gustatory hairs and the material filling the pore and the pore filaments of the olfactory hairs. The well-defined site of these two substances suggests that both are homologous and serve the same function. The texture of both is vesicular (Figs. 4 and 7). A material with similar vesicular texture has been mentioned by Noble-Nesbitt (1963) in the collembolan, *Podura*. Based on electron-microscopic studies, he found that the cytoplasm of epidermal cells forms a vesicular substance as the first step of the molt. He describes vesicles of 20 to 50 mμ, and some of up to 200 mμ, which were formed "as outpushings of the plasma membrane, which is then pinched off, leaving vesicles bounded by smooth membranes free of the cell surface." In chemoreceptive sensilla, no cytoplasm is found at locations where the vesicular structure occurs, except for the dendritic fibers. The large amount, in which this viscous material occasionally has been found in the taste hair of the blowfly (Stürckow, 1967), makes it unlikely that the substance originates as outpushings of the dendritic fibers. If this should be the case, they probably would enlarge later on through uptake of moisture from the vacuolar fluid. However, the vacuolar fluid itself could form this oozing substance. The coarse porous system and the coarse vesicles at the tip of the taste hair of the blowfly (Figs. 6 and 7) and the fine porous system and the fine vesicular structure in olfactory hairs (Figs. 3 and 4) suggest that this substance is an extrusion of the vacuolar fluid through the porous structures of the cuticle.

Regardless of the formation and origin of the vesicular substance, its characteristics strongly suggest this working hypothesis: the vesicular and viscous substance acceptor-donor type, which, after Rosenberg et al. (1968), might be β-carotene, offers a suitable membrane for the embedding of chemoreceptive molecules of an vitamin A, and other high molecular substances. These receptive molecules develop a current flow at the moment of absorption (compare Rosenberg et al., 1968; and the considerations of Rees, 1967). In this way, the dendritic endings become activated within a time interval that explains latencies of 1 and 5 msec between the application of a stimulus and the first impulse of sensory neurons as found by Barton Browne and Hodgson (1962). A careful analysis of the electroantennogram or the corresponding overall response from gustatory sensilla should show in analogy to studies of vertebrate vision an early receptor potential (Cone, 1965; and Pak, 1965), which reflects the whole or a portion of the electrical changes occurring at the moment of impingement of molecules on the receptive membrane. This chemoreceptive membrane degenerates on the exposed surface and blocks the reception of stimuli, if present in too large an amount (Stürckow, 1967); but generally it is probably replaced continually by a new one. When insects are cleaning their mouthparts, antennae, or legs by pulling them through a special cleaning spine (Dahl, 1884; Gemmerich, 1922) or through their mandibles, they probably pull off the degenerated portion of the receptive substance, which at the same time is replaced by fresh material due to the viscous characteristics of the substance and maybe a pressure from the inside.

To facilitate the supply of receptive material, the extracellular vacuole in the extension of the sensillum might be under a higher turgor than the neighboring tissue, which could be sustained by the tormogen and trichogen cells. Hodgson (1967) found a steady transport of tritium-labeled water from the labellum of the blowfly, *Phormia*, into the lumen of the labellar taste hairs during 2½ hours of observation, which suggests a fluid draft in distal direction.

METHODOLOGY: THE FLOW CAPILLARY USED IN STUDIES ON GUSTATION OF INSECTS

There has been considerable interest in the flow capillary used in studies with the black and blue blowflies, *Phormia regina* and *Calliphora erythrocephala*. This is a timely opportunity to describe its fabrication and discuss its advantages and disadvantages.

The development of the flow capillary (Fig. 17) extended over more than one year, during which five variations were employed to rinse the tip of a single labellar taste hair with alternating stimulating solutions. The capillary described gave the most reliable results and offered the quickest exchange between solutions. It was used for recordings of the impulse activity of gustatory neurons (Stürckow, 1963, 1967; Stürckow et al., 1967) and for behavioral experiments (Stürckow, 1967). The advantage of a flowing stimulus is the control of a constant concentration and temperature at the tip of a filled capillary. In a static liquid, both change rapidly through heat loss by evaporation (Gillary, 1966). In addition, a multichanneled

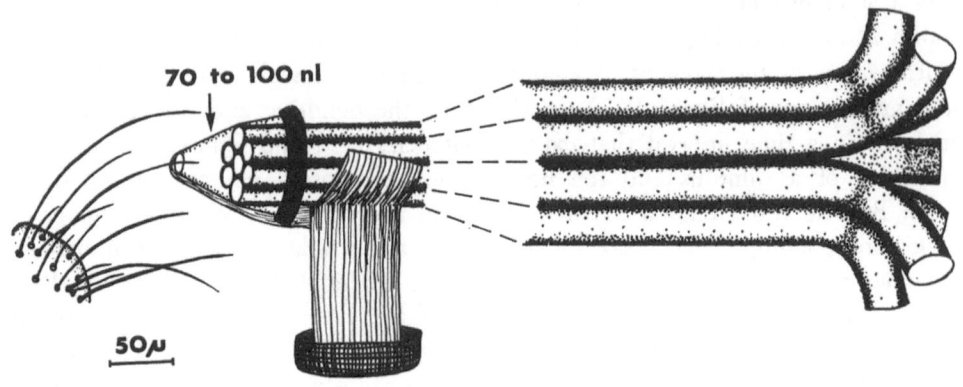

FIG. 17. Flow capillary used to stimulate single taste hairs of flies in neurophysiologic and behavioral experiments.

capillary allows one to change from one solution to another without threading the chemosensitive tip of the taste hair through the surface tension of a new solution.

The capillary is more rigid with an odd number of channels (three, five, or seven) than with an even number of channels. Pyrex tubing (10 cm long and 2 mm in outer diameter) of equal thickness is the initial material. As many tubes, as equal the number of channels desired, are bent 1.5 cm distant from each end in an angle of 45 degrees and are bundled at these points with strings. The ends should extend from the bundle in a whirl-like fashion. The middle of the bundle is then drawn out in the flame of a torch supplied with oxygen or compressed air. The bundle thus furnishes two raw capillaries. An important requirement in the fabrication of the capillary is the *regular rotation* of the bundle in the flame. During melting, only the distal ends of the channels fuse; therefore, and because of the close proximity of thick and thin glass, breakage easily occurs during cooling. This danger is considerably reduced if the freshly drawn capillaries are held by hand for a few minutes. The drawn out ends are broken off at a height where each channel has a width of 20 to 25 μ.

A *careful selection* on the basis of suitability of the unfinished product is now made with a stereo microscope at a magnification of 20 to 40 times or more. Capillaries with unequal tip diameters of the individual channels or with oval-shaped channels must be rejected, since they do not ensure an equal flow of solutions. A capillary with equal and circular cross sections for all channels is glued throughout its entire length with epoxy resin. Care must be taken that the resin, which is partly distributed by capillary forces, does not enter the channels at the open ends.

After drying, a regular capillary is fitted and carefully glued upon the tapering end of the fused channels to serve as the mixing chamber. This capillary must decrease in diameter to 20 or 30 μ in the *steepest taper* that can be obtained, since a *quick exchange between solutions* will occur only in a small mixing chamber. My smallest mixing chambers had a volume of 70 to 100 nl ($=70$ to 100×10^{-9}

liter) and allowed a complete exchange between two solutions within 1 to 1.5 sec, judged by calculations of the flow rate from the volume ejected during 3 and 5 hours (measured in a moist chamber). Such a rise time of the stimulus is among the slowest of applied sensory stimuli; however, rinsing a flexible tip of 1.5μ in diameter did not allow a faster exchange of solutions. If the flow rate is excessive, the tip of the hair will be forced out of the stream.

The ejected solution, which runs off, drawn by capillary forces, along the outside of the mixing chamber, must be taken up with a strip of filter paper, since accumulation and falling of drops will result in an irregular flow. The tip of the hair can be threaded into the capillary only during slow flow or more easily at no flow. Care should be taken that the tip floats in the middle of the stream and does not stick to the glass wall inside the capillary.

The connections between the different channels of the capillary and the bottles containing the stimulating solutions were made with polyethylene tubes. They had been flushed with water for over six months and did not deliver detectable amoun:s of ions, judged by responses without impulse activity from taste hairs that lacked a water receptor cell. The polyethylene tubes connected each channel with its bottle over a switchboard made of a 3×3 cm Plexiglas rod, into which stopcocks were fitted. At two marked positions, these stopcocks connected or disconnected channels of 1 mm width, which were drilled through the rod and the stopcocks. The flow rate was easily controlled through a bottle stand, the height of which was adjustable in a sturdy frame one meter high.* All bottles were filled to the same height and furnished with an air drawing pipette reaching into the solution.

Spectroscopic analysis of expelled fluid showed that only traces of the nonflowing solutions diffused into the flowing solution. If, for example, the test solution was changed from 1 M NaCl to distilled water, the final water contained NaCl equivalent to 1 mM or less. Salt and sugar solutions used had concentrations of 1 M or less; sugars furnished more viscous solutions than salts. One channel was always connected to distilled water and kept at steady flow, when the capillary was not used, to prevent diffusion through contact of the solutions in the mixing chamber.

In neurophysiologic experiments, the capillary was simultaneously used for stimulation and as an electrode. Platinum wires were inserted into each channel at the junction of the glass and polyethylene tubes. These electrodes were connected via a capacitor to the amplifying system to avoid interferences due to diffusion-, contact-, and electrode-potentials, which could amount to as much as 40 mV. If the capillary serves at the same time as an electrode, it can only be used for recordings of impulses, since the disturbing potentials cannot be eliminated in recordings of slow potential changes.

The capillary has been used successfully to study the response of taste hairs over more than half an hour. The initial impulse pattern in response to a new solution has been recorded without artifacts. Although the slow rise time of the

* I thank Mr. H. Kilb, Institute of Physiology of the University of the Saarland, Homburg/Saar, West Germany very much for assistance in the construction of the switchboard and the bottle stand.

stimulus causes the initial impulse frequency to ascend only to a portion of the height known from taste hairs stimulated through regular capillaries (Stürckow, 1967), the change from a salt to a sugar solution elicits a behavioral response and is worth being studied neurophysiologically.

REFERENCES

Abbott, C. E. 1937. The physiology of insect senses. *Entomologica Americana*, 26:225-285.

Abushama, F. T. 1966. Electrophysiological investigations on the antennal olfactory receptors of the damp-wood termite *Zootermopsis angusticollis*. *Entomol. Exp. Appl.*, 9:343-348.

Adams, J. R. 1961. The location and histology of the contact chemoreceptors of the stable fly, *Stomoxys calcitrans* L. Doctoral Dissert., Rutgers—The State University, New Brunswick, New Jersey.

Barber, S. B. 1956. Chemoreception and proprioception in *Limulus*. *J. Exp. Zool.*, 131:51-73.

Barrows, W. M. 1907. The reactions of the pomace fly, *Drosophila ampelophila* Loew, to odorous substances. *J. Exp. Zool.*, 4:515-537.

Barton Browne, L., and Hodgson, E. S. 1962. Electrophysiological studies of arthropod chemoreception. IV. Latency, independence, and specificity of labellar chemoreceptors of the blowfly, *Lucilia*. *J. Cell. Comp. Physiol.*, 59:187-202.

Beetsma, J., and Schoonhoven, L. M. 1966. Some chemonsensory aspects of the social relations between the queen and the worker in the honeybee (*Apis mellifica* L.). *Proc. Kon. Nederl. Akad. Wet.* [*Biol. Med.*], 69:645-647.

Beidler, L. M., and Smallman, R. L. 1965. Renewal of cells within taste buds. *J. Cell Biol.*, 27:263-272.

Bliss, C. I. 1935. The calculation of the dosage-mortality curve. *Ann. Appl. Biol.*, 22:134-167.

Boeckh, J. 1962. Elektrophysiologische Untersuchungen an einzelnen Geruchsrezeptoren auf den Antennen des Totengräbers (*Necrophorus*, Coleoptera). *Z. Vergl. Physiol.*, 46:212-248.

———— 1967a. Inhibition and excitation of single insect olfactory receptors, and their role as a primary sensory code. *In* Hayashi, T., ed. Olfaction and Taste. Oxford, Pergamon Press, Inc., Vol. 2, pp. 721-735.

———— 1967b. Reaktionsschwelle, Arbeitsbereich und Spezifität eines Geruchsrezeptors auf der Heuschreckenantenne. *Z. Vergl. Physiol.*, 55:378-406.

———— Kaissling, K.-E., and Schneider, D. 1960. Sensillen und Bau der Antennengeissel von *Telea polyphmeus*. (Vergleiche mit weiteren Saturniden: *Antheraea*, *Platysamia* und *Philosamia*). *Zool. Jahrb. Anat.*, 78:559-584.

———— Kaissling, K.-E., and Schneider, D. 1965. Insect olfactory receptors. Cold Spring Harbor Symposia Quant. Biol., 30:263-280.

———— Priesner, E., Schneider, D., and Jacobson, M. 1963. Olfactory receptor response to the cockroach sexual attractant. *Science*, 141:716-717.

Boistel, J. 1953. Etude fonctionelle des terminaisons sensorielles des antennes d'Hyménoptères. *C. R. Soc. Biol.*, 147:1683-1688.

———— and Coraboeuf, E. 1953. L'activité électrique dans l'antenne isolée de Lépidoptère au cours de l'étude de l'olfaction. *C. R. Soc. Biol.*, 147:1172-1175.

Borden, J. H. 1968. Antennal morphology of *Ips confusus* (Coleoptera:Scolytidae). *Ann. Entom. Soc. Amer.*, 61:10-13.

Borg, G., Diamant, H., Oakley, B., Ström, L., and Zotterman, Y. 1967a. A comparative study of neural and psychophysical responses to gustatory stimuli. *In* Hayashi, T., ed. Olfaction and Taste. Oxford, Pergamon Press, Inc., Vol. 2, pp. 253-264.

——— Diamant, H., Ström, L., and Zotterman, Y. 1967b. The relation between neural and perceptual intensity: A comparative study on the neural and psychophysical response to taste stimuli. *J. Physiol. (London)*, 192:13-20.

von Buddenbrock, W., ed. 1952. Chemorezeption. *In* Vergleichende Physiologie. Basel, Birkhäuser Verlag, Vol. 1, 338-464.

Butenandt, A., Hecker, E., Hopp, M., and Koch, W. 1961. Konstitutionsermittlung und Synthese des Bombykols, des Sexual-Lockstoffs des Seidenspinners *Bombyx mori. Sitzungsber. Bayr. Akad. Wiss. Math. Naturwiss., Kl.* 5, Mai 1961.

Chapman, J. A., and Craig, R. 1953. An electrophysiological approach to the study of chemical sensory reception in certain insects. *Canad. Entom.*, 85:182-189.

Cone, R. A. 1965. The early receptor potential of the vertebrate eye. *Cold Spring Harbor Symp. Quant. Biol.*, 30:483-491.

Cushing, J. E., Jr. 1941. An experiment on olfactory conditioning in *Drosophila guttifera. Proc. Nat. Acad. Sci. (U.S.A.)*, 27:496-499.

Dahl, F. 1884. Beiträge zur Kenntnis des Baues und der Funktion der Insektenbeine. *Arch. Naturgeschichte*, 1 (50):146-193.

Dethier, V. G. 1941. The function of the antennal receptors in lepidopterous larvae. *Biol. Bull.*, 80:403-414.

——— 1943. Testing attractants and repellents. Laboratory procedures in studies of the chemical control of insects. *Publ. Amer. Ass. Advan. Sci.*, 20:167-172.

——— ed. 1947. Chemical Insect Attractants and Repellents. Philadelphia, Blakiston Company.

——— 1953. Chemoreception. *In* Roeder, K. D., ed. Insect Physiology. New York, John Wiley & Sons, Inc., pp. 544-576.

——— 1954. The physiology of olfaction in insects. *Ann. N.Y. Acad. Sci.*, 58:139-157.

——— 1955. The physiology and histology of the contact chemoreceptors of the blowfly. *Quart. Rev. Biol.*, 30:348-371.

——— 1956. Chemoreceptor mechanisms. Molec. Structure and Funct. Activity of Nerve Cells, Amer. Inst. Biol. Sci., Publ. 1:1-30.

——— 1962. Chemoreceptor mechanisms in insects. *Sympos. Soc. Exp. Biol.*, 16:180-196.

——— ed. 1963. The Physiology of Insect Senses. New York, John Wiley & Sons, Inc.

——— and Arab, Y. M. 1958. Effect of temperature on the contact chemoreceptors of the blowfly. *J. Insect Physiol.*, 2:153-161.

——— and Chadwick, L. E. 1948a. Chemoreception in insects. *Physiol. Rev.*, 28:220-254.

——— and Chadwick, L. E. 1948b. The stimulation effect of glycols and their polymers on the tarsal receptors of blowflies. *J. Gen. Physiol.*, 32:139-151.

——— Hackley, B. E., Jr., and Wagner-Jauregg, T. 1952. Attraction of flies by iso-valeraldehyde. *Science*, 115:141-142.

——— Larsen, J. R., and Adams, J. R. 1963. The fine structure of the olfactory receptors of the blowfly. *In* Zotterman, Y., ed. Olfaction and Taste. Oxford, Pergamon Press, Inc., Vol. 1, pp. 105-110.

———— and Yost, M. T. 1952. Olfactory stimulation of blowflies by homologous alcohols, *J. Gen. Physiol.*, 35:823-839.

Dostal, B. 1958. Riechfähigkeit und Zahl der Riechsinneselemente bei der Honigbiene. *Z. Vergl. Physiol.*, 41:179-203.

DuBose, W. P., Jr. 1967. Sensory structures of the antennal flagella of *Hippelates* eye gnats (Diptera: Chloropidae). Doctoral Dissert., North Carolina State University at Raleigh, North Carolina.

Eiter, K., Truscheit, E., and Boness, M. 1967. Neuere Ergebnisse der Chemie von Insektensexuallockstoffen. Synthesen von D,L-10-Acetoxy-hexadecen-(7-*cis*)-ol-(1). 12-Acetoxy-octaden-(9-*cis*)-ol-(1) ("Gyplure") und 1-Acetoxy-10-propyl-tricadien-(5-*trans*. 9). *Liebigs Ann. Chem.*, 709:29-45.

Engen, T. 1960. Effects of practice and instruction on olfactory thresholds. *Perceptual and Motor Skills*, 10:195-198.

———— and Pfaffmann, C. 1959. Absolute judgments of odor intensity. *J. Exp. Psychol.*, 58:23-26.

Erickson, R. P., Doetsch, G. S., and Marshall, D. A. 1965. The gustatory neural response function. *J. Gen. Physiol.*, 49:247-263.

Evans, D. R. 1961. Depression of taste sensitivity to specific sugars by their presence during development. *Science*, 133:327-328.

———— 1963. Chemical structure and stimulation by carbohydrates. *In* Zotterman, Y., ed. Olfaction and Taste. Oxford, Pergamon Press, Inc., Vol. 1, pp. 165-176.

———— and Mellon, DeF., Jr., 1962a. Electrophysiological studies of a water receptor associated with the taste sensilla of the blowfly. *J. Gen. Physiol.*, 45:487-500.

———— and Mellon, DeF., Jr. 1962b. Stimulation of a primary taste receptor by salts. *J. Gen. Physiol.*, 45:651-661.

Fischer, W. 1957. Untersuchungen über die Riechschärfe der Honigbiene. *Z. Vergl. Physiol.*, 39:634-659.

Forel, A., ed. 1908. The Senses of Insects, trans. by Yearsley, M. London, Methuen & Company.

Frings, H., and Frings, M. 1949. The loci of contact chemoreceptors in insects. A review with new evidence. *Amer. Midl. Natur.*, 41:602-658.

———— and Frings, M., 1956. The loci of contact chemoreceptors involved in feeding reactions in certain Lepidoptera. *Biol. Bull.*, 110:291-299.

———— and Frings, M. 1959. Studies on antennal contact chemoreception by the wood nymph butterfly, *Cercyonis pegala*. *N.Y. Entom. Soc.*, 67:97-105.

von Frisch, K. 1919. Über den Geruchssinn der Biene und seine blütenbiologische Bedeutung. *Zool. Jahrb. Physiol.*, 37:1-238.

———— 1921. Über den Sitz des Geruchssinnes bei Insekten. *Zool. Jahrb. Physiol.*, 38:449-516.

———— 1934. Uber den Geschmackssinn der Bienen. *Z. Vergl. Physiol.*, 21:1-156.

———— 1967. Honey bees: Do they use direction and distance information provided by their dances? *Science*, 158:1072-1096.

———— 1968. Verstehen die Bienen wirklich ihre eigene Sprache nicht? *Allg. Deut. Imkerzeitung*, 2:3-8.

Gemmerich, J. 1922. Morphologische und biologische Untersuchungen der Putzapparate der Hymenopteren. *Arch. Naturgeschichte Abt. A*, 88 (12):1-63.

Gesteland, R. C. 1966. The mechanics of smell. *Discovery*, 27:29-34.

———— Lettvin, J. Y., and Pitts, W. H. 1965. Chemical transmission in the nose of the frog. *J. Physiol. (London)*, 181:525-559.

Ghiradella, H., Case, J., and Cronshaw, J. 1968. Fine structure of the aesthetasc hairs of *Coenobita compressus* Edwards. *J. Morphol.*, 124:361-386.

Gillary, H. 1966a. Quantitative electrophysiological studies on the mechanism of stimulation of the salt receptor of the blowfly. Doctoral Dissert., The Johns Hopkins University, Baltimore, Maryland.

—— 1966b. Stimulation of the salt receptor of the blowfly. I. NaCl. *J. Gen. Physiol.*, 50 (2):337-350.

—— 1966c. Stimulation of the salt receptor of the blowfly. II. Temperature. *J. Gen. Physiol.*, 50 (2):351-357.

Hassett, C. C., Dethier, V. G., and Gans, J. 1950. A comparison of nutritive values and taste thresholds of carbohydrates for the blowfly. *Biol. Bull.*, 99:446-453.

Hauser, G. 1880. Physiologische und histologische Untersuchungen über das Geruchsorgan der Insekten. *Z. Wiss. Zool.*, 34:367-403.

Hecker, E. 1960. Chemie und Biochemie des Sexuallockstoffes des Seidenspinners (*Bombyx mori* L.). *Proc. XI Intern. Congr. Entomol.*, 3:69-72.

Henke, K., and Rönsch, G. 1951. Über Bildungsgleichheiten in der Entwicklung epidermaler Organe und die Entstehung des Nervensystems im Flügel der Insekten. *Naturwissenschaften*, 38:335-336.

Hodgson, E. S. 1955. Problems in invertebrate chemoreception. *Quart. Rev. Biol.*, 30:331-347.

—— 1956. Temperature sensitivity of primary chemoreceptors of insects. *Anat. Rec.*, 125:560-561.

—— 1957. Electrophysiological studies of arthropod chemoreception. II. Responses of labellar chemoreceptors of the blowfly to stimulation by carbohydrates. *J. Insect Physiol.*, 1:240-247.

—— 1958. Chemoreception in arthropods. *Annu. Rev. Entomol.*, 3:19-36.

—— 1964. Chemoreception. *In* Rockstein, M., ed. The Physiology of Insecta. New York, Academic Press, Inc., Vol. 1, pp. 363-396.

—— 1965. The chemical senses and changing viewpoints in sensory physiology. *In* Carthy, J. D., and Duddington, C. L., eds. Viewpoints in Biology. London, Butterworths, pp. 83-124.

—— 1967. Chemical senses in the invertebrates. *In* Kare, M. R., and Maller, O., eds. The Chemical Senses and Nutrition. *Proc. Sympos.* Ithaca, N.Y. Baltimore, The Johns Hopkins Press, pp. 7-18.

—— Lettvin, J. Y., and Roeder, K. D. 1955. Physiology of a primary chemoreceptor unit. *Science*, 122:417-418.

—— and Roeder, K. D. 1956. Electrophysiological studies of arthropod chemoreception. I. General properties of the labellar chemoreceptors of Diptera. *J. Cell. Comp. Physiol.*, 48:51-75.

—— and Steinhardt, R. A. 1967. Hydrocarbon inhibition of primary chemoreceptor cells. *In* Hayashi, T., ed. Olfaction and Taste. Oxford, Pergamon Press, Inc., Vol. 2, pp. 737-748.

Hoffmann, C. 1961. Vergleichende Physiologie des Temperatursinnes und der chemischen Sinne. *Fortschr. Zool.*, 13:190-256.

Horridge, G. A. 1955. Chemoreceptors. *In* Bullock, T. H., and Horridge, G. A., eds. Structure and Function in the Nervous Systems of Invertebrates. San Francisco, W. H. Freeman and Co. Publishers, Vol. 2, pp. 1051-1054.

Ishikawa, S. 1963. Responses of maxillary chemoreceptors in the larva of the silkworm, *Bombyx mori*, to stimulation by carbohydrates. *J. Cell. Comp. Physiol.*, 61:99-107.

———— 1966. Electrical response and function of a bitter substance receptor associated with the maxillary sensilla of the larva of the silkworm, *Bombyx mori* L. *J. Cell. Physiol.*, 67:1-11.

Jahn, T. L., and Wulff, V. J. 1950. Chemoreception. *In* Prosser, C. L., ed. Comparative Animal Physiology. Philadelphia, W. B. Saunders Company, pp. 447-470.

Johnson, D. L. 1967. Honey bees: Do they use the direction information contained in their dance maneuver? *Science*, 155:844-847.

Johnston, J. W., Jr. 1967. Quantification of olfactory stimuli. *In* Brown, C. C., ed. Methods in Psychophysiology. Baltimore, The Williams & Wilkins Co., pp. 192-220.

Kay, R. E., Eichner, J. T., and Gelvin, D. E. 1967. Quantitative studies on the olfactory potentials of *Lucilia sericata*. *Amer. J. Physiol.*, 213:1-10.

Kraepelin, K. M. F., ed. 1883. Ueber die Geruchsorgane der Gliederthiere; eine historisch-kritische Studie. Hamburg, Osterprogramm der Realschule des Johanneums.

Krause, B. 1960. Elektronenmikroskopische Untersuchungen an den Plattensensillen des Insektenfühlers. *Zool. Beitr.*, 6:161-205.

Lacher, V. 1964. Elektrophysiologische Untersuchungen an einzelnen Rezeptoren für Geruch, Kohlendioxyd, Luftfeuchtigkeit und Temperatur auf den Antennen der Arbeitsbiene und der Drohne. *Z. Vergl. Physiol.*, 48:587-623.

———— 1967a. Verhaltensreaktionen der Bienenarbeiterin bei Dressur auf Kohlendioxid. *Z. Vergl. Physiol.*, 54:75-84.

———— 1967b. Elektrophysiologische Untersuchungen an einzelnen Geruchsrezeptoren auf den Antennen weiblicher Moskitos (*Aëdes aegypti* L.). *J. Insect Physiol.*, 13:1461-1470.

———— and Schneider, D. 1963. Elektrophysiologischer Nachweis der Riechfunktion von Porenplatten (Sensilla placodea) auf den Antennen der Drohne und der Arbeitsbiene (*Apis mellifica* L.). *Z. Vergl. Physiol.*, 47:274-278.

Larsen, J. R. 1962. The fine structure of the labellar chemosensory hairs of the blowfly, *Phormia regina* Meig. *J. Insect Physiol.*, 8:683-691.

Lefebvre, A. 1838. Note sur le sentiment olfactif des antennes. *Ann. Soc. Entomol.* (*France*), 7:395-399.

Lewis, C. T. 1954a. Contact chemoreceptors of blowfly tarsi. *Nature* (*London*), 173:130-131.

———— 1954b. Studies concerning the uptake of contact insecticides. I. The anatomy of the tarsi of certain diptera of medical importance. *Bull. Entomol. Res.*, 45:711-722.

Litchfield, J. T., Jr., and Wilcoxon, F. 1949. A simplified method of evaluating dose-effect experiments. *J. Pharmacol. Exp. Ther.*, 96:99-113.

Loftus, R. 1966. Cold receptor on the antenna of *Periplaneta americana*. *Z. Vergl. Physiol.*, 52:380-385.

Marler, P. R., and Hamilton, W. S., III. 1966. Stimulus filtering: Chemoreception. *In* Marler, P. R., and Hamilton, W. S., III, eds. Mechanisms of Animal Behavior. New York, John Wiley & Sons, Inc., pp. 270-315.

Matthews, D. F., and Tucker, D. 1966. Single unit activity in the tortoise olfactory mucosa. *Fed. Proc.*, 25:329.

McIndoo, N. E. 1914. The olfactory sense of the honey bee. *J. Exp. Zool.*, 16:265-346.

Mellon, DeF., Jr. 1961. Quantitative electrophysiological studies on the contact chemoreceptors of the blowfly. Doctoral Dissert., The Johns Hopkins University, Baltimore.

Minnich, D. E. 1921. An experimental study of the tarsal chemoreceptors of two nymphalid butterflies. *J. Exp. Zool.*, 33:173-203.

——— 1922a. The chemical sensitivity of the tarsi of the red admiral butterfly, *Pyrameis atalanta* Linn. *J. Exp. Zool.*, 35:57-81.

——— 1922b. A quantitative study of tarsal sensitivity to solutions of saccharose in the red admiral butterfly, *Pyrameis atalanta* Linn. *J. Exp. Zool.*, 36:445-457.

——— 1926a. The chemical sensitivity of the tarsi of certain muscid flies. (*Phormia regina* Meigen, *Phormia terrae-novae* R. D. and *Lucilia sericate* Meigen). *Biol. Bull.*, 51:166-178.

——— 1926b. The organs of taste on the proboscis of the blowfly, *Phormia regina* Meigen. *Anat. Rec.*, 34:126.

——— 1929. The chemical sensitivity of the legs of the blow-fly, *Calliphora vomitoria* Linn., to various sugars. *Z. Vergl. Physiol.*, 11:1-55.

——— 1932. The contact chemoreceptors of the honey bee, *Apis mellifera* Linn. *J. Exp. Zool.*, 61:375-393.

Moeck, H. A. 1968. Electron microscopic studies of antennal sensilla in the ambrosia beetle *Trypodendron lineatum* (Olivier) (Scolytidae). *Canad. J. Zool.*, 46:521-556.

Morita, H. 1967. Effects of salt on the sugar receptor of the fleshfly. *In* Hayashi, T., ed. Olfaction and Taste. Oxford, Pergamon Press, Inc., Vol. 2, pp. 787-798.

——— Doira, S., Takeda, K., and Kuwabara, M. 1957. Electrical response of contact chemoreceptor on tarsus of the butterfly, *Vanessa indica*. *Mem. Fac. Sci., Kyushu Univ., Ser. E. Biol.*, 2:119-139.

——— and Shiraishi, A. 1968. Stimulation of the labellar sugar receptor of the fleshfly by mono- and disaccharides. *J. Gen. Physiol.*, 52(4):559-583.

——— and Yamashita, S. 1959. Generator potential of insect chemoreceptor. *Science*, 130:922.

——— and Yamashita, S. 1961. Receptor potentials recorded from sensilla basiconica on the antenna of the silkworm larvae, *Bombyx mori*. *J. Exp. Biol.*, 38:851-861.

——— and Yamashita, S. 1966. Further studies on the receptor potential of chemoreceptors of the blowfly. *Mem. Fac. Sci., Kyushu Univ., Ser. E. Biol.*, 4(2):83-93.

Moulton, D. G., and Beidler, L. M. 1967. Structure and function in the peripheral olfactory system. *Physiol. Rev.*, 47:1-52.

Mozell, M. M. 1961. Olfactory neural and epithelial responses in the frog. *Fed. Proc.*, 20:339.

Nedel, J. O. 1960. Morphologie und Physiologie der Mandibeldrüse einiger Bienen-Arten (Apidae). *Z. Morphol. Oekol. Tiere*, 49:139-183.

Noble-Nesbitt, J. 1963. The cuticle and associated structures of *Podura aquatica* at the moult. *J. Cell. Sci.*, 104:369-391.

Oakley, B., and Benjamin, R. M. 1966. Neural mechanisms of taste. *Physiol. Rev.*, 46:173-211.

den Otter, C. J. 1967. Specificities of the sense cells of taste hairs of *Calliphora erythrocephala* Mg. *Acta Physiol. Pharmacol. Neer.*, 14:389-390.

——— and von der Poel, A. M. 1965. Stimulation of three receptors in labellar chemosensory hairs of *Calliphora erythrocephala* Mg. by monovalent salts. *Nature (London)*, 206:31-32.

——— and van der Starre, H. 1967. Responses of tarsal hairs of the bluebottle, *Calliphora erythrocephala* Meig., to sugar and water. *J. Insect Physiol.*, 13:1177-1188.

Pak, W. L. 1965. Some properties of the early electrical response in the vertebrate retina. *Cold Spring Harbor Symp. Quant. Biol.*, 30:493-499.

Peters, W. 1965. Die Sinnesorgane an den Labellen von *Calliphora erythrocephala* Mg. (Diptera). *Z. Morphol. Oekol. Tiere*, 55:259-320.

——— and Richter, S. 1963. Morphological investigations on the sense organs of the labella of the blowfly, *Calliphora erythrocephala* Mg. *Proc. XVI Int. Congr. Zool.*, 3:89-92.

Prosser, D. L. 1961. Chemoreception. *In* Prosser, C. L., and Brown, F. A., Jr., eds. Comparative Animal Physiology. Philidaephia, W. B. Saunders Company, pp. 319-334.

Randebrock, R. E. 1968. Molecular theory of odour. *Nature (London)*, 219:503-505.

Rees, C. J. C. 1967. Transmission of receptor potential in dipteran chemoreceptors. *Nature (London)*, 215:301-302.

Ribbands, C. R. 1955. The scent perception of the honeybee. *Proc. Roy. Soc. [Biol.]*, 143:367-379.

Richter, S. 1962. Unmittelbarer Kontakt der Sinneszellen cuticularer Sinnesorgane mit der Aussenwelt. *Z. Morphol. Oekol. Tiere*, 52:171-196.

Ritter, E. 1936. Untersuchungen über den chemischen Sinn beim schwarzen Kolbenwasserkäfer *Hydreus piceus*. *Z. Vergl. Physiol.*, 23:543-570.

Rosenberg, B., Misra, T. N., and Switzer, R. 1968. Mechanism of olfactory transduction. *Nature (London)*, 217:423-427.

Roys, C. 1954. Olfactory nerve potentials, a direct measure of chemoreception in insects. *Ann. N. Y. Acad. Sci.*, 58:250-255.

Saxena, K. N. 1967. Some factors governing olfactory and gustatory responses of insects. *In* Hayashi, T., ed. Olfaction and Taste. Oxford, Pergamon Press, Inc., Vol. 2, pp. 799-819.

Schallenberger, R. S., and Acree, T. E. 1967. Molecular theory of sweet taste. *Nature (London)*, 216:480-482.

Schenk, O. 1903. Die antennalen Hautsinnesorgane einiger Lepidopteren und Hymenopteren. *Zool. Jahrb. Anat.*, 17:573-618.

Schneider, D. 1955. Mikro-Elektroden registrieren die elektrischen Impulse einzelner Sinneszellen der Schmetterlingsantenne. *Industrie-Elektronik*, 3(3,4):3-7.

——— 1957a. Electrophysiological investigation on the antennal receptors of the silk moth during chemical and mechanical stimulation. *Experientia*, 13:89-91.

——— 1957b. Elektrophysiologische Untersuchungen von Chemo- und Mechanorezeptoren der Antenne des Seidenspinners *Bombyx mori* L. *Z. Vergl. Physiol.*, 40:8-41.

——— 1961a. Untersuchungen zum Bau und zur Funktion der Riechorgane von Schmetterlingen und Käfern. *Berichte Physikal.-Medizin. Ges. Würzburg*, 70:158-168.

——— 1961b. Der Geruchssinn bei den Insekten. *Dragoco Bericht*, 2:27-38.

——— 1962. Electrophysiological investigation on the olfactory specificity of sexual attracting substances in different species of moths. *J. Insect Physiol.*, 8:15-30.

——— 1963a. Electrophysiological investigation of insect olfaction. *In* Zotterman, Y., ed. Olfaction and Taste. Oxford, Pergamon Press, Inc., Vol. 1, pp. 85-103.

——— 1963b. Vergleichende Rezeptorphysiologie am Beispiel der Riechorgane von Insekten. *Jahrb. der Max-Planck-Ges.*, pp. 150-177.

——— 1964. Insect antennae. *Annu. Rev. Entomol.*, 9:103-122.

——— 1965. Chemical sense communication in insects. *Symp. Soc. Exp. Biol.*, 20:273-297.

—— 1967. Wie arbeitet der Geruchssinn bei Mensch und Tier? *Naturwiss. Rundsch.*, 20:319-326.

—— Block, B. C., Boeckh, J., and Priesner, E. 1967. Die Reaktion der männlichen Seidenspinner auf Bombykol und seine Isomeren: Elektroantennogramm und Verhalten. *Z. Vergl. Physiol.*, 54:192-209.

—— and E. Hecker. 1956. Zur Elektrophysiologie der Antenne des Seidenspinners *Bombyx morie* bei Reizung mit angereicherten Extrakten des Sexuallockstoffes. *Z. Naturforsch.*, 11 (b):121-124.

—— and Kaissling, K.-E. 1957. Der Bau der Antenne des Seidenspinners *Bombyx mori* L. II. Sensillen, cuticulare Bildungen und innerer Bau. *Zool. Jahrb. Anat.*, 76:223-250.

—— Lacher, V., and Kaissling, K.-E. 1964. Die Reaktionsweise und das Reaktionsspektrum von Riechzellen bei *Antheraea pernyi* (Lepidoptera, Saturnidae). *Z. Vergl. Physiol.*, 48:632-662.

—— Steinbrecht, R. A., and Ernst, K.-D. 1966. *Naturwiss. Rundsch.*, 19(3): Title page.

Schoonhoven, L. M. 1966. Some cold receptors in larvae of three lepidoptera species. *J. Insect Physiol.*, 13:821-826.

—— 1967a. Loss of host plant specificity by *Manduca sexta* after rearing on an artificial diet. *Entomol. Exp. Appl.*, 10:270-272.

—— 1967b. Chemoreception of mustard oil glucosides in larvae of *Pieris brassicae*. *Proc. Kon. Nederl. Akad. Wet.* [*Biol. Med.*] 70:556-568.

—— 1968. Chemosensory bases of host plant selection. *Annu. Rev. Entomol.*, 13: 115-136.

—— and Dethier, V. G. 1966. Sensory aspects of host-plant discrimination by lepidopterous larvae. *Neth. J. Zool.*, 16:497-530.

Schwarz, R. 1955. Über die Riechschärfe der Honigbiene. *Z Vergl. Physiol.*, 37:180-210.

von Skramlik, E. ed. 1926. Handbuch der Physiologie der niederen Sinne. I. Die Physiologie des Geruchs- und Geschmackssinnes. Leipzig, Georg Thieme Verlag.

—— 1948. Über die zur minimalen Erregung des menschlichen Geruchs beziehungsweise Geschmackssinnes notwendigen Molekülmengen. *Pflueger Arch. Ges. Physiol.*, 250:702-716.

Slifer, E. H. 1961. The fine structure of insect sense organs. *Int. Rev. Cytol.*, 11:125-159.

—— 1966. Sense organs on the antennal flagellum of a walkingstick *Carausius morosus* Brünner (Phasmida). *J. Morphol.*, 120:189-202.

—— 1967. The thin-walled olfactory sense organs on insect antennae. *In* Beament, J. W. L., and Treherne, J. E., eds. Insects and Physiology. Edinburgh, Oliver & Boyd, pp. 233-245.

—— 1968. Sense organs on the antennal flagellum of a praying mantis, *Tenodera angustipennis*, and of two related species (Mantodea). *J. Morphol.*, 124:105-116.

—— Prestage, J. J., and Beams, H. W. 1957. The fine structure of the long basiconic sensory pegs of the grasshopper (Orthoptera, Acrididae) with special reference to those on the antennae. *J. Morphol.*, 101:359-381.

—— Prestage, J. J., and Beams, H. H. 1959. The chemoreceptors and other sense organs on the antennal flagellum of the grasshopper (Orthoptera; Acrididae). *J. Morphol.*, 105:145-191.

—— and Sekhon, S. S. 1961. Fine structure of the sense organs on the antennal flagellum of the honey bee, *Apis mellifera* Linnaeus. *J. Morphol.*, 109:351-381.

——— and Sekhon, S. S. 1963. Sense organs on the antennal flagellum of the small milkweed bug, *Lygaeus kalmii* Stal (Hemiptera, Lygaeidae). *J. Morphol.*, 112:165-193.

——— and Sekhon, S. S. 1964a. Fine structure of the sense organs on the antennal flagellum of a flesh fly, *Sarcophaga argyrostoma* R.-D. (Diptera, Sarcophagidae). *J. Morphol.*, 114:185-208.

——— and Sekhon, S. S. 1964b. The dendrites of the thin-walled olfactory pegs of the grasshopper (Orthoptera, Acrididae). *J. Morphol.*, 114:393-410.

——— Sekhon, S. S., and Lees, A. D. 1964. The sense organs on the antennal flagellum of aphids (Homoptera), with special reference to the plate organs. *J. Cell. Sci.*, 105:21-29.

Smyth, T., and Roys, C. C. 1955. Chemoreception in insects and the action of DDT. *Biol. Bull.*, 108:66-76.

Snedecor, G. W., and Cochran, W. G., eds. 1967. Statistical Methods. 6th ed. Ames, Iowa State University Press.

Snodgrass, R. E., ed. 1935. Principles of Insect Morphology. New York, McGraw-Hill Book Company.

Soliman, S. A., and Cutcomp, L. K. 1963. A comparison of chemoreceptor and whole-fly responses to DDT and parathion. *J. Econ. Entomol.*, 56:492-494.

Steinbrecht, R. A. 1964. Die Abhängigkeit der Lockwirkung des Sexualduftorgans weiblicher Seidenspinner *(Bombyx mori)* von Alter und Kopulation. *Z. Vergl. Physiol.*, 48:341-356.

Steinhardt, R. A. 1966. Physiology of labellar electrolyte receptors of the blowfly, *Phormia regina*. Doctoral Dissert., Columbia University, New York.

——— Morita, H., and Hodgson, E. S. 1966. Mode of action of straight chain hydrocarbons on primary chemoreceptors of the blowfly, *Phormia regina*. *J. Cell. Physiol.*, 67:53-62.

Steward, C. C., and Atwood, C. E. 1963. The sensory organs of the mosquito antenna. *Canad. J. Zool.*, 41:577-594.

Stürckow, B. 1959. Über den Geschmackssinn und den Tastsinn von *Leptinotarsa decemlineata* Say (Chrysomelidae). *Z. Vergl. Physiol.*, 42:255-302.

——— 1960. Elektrophysiologische Untersuchungen am Chemorezeptor von *Calliphora erythrocephala* Meigen. *Z. Vergl. Physiol.*, 43:141-148.

——— 1961a. Histologische Untersuchungen am labellaren Chemorezeptor von *Calliphora* und *Phormia*. *Proc. XI Int. Congr. Entomol.*, 1:410-411.

——— 1961b. Zum Test mit *Solanum*-Alkaloidglykosiden am Chemorezeptor von *Leptinotarsa decemlineata* Say. In Schreiber, K., ed. Chemie und Biochemie der Solanum-Alkaloide. Berlin, Deut. Akad. Landwirtschaftswiss. Tagungsber., 27:17-21.

——— 1963. Electrophysiological studies of a single taste hair of the fly during stimulation by a flowing system. *Proc. XVI Int. Congr. Zool.*, 3:102-104.

——— 1965. The electroantennogram (EAG) as an assay for the reception of odours by the gypsy moth. *J. Insect Physiol.*, 11:1573-1584.

——— 1967. Occurrence of a viscous substance at the tip of the labeller taste hair of the blowfly. *In* Hayashi, T., ed. Olfaction and Taste. Oxford, Pergamon Press, Inc., Vol. 2, pp. 707-720.

——— Adams, J. R., and Wilcox, T. A. 1967a. The neurons in the labellar nerve of the blow fly. *Z. Vergl. Physiol.*, 54:268-289.

——— Holbert, P. E., and Adams, J. R. 1967b. Fine structure of the tip of chemosensitive hairs in two blow flies and the stable fly. *Experientia*, 23:780-782.

—— and Löw, I, 1961. Die Wirkung einiger *Solanum*-Alkaloidglykoside auf den Kartoffelkäfer, *Leptinotarsa decemlineata* Say. *Entomol. Exp. Appl.*, 4:133-142.

Takeda, K. 1961. The nature of impulses of single tarsal chemoreceptors in the butterfly, *Vanessa indica*. *J. Cell. Comp. Physiol.*, 58:233-245.

Thorpe, W. H. 1939. Further experiments on olfactory conditioning in a parasitic insect: The nature of the conditioning process. *Proc. Roy. Soc. [Biol.]*, 126-370-397.

—— ed. 1963. Learning and Instinct in Animals. 2nd ed. London, Methuen and Co. Ltd.

—— and Jones, F. G. W. 1938. Olfactory conditioning in a parasitic insect and its relation to the problem of host selection. *Proc. Roy. Soc. [Biol.]* 124:56-81.

Vogel, R. 1923. Zur Kenntnis des feineren Baues der Geruchsorgane der Wespen und Bienen. *Z. Wiss. Zool.*, 120:281-324.

Warnke, G. 1931. Experimentelle Untersuchungen über den Geruchssinn von *Geotrupes silvaticus* Panz. und *Geotrupes vernalis* Lin. Zugleich ein Beitrag zum Problem der Orientierung der Tiere im Raum. *Z. Vergl. Physiol.*, 14:121-199.

Weber, H., ed. 1933. Lehrbuch der Entomologie. Jena, Gustav Fischer Verlag.

—— ed. 1954. Grundriss der Insektenkunde. 3rd ed. Stuttgart, Gustav Fischer Verlag.

Weide, W. 1960. Einige Bemerkungen über die antennalen Sensillen sowie über das Fühlerwachstum der Stabheuschrecke *Carausius (Dixippus) morosus* Br. (Insecta:Phasmida). *Wiss. Z. Martin Luther Univ. Halle-Wittenberg*, 9:247-250.

Wenner, A. M. 1967. Honey bees: Do they use the distance information contained in their dance maneuver? *Science*, 155:847-849.

Wharton, D. R. A., Miller, G. L. and Wharton, M. L. 1954a. The odorous attractant of the American cockroach, *Periplaneta americana* (L.). I. Quantitative aspects of the response to the attractant. *J. Gen. Physiol.*, 37:461-469.

—— Miller, G. L., and Wharton, M. L. 1954b. The odorous attractant of the American cockroach, *Periplaneta americana* (L.). II. A bioassay method for the attractant. *J. Gen. Physiol.*, 37:471-481.

Wieting, J. O. G., and Hoskins, W. M. 1939. The olfactory responses of flies in a new type of insect olfactometer. II. Responses of the housefly to ammonia, carbon dioxide and ethyl alcohol. *J. Econ. Entomol.*, 32:24-29.

Wigglesworth, V. B. 1941. The sensory physiology of the human louse *Pediculus humanus corporis* De Geer (Anoplura). *Parasitology*. 33:67-109.

—— 1953. The origin of sensory neurones in an insect, *Rhodnius prolixus* (Hemiptera). *J. Cell. Sci.*, 94:93-112.

—— ed. 1965. The Principles of Insect Physiology. 6th ed. London, Methuen & Co. Ltd.

Will, F. 1885. Das Geschmacksorgan der Insekten. *Z. Wiss. Zool.*, 42:674-707.

Wirth, W. 1928. Untersuchungen über Reizschwellenwerte von Geruchsstoffen bei Insekten. *Biol. Zentralbl.*, 48:567-576.

Wolbarsht, M. L. 1957. Water taste in *Phormia*. *Science*, 125:1248.

—— and Dethier, V. G. 1958. Electrical activity in the chemoreceptors of the blowfly. I. Responses to chemical and mechanical stimulation. *J. Gen. Physiol.*, 42:393-412.

—— and Hanson, F. E. 1967. Electrical and behavioral responses to amino acid stimulation in the blowfly. *In* Hayashi, T., ed. Olfaction and Taste. Oxford, Pergamon Press, Inc., Vol. 2, pp. 749-760.

Yamada, M. 1968. Extracellular recording from single neurones in the olfactory centre of the cockroach. *Nature (London)*, 217:778-779.

——— Ishii, S., and Kuwahara, Y. 1968. Preliminary report on olfactory neurons specific to the sex pheromone of the American cockroach. *Botyu-Kagaku*, 33:37-39.

Zacharuk, R. Y. 1962. Exuvial sheaths on sensory neurons in the larva of *Ctenicera destructor* (Bown) (Coleoptera, Elateridae). *J. Morphol.*, 111:35-47.

METHODOLOGY

Gillary, H. 1966. Stimulation of the salt receptor of the blowfly. II. Temperature. *J. Gen. Physiol.*, 50(2):337-350.

Stürckow, B. 1963. Electrophysiological studies of a single taste hair of the fly during stimulation by a flowing system. *Proc. XVI Intern. Congr. Zool.*, 3:102-104.

——— 1967. Occurrence of a viscous substance at the tip of the labellar taste hair of the blowfly. *In* Hayashi, T., ed. Olfaction and Taste. Oxford, Pergamon. Press, Inc., Vol. 2, pp. 707-720.

——— Adams, J. R., and Wilcox, T. A. 1967. The neurons in the labellar nerve of the blow fly *Z. Vergl. Physiol.*, 54:268-289.

DISCUSSION

STUART: Dr. Stürckow, what do you think is the function of the microtubules in the chemoreceptor hair in the light of the contention of Porter and Roth that microtubules are structural? Do you think secretion passes through them?

STÜRCKOW: Well, I think that the microtubules or pore filaments in the cuticle of olfactory hairs are filled with a substance produced by the plasma of the trichogen cell and that this substance, as in gustatory hairs, is a medium that receives the stimulus and transmits by current flow a message to the dendritic fibers. The longitudinal section could not show this material because an ultrathin section is made from fixed material, and fixatives collapse this substance. However, the replica that we made of the tip of the taste hair has shown that in the living state, this substance (and probably also the substance in the pore and microtubules of olfactory hairs) extends very slightly out of the pore, like an antenna to receive the stimulus. When the insect cleans the antennae, it pulls this substance off and a new one is supplied immediately, through its viscosity and maybe a turgor from within the hair.

STUART: Do you know whether the secretion is actually going out? It may be a particular secretion connected with the cuticle of the insect.

MASON: You commended on the relationship of this material to that isolated by Dastoli and Price. Are you indicating that this is a protein substance?

STÜRCKOW: No. We cannot tell from electron micrographs whether it is a protein, a carbohydrate, or something else. Nevertheless, I think future biochemical work should be able to determine whether or not this is a protein. If we were to milk about a thousand of the taste hairs, such as those which had a huge ball of substance at the tip, a biochemist could work with this substance. It should then be possible to determine whether or not it is a protein.

WILSON: Very good. I want to mention parenthetically with reference to the rather poor results obtained by Dr. Stuerckow using gyplure and gyptol, that a group of German chemists has recently reported the structural formulas of Jacobson and his co-workers to be in error [Eiter, K., Truscheit, E., and Bonnes, M., *Liebigs Ann. Chem.*, 709:29-45 (1967)]. The authors synthesized the putative pheromones and found them to be inactive. They also report that synthetic "propylure", the supposed female sex attractant of *Pectinophora gossypiella*, is inactive. Thus, the structural formulas of a large percentage of the insect sex attractants are suspect. If nothing else, this demonstrates the great need for close future collaboration of chemists and biologists in pheromone studies.

JOHNSTON: Dr. Roth, has anyone speculated upon the minimum detectable concentration of the pheromone in air?

ROTH: Jacobson and Beroza have estimated that 30 molecules would elicit a sex response in male American cockroaches. However, this was based on the wrong chemical structure of the pheromone. I think even if it were 3,000, it would be a pretty small number of molecules.

TURK: These molecular counts bother me sometimes. When we count molecules as few as 30, or even as few as 10^5, I presume we are diluting something that is so concentrated that we have some *confidence in its original value,* and that we then extrapolate according to the ratio of dilutions. The only occasions that I know that such a dilution procedure has been tested is with tracer materials, such as sulphur hexafluoride. We find that in some systems we can have confidence in concentrations down to 10^{-11} or perhaps 10^{-14} (volume fraction). When I say confidence, I mean that we can reconstitute to a higher concentration by enhancement factors of 10^3 to 10^6 and get back most of the analytical signal. Professor Johnston once gave me a sample of Exaltone to dilute, and I used the same reconcentration system, which was successful with SF_6, but I never found Exaltone again. So I lost it either on the way down in dilution or on the way up in concentration. However, I think that in most cases, investigators who cite 30 or 10^5 molecules or something like that have not tried reconstitution, so there is no real independent confidence in what these numbers may be.

WILSON: Yes, that certainly is an excellent point. We have not yet made a cross-check on threshold concentration estimations using different methods, e.g., the stream-dilution method and the diffusion model method.

7

Researches on Trail and Alarm Substances in Ants*

A. GABBA and M. PAVAN

Istituto di Entomologia Agraria
dell'Università di Pavia
Pavia, Italy

INTRODUCTION 161
THE TRAIL SUBSTANCE IN THE FORMICIDAE 162
THE ALARM SUBSTANCES IN THE FORMICIDAE 176
CONCLUSION 189
REFERENCES 191
DISCUSSION 195

INTRODUCTION

The world of Arthropods, especially of insects, is becoming more and more interesting to general biologists because studies of insect behavior, ecology, and physiology of the sense organs have progressed remarkably over the last few years, and new possibilities for research are continuously opening up. Research on behavior conditioned by reactions to chemical substances is among the most interesting material offered by insect biology. The study of behavior reactions in insects to substances they produce and employ to regulate particular aspects of their social life is especially important. We therefore stress research on the trail and alarm substances of social insects, and the social behavior determined by these substances—studies which have developed in a modern direction only in the last few years. It is quite possible that, as these researches progress, we may also find useful indices for the study of behavior physiology in higher animals. Studies of the intimate chemical mechanisms of elaboration and transmission of mass information to obtain a social effect, may contribute to the solution of analagous problems in higher animals (where these problems are more difficult due to the greater complexity of the nervous organs).

The relative simplicity of the central and peripheral nervous systems of insects

* This work was sponsored by a grant from the Consiglio Nazionale della Ricerche, Rome.

as compared with warm-blooded animals may perhaps contribute towards a satisfying solution of the complex and controversial problems involved in the theory of smells, their perception, and the reactions they induce in animal organisms.

Therefore, we consider highly interesting the proposal to deal with the problem of trail and alarm substances in the Family Formicidae of social Hymenoptera as part of a monograph which will examine these facts as they appear throughout the whole zoological range.

THE TRAIL SUBSTANCE IN THE FORMICIDAE

Among the most interesting aspects of ants' social life are collaboration, the division of labor, exploration of the ambient environment, food-gathering, and the road network which links the main nest to the smaller ones, and which links the nest with the food and hunting reserves. During the period of activity there is continuous movement around the nest, in which columns of ants come and go, usually following the tracks unhesitatingly chosen by single insects. However, if the track is interrupted at a certain point, for example by placing an extraneous object across it or digging a small furrow, the ants stop in confusion. What guides them along the tracks is not simply a topographic knowledge of the area or a particular sense of direction. They often follow an odorous trail left on the ground by the continuous passing of their fellow ants, the smell of which, often characteristic of the species, is picked up by extremely sensitive organs localized in the antennae. Except for a few cases when it is especially strong and characteristic, the smell cannot be perceived by man.

Close observation of the structure and characteristics of these trails and the anatomic origin and chemical nature of the substances marking them, has been recently made possible first of all by the development of artificial trail techniques, by which we can study laboratory colonies. These techniques consist in making extracts from ant organs and glands and then laying trails with them along lines of various shapes, intersecting with other unmarked lines. By checking whether and by which extract the ants are attracted, it is possible to single out the presence of a trail substance, and the organ producing it, and to understand how it works and how it is used.

One faces various problems when trying to study the trail-laying phenomenon. First, one must ascertain that the phenomenon applies to the species under consideration; then one must investigate the origin of the trail substances and determine the producing organ or gland, the modalities of the secretion, the range of action (specific, interspecific, intergeneric), the chemicophysical properties, and the chemical structure of the substances considered.

Several subfamilies of Formicidae lay odorous trails which vary considerably in their origin and the organs producing them.

Experiments on the trail substance in the various subfamilies

The literature on this subject is considerable, and the species studied are numerous. Trail-laying behavior may be observed in workers returning to their nest either "empty-handed" after exploration or bearing insects, larvae, or other food. Many species have been observed to intermittently touch the ground with their abdomen (thus leaving a signal) when they return from hunting.

IN FORMICINAE.* The trail layed by *Lasius* (*Dendrolasius*) *fuliginosus*† consists of a series of small lines which narrow down in the direction in which the ant is moving. This trail-laying, which happens especially during their return to the nest, was observed both in workers carrying food and in ants carrying larvae (Carthy, 1951b).

During the laying of the trail the workers assume a special position in order to bring the point of the gaster as near as possible to the substratum. They raise their bodies by stretching out their legs, and then by bending the abdomen, bring the anal aperture into contact with the ground. The workers do not follow these odorous trails closely but keep in contact with it, exploring the ground with their antennae. If the trail is interrupted the ants lose their sense of direction.

Chemical analyses by Carthy (1951a) showed that in *Lasius fuliginosus* the product laid to mark the trail was an anal emission of a clear viscous liquid, with additions which could be seen after discoloration. The additions are bits of peritrophic membrane which are discharged with the secretion (Carthy, 1951b). The liquid also contained uric acid, polysaccharides, and proteins: that is, fecal matter rather than a special glandular secretion. This, however, did not rule out the possibility that particular attractive secretion products could be present in small quantities, or that a proper trail substance was produced by another organ (for example the venom gland) and that this could be mixed with the fecal emission. Nevertheless, Carthy's first conclusion seems very probable. It was, in fact, observed that in other species, for example *Myrmelachista ramulorum* and *Paratrechina longicornis,* the origin of the substance is the hind intestine (Blum and Wilson, 1964). Regarding these two species, however, the possible presence of a particular trail substance—whether it is a digestive product or a special secretion of the intestinal wall—is still unsolved.

Interesting observations on the characteristics of the *L. fulginosus* odorous trail were made also by Hangartner and Bernstein in 1964. They saw that active trails could be obtained by dissolving the contents of the hind intestine or stomach of this Formicina in water. The trails were not effective when dry, but could be reactivated by wetting.

Hangartner (1967) found that the trail substance was in the rectal ampulla in three *Lasius* species (*emarginatus, flavus, niger*). The trail substance in the rectal ampulla of two other species, *Formica cinerea rufibarbis* and *Formica rufa,* also proved effective on *L. fuliginosus.* Hangartner's studies furnish interesting information on trail-substance specificity, on its chemical and physical characteristics, and on the behavior and sense of direction of variously treated workers as they follow an odorous trail. While *L. fuliginosus* and *L. flavus* seem to have a strictly specific trail substance, the other species examined produce an interspecifically active trail substance. *L. fuliginosus* is attracted by the artificial trails of the other species (except that of *L. flavus,* which is species-specific). *L. flavus* follows only its own trails—and even these not very actively.

_____

* See Table 1.
† The authors' names of the species mentioned appear in Tables 1 through 5. For the species indicated in the text but not listed in the tables, the author's name is indicated only in the first quotation.

TABLE 1

Formicidae species producing a trail substance, from a known organ.

KNOWN SPECIES PRODUCING THE TRAIL SUBSTANCE	GLANDULAR ORGAN PRODUCING THE TRAIL SUBSTANCE	TRAIL SUBSTANCE EMITTED FROM:	REFERENCES
Ponerinae			
Termitopone laevigata (Fr. Smith)	hindgut	anus	3, 83, 85
Dorylinae			
Eciton hamatum (Forel)	hindgut	anus	6
Neivamyrmex nigrescens (Cresson)	hindgut	anus	72
Myrmicinae			
Acromyrmex octospinosus (Reich)	poison gland	sting	5
Atta cephalotes (L.)	poison gland	sting	5
Atta texana (Buckley)	poison gland	sting	37
Crematogaster peringueyi (Emery)	metathoracic legs, distal segment	probably terminal tarsal subsegment of hind legs	24
Cyphomyrmex rimosus (Spinola)	poison gland	sting	5
Monomorium floricola (Jerdon)	poison gland	sting	3
Monomorium minimum (Buckley)	poison gland	sting	3
Monomorium pharaonis (L.)	poison gland	sting	3
Myrmica brevinodis (Emery)	poison gland	sting	3
Pheidole fallax (Mayr)	Dufour's gland	sting	81
Sericomyrmex urichi (Forel)	poison gland	sting	6
Solenopsis geminata (Fabr.)	Dufour's gland	sting	77
Solenopsis saevissima (Fr. Smith)	Dufour's gland	sting	75
Solenopsis xiloni (McCook)	Dufour's gland	sting	77
Tetramorium caespitum (L.)	poison gland	sting	7
Tetramorium guineense (Fabr.)	poison gland	sting	7
Trachymyrmex septentrionalis (McCook)	poison gland	sting	5
Dolichoderinae			
Iridomyrmex humilis (Mayr)	Pavan's gland	outlet between IV and V gastral urosternum	86
Iridomyrmex pruinosus (Roger)	Pavan's gland	outlet between IV and V gastral urosternum	86
Monacis bispinosa (Oliv.)	Pavan's gland	outlet between IV and V gastral urosternum	86
Tapinoma sessile (Say)	Pavan's gland	outlet between IV and V gastral urosternum	83
Formicinae			
Formica cinerea rufibarbis (For.)	hindgut	anus	27
Formica rufa (L.)	hindgut	anus	27
Lasius emarginatus (Oliv.)	hindgut	anus	27
Lasius flavus (Fabr.)	hindgut	anus	27
Lasius fuliginosus (Latr.)	hindgut	anus	16, 27
Lasius niger (L.)	hindgut	anus	27
Myrmelachista ramulorum (Wheeler)	hindgut	anus	11
Paratrechina longicornis (Latr.)	hindgut	anus	11, 81

In order to explain these complex interspecific relationships, Hangartner forms a hypothesis similar to that of Blum and Ross (1965) regarding the attraction exercised by the venom gland secretion in various Myrmicinae. After demonstrating experimentally that this interspecific attraction (always lower than that of the producing species) is not due to a different concentration of the trail substance, he hypothesizes the existence of several chemical substances present simultaneously with the true pheromone; in the trail substance these substances are apparently analogous, but not identical, to the trail pheromone of the other species.

As regards the trail substance interspecificity, some Formicidae species regularly use odorous trails left by other species. For example, *Crematogaster limata parabiotica* and *Monacis debilis* usually form nests in close association, and foraging activities take place along common trails.

A similar association was also found by Wheeler between *Crematogaster parabiotica* and *Camponotus femoratus* and between the European species *Crematogaster scutellaris* and *Camponotus lateralis*; the latter (*camponotus* Spp.) follow the former's (*Crematogaster* Spp.) odorous trails.

Wilson (1965a) describes the common trails between Dolichoderine *Azteca chartifex* and Formicine *Camponotus beebei*: the *Azteca* odorous trails are followed both by *Azteca* and *Camponotus*; moreover, the latter seems to prefer the *Azteca* trails to its own. *Camponotus* workers use *Azteca* trails during the day, when the *Azteca's* foraging activity is limited, so they trouble the other species but little. However, when individuals of the two species meet, *Azteca* adopts a hostile attitude to *Camponotus*. *Camponotus* workers have also kept their own trail system. The *Camponotus* trails last for about 15 minutes and do not have any noticeable effect on *Azteca*.

IN DOLICHODERINAE. In 1955b Pavan found a new organ in the Dolichoderine *Iridomyrmex humilis*—an exocrine gland opening between the IV and V gastral urosternum—which he called the "ventral organ." He assumed that it might be used for producing the trail substance, as it opens towards the outside in a manner and in a position suitable for that function. Later, Wilson and Pavan (1959) used the artificial-trail technique in a first attempt to localize the organ producing the trail substance in that species. Tracks marked with extracts from various parts of the body were found inactive, except for the gaster which proved active. There are five organs in the gaster which are able to emit glandular secretions, and further experiments attempted to localize the glandular apparatus responsible for trail-laying. It was found that not only in *I. humilis*, but also in *I. pruinosus* and *Monacis bispinosa*, which represent two phylogenetically distant genera of Dolichoderinae, the main or single source of the odorous trail was the ventral organ. This proved the 1955 hypothesis to be true.

This gland, later renamed by American authors "Pavan's gland" (present only in Dolichoderinae, but not in every species) is therefore an organ specialized in the pheromone secretion of the odorous trail. It is the first organ known in entomology with the specialized function of producing trail pheromones. The outlet usually has a particular protruding shape which seems meant to facilitate the laying of the secretion. Results up to now show that not all Dolichoderinae having this particular conformation have the corresponding Pavan's gland.

As Pavan's gland is a typical apparatus for the production of the trail substance, we are still left with the problem of whether a trail substance exists—and what its origin is—in those Dolichoderinae species which do not have this organ, as, for example, *Tapinoma nigerrimum, Liometopum microcephalum*, and many others, which travel in columns all the same. In the Dolichoderinae which do not have that organ, it is a matter of checking whether the trail substance is produced by the digestive apparatus, as in Formicine, Doriline, and Ponerine ants.

The literature (Pavan, 1955b; Miradoli et al., 1957; Wilson and Pavan, 1959; Gabba, 1967), concerning the distribution of Pavan's gland is summarized in Table 2.

Wilson and Pavan (1959), in their experiments with odorous trails of *I. humilis, I. pruinosus, M. bispinosa, Liometopum occidentale* and *Tapinoma sessile* odorous trails, carried out with these species to determine the interspecific and intergeneric responses, found that the trail substances were species specific.

IN DORYLINAE.* The biology, activity cycles, and social organization systems in many species of army ants have been extensively studied and described by Schneirla (1940, 1956, 1960). Some genera (*Eciton, Aenictus, Neivamyrmex*) form societies in which static and nomadic stages alternate, depending on the growth level of the brood, and the physogastric conditions of the queen.

The workers are warlike and present a marauding activity which, from a minimum during the egg-laying period, increases gradually and proportionately with the colony's nutritive needs. Schneirla showed that in both activities, exodus and raiding, they moved along well-developed and long-lasting chemical trails which could be followed even weeks after they had been laid (Schneirla and Brown, 1950; Schneirla, 1963). Although rain did not efface them rapidly, they lasted much longer during the dry season, showing that the trail pheromone is relatively photostable, thermostable, insoluble in water, and has a low volatility.

These marked trails are formed by the workers by dragging their abdomens on the ground.

Schneirla mentions the existence of chemical trails with regards to *Aenictus laeviceps, A. gracilis, Eciton hamatum, E. burchelli*, and *Neivamyrmex nigrescens* species.

Schneirla and Brown also noticed that *E. burchelli* followed a trail of about 40 meters laid more than three weeks before by *E. hamatum*, and vice versa. Workers of this last species were seen to move along a trail laid by *E. burchelli*.

Regarding *E. hamatum*, Blum and Portocarrero (1964), after several experiments, concluded that the hind intestine is the origin of the trail pheromone. However, none of the trail pheromones has ever been chemically characterized, nor do we know whether it is produced by special gland cells of the hind intestine, or if it is a digestive product formed further up in the digestive tract.

The intestinal content must have an attractive function, since artificial trails prepared with extracts of empty hind gut, following fasting, showed little or no activity.

Watkins (1964) went deeper into the research work carried out on *Neivamyrmex*, a primitive genus of Dorylinae, paying particular attention to *N. opacithorax, N. nigre-*

* See Table 3.

TABLE 2

Data on Pavan's gland in Dolichoderinae

SPECIES OF DOLICHODERINAE	CARINOFORM TIP BETWEEN THE IV AND V GASTRAL URO-STERNUM	PRESENCE OF PAVAN'S GLAND	REFERENCES
Aneuretus simoni Em.	+	+	36
Azteca aurita Em.	+	?	46
A. instabilis Sm.	+	?	46
A. mulleri Em.	+	?	46
A. velox For.	+	?	46
Dolichoderus attelaboides F.	+	?	46
D. bidens L.	+	?	46
D. capitatus Santschi	+	?	46
D. decollatus Sm.	+	?	46
D. feae Em.	+	?	46
D. germaini Em.	+	?	46
D. gibbifer Em.	+	?	46
D. (Hypoclinea) doriae Em.	+	+	25, 86
D. imitator Em.	+	?	46
D. lugens Em.	+	?	46
D. quadripunctatus L.	+	+	25
D. rugosus Sm.	+	?	46
D. scrobiculatus Mayr	+	?	46
D. sulcaticeps Mayr	+	?	46
Dorymyrmex pyramicus (Roger)	+	?	46
Iridomyrmex detectus (Sm.)	−	?	46
I. humilis Mayr	+	+	46
I. nitidus Mayr	−	?	46
I. pruinosus (Roger)	?	+	86
Leptomyrmex erithrocephalus F.	+	?	46
Le. nigriventris Guerin	+	?	46
Le. puberula Wheeler	+	+	36
Le. unicolor Em.	+	?	46
Liometopum apiculatum Mayr	+	?	46
L. microcephalum Panz.	+	−	46
Monacis bispinosa (Oliv.)	?	+	86
Tapinoma erraticum Latr.	+	?	46
T. flavidum Andre	+	?	46
T. nigerrimum Nyl.	+	−	46
T. sessile (Say)	?	+	83
Technomyrmex andrei Em.	+	?	46

scens and *N. carolinensis*. The trail interspecificity tests between these three species were all positive, which leads to the conclusion that there is no strict trail pheromone specificity in the species examined.

It was also seen that is was possible to lay artificial trails with the contents of *N. nigrescens* stomach and rectum and that they were active on *N. opacithorax* and *N. nigrescens* workers. Here, too, the trail substance seems to be fecal matter with or without the addition of glandular substance.

The recent experiments by Watkins et al. (1967) confirmed, on the whole. the previous data on the lack of trail-substance specificity in the genus *Neivamyrmex*. The species examined included five species of *Neivamyrmex* (*nigrescens, opacithorax, pilosus, pauxillus, wheeleri*) and *Labidus coecus*. The trails laid by these six species were active on one another with only one exception: *N. pilosus* followed only its own trail, which, however, was not specific. Preference tests were also made where the workers of a species could choose their own trails or those laid by another species. Although, from what we have already seen, many army ants respond to odorous trails laid by related species, generally, when given the chance, they preferred to follow their own trail.

IN MYRMICINAE. The numerous species of the Attine tribe show different behavior patterns in their foraging activity. An individual search for food was observed in the less-advanced species, and an organized food-gathering activity in the most advanced species (Weber, 1958). Among the latter are the *Acromyrmex* and *Atta* genera, whose workers form long columns along odorous trails, the *Monomorium* genus, and the *Trachymyrmex* genus which seem to have an intermediate degree of development as regards trail-laying behavior (they may hunt either alone or in columns). Sudd (1959) reports that in *Monomorium pharaonis*, a chemical-trail-laying species, the workers have a habit of moving in close contact, even if the tracks are not crowded. At a fork, *Monomorium* has a tendency to follow the tracks chosen by the worker moving in front of it.

Weber (1958), on the basis of the morphology, behavior and structure of the nest, thought that *Cyphomyrmex rimosus* was the most primitive of Attini. Noticing the occurrence of individual foraging, he concluded that this species does not leave odorous trails. Nevertheless, Blum et al. (1964), observing two columns of *C. rimosus* in nature, proved the existence of odorous trails also in this primitive species.

Moser and Blum in 1963 had already carried out interesting studies on trail laying by *Atta texana*. The workers of this species, while foraging, extrude their stinger and periodically rest their abdomen on the ground. This behavior is more noticeable during the return to the nest from the source of food than in the opposite direction, when trail-laying is much more occasional. Research showed that the *A. texana* lays long-lasting odorous trails by its stinger. The venom gland secretes a substance which accumulates in a reservior; this is half empty in the young workers, full in the older workers. With extracts from young workers with half empty reservoirs, the response was moderate or strong, but the reaction obtained with the contents of older workers' full reservoirs was much stronger.

The trail substance of this species appears as a clear, viscous, strongly alkaline liquid. It forms a white suspension in acetone, alcohol, or water. When exposed to the air the liquid turns into a semi-solid mass.

Well-drawn trails may be found which extend for dozens of yards. The same track may be followed for months, indicating that the trail substances have a low vapor tension. The foraging activity goes on also when the track is wet, indicating that the odorous substance is not water-soluble.

The trail substance was fully effective after five months at $-12°C$ and did not freeze at this temperature. It was also seen that the venom reservoirs were intact in

TABLE 3

Specificity of trail substance in Formicidae

SPECIES NO.	FORMICIDAE TESTED FOR TRAIL-LAYING AND FOLLOWING RESPONSE	FOLLOWS TRAILS OF: (REFERENCE TO THE NUMBER OF 1ST COLUMN)	DOES NOT FOLLOW TRAILS OF: (REFERENCE TO THE NUMBER OF 1ST COLUMN)	REFERENCES
	Ponerinae			
1	*Termitopone laevigata* (Fr. Smith)	1	18	3, 4, 83, 84
	Dorylinae			
2	*Aenictus gracilis* Emery	2	—	63
3	*Aenictus laeviceps* F. Smith	3	—	63
4	*Eciton burchelli* (Westwood)	—	18	4, 59, 60
5	*Eciton hamatum* (Forel)	5	18	4, 5, 59, 60
6	*Labidus coecus* (Latreille)	6, 8, 9, 10, 11, 12	—	73
7	*Neivamyrmex carolinensis* (Emery)	7, 8, 9	—	72
8	*Neivamyrmex nigrescens* (Cresson)	6, 7, 8, 9, 10, 11, 12	52	62, 72, 73
9	*Neivamyrmex opacithorax* (Emery)	6, 7, 8, 9, 10, 11, 12	52	72, 73
10	*Neivamyrmex pauxillus* (Wheeler)	6, 8, 9, 10, 11, 12	—	73
11	*Neivamyrmex pilosus* (F. Smith)	11	6, 8, 9, 10, 12	73
12	*Neivamyrmex wheeleri* (Emery)	6, 8, 9, 10, 11, 12	—	73
	Myrmicinae			
13	*Acromyrmex coronatus* (F.)	13	—	6
14	*Acromyrmex nr. coronatus* (F)	14	—	6
15	*Acromyrmex octospinosus* (Reich)	15, 16, 17, 47	—	4
16	*Atta cephalotes* (L.)	15, 16, 47	—	4
17	*Atta texana* (Buckley)	15, 16, 17, 24, 40($\pm$), 46, 47	43, 45	4, 6, 7, 21, 37
18	Species of Attines			4
19	*Crematogaster limata parabiotica* Forel	19, 58, 65	—	82
20	*Crematogaster lineolata* (Say)	—	18, 43, 57	4, 77, 86
21	*Crematogaster peringueyi* Emery	21	—	24
22	*Crematogaster scutellaris* (Oliv.)	22	66(?)	82
23	*Crematogaster sordidula* Nyl.	23	—	68
24	*Cyphomyrmex rimosus* (Spinola)	17, 24, 47	—	4
25	*Huberia striata* (F. Smith)	25	31, 32	3
26	*Manica rubida* Latr.	26	—	68
27	*Myrmica brevinodis* Emery	27	—	3
28	*Myrmica laevinodis* Nyl.	—	72	27
29	*Myrmica rubra* (L.)	—	43	77
30	*Myrmica ruginodis* Nyl.	30	—	17, 34

TABLE 3 (*cont.*)

SPECIES NO.	FORMICIDAE TESTED FOR TRAIL-LAYING AND FOLLOWING RESPONSE	FOLLOWS TRAILS OF: (REFERENCE TO THE NUMBER OF 1ST COLUMN)	DOES NOT FOLLOW TRAILS OF: (REFERENCE TO THE NUMBER OF 1ST COLUMN)	REFERENCES
31	Monomorium minimum (Buckley)	31, 32	18, 25, 27, 33, 46, 47	3, 4
32	Monomorium pharaonis (L.)	32	25, 27, 31, 33, 46, 47	3, 67
33	Monomorium floricola (Jerdon)	33	27, 31, 32, 46, 47	3,
34	Pheidole antillensis	34	—	21
35	Pheidole crassinoda	35	—	68
36	Pheidole ecitonodora Wheeler	36	—	21
37	Pheidole fallax Mayr	37	74	11, 21
38	Pheidole pallidula Nyl.	38	—	68
39	Pogonomyrmex badius (Latr.)	—	43	77
40	Sericomyrmex urichi Forel	40	17, 47	6
41	Solenopsis geminata (Fabr.)	41, 42(±), 44	57, 74	11, 21, 77, 86
42	Solenopsis (Diplorhoptrum) molesta (Say)	—	43	77
43	Solenopsis saevissima (Fr. Smith)	43, 44	17, 18, 41, 45, 46, 47, 57	4, 7, 21, 75, 77, 78, 79, 86
44	Solenopsis xyloni (McCook)	41(±), 44	—	21, 77
45	Tetramorium caespitum (L.)	45	17, 43, 46, 47	7
46	Tetramorium guineense (Fabr.)	17, 46, 47	43, 45	7
47	Trachymyrmex septentrionalis (McCook)	15, 16, 17, 24, 46, 47	40, 43, 45	4, 6, 7
48	Xenomyrmex floridanus Emery	—	43	77
	Dolichoderinae			
49	Azteca chartifex Forel	49	64	82
50	Conomyrma pyramica (Roger)	—	18	4
51	Dorymyrmex goetschi	51	—	68
52	Iridomyrmex analis (André)	52	8, 9	72
53	Iridomyrmex humilis Mayr	53	43, 55, 56, 57, 62	21, 77, 86
54	Iridomyrmex melleus Wheeler	—	74	11
55	Iridomyrmex pruinosus (Roger)	55	18, 53	4, 21, 86
56	Liometopum occidentale (Emery)	56	53	86
57	Monacis bispinosa (Olivier)	57	43, 53, 62	21, 77, 86
58	Monacis debilis (Emery) [=Dolichoderus debilis var. parabiotica Forel]	19, 58	—	82
59	Tapinoma antarcticum For.	59	—	68
60	Tapinoma erraticum Latr.	60	72	27, 68
61	Tapinoma melanocephalum (Fabr.)	—	74	11
62	Tapinoma sessile (Say)	62	43, 53, 55, 56, 57	77, 83, 84, 86
	Formicinae			
63	Acantholepis frauenfeldi (Mayr)	63	—	68
64	Camponotus beebei Wheeler	49, 64	—	82

TABLE 3 (*cont.*)

SPECIES NO.	FORMICIDAE TESTED FOR TRAIL-LAYING AND FOLLOWING RESPONSE	FOLLOWS TRAILS OF: (REFERENCE TO THE NUMBER OF 1ST COLUMN)	DOES NOT FOL- LOW TRAILS OF: (REFERENCE TO THE NUMBER OF 1ST COLUMN)	REFERENCES
65	*Camponotus femoratus* (Fabr.)	19, 65	—	82
66	*Camponotus lateralis* (Oliv.)	22, 66	—	82
67	*Camponotus maculatus barbaricus* Emery	67	—	27
68	*Formica cinerea rufibarbis* Forel	68	72	27
69	*Formica rufa* L.	69	72	27
70	*Lasius emarginatus* Oliv.	70, 73(±)	71, 72	27
71	*Lasius flavus* Fabr.	71(±)	70, 72, 73	27
72	*Lasius fuliginosus* (Latr.)	68, 69, 70, 72, 73	71	15, 16, 27, 28
73	*Lasius niger* L.	70, 73	71, 72	17, 27
74	*Myrmelachista ramulorum* Wheeler	74	—	11
75	*Paratrechina longicornis* (Latr.)	75	74	11
76	*Plagiolepis pigmaea* Latr.	76	—	68

dead workers after the rest of the abdomen was decomposed, and the substance remained active.

In 1964 Blum et al. continued their observations of the Attine *Trachymyrmex septentrionalis, Acromyrmex octospinosus, Atta cephalotes, A. texana,* and *Cyphomyrmex rimosus.* As already found for *A. texana,* the venom glands may be considered as the producers of the trail pheromone also for the other four species. It was seen also that there is no specificity in the odorous trail pheromones between these four Attine genera.

In 1959 Wilson made a series of experiments on the Myrmicine *Solenopsis saevissima,* whose workers drag the end of their abdomen on the ground, with their stinger extruded. Artificial trails with venom extracts were drawn in order to verify whether the substance emitted through the stinger was effective as a trail pheromone.

Although the workers' response was positive, as shown by their attraction to the artificial trails, they nonetheless preferred the natural ones when a choice was available.

Numerous artificial trails, made in the same conditions from the wall and contents of the principal parts of the intestine or from the pulped tissues and hemolymph, caused various reactions, but did not exercise an overwhelming call. Later experiments showed that the essential substance of the trail is a secretion of the Dufour's gland emitted through the stinger.

Also in the Myrmicine *Tetramorium guineense* the food-gathering activities take place along well-developed odorous trails. Trail-laying behavior was observed by Blum and Ross (1965) in workers carrying insects back to the nest. The loaded ants periodically touched the ground with the gaster, in the way described for other trail-laying Myrmicine ants (Wilson, 1959; Moser and Blum, 1963). Photographs showed that the stingers of workers loaded with food were thrust outwards when their abdomens were periodically near the ground.

Blum and Ross (1965) prepared artificial trails by applying methylene chloride extracts of hind gut, venom gland, and Dufour's gland on circles 15 cm in diameter drawn on white paper. After the solvent evaporated, groups of 12 workers were put in the middle of the circle, and a positive answer was recorded if, after meeting the circle, a worker followed it all round. This experiment (190 positive answers out of 200) showed that the venom gland was the origin of the trail pheromone in *T. guineense*. This pheromone was not found in the young workers' empty glands.

The content of the venom gland is transparent. When a venom sack is broken in the air, its content rapidly turns into an amorphous semi-solid substance which liquifies easily in distilled water. The rapid "solidification" of the venom may depend simply on the rapid evaporation of the small quantity of solvent (probably water) in which the components of the venom are dissolved. The presence of water in the venom was confirmed by gas-chromatographic analysis. The greater part of the material produced by the venom gland was insoluble in all the organic solvents tried.

The trail substance discharging through the protruding stinger probably comes almost exclusively from the venom gland. This is shown by the fact that, after emitting numerous drops, the venom gland vesicle is empty, while the Dufour's gland is always swollen. At most, only traces of the content of the Dufour's gland are secreted with the products of the venom gland.

The chemical substance of the *T. guineense* trail is not very volatile. The artificial trails prepared on nonabsorbent paper and kept at 28°C. were still active after 168 hours.

Therefore, the pheromone does not seem to have a high vapour pressure, which is consistent with the fact the natural odorous trails of this Myrmicine are long lasting.

The odorous trail pheromone specificity of *T. guineense* was studied employing other Myrmicinae species belonging to four different genera (Blum and Ross, 1965) (*T. caespitum, A. texana, T. septentrionalis,* and *S. saevissima*). The trail substance is produced in both *Tetramorium* species by the venom glands, but the odorous trail is entirely specific for the producing species. In contrast with this specificity, *T. guineense* followed artificial trails prepared with the venom glands of two different Myrmicine, *A. texana* and *Tr. septentrionalis*. Similarly, transposition studies in which *T. guineense* was the originating species showed that the venom gland of this Myrmicine contained a trail substance which was active for the two Attine species, like the latter's own substance. On the other hand, *T. caespitum,* which did not follow artificial trails prepared with *T. guineense* glands, neither followed trails prepared with *A. texana* or *Tr. septentrionalis* venom glands.

The *S. saevissima* Dufour's gland secretion proved inactive as a trail substance when tried on other Myrmicinae.

The foraging activity in *Monomorium pharaonis* is relatively simple and well organized: it follows some well-determined tracks leading to the probable supply areas (Sudd, 1960). These roads are long lasting and partially determined by topography. Some have been described which are more than 10 meters long.

After finding food, explorers lay out branching trails from these main roads. In fact, in this Myrmicine species, only a small number of scout workers are individually engaged in the research for supply areas. After finding the food they do not carry it to the nest, but pick up a bit in their mandibles and go back laying a trail, thus activating their companions to look for food.

The trail left by a single *M. pharaonis* worker is sufficient to call other workers, but can communicate only the direction in which the food lies, not the distance or its quality (Sudd, 1960). The workers follow the odorous trail, which represents their only information about the presence and location of food, and, on their way back to the nest, each in their turn strengthens the trail, so that the recruitment continually increases.

The residues of the trails in their proximal part remain and are incorporated, as we have seen, into a road network which serves as a first, approximate guide for the scouts. Therefore their position is fairly constant, and their distribution around the nest depends on the earlier forage areas. Further research regarding the Myrmicine genus have been carried out recently by Blum (1966) for *M. pharaonis*, *M. minimum*, *M. flavicola*, and *M. antarcticum* Wheeler and *Huberia striata*.

In *M. Pharaonis* the organ producing the trail substance is the venom gland.

Pheromone is laid on the ground through the stinger in the same way as described for *Solenopsis saevissima*. The only behavioral difference between these two species when they lay their trail is that *Monomorium* withdraws its stinger less frequently, thus forming a dotted line whose single trails are longer. Also in other species examined (*M. minimum*, *M. floricola*, *Huberia striata*) the origin of the pheromone trail has been found in the poison gland. On the other hand, in *M. antarcticum*, the hind-gut, venom gland, and Dufour's gland extract do not provoke any reaction. Furthermore, the other species examined did not respond to the latter's glandular extracts.

Artificial trails made with the venom-gland content of the three species of *Monomorium* and *Huberia striata* proved to be strictly specific, with only one exception: *M. minimum* was equally attracted both by extracts from its own venom gland and from those of *M. pharaonis* (the contrary was not true).

The specificity of the *Monomorium* trail pheromone was studied also by laying artificial trails with extracts of the *Trachymyrmex septentrionalis*, *Myrmica brevinodis*, and *Tetramorium guineense* venom glands. None of the *Monomorium* species followed these trails.

The venom in all the *Monomorium* species examined is a transparent, alkaline liquid, which rapidly solidifies into an amorphous substance in air. The trails of *M. pharaonis* were fairly long lasting, remaining active for about 24 hours, while those of *M. minimum* were hardly active after 2½ hours. Both substances were soluble in organic solvents.

An interesting biologic fact is reported in a recent paper by Blum and Portocarrero (1966) about a primitive Myrmicine ant, *Daceton armigerum* (Latreille), whose foraging activity is carried out by single workers who do not rely on odorous trails. This fact was observed both in nature and in laboratory colonies.

However, the *Daceton armigerum* venom is highly active as a trail substance for the workers of three Attine species. Positive responses were obtained in the *Trachymyrmex septentrionalis*, *Acromyrmex coronatus*, *Acromyrmex* nr. *coronatus*,

Atta texana, and *Atta cephalotes.* Only one of the species examined, *Sericomyrmex urichi,* did not follow the odorous trail. On the other hand, it was found that this same species did not respond to artificial trails prepared with the venom gland secretion of *Trachymyrmex* and *Atta texana* either, which seems to show a difference of *sericomyrmex* trail pheromone with respect to that of the other Attine.

Another Myrmicine which does not lay trails, *Cordiocondyla nuda minutor* Forel, has a substance which is fairly active for *Monomorium minimum* in its venom gland (Blum, 1966).

In the Myrmicinae, whose trail-laying activities have been studied, it was found that the venom gland is widely used in the production of the trail substance. The lack of specificity, according to Blum and Portocarrero (1966), could be explained by supposing that in the different venons, whether used for marking the trail or not, (as in the case of *Daceton* and *Cardiocondyla*) very small quantities of various common substances are present which possess the property of pheromones for the various species. In this way, even if the members of different genera have different trail substances, they will be attracted by other species' venoms, as their own trail pheromone is also present in small quantities in the secretions. An analogous fact to that described by Blum for *Daceton* had been noticed also by Wilson and Pavan (1959) regarding the *Monacis bispinosa* Dufour's-gland secretion. In this Dolichoderine, as we have seen, the trail substance is secreted by Pavan's gland, but artificial trails made with extracts from Dufour's gland attracted *Solenopsis saevissima* as effectively as extracts of Dufour's gland from the same *Solenopsis.*

Fletcher and Brand (1968), watching *Crematogaster peringueyi* Emery workers during their foraging activity, saw this species, after discovering a new source of food, go back to the nest without touching the ground with their abdomen—a behavior characteristic of all trail-laying Formicidae species so far examined. However, *Crematogaster* did lay odorous trails capable of attracting their companions. Observations showed that the food-gathering ants walked in an unusual way. The gaster was slightly raised, the back legs much more close together than in the normal walking position. The tarsi, held more or less parallel with respect to the body's longitudinal axis, were periodically struck on the substratum while the worker slowly advanced. These last observations, made a posteriori, confirmed the results of experimental research.

In fact, extracts from various parts of the body, tried out with the usual artificial trail technique, showed that in this species the trail substance originates in the metathoracic leg distal segments, and probably comes out at the terminal tarsal subsegment, since this is the only point which comes into contact with the ground while laying the trail.

In order to explain this unusual behavior Fletcher and Brand hypothesize that in this species a specific trail pheromone may have evolved from a nonspecific secretion, formerly used only to increase the adhesive power of the tarsal pads, a fact common in insects.

A third organ suitable for the production of the trail substance has been found in Myrmicinae, that may indicate a systematic revision of the subfamily, as suggested by Fletcher and Brand.

As to the chemical characteristics of the various substances used to mark the trail, few data are available besides those we have seen in this paper.

Walsh et al. (1965) purified a tiny, apparently homogeneous, quantity of S. *saevissima* trail substance. By means of gas-liquid chromatography a peak that contained all the active substance was observed. Of course this does not rule out the possibility that the active fraction is contaminated with an inactive compound having similar chromatographic properties. Remarkable losses of active material take place during the purification process. Also, the material, both in a raw and purified state, rapidly loses its activity even if kept at $-20°C$ in dichloromethane, making chemical identification of the active substance difficult (it is still unknown).

Characteristics of olfactory communication and mass response

Interesting experiments on the laying of odorous trails and on their properties were carried out by Wilson (1962a, 1962b, 1962c), on S. *saevissima*. If a worker finds food that is small and solid, it picks it up in its mandibles and drags it towards the nest; however, if it is liquid or unmovable, after a quick inspection it goes back to the nest, laying an odorous trail. The resulting chemical trail resembles a more or less sinuous dotted line connecting the food to the nest. The workers who come across a newly laid trail smell its presence at a distance of about 10 mm. The track is not unidirectional because it can be followed in both directions.

Wilson's observations on S. *saevissima* showed that a single worker cannot lay a trail over 50 cm long. Longer trails must be due to chain activity by numerous individuals.

It is almost exclusively by the trail substance that the ants are called on for a mass gathering at the food. It strongly attracts the workers and induces them to leave the nest for the odorous trail. The initial increase in number of workers arriving at the food source is exponential, until the area becomes over crowded—then the increase levels off. Unable to get at the food source, they go back without laying a trail; on the other hand, the trail laid by other single workers evaporates within a few minutes. Consequently, the number of workers around the food tends to be a function of the areas and/or mass of the food. Again, according to Wilson, trail laying, measured from the time when the stinger is brought forward, is of the "all or nothing" type of response. The individual trail does not designate the quantity or quality of food. The quantity of the food is instead indicated by the numbers of workers, since, as we have already seen, their gathering is proportional to the area of available food. Quality is reported in the same way, that is by means of a mass response: the individuals reaching the food source decide whether to leave a trail or not after an inspection of the discovered food. The more desirable the food discovered, the higher the percentage of positive answers, the greater the quantity of trail substance presented to the colony, and thus the more numerous, proportionally, the ants leaving the nest.

The trail pheromone is a chemical communicator. The real information transmitted was measured in the case of the odorous trail left by S. *saevissima* (Wilson, 1962b). When the workers find a new source of food they return to their nest leaving an odorous trail made of tiny quantitites of a substance from the Dufour's gland. This volatile substance, spreading out in the air, forms a semi-ellipsoidal active space, with a maximum transversal radius of about 1 cm, which persists for about 100 seconds. The ants attracted by this trail follow the track. The longer the trail, the greater the quantity of directional information, but obviously the information is much reduced owing to the time needed to mark and follow this trail. Even so, both the quantity of information transmitted and the rate of transmission are considerable. However, the odorous trails do not assuredly lead to the object, and the ants do not follow the track closely.

Both stages, laying and following the trail, include errors which can be evaluated and represented approximately with a Gaussian curve. This distribution of the errors in its turn serves to determine the quantity of information transmitted.

A general theory of information can be applied to the study of animal communication. The essential step is breaking down the communication system into bits, the basic units of information. A bit is defined as the quantity of information required to make a choice between two equiprobable alternatives. If n alternatives are present, a choice gives the following quantity of information:

$$H = \log_2 n$$

where H = quantity of information in bits.

Wilson's analysis showed that the odorous trails left by a worker transmit four bits of information about direction and two bits about distance, which is equivalent to determining one of the 16 equiprobable sectors of a circle and one of the four equiprobable intervals of a distance scale.

The Alarm Substances in the Formicidae

In the complex social behavior of ants, alarm transmission and reactions are important mass phenomena affected by various means of communication: tactile, acoustic, and chemical (pheromones). Sounds and tactile stimuli may start and spread the alarm, but generally massive diffusion of the alarm is obtained through the emission of glandular chemical secretions by aroused workers. Alarm behavior is the initial stage of a complex of coordinated, defensive acts.

We have observed that all Formicidae species forming large colonies possess alarm secretions, whereas chemical alarm mechanisms are not traceable in relatively primitive species such as *Ponera coarctata* (Latr.) (Ponerinae) and *Myrmicina graminicola* (Latr.) (Myrmincinae), which form very small colonies generally of less than 100 workers. In these small colonies a disturbance can be detected directly by the single workers, therefore a mass information system does not appear to be strictly necessary.

There are several causes of alarm behavior, but it is generally a reaction to extraneous individuals entering the nest or in any way representing a disturbance or menace to the colony. The workers have alarm reactions and manifest their excitement with typical quick-oriented movements. It is still unknown whether sexuals react in the same way. However, it has been noticed that the queens possess alarm substances, but they do not participate in the immediate subsequent stages of defense and attack.

The workers' behavior is not uniform when faced by a cause for alarm. Their reactions vary, depending particularly on the species they belong to: some react passively and hide inside the nest or abandon it, others become decidedly aggressive. Reactions vary also according to the site of the alarm, the concentration and the length of time the alarm substance is exposed. Likewise the strength of the population, the age, and the state of activity or inactivity of the workers may influence the intensity and type of reaction.

For example, behavior inside the nest differs from behavior in the foraging area. In the foraging areas the response usually comes only from the bigger workers who are capable of defending themselves [e.g., various species of *Formica* and *Tapinoma nigerrimum* (Maschwitz, 1964a)] while the others usually flee. The reactions inside the nest are more violent: even the small and weak ants are attracted by the alarm substance to the disturbed area to defend the queen and the brood.

*Experiments on the alarm substances in
the various subfamilies*

IN PONERINAE. In the Ponerinae subfamily the alarm pheromone (of the species examined) originates in the mandibular glands.

Several authors have mentioned the existence of chemical communication in *Paraponera clavata, Myrmecia gulosa, Onychomyrmex hadley* Emery, and *Pachycondila harpax*, but the pheromones have not been chemically identified.

It is known that the African ant, *Paltothyreus tarsatus* (Fabr.), when in a state of defense, emits a strong unpleasant odor; *Megaponera foetens* (F.) behaves in the same way. The smell, which Pavan (Casnati et al., 1967) showed as coming from the mandibular glands,* is perceivable from a distance of several meters; when it is emitted, the ant colony enters into a state of alarm.

Recent chemical research by Casnati et al. (1967) on *P. tarsatus* showed that the mandibular secretion contains dimethylsulfide and dimethyltrisulfide. This was the first time that sulfurated substances were found as alarm and defense products in insects.

Thus, there are interesting possibilities for a comparitive study of other Ponerines' secretions in relation to the primitivity of this family compared with the other Formicidae.

IN DORYLINAE. The alarm behavior in Dorylinae was studied by Brown (1960) observing *Eciton hamatum, Labidus praedator,* and *Nomamyrmex esenbecki* colonies. On disturbing a column of *Eciton hamatum* Brown perceived a particular smell which gradually grew in proportion to the number of workers disturbed who, at the same time, showed aggressive reactions to the extraneous object.

When heads of workers or soldiers of the three above-mentioned species were

* The literature often attributes the origin of the smell to other glands.

put on the respective foraging columns, the companions reacted in an alarmed manner and attacked violently, biting the heads. Also, objects smeared with crushed heads provoked an intense attack. Beheaded bodies did not provoke a similar behavior; at the most, they attracted only slight attention. According to Brown, the existence of an alarm pheromone in the mandibular glands of these three genera is very probable.

In order to explain why, in the alarm phase, the individuals emitting the pheromone are not themselves attacked by the other workers, Brown makes two hypotheses: either the quantity of substance normally emitted is much less than what is perceived on presenting the colony with a detached or crushed head, or the postcephalic part of the body secretes substances which prevent an attack, thus protecting the alarmed individuals.

IN DOLICHODERINAE. In the species of the Dolichoderinae subfamily so far examined, the alarm substances are produced exclusively by the supra-anal glands;* in many species the secretion from these glands also has a defense function. In the case of *Iridomylrmex humilis*, Pavan in 1947 isolated a new substance from the defense secretion of the supra-anal gland. This is the insecticide principle: iridomyrmecin.

After formic acid, isolated from Formicidae in the seventeenth century, iridomyrmecin was the first active principle of the Formicidae defensive secretions to be isolated and chemically defined as α-(2-hydroxymethyl-3-methylcyclopentane) propionic acid 8-lactone; numerous researches derived from this, yielding interesting data on the composition of the venoms of many Dolichoderinae and other Formicidae, have opened a new chapter in chemistry—that of iridoids. In Table 4 there is a summary of the data for Dolichoderinae. However, iridomyrmecin does not act as an alarm pheromone.

The dual function of defense and alarm is found in the secretions emitted by *Tapinoma nigerrimum* supra-anal glands, whose chemical composition was studied by Trave and Pavan (1956). *Tapinoma* workers employ the venom actively for defensive and offensive purposes against competing insects. The release of the secretion provokes violent excitement among their companions.

The substance produced by the supra-anal glands is stored in a liquid state in its specific reservoir until the time of emission. If this reservoir is opened while immersed in water the liquid released floats and gives off the typical smell which is perceived when pressing on the abdomen of a living worker. The secretion is composed of two ketones (methylheptanone and propylisobutylketone) and a dialdehyde (iridodial)† that may function to partially retain the other two volatile substances, and thus prolong the active time span. *Tapinoma sessile*, after being exposed for a long time to *T. nigerrimum* anal secretion, starts mass exodus movements leaving the old nest. Small quantities of propylisobutylketone and methylheptanone presented in separate tests provoked intense alarm in *T. sessile* (Wilson and Pavan, 1959). Also *I. pruinosus*, *T. sessile*, *Liometopum occidentale*, and *Monacis bispinosa* emit alarm substances from the supra-anal glands. The pheromone, whose smell can easily be perceived, is

---

* Also improperly called "anal glands."
† Iridodial was isolated independently in Italy by Trave and Pavan, 1956, and by Cavill and coworkers from species of the *Iridomyrmex* and *Dolichoderus* genera of Australia.

TABLE 4

Chemicals produced by supra-anal glands of Formicidae Dolichoderinae

SPECIES OF FORMICIDAE DOLICHODERINAE	CHEMICALS OF SUPRA-ANAL GLANDS
Conomyrma pyramicus (Roger)	2-heptanone; iridodial; 2-methyl-2-hepten-6-one
Dolichoderus (Acanthoclinea) clarcki (Wheeler)	dolichodial; 4-methyl-2-hexanone
D. (A.) dentata (Forel)	dolichodial
D. (Diceratoclinea) scabridus (Roger)	iridodial; dolichodial; isoiridomyrmecin; 2-methyl-2-hepten-6-one
Iridomyrmex conifer For.	iridodial; 2-methyl-2-hepten-6-one
I. detectus (Smith)	iridodial; 2-methyl-2-hepten-6-one
I. humilis Mayr	iridomyrmecin
I. myrmecodiae (Em.)	dolichodial
I. nitidiceps (André)	iridodial; 2-methyl-2-hepten-6-one
I. nitidus Mayr	isoiridomyrmecin; isodihydronepetalactone
I. pruinosus (Roger)	2-heptanone
I. rufoniger Lowne	dolichodial; iridodial; 2-methyl-2-hepten-6-one
Liometopum microcephalum Panz.	2-methyl-2-hepten-6-one
Tapinoma nigerrimum Nyl.	iridodial; 2-methyl-2-hepten-6-one; propyl-isobutyl-ketone

secreted in large quantities every time the workers are disturbed; this provokes an immediate and typical alarm reaction in the nest population, characterized by accelerated movements.

Except for *I. pruinosus*, the supra-anal gland substances of these species have not been chemically identified. Although the substances smell different, all interspecificity tests gave positive answers.

The lack of specificity can be explained here (as already seen for the trail) if we admit the presence of common components, or more probably, as the authors suggest, if we assume that the workers reactions are aspecific. The Myrmicine *Solenopsis saevissima* Dufour's gland secretion though, which represents the alarm substance for this species, had no effect or only induced slight excitement in *T. sessile* workers. The same weak reaction was observed with exposure to formic acid.

Blum et al. (1963) isolated 2-heptanone from *I. pruinosus* supra-anal glands. The substance is stored in the well-developed anal reservoirs of this species and has a characteristic strong odor which is easily detected on crushing a worker or disturbing a nest. In the latter case, the smell gradually increases in intensity, corresponding to the spreading of the alarm inside the colony, and therefore to the number of excited workers secreting the pheromone. In this case, the ketone represents a double-function secretion: besides being a very effective alarm pheromone with a strong attraction for the workers, it is used by *I. pruinosus* as a defense substance as well.

Experiments in nature, and on laboratory colonies, showed the important biologic effects that this ketone induced in the producing species (Blum et al., 1966): low concentrations of 2-heptanone attracted the workers and induced them to leave the nest; when the concentration was increased, the ants attracted appeared to be much more excited and moved much faster. The effect was sufficiently strong to induce the workers to move onto surfaces heated to a lethal temperature.

2-Heptanone is also the alarm pheromone of the Dolichoderine *Conomyrma pyramica* whose alarm behavior was recently studied by Blum and Warter (1966). *C. pyramica* is an extremely active species. The workers are highly excitable: when disturbed they move frantically, emitting, from well-developed supra-anal glands, a volatile secretion containing 2-heptanone which provokes typical alarm behavior in their companions. This ketone strongly attracts *Conomyrma* both in high and low concentrations. Placed at the entrance of the nest it draws out many workers in a short time. The number of workers at the nest entrance increases at least tenfold in 60 seconds. The spreading of alarm in the colony is just like that described for *Pogonomyrmex badius* (Wilson, 1958): there is a widening wave-transmission around the initial stimulation center, produced by the centrifugal movements of the excited workers who, upon meeting other individuals, transmit the alarm through tactile stimuli and the emission of the anal secretion.*

Near the pheromone source, the frantic movements of the workers turn into staggering, almost rolling, movements. When the concentration of 2-heptanone is very high, the workers find it repellent.

A high, 2-heptanone concentration is also capable of provoking a digging reaction in the workers, which lasts as long as they stay exposed to the pheromone. Similarly, Blum and Warter observed that pieces of paper or other objects impregnated with the ketone were generally buried by the ants, who thus eliminated the cause of alarm. This phenomenon might explain the usefulness of the digging behavior for eliminating a source of disturbance.† In *Conomyrma pyramica*, (as in *Tapinoma nigerrimum* and *I. pruinosus*) the supra-anal gland secretion may also be employed for defensive purposes.

Maschwitz (1964a) discovered an alarm function in the supra-anal gland secretion of two other species of Dolichoderinae, *Tapinoma erraticum* and *Botriomyrmex meridionalis*.

IN FORMICINAE. In 1956 Pavan isolated a new substance, dendrolasin, with a molecular weight of 218, from the mandibular glands of *Lasius (Dendrolasius) fuliginosus* Latr. In the same year, Quilico, Piozzi and Pavan made known the structure of the substance, β-(4,8-dimethylnona-3,7-dienyl) furan. This was the first furan known in the animal kingdom, and the first chemically known alarm pheromone. The

---

* This kind of mass information transmission reminds us of a very efficient news communication system of the human populations in great equatorial forests, for example the Congo. People there use the tam-tam, a large drum consisting of an empty trunk with an opening, beating the rim with two sticks. The resulting deep sound propagates over a great distance. With this excellent mass information system, the tam-tams beaters, upon receiving the news, broadcast it immediately, each in turn. In a very short time the information spreads rapidly with a parabolic pattern (for example, in 20 minutes a good tam-tam network can spread information at a distance of 100 km. in all directions, covering an area of about 32,000 km²). In our analogy, in the world of ants the workers releasing the chemical-informative substance in the air correspond to the tam-tams which spread the news with sound vibrations: the broadcast mechanism in both cases follows the same parabolic pattern even though the means are completely different.
† This reminds us of Ghidini and Pavan's observations (Pavan, 1962). The *I. humilis* workers were given, in nature, poisonous sugary liquids to drink, in small bowls placed on the ground; after numerous deaths by poisoning, sand was carried to the bowls in organized mass action until they were completely buried, thus stopping the poisoning process.

mandibular gland secretion, whose odor can be detected by man, elicits an alarm reaction in L. fuliginosus workers. Objects contaminated with raw secretion are repellent for other Formicidae species. Formica rufa L. workers, experimentally contaminated with raw secretion and also with pure dendrolasin, are either avoided or attacked by companions of the same nest who do not recognize them, as their identity is masked by the smell of the Dendrolasius secretion. In nature, ants who escaped from Dendrolasius mandibles in a fight, are therefore contaminated with their secretion, are not recognized, and are attacked by their sister ants and chased from their own nest.

It has been known since 1956 that dendrolasin is not the only constituent of the mandibular gland secretion. Recently, Bernardi et al. (1967) showed that the secretion contains also perillen, farnesal, cis- and trans-citral. It is interesting to see the relationship that may exist between some of these substances.

In various Lasius species (brunneus, flavus, fuliginosus, and niger), alarm substances are secreted from the Dufour's gland, as shown by Maschwitz, 1964a.

In interspecific experiments in the Lasius genus, the Dufour's gland secretion of L. fuliginosus proved effective as an alarm-inducing substance on all three of the species tested (L. niger, flavus, brunneus). The L. niger pheromone proved fairly active, while those of L. flavus and brunneus were less active. The Lasius fuliginosus Dufour's gland induced alarm also in other Formicinae of two different genera, Formica polyctena and Plagiolepis pigmaea Latr. Reciprocally, the latter's secretion was active on L. fuliginosus.

In the Formica, it was seen (Maschwitz, 1964a) that F. cinerea, F. fusca, and F. polyctena have a multiple alarm system with a threefold glandular source: the mandibular, Dufour's, and venom glands. The last of these, which secretes formic acid, is known specially for its defensive function. The action of Formica alarm substances, regardless of their origin, is clearly interspecific. The F. polyctena Dufour's gland secretion is also active on L. niger.

Recently Bergström and Löfquist (1968) chemically identified numerous substances in the Dufour's gland of three Formica species (sanguinea, fusca, and rufibarbis) and in Polyergus rufescens, but did not test whether the substances act as alarm pheromones.

The Acanthomyops claviger mandibular glands have large reservoirs, the contents of which are emitted by the roused workers after stimulation, as in L. (D.) fuliginosus. Ghent (1961) found that citronellal, present in the mandibular secretion, acted as alarm pheromone for this species.

Chemical researches by Chada et al. (1962) identified the volatile components of the secretion as consisting essentially of a mixture of citral and citronellal in a rough proportion of 1:9.

Furthermore, the alarm communication function of Acanthomyops is not limited to the mandibular glands alone. The Dufour's gland plays an equally important role here also, as shown recently by Regnier and Wilson (1968). In the mandibular glands secretion they identified two other substances together with citronellal and citral: 2,6-dimethyl-5-hepten-1-al and the corresponding alcohol. In the Dufour's gland undecane, tridecane, 2-tridecanone, pentadecane, and 2-pentadecanone were identified chemically. Citronellal, citral, 2-tridecanone, and undecane seem to have the function of alarm pheromones. Undecane also seems to have a strong defensive action. The other substances found are probably not important in the communication system, as their quantity is too small and, in the case of tridecane and 2-pentadecanone, the vapour pressure is too low to allow an adequate diffusion of the signal.

IN MYRMICINAE. Sudd's observations (1957) on *Monomorium pharaonis* are among the first on the Myrmicine alarm reactions. The workers of this species stop at a crushed ant lying across their trail, or at a spot where the body of one of their companions has been crushed and then removed. After a quick reconnaissance with their antennae they flee. This reaction does not take place with bodies of other species normally used for food. The flight reaction is not due to an interruption of the trail since artificial interruptions (such as effacing part of the chemical trail or laying an obstacle upon it) do not induce the workers to flee. Sudd therefore thinks that they perceive some volatile substance constained in the ants' bodies and this serves as an alarm.

Interesting systematic researches were carried out by Wilson (1958) on *Pogonomyrmex badius* colonies in artificial nests. The reaction to the initial stimulus varies according to an intensity gradient which is proportional to the strength of the initial stimulus. The workers also raise their heads and antennae and lower their abdomen to a position roughly perpendicular to the substratum. In the case of a high alarm intensity the workers also hold their mandibles partially open. However, when far away from the entrance of the nest, they may simply exhibit a flight reaction. The excited workers, on meeting other companions, communicate the state of alarm, which spreads in centrifugal waves throughout the colony. Wilson showed that the transmission takes place through the release of cephalic volatile substances, produced by the mandibular glands, which can also be smelled by man.

Workers exposed to these glandular secretions for a prolonged period of time also reveal a digging reaction, oriented toward the stimulus, which stops when the pheromone is removed. Wilson hypothesizes that this digging behavior may have a biologic meaning: a continuous emission of the glandular secretion might cause the workers to run to the rescue of companions accidentally buried inside the nest.

Other species of this genus were recently studied by McGurk et al. (1966) who isolated 4-methyl-3-heptanone from the mandibular glands of *Pogonomyrmex barbatus*. The same ketone was found in the heads of *P. californicus, P. desertorum. P. occidentails* and *P. rugosus*. It has been demonstrated that 4-methyl-3-heptanone is the alarm pheromone in *P. barbatus*. The workers behaved exactly like *P. badius*. In this species it was possible to note digging movements after prolonged exposure. The ketone also provoked an aggressive reaction to crushed heads, objects, or living workers contamimated with the substance under examination, as described by Brown (1959) for Dorylinae.

McGurk et al. (1966) identified also the corresponding alcohol, 4-methyl-3-heptanol, together with 4-methyl-3-heptanone, in *P. badius* heads.

4-Methyl-3-heptanone is also the alarm pheromone for *Atta texana*. In the mandibular gland of *Atta texana* workers, Moser et al. (1968) identified 4-methyl-3-heptanone, 2-heptanone and detected other unidentified substances. 2-Heptanone proved about 1000 times less effective than 4-methyl-3-heptanone. The workers reactions went from attraction to alarm to repulsion, as the concentration of the presented substances varied. Activity gradually decreased until it returned to normal about three minutes from the beginning of the alarm. A queen's head contains a substance capable of provoking intense and lasting alarm in the workers; the queen, however, shows no alarm either at the workers' glandular substances or to the crushed heads of males, workers, or queens. A male's crushed head induces in workers a behavior similar to that induced by 4-methyl-3-heptanone (1:100,000) or by workers' heads. On the contrary, males did not show any alarm reaction whatever.

Butenandt et al. (1959) found citral in the mandibular glands of *Atta sexdens rubropilosa*. This aldehyde functions as an alarm pheromone.

Wilson (1962c) observed typical alarm reactions in *Solenopsis saevissima*, caused with low and high concentrations of a cephalic substance produced by the mandibular glands (Wilson, 1965ab). Maschwitz (1964a; 1964b; 1966) carried out interspecificity experiments verifying the occurrence of the alarm behavior in numerous Myrmicinae. Various species of *Myrmica* (*laevinodis, rubida, ruginodis,* and *sulcinodis*) possess alarm substances in both the mandibular and venom glands. Both secretions generally act interspecifically. *M. ruginodis* responds well also to the *Pheidole pallidula* abdominal substance, whereas there is no alarm reaction with the cephalic substance of the same species. Tests in which the *Leptothorax nigrescens* Mayr, *Tetramorium caespitum,* and *Messor barbarus* alarm substance was presented to *Myrmica* were also negative.

Crushed heads of *Messor barbarus* and *Crematogaster auberti, C. scutellaris, Pheidole pallidula,* and *Tetramorium caespitum* had an intraspecific but not interspecific effect.

Characteristics of the alarm transmission system

Adult workers may present different kinds of alarm reactions, varying from simple attraction and agitation to attack and fighting. The typical behavior of individuals in a state of alarm includes a first stage in which the worker attracted contacts the source of the emission of the substance, receiving the message with its antennae. Then the reaction stage follows, characterized by strong excitement, an increase of the preambulatory activity in the form of circular or zig-zag lines and emission of the alarm pheromone. The workers often assume a defensive attitude, running around the place of emission with open mandibles. Sometimes they throw themselves on the odorous object, seize it, drag is away, or bury it.

The physical and chemical characteristics and the properties relative to the alarm communication system were analyzed in *Pogonomyrmex badius* by Wilson and Bossert (1963). The quantitative study carried out by these authors on this Myrmicine is fundamental for an understanding and interpretation of the same phenomenon in other species. The volatile substances emitted by the mandibular glands of *P. badius* form a sphere which expands rapidly and immediately, reaching a radius of about 6 cm in 13 seconds, at which time the alarm pheromone concentration is equal to or higher than the minimum concentration necessary to trigger off a reaction in the receptor (threshold-concentration). Supposing an environment where no air moves, the density of the volatile substance decreases from the centre of the sphere outwards: in this case the low concentration of the outer layer attracts the workers. In the central area, on the other hand, the pheromone is concentrated enough to provoke true alarm behavior. However, since the substance is quite volatile, the sphere dissolves almost completely within 35 seconds, and the signal is extinguished.

Biologically it is very important and useful for the regular functioning of a colony's life that the ants in question be able to receive the alarm signal within very short limits of time and space. An ants' nest is continuously subject to various disturbances—often unimportant and not dangerous to the community—which can be quickly eliminated by small groups of workers. If the alarm-producing phenomenon is localized and not very important the number of individuals directly stimulated is small, and therefore the diffusion of the message is limited. The number of workers directly alarmed increases with increasing duration and strength of the disturbance, and therefore the danger to the colony increases. This leads to the emission of a large quantity of the alarm pheromone, that reaches areas farther away from the nest and lasts for a longer period of time. When the danger stimulus stops, the signal disappears rapidly.

Structural Characteristics and Chemical Specificity of Alarm Substances*

The attraction of a chemical substance to insects generally increases with increase in molecular size. On the other hand, if the size and complexity of the molecule increases there is a corresponding decrease in volatility and therefore rapidity of diffusion; these, as we have seen, are properties essential to an effective alarm system. Moreover, it is biologically useful for insects to use alarm substances with a relatively simple structure that can be synthesized rapidly. An alarm substance acts mostly inside the nest, so that the message does not have to go far: consequently, a great stimulating power is not necessary. Considering the above factors, Wilson and Bossert's (1963) predicted the molecular weight of the insects' alarm substance to be between 100 and 200, with 5 to 10 carbon atoms.

Although our knowledge of the composition of chemical alarm substances is still very incomplete, the pheromones hitherto identified confirm this hypothesis. The only exceptions are to be found in dendrolasin which has 15 carbon atoms and a molecular weight of 218, and formic acid with 1 carbon atom and a molecular weight of 46. However, in both these substances it seems the alarm function is secondary to the defensive function.

Apart from natural alarm secretions, many other odorous volatile substances cause agitation, digging, and exodus reactions in ants. The effects induced by extraneous substances are usually less than those provoked by specific alarm substances. Wilson (1958) observed that P. badius workers reacted strongly to nu-

* See Table 5.

TABLE 5

Alarm substances (known or presumed) and their source in Formicidae

FORMICIDAE	SOURCE	CHEMICAL IDENTITY	FUNCTION	REFERENCES
Ponerinae				
Myrmecia gulosa (Fabr.)	mandibular glands	—	alarm	21
Pachycondila harpax (Fabr.)	mandibular glands	—	alarm	81
Paltothyreus tarsatus (Fabr.)	mandibular glands	dimethyldisulfide dimethyltrisulfide	defense, alarm	20
Paraponera clavata (Fabr.)	mandibular glands	—	alarm	81
Dorylinae				
Eciton sp.	mandibular glands	—	alarm	81
Eciton hamatum (Forel)	head (mandibular glands?)	—	alarm, attack	13
Labidus praedator (Fr. Smith)	head (mandibular glands?)	—	alarm, attack	13
Nomamyrmex esenbecki (Westwood)	head (mandibular glands?)	—	alarm, attack	13
Dolichoderinae				
Bothriomyrmex meridionalis Rog.	supra-anal glands	2-heptanone	alarm	31
Conomyrma pyramica (Roger)	supra-anal glands	—	alarm, digging, attraction, defense (?)	8
Iridomyrmex pruinosus Roger	supra-anal glands	2-heptanone	alarm, attraction, defense	9, 10
Liometopum microcephalum Panz.	supra-anal glands	methylheptenone	alarm	19
Liometopum occidentale Emery	supra-anal glands	—	alarm, defense	86
Monacis bispinosa (Olivier)	supra-anal glands	—	alarm, defense	86
Tapinoma erraticum Latr.	supra-anal glands	—	alarm	31
Tapinoma erraticum ssp. *nigerrima* (Nyl.)	supra-anal glands	—	alarm	31
Tapinoma nigerrimum (Nyl.)	supra-anal glands	methyleptenone, propyl-isobutyl-ketone, iridodial	defense, alarm	70
Tapinoma sessile (Say)	supra-anal glands	—	? alarm	86

TABLE 5 (*cont.*)

Alarm substances (known or presumed) and their source in Formicidae

FORMICIDAE	SOURCE	CHEMICAL IDENTITY	FUNCTION	REFERENCES
Formicinae				
Acanthomyops claviger (Roger)	mandibular glands	citronellal	defense, alarm	22, 58
		citral		
		2, 6-dimethyl-5-hepten-1-al	?	
		2, 6-dimethyl-5-hepten-1-ol	?	
	Dufour's gland	n-C$_{11}$-ane	defense, alarm	58
		2-tridecanone	alarm	
		n-C$_{13}$-ane	?	
		2-pentadecanone	?	
		n-C$_{15}$-ane	?	
Mormica cinerea Mayr	Dufour's gland	formic acid	alarm	31
	poison gland	—	alarm, defense	31
	mandibular glands	—	alarm	31
Formica fusca L.	abdomen (Dufour's gland?)	farnesene	alarm (?)	1
		n-C$_9$-ane		
		n-C$_{10}$-ane		
		n-C$_{11}$-ane		
		n-C$_{12}$-ane		
		n-C$_{13}$-ane		
		n-C$_{15}$-ane		
		n-C$_{17}$-ane		
		C$_{12}$-ane branched		
		C$_{13}$-ene		
Formica fusca L. *lemani* Bondr.	Dufour's gland	formic acid	alarm	31
	poison gland	—	alarm, defense	31
	mandibular glands	—		
Formica polyctena (Förster)	mandibular glands	—	alarm	31
	Dufour's gland	—	alarm	31
	poison gland	formic acid	alarm	31
			alarm, defense	31

Species	Source	Compound	Function	Reference
Polyergus rufescens (Latr.)	abdomen (Dufour's gland?)	farnesene	alarm (?)	1
Formica rufibarbis F.	Dufour's gland	n-C$_{11}$-ane n-C$_{17}$-ane n-C$_{10}$-acetate n-C$_{11}$-acetate n-C$_{12}$-acetate 2-tridecanone n-C$_9$-ane n-C$_{10}$-ane n-C$_{11}$-ane n-C$_{12}$-ane n-C$_{13}$-ane n-C$_{15}$-ane n-C$_{17}$-ane n-C$_{19}$-ane C$_{17}$-ene	alarm (?)	1
Formica sanguinea Latr.	Dufour's gland	n-C$_{10}$-acetate n-C$_{11}$-acetate n-C$_{12}$-acetate n-C$_{10}$-alcohol n-C$_{11}$-alcohol n-C$_{12}$-alcohol farnesene n-C$_9$-ane n-C$_{10}$-ane n-C$_{11}$-ane n-C$_{13}$-ane	alarm (?)	1
Lasius brunneus Latr.	Dufour's gland	—	alarm	31
Lasius flavus Fabr.	Dufour's gland	—	alarm	31
Lasius fuliginosus Latr.	mandibular glands	farnesal citral perillen dendrolasin 6-methyl-hept-5-en-2-one	alarm, defense	2, 47
Lasius niger niger L.	Dufour's gland	—	alarm	31
Plagiolepis pygmaea Latr.	Dufour's gland	—	alarm	31
Serviformica sp.	Dufour's gland	—	alarm	31
	poison gland (?)	—	alarm, defense	32

TABLE 5 (cont.)

Alarm substances (known or presumed) and their source in Formicidae

FORMICIDAE	SOURCE	CHEMICAL IDENTITY	FUNCTION	REFERENCES
Myrmicinae				
Aphaenogaster testacea pilosa ssp. spinosa	head (mandibular glands?)	citral	alarm	32
Atta sexdens rubropilosa Forel	mandibular glands	—	alarm	14
Atta texana (Buckley)	mandibular glands	4-methyl-3-heptanone 2-heptanone	alarm, attrac- tion attack (?)	38
Crematogaster auberti Em.	head (mandibular glands?)	—	alarm	32
Crematogaster scutellaris Oliv.	head (mandibular glands?)	—	alarm	32
Messor Barbarus L.	head (mandibular glands?)	—	alarm	32
	abdomen (poison glands?)	—	alarm	31
Messor instabilis ssp. hispanica	head (mandibular glands?)	—	alarm	32
	abdomen (poison glands?)	—	alarm	31
Myrmica laevinodis (Nyl.)	head (mandibular glands?)	—	alarm	32
	poison gland	—	alarm	31
Myrmica rubida L.	head (mandibular glands?)	—	alarm	32
	poison gland	—	alarm	31
Myrmica ruginodis Nyl.	head (mandibular glands?)	—	alarm	32
	poison gland	—	alarm	31
Myrmica sulcinodis Nyl.	head (mandibular glands?)	—	alarm	32
	poison gland	—	alarm	31
Pheidole pallidula Nyl.	head (mandibular glands?)	—	alarm	32
	abdomen (poison glands?)	—	alarm	32
Pogonomyrmex badius (Latr.)	mandubular glands	4-methyl-3-heptanone	alarm, digging, (?)	74
Pogonomyrmex barbatus (Fr. Smith)	head (mandibular glands?)	4-methyl-3-heptanol	alarm, attack, digging	35
	mandibular glands	4-methyl-3-heptanone		35
Pogonomyrmex californicus (Buckley)	head (mandibular glands?)	4-methyl-3-heptanone	alarm (?)	35
Pogonomyrmex desertorum (Wheeler)	head (mandibular glands?)	4-methyl-3-heptanone	alarm (?)	35
Pogonomyrmex occidentalis (Cresson)	head (mandibular glands?)	4-methyl-3-heptanone	alarm (?)	35
Pogonomyrmex rugosus Emery	head (mandibular glands?)	4-methyl-3-heptanone	alarm (?)	35
Solenopsis saevissima (Fr. Smith)	mandibular glands, Dufour's gland	—	trail, alarm	77
Strongylognathus huberi For.	mandibular glands, abdomen	—	alarm	31 32
Tetramorium caespitum (L.)	head (mandibular glands?)	—	alarm	32
	poison gland	—	alarm	31

merous organic substances such as formic acid, butyric acid, caproic acid, and ethylamine; they showed typical alarm and digging behaviour, though the response was weaker than that induced by equal concentrations of the pheromone secreted by the mandibular glands.

Exhaustive studies on the specificity of the alarm substances were recently carried out by Blum et al. (1966) and Regnier and Wilson (1968). Blum et al. (1966) tried out a series of 49 ketones on *Iridomyrmex pruinosus* colonies, both in nature and in the laboratory, to find out the relationship between chemical structure and alarm-inducing power. By increasing the number of carbon atoms from 3 to 13 a lack of activity was shown for the first (C_3–C_4) and the last members (C_{11}–C_{13}) of the 2-alkanone series. Optimal acitivity occurred between C_6–C_9. The natural alarm pheromone, 2-heptanone, was more effective than the other ketones.

Structural variations, such as a displacement of the carbonyl group, the introduction of a second ketone group or the presence of side-chain methyl groups, usually weakened the effect of the substance. While cycloalkanones were ineffective, the introduction of an aromatic ring in the place of a portion of aliphatic chain preserved the activity.

The same relationship between chemical structure and activity was found by Blum and Warter (1966) by trying out a series of ketones on the Dolichoderine *Conomyrma pyramica*, for which 2-heptanone is the natural alarm pheromone.

For *Acanthomyops claviger*, Regnier and Wilson (1968) found that the alkanes falling between C_{10} and C_{13} usually elicited good responses from the workers, and showed excellent properties as alarm substances. Since they have a vapor pressure high enough to generate a sufficiently large active space in a few seconds, they act at low concentrations. (As previously mentioned, one natural pheromone of *A. claviger* is undecane.)

CONCLUSION

The hymenopteran family Formicidae includes about 7,000 species. These social insects show various kinds of development in the elaboration, transmission, and reception of information capable of determining particular social behavior patterns. In this chapter, particular attention has been given to the systems for indicating the trails used to reach hunting and foraging areas, communicating nests (branches), and so on, and to the alarm system as the first factor in a series of reactions leading to complex defensive and offensive action.

In the literature there are more or less detailed studies of at least 76 Formicidae species regarding trail problems, 45 species for alarm problems. However, these reactions seem to be very widely developed throughout the whole group.

To indicate the trail, the chemical information systems use substances (pheromones) elaborated by various organs, for example the digestive tube,* Dufour's

---

* This calls to mind the case of attraction substances for adults, males and females, produced by the intestine of the Coleoptera of the *Ips* genus and *Dendroctonus* living in the trunks of conifers: the presence of attraction substances in the excrement is interpreted as a general call to adults to assemble for mating in a place suitable for the feeding and rearing of the offspring (Pavan and Quilico, 1968).

gland, and Pavan's gland (a typical specific trail organ in Dolichoderinae). The problem of the vicarious organ for the trail substance is still unsolved as regards those Dolichoderinae which have a habit of moving in columns along well-defined trails, even though they do not possess such glands.

The glands producing informative substances with an alarm function are the mandibular glands, the supra-anal glands (Dolichoderinae), the Dufour's gland, and the venom apparatus.*

As yet, the trail-laying function seems to be fulfilled by just one organ per species (either the digestive tube, or the Dufour's gland, or the Pavan's gland, etc.)

The alarm function can be developed contemporaneously, in one species, by different organs (for example the mandibular glands, the Dufour's gland and the venom gland in some Formicinae species).

Experiments carried out by various authors to check on the attraction validity of the trail substance for the producing species and possibly for other species, revealed that sometimes the trail is strictly specific, other times it is followed by species of the same genus or even of a different genus. However, the normal condition is that the trail is specific.

On the other hand, alarm pheromones are usually not specific.

Precise data on trail pheromone chemical structure are not available.† However, the producing organs have been identified and we expect their determination soon. We do not know whether the substances emitted from the digestive tube which function as trail pheromones are simply excrements or particular secretions produced to this end.

Also the trail pheromone produced by the Dufour's gland of three *Solenopsis* species is not chemically known. On the other hand, various substances secreted by the Dufour's gland of other species are chemically known, but in their case no tests have been carried out on their possible function as trail pheromones.‡ However, in various cases the Dufour's gland has been proved to produce alarm substances.

Fifteen ant alarm substances have been chemically identified and derive from 13 species. The substances include aldehydes, ketones, sulfides, furans, alkanes, and acids. Some of them have not been experimented in isolation, but only as components of the raw secretion (e.g., farnesal, perillene, for *Lasius (Dendrolasius) fuliginosus*). The alarm pheromones, as we have seen, may be produced not only by the Dufour's gland but also by the mandibular glands, the venom gland, and the supra-anal glands. In several cases, the products from these glands have been shown to possess other social properties as well (e.g., for defense, trail-laying).

---

* It should be remembered that in various Apidae (e.g., *Apis mellifera* L.) it is the Nassanoff gland which functions typically as a trail substance elaborator, in others (as in some *Trigona* species) it is the *mandibular* glands.
† In Termitidae the trail substance was studied in *Microcerotermes edentatus* Was. and in *Calotermes flavicollis* (F.) and identified as cis-3-exen-1-ol.
‡ See Bergström and Löfquist, 1968.

REFERENCES

1. Bergström, G., and Löfquist, J. 1968. Odour similarities between the slave-keeping ants *Formica sanguinea* and *Polyergus rufescens* and their slaves *Formica fusca* and *Formica rufibarbis*. *J. Insect Physiol.*, 14(7):995-1011.

2. Bernardi, R., et al. 1967. On the components of secretion of mandibular glands of the ant *Lasius* (*Dendrolasius*) *fuliginosus*. *Tetrahedron Lett.*, (40):3893-3896.

3. Blum, M. S. 1966. The source and specificity of trail pheromones in *Termitopone*, *Monomorium*, and *Huberia* and their relation to those of some other ants. *Proc. Roy. Entomol. Soc.* (Lond.), 41(10-12):155-160.

4. —— Moser, J. C., and Cordero, A. D. 1964. Chemical releasers of social behavior. II. Source and specificity of the odor trail substances in four Attine genera. (*Hymenoptera Formicidae.*). *Psyche*, 71(1):1-7.

5. —— and Portocarrero, C. A. 1964. Chemical releasers of social behavior. IV. The hindgut as the source of the odor trail pheromone in the neotropical army ant Genus *Eciton*. *Ann. Entomol. Soc. Amer.*, 57(6):793-794.

6. —— and Portocarrero, C. A. 1966. Chemical releasers of social behavior. X. An Attine trail substance in the venom of a non-trail laying Myrmicine, *Daceton armigerum* (Latreille). *Psyche*, 73(2):150-155.

7. —— and Ross, G. N., 1965. Chemical releasers of social behaviour. V. Source, specificity, and properties of the odour trail pheromone of *Tetramorium guineense* (F.) (*Formicidae: Myrmicinae*). *J. Insect Physiol.*, 11:857-868.

8. —— and Warter, S. L. 1966. Chemical releasers of social behavior. VII. The isolation of 2-heptanone from *Conomyrma pyramica* (*Hymenoptera: Formicidae: Dolichoderinae*) and its *modus operandi* as a releaser of alarm and digging behavior. *Ann. Entomol. Soc. Amer.*, 59(4):774-779.

9. —— Warter, S. L., Monroe, R. S., and Chidester, J. C. 1963. Chemical releasers of social behaviour. I. Methyl-*n*-amyl ketone in *Iridomyrmex pruinosus* (Roger) (*Formicidae: Dolichoderinae*). *J. Insect Physiol.*, 9:881-885.

10. —— Warter, S. L., and Traynham, J. G. 1966. Chemical releasers of social behaviour. VI. The relation of structure to activity of ketones as releasers of alarm for *Iridomyrmex pruinosus* (Roger). *J. Insect Physiol.*, 12:419-427.

11. —— and Wilson, E. O. 1964. The anatomical source of trail substances in Formicine ants. *Psyche*, 71(1):28-31.

12. Bossert, V. H., and Wilson, E. O. 1963. The analysis of olfactory communication among animals. *J. Theor. Biol.*, 5:443-469.

13. Brown, W. L., Jr. 1960. The release of alarm and attack behavior in some new world army ants. *Psyche*, 25-27.

14. Butenandt, A., Linzen B., and Lindauer, M. 1959. Über einen duftstoff aus der mandibeldrüse der blattschneiderameise *Atta sexdens rubropilosa* Forel. *Arch. Anat. Micr. Morph. Exp.*, 48:13-20.

15. Carthy, J. D. 1950. Odour trails of *Acanthomyops fuliginosus*. *Nature*, 166:154.

16. —— 1951a. The orientation of two allied species of British ant. II. Odour trail laying and following in *Acanthomyops* (*Lasius*) *fuliginosus*. *Behaviour*, 3(4):304-318.

17. —— 1951b. The return of ants to their nest. *Trans. 9th Int. Congr. Entom.*, 1:365-369; 1952.

18. —— 1966. Insect communication. *Sympos. Roy. Entom. Soc.* (Lond.) (3) *In* Insect Behaviour. Haskell, P. T., ed. London, pp. 69-80.

19. Casnati, G., Pavan, M., and Ricca, A. 1964. Ricerche sul secreto delle glandole anali di *Liometopum microcephalum* Panz. (*Hymenoptera: Formicidae*). *Boll. Soc. Entom. Ital.*, 94(9-10):147-152.

20. —— Ricca, A., and Pavan, M. 1967. Sulla secrezione difensiva delle glandole mandibnolari di *Paltothyreus tarsatus* (Fabr.) (*Hymenoptera: Formicidae*). *Chim. Ind.* (Milano), 49(1):57-58.

21. Cavill, G. W. K., and Robertson, P. L. 1965. Ant venoms, attractants, and repellents. *Science*, 149 (3690): 1337-1345.

21a. —— Ford, D. L., Locksley, H. D. 1956. Iridodiae and iridolactone. *Chem. and Industry*, pp. 465.

22. Chadha, M. S., Eisner, T., Monro, A., and Meinwald, J. 1962. Defence mechanisms of Arthropods. VII. Citronellal and citral in the mandibular gland secretion of the ant *Acanthomyops claviger* (Roger). *J. Insect Physiol.*, 8:175-179.

23. Eisner, T., and Meinwald, J. 1966. Defensive secretions of Arthropods. *Science*, 153(3742):1341-1350.

24. Fletcher, D. J. C., and Brand, J. M. 1968. Source of the trail pheromone and method of trail laying in the ant *Crematogaster peringueyi*. *J. Insect Physiol.*, 14(6):783-788.

25. Gabba, A. 1967. Aspetti dell'organizzazione negli insetti sociali. II. La sostanza della traccia nei *Formicidae*. *Natura*, 58(2):150-172.

25a. Ghent, R. L. 1961. Adaptive refinements in the chemical defense mechanism of certain Formicinae. Thesis, Cornell Univ., Ithaca, N. Y., pp. 1-88.

26. Ghidini, G. M. 1965. Le armi chimiche nelle formiche. *Atlante*, (7):46-55.

27. Hangartner, W. 1967. Spezifität und Inaktivierung des Spurpheromons von *Lasius fuliginosus* Latr. und Orientierung der Arbeiterinnen im Duftfeld. *Z. Vergl. Physiol.*, 57:103-136.

28. Hangartner, W., and Bernstein, S. 1964. Über die Geruchsspur von *Lasius fuliginosus* zwischen Nest und Futterquelle. *Experientia*, 20:392-393.

29. Law, J. H., Wilson, E. O., and McCloskey, J. A. 1965. Biochemical polymorphism in ants. *Science*, 149(3683):544-546.

30. Lindauer, M. 1962. Ethology. *Ann. Rev. Psychol.*, 13:35-70.

31. Maschwitz, U. W. 1964a. Gefahrenalarmstoffe und gefahrenalarmierung bei sozialen hymenopteren. *Z. Vergl. Physiol.*, 47:596-655.

32. Maschwitz, U. W. 1964b. Alarm substances and alarm behaviour in social Hymenoptera. *Nature*, 304 (4956):324-327.

33. Maschwitz, U. W. 1966. Alarm substances and alarm behavior in social insects. *Vitamins, Hormones*, 24:267-290.

34. McGregor, E. G., 1948. Odorous as a basis for oriented movement in ants. *Behavior*, 1:267-296.

35. McGurk, D. J. et al. 1966. Volatile compounds in ants: identification of 4-methyl-3-heptanone from *Pogonomyrmex* ants. *J. Insect Physiol.*, 12:1435-1441.

36. Miradoli Zatti, M. A., and Pavan, M. 1957. Studi sui *Formicidae*. III. Nuovi reperti dell'organo ventrale nei *Dolichoderinae*. *Boll. Soc. Entom. Ital.*, 87 (5-6):84-87.

37. Moser, J. C., and Blum, M. S. 1963. Trail marking substance of the Texas leaf-cutting ant: source and potency. *Science,* 140(3572):1228.

38. Moser, J. C., Brownlee, R. C., and Silverstein, R. 1968. Alarm pheromones of the ant *Atta texana. J. Insect Physiol.,* 14(4):529-535.

39. Pavan, M. 1948a. Iridomirmecine, antibiotique nouveau extrait de la Fourmi argentine Paris, *XIII Congr. Int. Zool.,* pp. 500-501.

40. ———— 1948b. Iridomyrmecin, an antibiotic substance extracted from the argentine ant (*Iridomyrmex pruinosus humilis* Mayr). *VIII Int. Congr. Entom.,* Stockholm: 863-865.

41. ———— 1949. Ricerche sugli antibiotici di origine animale: Nota riassuntiva. *Ric. Sci.,* 19(9):1011-1017.

42. ———— 1950a. Potere insetticida della "iridomirmecina" e significato della sostanza nella biologia di *Iridomyrmex humilis* Mayr (Formica argentina). *Ric. Sci.,* 20(12):1853-1855.

43. ———— 1951. "Iridomyrmecin" as insecticide. *IX Int. Congr. Entom.,* 1:321-327.

44. ———— 1952. Primo contributo sperimentale allo studio farmacologico della iridomirmecina. *Arch Int. Pharmacodyn.,* 89(2):223-228.

45. ———— 1955a. Gli insetti come fonte di prodotti biologicamente attivi. *Chim. Ind.* (Milano), 37(9):714-725.

46. ———— 1955b. Studi sui *Formicidae.* I. Contributo alla conoscenza degli organi gastrali nei *Dolichoderinae. Natura* (Milano), 46:135-145.

47. ———— 1956. Studi sui *Formicidae.* II. Sull'origine, significato biologico e isolamento della dendrolasina. *Ric. Sci.,* 26(1):144-150.

48. ———— 1950b. Significato chimico e biologico di alcuni veleni di insetti. Ed. Tipografia Artigianelli, Pavia: 1-75.

49. ———— 1958, 1959. Biochemical aspects of insect poisons. IV. Int. Congr. Biochem., XII: Biochem. of Insects, Pergamon Press, London: 15-36.

50. ———— 1960a. Estrazione e purificazione di alcuni componenti delle secrezioni difensiva degli Artropodi. *XI Int. Congr. Entom.,* Wien, Verh. Bd. III: 276-283.

50a. ———— 1960b. Sviluppi delle ricerche sulle secrezioni di insetti. *Atti Accad. Naz. It. Entom., Rend.,* 8:228-242.

51. ———— 1961. Données chimiques et biologiques sur les secretions des *Formicidae* et *Apidae. Atti IV Congr. U.I.E.I.S.,* 12:19-37.

51a. ———— 1962. Aspectes du comportement chez les fourmis. *Atti IV Congr. U.I.E.I.S.,* 12:122-131.

52. ———— and Quilico, A. 1968. Prospettive di controllo degli insetti nocivi con sostanze naturali agenti sul loro comportamento. Convegno Internazionale su "Nuove prospettive nella lotta controgli Insetti nocivi," Roma, Accad. Naz. dei Lincei, 16-18 settembre 1968 (in press).

53. ———— and Ronchetti, G. 1955. Studi sulla morfologia esterna e anatomia interna dell'operaia di *Iridomyrmex humilis* Mayr e ricerche chimiche e biologiche sulla iridomirmecina. *Atti Soc. Ital. Sci. Nat.,* 94(3-4):379-477.

54. ———— and Trave, R. 1958. Etudes sur les *Formicidae.* IV. Sur le venin du Dolichodéride *Tapinoma nigerrimum* Nyl. *Insectes Sociaux,* 5(3):299-308.

55. Quilico, A., Piozzi, F., and Pavan, M. 1956. Sulla dendrolasina. *Ric. Sci.,* 26(1):177-180.

56. ———— Piozzi, F., and Pavan, M. 1957a. Ricerche chimiche sui *Formicidae;*

sostanze prodotte dal *Lasius (Chthonolasius) umbratus* Nyl. *Rend. Ist. Lomb. Sc. Lett., Cl. Sci.,* 91:271-279.

57. ——— Piozzi, F. and Pavan, M. 1957b. The structure of dendrolasin. *Tetrahedron Lett.,* 1(3):177-185.
58. Regnier, F. E., and Wilson, E. O. 1968. The alarm-defence system of the ant *Acanthomyops claviger.* *J. Insect Physiol.,* 14(7):955-970.
59. Schneirla, T. C. 1940. Further studies on the army-ant behaviour pattern: Mass organization in the swarm-raiders. *J. Comp. Physiol. Psychol.,* 29:401-460.
60. ——— 1956. A preliminary survey of colony division and related processes in two species of terrestrial army ants. *Insectes Sociaux,* 3:49-69.
61. ——— 1960. The army ants. *Smithsonian Treasury Science.* New York, Simon and Schuster, Inc., pp. 664-696.
62. ——— 1963. The behaviour and biology of certain nearctic army ants: Spring-time resurgence of cyclic function—southeastern Arizona. *Anim. Behav.,* 11(348):583-595.
62a. ———, and Brown, R. Z. 1950. Army ant life and behaviour under dry season conditions. *Amer. Mus. Nat. Hist.,* 95:263-354.
63. ——— and Reyes, A. Y. 1966. Raiding and related behaviour in two surface-adapted species of the old world Doryline ant, *Aenictus. Anim. Behav.,* 14(571):132-148.
64. Sudd, J. H. 1957. A response of worker ants to dead ants of their own species. *Nature* (London), 179(4556):431-432.
65. ——— 1959. Interaction between ants on a scent trail. *Nature* (London), 183 (4675):1588.
66. ——— 1960. The foraging method of Pharaoh's ants, *Monomorium pharaonis* (L.). *Anim. Behav.,* 8:67-75.
67. ——— 1958. The foraging methods of some Myrmicine ants in Nigeria. XVth *Int. Congr. Zool.,* Sect. XI, Paper 18.
68. ——— 1967. An introduction to the behaviour of ants. London, Arnold E. Ltd. pp. 1-200.
69. Tanyolaç, N. ed. 1968. Theories of odors and odor measurement. Robert Coll. Res. Center, Bebek, Istanbul, Turkey 1-572.
70. Trave, R., and Pavan, M. 1956. Veleni degli Insetti. Principi estratti dalla formica *Tapinoma nigerrimum* Nyl. *Chim. Ind.* (Milano), 38:1015-1019.
71. Walsh, C. T., Law, J. H., and Wilson, E. O. 1965. Purification of the fire ant trail substance. *Nature* (London), 207(4994):320-321.
72. Watkins II, J. F. 1964. Laboratory experiments on the trail following of army ants of the genus *Neivamyrmex (Formicidae: Dorylinae). J. Kansas Entom. Soc.,* 37(1):22-28.
73. ——— Cole, T. W., and Baldridge, R. S. 1967. Laboratory studies on inter-species trail following and trail preference of army ants (Dorylinae). *J. Kansas Entom. Soc.,* 40(2):146-151.
73a. Weber, N. A. 1958. Evolution in fungous-growing ants. *Proc. 10th Int. Congr. Entom. 1956 Montreal.,* 2:459-473.
74. Wilson, E. O. 1958. A chemical releaser of alarm and digging behavior in the ant *Pogonomyrmex badius* (Latreille). *Psyche,* 65(2-3):41-51.
75. ——— 1959a. Source and possible nature of the odor trail of fire ants. *Science,* 129(3349):643-644.
76. ——— 1959b. Pheromones in the organization of ant societies. *Anat. Rec.,* 134 (3):653.

77. ——— 1962a. Chemical communication among workers of the fire ant *Solenopsis saevissima* (Fr. Smith). I. The organization of mass-foraging. *Anim. Behav.*, 10(1-2):134-147.

78. ——— 1962b. Chemical communication among workers of the fire ant *Solenopsis saevissima* (Fr. Smith). II. An information analysis of the odour trail. *Anim. Behav.*, 10(1-2):148-158.

79. ——— 1962c. Chemical communication among workers of the fire ant *Solenopsis saevissima* (Fr. Smith). III. The experimental induction of social responses. *Anim. Behav.*, 10(1-2):159-164.

80. ——— 1963a. Pheromones. *Sci. Amer.*, 208(5):100-114.

81. ——— 1963b. The social biology of ants. *Ann. Rev. Entom.*, 8:345-368.

82. ——— 1965a. Trail sharing in ants. *Psyche*, 72(1):2-7.

83. ——— 1965b. Chemical communication in the social insects. *Science*, 149(3688): 1064-1071.

84. ——— 1966. Gli insetti comunicano mediante sostanze chimiche. *Sapere*, 673:35-39.

85. ——— and Bossert, V. H. 1963. Chemical communication among animals. *Rec. Progr. Hormone Res.*, 19:673-716.

86. ——— and Pavan, M. 1959. Glandular sources and specificity of some chemical releasers of social behavior in Dolichoderine ants. *Psyche*, 66(4):70-76.

DISCUSSION

ROTH: Dr. Wilson will be glad to answer questions directed to Dr. Gabba's paper.

BURGHARDT: Have any trail substances been identified chemically?

WILSON: No. The problem has proven intractable. In the past several years, for example, we have extracted over 2 million fire ants. Last year we finally pinpointed the trail pheromone at a consistent identifiable peak. This revealed that the pheromone is present in the Dufour's gland of individual ants in less than nanogram quantities. Even more remarkable, when we forced an entire fire ant colony to migrate across a bridge, a movement involving trail-laying by hundreds of thousands of individuals for over an hour, we still recovered less than 1 nanogram of trail substance in an efficient odor trip. This is why no trail substances have been identified. Dr. B. P. Moore has had the same experience in his attempts to identify the *Nasutitermes* trail substance, mentioned by Dr. Stuart earlier. Moore found that the pheromone, a diterpene produced by the sternal glands, is present in nanograms or less per termite. But I am optimistic about the future. Chemists have now perfected the technique of coupled gas chromatography–mass spectrometry, which makes it possible to identify microgram quantities of low molecular-weight substances. So it should be possible to obtain enough of the trail pheromone to go ahead with chemical analysis. I predict that we will see the first identification of a trail substance very soon. Then such identifications will become as routine in the case of the alarm substances, which have hitherto been more accessible only because they are present in larger quantities.

UNIDENTIFIED VOICE: Are these substances stored in reservoirs?

WILSON: Yes, they are; that is the surprising thing. You can see liquid in the organs mentioned by Dr. Gabba. This is true for Pavan's gland and also for Dufour's gland. Evidently, the real trail substances themselves, the pheromones, while identifiable as a single peak in a chromatogram, make up only a small fraction of the visible liquid contents.

UNIDENTIFIED VOICE: Would it make sense to pick out the glands?

WILSON: No, not individually. I don't think you would get enough material without an inordinate amount of labor, in the system so far investigated.

STUART: Dr. Gabba, I wonder if you would like to comment on what the actual function of an alarm substance is, in the light of Maschwitz's idea that an alarm substance should not actually elicit attack; he says attack is reserved for the enemy.

GABBA: Could you elaborate, please?

STUART: Maschwitz (1967)* maintains that alarm substances should merely transmit excitation rather than induce an animal to snap, lunge, and attack. As Maschwitz says, it should just excite an animal, not induce it to attack.

GENERAL DISCUSSION
following chapters 4 through 7

STUART: I would like to ask Dr. Wilson what his idea of an alarm substance in a social insect really is. Can we formulate a good definition of it and consider the possible biologic significance of the fact that specific species have certain alarm substances? I ask this because no one has found, until recently, an alarm substance in termites, and I feel that it is not quite necessary in most termites to have a volatile alarm substance; contact chemoreception or contact reception seem more suited to the environment in which most termites live. With the ants and bees that forage at greater distance, perhaps there is more reason for volatile substances.

WILSON: The expression "alarm" has come to be used for a distressingly large variety of pheromone-mediated behavioral responses. It is used in an obviously appropriate way for the fright reactions of fish and the frenzied responses of many ant species to their alarm substances. It is also applied to responses that involve initial attraction to the source of the alarm substances, which may or may not be followed by frenzy and aggression. So, if we attempt to employ the ordinary definition of alarm as applied to human communication and human emotion, we find that some of the responses that have been placed in the category of alarm in animal behavior fit well, and others do not. The classification of substances into alarm substances or non-alarm substances is therefore a rather poor one. There are also some examples of species which employ the same pheromones to alarm, to recruit to a food source, or to something relatively agreeable. Yet we should probably wait until our knowledge of chemical communication is more complete before devising a new classification.

STUART: Are there any other comments on alarm substances?

BARDACH: Yes, but in a somewhat different way, in reference to something that Dr. Le Magnen said, talking about the alarm reaction of fish. In the large group of

---

* Alarm substances and alarm behaviour in social insects. *In* Harris, Loraine, and Wool, eds. Vitamins and Hormones. London, Academic Press, Inc.

fish where alarm substances (the *schreckstoff* of von Frisch), have been shown to exist, that is, in the Stariophysie, the material is released when the skin is broken following attack by a predator. This substance is not a stress product due to a change in the physiologic state of the animal. If it were, it could hardly be produced instantaneously so as to elicit immediate dispersal of a school when the subject is attacked.

LE MAGNEN: I am referring to your own work, Dr. Bardach, about the fact that a little fish after having contact with the big fish is now more recognized by its individual odor. Its individual odor is marked by a new odor consecutive to the fear provoked by the contact with the big fish, provided that I understand the true sense of your work.

BARDACH: Yes, but this change in status and associated substances, which takes some time to build up, are different from the very specific *schreckstoff* of von Frisch. There may be a connection between the *schreckstoff* which has only been found in one large group of fish and the reaction that we have observed (a change in status changes the odor of an animal) since the fish among which both phenomena were observed, belong to the same group. It is possible that the *schrecksoff sensu strictu* that leads to the dispersal of the school, developed from a much more versatile social substance, or substances of social significance, in such groupings as we observed them. Of course, this is speculation.

MASON: I would like to direct this to Dr. Wilson. Is it possible in the case of the ants, that the alarm reaction elicited by chemicals (which you listed on the board) is due to their interference with the reception of chemicals that are being produced constantly; that is, could they have their effect by interfering with a chemical that maintains social equilibrium? The reason that I ask that is that there has been considerable emphasis on specificity here and yet those chemicals you showed, indicate to me a lack of specificity insofar as their structures are concerned.

WILSON: The alarm substances shown of *Acanthomyops claviger*, (which I listed), are not species specific, and this is true for ants generally. Most of the *Acanthomyops* substances are very efficient alarm substances. They have thresholds of between 10^{10} and 10^{12} molecules per cc, which is a high threshold for an odorant. But the quantity of these substances released by individual workers (0.01 to 1 mg) make that level of threshold necessary in order to generate a signal on the order of one centimeter. If the threshold concentrations were much below 10^{10} and 10^{12} molecules per cm^3, then the signal would extend beyond a centimeter, and would be very long lasting, throwing most, or all, of the colony into tumult. On the other hand, higher threshold concentrations would create signals that would be less than a centimeter in reach, and only seconds in duration, hardly enough to permit communication to other workers. Thus, the system seems to have been well engineered in evolution.

Your other question could be restated as whether the effect is inhibitory rather than stimulatory. We do not know the answer to that. It could be settled by electro-antennogram analysis or some other more sensitive form of neurophysiologic study.

STUART: Perhaps I could make a comment here on termites. Dr. Mason was talking about whether there is a disruption in the colony odor or whether there is a disruption in the ants' perception. In termites, if a drop of distilled water or some other seemingly innocuous substance is introduced into a nest, then the change in odor will elicit a response which might vary from attack to a purely "jerk" reaction, and perhaps to the elimination by fecal disposition of whatever odor there was. In termites, I think you are getting a response to something that has a different odor than that of the normal environment. This is why I'm a little bit worried about what alarm is, because in some instances alarm is used synonymously with attack. An essay for alarm that has been described in some recent work is as follows: a piece of filter paper dipped in the substance to be tested is presented to a termite; if a frenzy and attack are elicited, then the test is judged positive. Now the trouble is that in termites, if one dips a piece of filter paper in some phenol solution, smears it by rubbing with one's hand (or contaminating it in some other way), and presents that to a termite, the elicited attack will be greater than that usually found in a control experiment using a piece of filter paper, either wetted by distilled water or dry. So phenol will induce a termite to go into a frenzy and attack; but I do not think phenol is an alarm substance. Are there any other comments?

MASON: A word of warning: I do not know how you got your distilled water, but chemists know it is a problem to get pure distilled water; we are talking about things down to the level of 10^{-15} or 10^{-18}. For this reason, I do not believe that there is such a thing as *pure distilled water*; you can de-ionize it and glass-distill it in the presence of oxidizing agents such as permanganate which will knock out all organic compounds and still get impure water. *Storage in glass containers also causes impurity.* Thus, your distilled water may well be impure.

STUART: Yes, but that is my point: any odor can induce attack.
Are there any questions specifically directed to Dr. Stürckow?

JOHNSTON: According to your hypothesis, Dr. Stürckow, the flow of neutral current is generated in the viscous material; is that correct?

STÜRCKOW: Yes, the morphologic investigations of Dr. Adams and myself, together with Miss Holbert, on the tip of the hair, demonstrated that the dendrites end 0.5μ from the outside of the hair. Thus, a molecule would have to diffuse over this distance through viscous substance, which would take quite a bit of time and would not explain the first impulse within 1 m/sec, as it has been found by Barton, Browne, and Hodgson. Within this 1 m/sec a receptor potential has to be built up; it must be transferred to the cell body (which in some cells is 300μ away from the tip) and the impulse must be generated. One m/sec is too short a time for an enzymatic mechanism and the transfer of a potential over 300μ. Depolarization or a resistant-change seem to be a more likely receptive mechanism. The work of Rosenberg and his associates supports my suggestion. His idea is that beta-carotene, vitamin A, or some other high-molecular substances, which are embedded in the receptive membrane, give a fast current change at the moment of absorption of ammonia or other odorous substances.

JOHNSTON: When one considers that there are two *known* parameters to the olfactory system, quality and quantity, how would you speculate that the flow of current codes both of these?

STÜRCKOW: That is a very difficult question.

JOHNSTON: Yes, it is. One variable may be the time of onset of flow.

STÜRCKOW: If this viscous substance proves to be the chemoreceptive substance, it might discriminate between different chemicals by having embedded different vitamins, which absorb different substances. How this differentiation is transmitted to the different dendrites, is difficult to tell, although several suggestions are possible. I mainly think of this one: we have two channels—the one is filled with dendrites, the other has no dendrites, except that at the base there is one denrite. This dendrite has been interpreted as a mechanoreceptive one, but I have evidence to doubt it. The third dendrite at the base might also receive chemical stimuli through a current, and this in a faster or slower way than the other dendrites. This might cause a time delay on either side, which could be used for differentiation. I think we are 10 or 20 years away from the answer, but we shall reach it.

JOHNSTON: I would like to refer to your [Stuart] social termites. That very interesting diagram towards the end of your lecture summarized two techniques which you thought were doubtful, and then you went on to explain a preferred technique for trail-laying. Will you please give me a review of the technique you approve?

STUART: Well, as far as trail-laying bioassays are concerned, I approve of the technique that I think was first instituted by Wilson for ants. You try to work on a whole colony of insects and disrupt the normal behavior by laying a trail from these insects away from them. If they follow it in a very immediate and characteristic manner, it is a trail pheromone. In ants, Wilson used a lure or bait (a meal worm I think it was). He got an aggregation around this meal worm so that any stimulus which was then imposed on these ants would have to have a positive effect to be noticed; the ants would have to give up feeding, as it were, and move along the trail. Now in termites one has to use a somewhat different techinque, since meal worms cannot be fed to termites. A little bit of the carton nest, or a little bit of the wood on which the termites have been feeding, however, causes aggregation, and trails can then be drawn from that. Control trails are also drawn and the response is observed. However, you are not giving the termite the choice of following only the one odor or the one substance that is being tested, and nothing else. You are actually giving it a choice between the experimental substance and all the other odors and orienting factors such as light, heat, and contact. Now, the two types of bioassay that I do not think are correct for trail-laying substances are the simple attractancy choice tests ("A or B"), with little pieces of filter paper with the extract on them; nor a drawn-out trail on a piece of very clean glass plate with the trail of 2 or 3 inches for the termite to follow individually up and down, once or twice. Other chemicals which are not repellent can give the same result. You cannot get it with the Wilson technique and also with my own technique with termites.

JOHNSTON: Thank you for the review. With respect to insect pheromones and the general subject of communication, I was impressed by one idea today which contrasts with something that was explained yesterday. Today, there was clear evidence that some insect pheromones are, in fact, a given single odorant. Now, it seems plausible that one chemical compound is functional, especially in a solitary insect species as it does not require identification of individual members. Dr. Whitten and I have discussed whether the mouse pheromone is a single substance or a small melange. I favor the notion that it may well be a mixture of odorants because this would provide the individualistic "signature" that would identify another member of the species. It seems to me that insects do not require the same quantity or elegance of information for their means of communication. They do not need to recognize individuals in many situations. Hence, the presence of a single odorant compound is functionally quite plausible, as well as an established fact for some species.

WILSON: I would like to add a little to Dr. Johnston's comment. He has touched on what I regard as a fundamental distinction between insect societies and vertebrate societies. All the evidence so far shows that vertebrate secretions tend to be complex mixtures, while invertebrate pheromones tend to be simple mixtures or comprised of one component. The difference may have the following significance: We know that organization in most vertebrate societies involves individual recognition based on memory. This recognition is the basis for the leadership phenomenon, for the strictly ordered dominance hierarchies which are so important in most vertebrate societies, and for prolonged parent–offspring relationships. In insect societies, on the other hand, there is seldom any true leadership, at least of an enduring kind. Leadership in foraging columns of army ants, for example, constantly changes. Dominance hierarchies occur widely in hymenopteran societies, but so far as we know they are based more on such physiologic conditions as ovarian development and size than on recognition of individuals. In fact, there is no evidence, of which I am aware, that individual recognition exists among the workers of an insect society. In fact, organization based on such recognition is improbable—even if it does occur—simply because most insect societies contain hundreds, thousands, or even millions of individuals each of which is rather shortlived. Parent–offspring relationships are never individualized in insect societies except during colony-founding. In larger colonies, it is better described as a giant crèche system, usually with the queen as the single progenitor of the whole lot. So I would agree with Dr. Johnston that the *vertebrates that need individual recognition tend to employ a very complex chemical signal towards this end. Insects, on the other hand, having neither the brains nor longevity for this degree of inter-individual relationship, find it quite adequate to employ single substances.*

STÜRCKOW: I cannot agree completely with this after having watched the behavior of male cockroaches. They differentiate between females and, apparently, not solely by means of different sex pheromone levels. Vision might also play a role, whereas auditory stimuli probably are not included; but is it important whether recognition occurs through the perception of sex pheromones, through vision, or through auditory signals? We think that because we have observed individual recognition, it is certain that in vertebrates (and higher animals) the whole sensory perception is considerably broader than in insects. In insects, the stimulation of a single neuron can elicit a behavioral response which would never be possible in vertebrates, but on the basis of this we cannot completely deny that there could be recognition in insects. Even if thousands of insects live in one state, they meet every day maybe 5 or 10 which are important to them. If I work, I meet 2 or 3 people a day who are important to me, and do not know the thousands of people in town. I do not see too much of a difference in personal recognition between certain insects and vertebrates. The relationship surely will be of different types!

WILSON: I think that is too much to believe for insect societies. There is no evidence that individual colony members have such personal relationships. Let me go further and say that the distinction between the vertebrate society, which is individualistic in construction, and the insect society, which is not, also involves personal relationships that are based on experience and must therefore be learned. That is, vertebrate A responds to vertebrate B on encountering it, not only because it recognizes it as a distinct individual, animal B, but also because it has had experiences with animal B which it remembers; this forms the basis for the relationship. In insect societies, there are differences in individual relations, but they are based mainly on graded responses determined in a general way by such parameters as size, ovarian development, life stage, pheromone productivity, and so on. There is no reason to believe that it can be based on experience in the past with single individuals who are labelled and recalled.

STÜRCKOW: No. We have watched cockroach males for two years, having about 20 males every day see different females for 10 minutes each. The work was done by a man who was very careful, and I watched occasionally. He learned to know his 20 males and about 30 females, and I learned to know them from the records and short observations. We could predict the behavior of a certain male to a certain female even if they had not seen each other for two months; therefore, memory must play a role.

WILSON: I am not familiar with this case. Has it been published?

STÜRCKOW: No. It was not allowed to be published because it was not statistically evaluated. Statistics could not have helped in this case, unless the work were continued for 10 years, for it was too individualistic. If you work with such fine behavioral observations you cannot put them easily in a scheme. This material was not collected to give proof that there is a memory; it had quite a different aim. The evidence for memory came as a secondary result. You know that bees can learn and have a memory over 8 or 10 days, as shown by von Frisch (flies do not learn, as Dethier has shown). Why should not certain insects have developed a memory of a certain individual?

WILSON: I must remain very skeptical. Remember, too, that there are 30 to 80 thousand bees in a bee colony and a bee has approximately 45 days to live during the busiest season. The same is true of ant colonies, which (in extreme cases) contain a million or more workers with no more than a year or two to live. A single ant can remember up to three or four visual objects for periods of up to four or five days. But this does not seem to be nearly enough to allow it to form individual relationships of any consequence, much less organize an entire colony. I concede that it is theoretically possible that a primitive insect society with long-lived individuals, such as roaches, might be organized to some degree in this fashion, but I hope you would agree with me that this has not been proven—at least there are no data to my knowledge in the literature which establish it. In the studies of individual behavioral ontogenies undertaken so far, that is day-by-day development of behavior of individual insects such as those in colonies of bees by Lindauer and in colonies of ants by myself, there was no indication of this sort of relationship. But I must admit that we did not label all the individuals in the colony, and subtle responses could have escaped us. Also, in the dominance hierarchies of *Bombus* and *Polistes*, it is impossible to say how much the responses are based on memory of individuals.

JOHNSTON: Dr. Stürckow, does your observation of male roaches imply that the olfactory organ of the roach has a very considerable number of receptors like the rough analogy of man and the honey bee? I ask this question because some time ago, Dr. Robert Wright of Vancouver, B.C., speculated that insects, at least the ones that he was particularly interested in (one species of true bug, certain parasitic hymenoptera, and mosquitoes), have relatively few kinds of receptor cells which can respond. Therefore, the insects would be anatomically or neurologically blind to what might be called "extraneous odors" from a biologically useful point of view. Now where do you stand? At one end or the other, or in the middle?

STÜRCKOW: At one end, insects have been shown to develop very strongly either vision, hearing, olfaction, taste, or mechanoreception, or two of these senses. Those insects which have very long antennae or antennal combs with many branches have also several thousand sensory neurons, and should receive a wide spectrum of olfactory stimuli, although the number of neurons is probably much smaller than the number of olfactory neurons in man. Mostly these insects with well-developed olfactory systems are poor in vision, as the cockroach, for example. The cockroach lives in the dark or semi-dark and is guided mainly by olfactory stimuli and also by mechanical stimuli. To come back to the behavior of the cockroach, I was surprised how many different pictures you can get from a cockroach which has two such long antennae (each one is about 5 cm long). They are used for balance and for expressions, which might be expressions of the face or the body, as in some other insect groups. The expressions of the antennae go into the behavioral pictures which were similiar to those which I know from the African toad *Xenopus*. We know that fish have affection for one another—at least Dr. Bardach claimed it—and I would believe that toads have it towards each other. They are very tender and delicate in showing it, moving their bodies in special ways to attract the opposite sex. In the cockroach, the antennae are used not only for olfaction and for the reception of mechanical or temperature stimuli, but also to give behavioral expressions. The grasshopper has a very well-developed auditory system, the bee has very well-developed vision and olfaction, and the cockroach is ideal for olfactory studies. Apparently, there are no cockroaches with short antennae and poor olfaction.

JOHNSTON: You might say that this use of the long, slender antennae in cockroaches reminds us of dancers waving their arms in a disciplined cadence. I have observed the waving of antennae by these insects.

STUART: Well, I do not know if we have resolved whether insects really do recognize one another. There does seem to be a difference of opinion. Dr. Stürckow, by using the electroantennogram, can you find individual differences in an individual cockroach to the odor of different individual cockroaches?

STÜRCKOW: These antennograms are sum phenomena and their height depends on the stimulus and on the site of your electrodes. If your skill is poor, you make poor antennograms. If you try often enough you might have luck and record some better ones, but you should never think you record everything that is going on. You get a poor overall tracing, but still of some value as the visual analogue, the dissection microscope, gives the histologist. If we were to use the electron microscope alone, we would not know where we were working. I consider the antennogram as one clue of many and know very well its limitations.

STUART: I would just like to ask what the variation is, then, with an individual odor in an individual preparation with an electroantennogram—in other words, just how reproducible are the results from individual to individual, keeping most conditions and the odor constant?

STÜRCKOW: I showed one slide in which the same stimulus was given four times, and you have seen that these superimposed antennograms did not differ from each other, but it was only under the condition that neither of the two electrodes were moved. You always can refer only to your present site of electrodes and must work with different antennae often enough to make sure that this concentration gives this effect. Even under such constant conditions, the antennogram is not explicit enough to differentiate between odors of different individuals, since antennograms of the same height can evoke the same, as well as different, sensations. They record only the amount of reception of an odor, but not its quality.

Chemical Communication in Fish*

JOHN E. BARDACH

School of Natural Resources
University of Michigan
Ann Arbor, Mich. 48104

JOHN H. TODD

Woods Hole Oceanographic Institution
Woods Hole, Mass. 02543

INTRODUCTION 205
THE IMPORTANCE OF THE CHEMICAL SENSES IN THE LIVES OF FISH 206
TESTS OF CHEMOSENSORY ACUITY OF FISH 211
EVIDENCE FOR INTRASPECIFIC SIGNALS (PHEROMONES) IN CHEMICAL
    COMMUNICATION IN FISH 215
INTERSPECIFIC CHEMICAL COMMUNICATION 232
DISCUSSION 234
REFERENCES 237

INTRODUCTION

A large variety of fish have successfully invaded and occupied aquatic habitats where vision would be of little assistance in regulating their behavior. In the lives of nocturnal fish and residents of caves, turbid waters, and the ocean depths, the chemical senses may be expected to dominate. Although almost nothing is known of their behavior, the highly developed olfactory and gustatory regions of the brains of many of them suggest that pheromones or chemical signals may be involved in their social behavior and that their chemical senses are important in assessing the environment.

* Certain portions of research described here were supported by P.H.S. Grant NB. 04687.

THE IMPORTANCE OF THE CHEMICAL
SENSES IN THE LIVES OF FISH

Those fish whose behavior has been extensively studied have been primarily visual animals like the salmon, sunfish, and cichlids. Even these fish have been shown to utilize their sense of smell in a variety of ways, from recognizing their home stream during spawning migrations to distinguishing their young from other parents' broods.

Reactions To Distant Water-Borne
Stimuli

The spawning migrations of salmon are among the most spectacular behavioral feats performed by animals. While all of the salmons' senses may act in concert, enabling them to reach their spawning sites, chemosensory information may be the most important factor in guiding the salmon to their natal stream (Hasler, 1966).

In order to demonstrate that olfactory stimuli are of vital importance to migratory silver salmon (O. kisutch), test salmon from two branches of a spawning stream were captured and their sense of smell eliminated by plugging the nostrils. They were then displaced downstream. Hasler (1966) also reports displacing a control group of salmon whose sense of smell was unaltered. Of the recaptured fish from branch A, 77 percent of the 51 test fish and 100 percent of the control (46 fish) had made the correct choice between the two branches. Of the recaptured fish from branch B, 70 percent of the 27 control fish, but only 16 percent of the 19 test fish had chosen correctly. Hasler (1966) concluded that "The data indicate that normal fish were readily able to repeat their original choice at the stream juncture, thus furnishing additional support for the home stream theory. Those with olfactory occlusion, however, were unable to select accurately."

It should be noted that branch A was considerably larger than branch B, suggesting that the salmon were guided by current as well as by olfactory stimuli. Although Hasler's particular conclusions seem sweeping in view of the small numbers of fish involved in his work, it would appear that the tendency of the stream B controls to home was not shared by their anosmic streammates.

Evidence from a variety of experiments supports Hasler's olfactory theory of stream homing. Sockeye salmon respond behaviorally to water from their home stream without any training while remaining inattentive to water from other sources (Fagerlund et al., 1963). Electroencephalographic recordings from olfactory bulbs of homing salmon showed marked increases in activity following the infusion of "home" water into the olfactory sac. Water from sources other than the natal stream produced little or no change in the level of activity of the olfactory bulbs. The optic lobes in the spawning salmon were inactive compared to those of non-migratory rainbow trout (Hara et al., 1965). These findings lend considerable support to the contention that olfactory cues derived from the environment are of overriding importance in the last phases of the Pacific salmon's life.

In the lives of other predominantly visual fish, the sense of smell may be significant. After one month of training to a single positive- and negative-odor test per day, blunt nose minnows, *Pimephalas notatus* (*Hyborhynchus notatus*), were able to discriminate between streams. They continued to make the correct choices throughout the year, suggesting that streams do not lose their chemical identity with seasonal changes. When the olfactory epithelia were destroyed, the minnows no longer responded correctly (Hasler and Wisby, 1951).

To determine the nature of the odorous substance, stream water was collected and evaporated. The residue was ignited, thereby driving off the organic fraction. The ash, consisting of the inorganic fraction of the residue, was presented to the fish, but they were unable to discriminate between the streams. Subsequently more stream water was distilled at 100°C, and neither the resulting distillate nor the rediluted residue elicited a response. Vacuum distillation at 25°C., however, led to recovery of most volatile organic compounds. Rediluted residues of the vacuum process did not cause the fish to react, but the distillate was found to contain the volatile organic substance which provided the fish with the odors representing the streams.

Plants may serve as "signs" or guideposts, particularly for fish that forage at night and return to occupy territories during the day. Walker and Hasler (1949), using techniques similar to the stream-discrimination study mentioned above, found that blunt nose minnows could be trained to choose between two kinds of aquatic plants (*Myriophyllum exalbescens* and *Ceratophyllum demersum*). They were also trained to distinguish a variety of plant combinations. When the sense of smell was destroyed, the minnows no longer responded to plant odors. Even after the minnows were habituated to a specific plant odor they still remained extremely sensitive to others.

A wide variety of fish restrict their movements and occupy home ranges or territories throughout most of their lives. When they are displaced they exhibit an excellent ability to return to their home territories (Gerking, 1959). The longear sunfish (*Lepomis megalotis*), is one such animal. They reside in the pools of small streams and lakes in the midwestern regions of the United States, and remain within their home ranges for several years (Gunning, 1959). If the longear sunfish are displaced several hundred feet, almost all of them are capable of homing, a feat that they continue to achieve even after they are blinded. With their sense of smell impaired, their success in homing is drastically reduced. Blind and anosmic longear sunfish were entirely disoriented (Gunning, 1959).

Home ranges or territories must have characteristic odors in order for blinded sunfish to be able to locate them with precision. It is possible that many fish recognize their home ranges as one section of an olfactory mosaic and that the topography of their environment is learned from chemical rather than from visual or tactile information.

Reaction to Nearby Water-Borne Stimuli

The pathways in the central nervous system of fish, coordinating smell and taste with muscular movements (Herrick, 1907), illustrate that the chemical senses in many groups of fish are the dominant sensory modalities. Responses to chemicals differ from species to species and depend on the nature of the stimulus, e.g., food

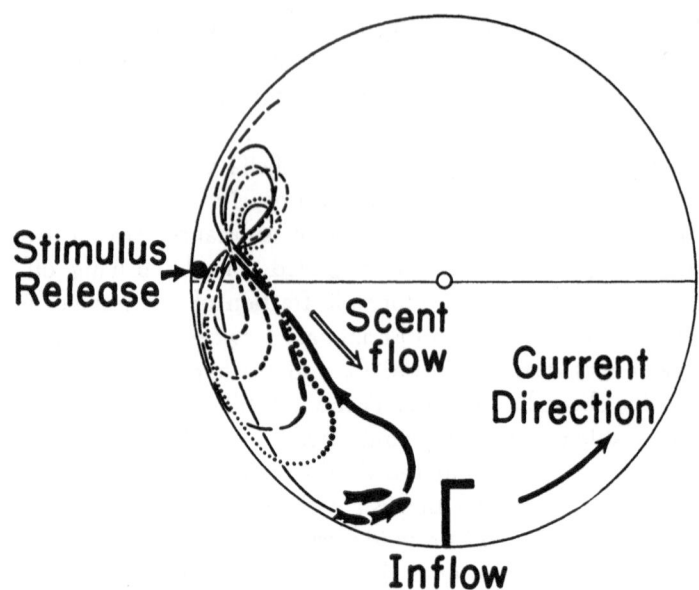

FIG. 1. Scent searching pattern of the tuna (*Euthynnus pelamis*). The lower leg of the figure becomes progressively longer as the scent drifts with the steady current while the speed into the current increases; the thickness of the lines is roughly proportional to the swimming speeds in the different portions of the figure.

scent versus species-specific odor or warning substance. The pattern of responses also differs depending on whether smell or taste is the primary mediator. Responses to food scents are sufficiently varied to warrant a general description as an illustration of a fish's reactions to chemical stimuli.

Where smell rather than taste is employed in food search, traces of a food scent released into a slow, weak current will cause fish to speed up their swimming movements, and to modify their paths (Kleerekoper, 1967). For example, a skipjack tuna (*Euthynnus pelamis*), approximately 50 cm in length, searching for a scent source without visual clues, but in a slow, steady circular current, doubles or triples its normal cruising speed of about 0.75 m/sec. The fish begins to move in and out of the scent cloud in a figure-8 pattern, with the location of the crossing of the loops roughly coinciding with the area where the scent was released (Fig. 1). Tuna swim faster as they approach an odor source, and their speed declines as they swim away from it. Comparison of stimulus strength encountered from moment to moment may be responsible for the changes in swimming speed (Bardach and Gooding, in preparation). It would appear that tuna, without visual cues, are capable of orienting to the source of a scent with considerable efficiency.

Tuna are predominantly visually oriented fish that take moving prey, the final capture of which depends on eyesight. Trout and salmon are also sight feeders though smell plays a strong role in some aspects of their behavior (see above). It will therefore not come as a surprise to trout biologists and fly fishermen to hear that stream dwelling trout have a typical searching reaction to a food scent which differs drastically from that described for the pelagic tuna.

Small rainbow trout (20 cm total length) were kept in a 4 by 2 by 0.5 m tank (the water was held at 0.2 m depth) with a slow flow through it. Some eddies formed, as ascertained by the release of dyes. The trout chose the quietest down-stream corner of the tank as a resting place, somewhat as they would do in a stream. When a trace of filtered liver juice was released upstream, the scent eventually reached the resting trout. As soon as this occurred, the fish faced towards the center of the tank. Lying in a corner of the tank, they selected a number of posi-tions in subsequent moves through a 90-degree arc, staying for from several seconds to almost half a minute at each station, pointing their heads towards the scent and the open water of the tank, as if scanning for the center of the concentration of the scent. While "pointing," as it were, the pectoral fins were moved at dif-ferential speeds in a pattern that caused the water to pass at different velocities over right and left narial openings. We do not know whether or not the trout compares between right and left odor inputs, but catfish, which transport water over their olfactory lamellae by means of cilia, in contrast to the trout's predominantly passive narial water exchange, can regulate the flow through one naris or the other (Chen and Bardach, in preparation). We do know, however, that the trout were able to locate the scent by the maneuvers described above because, after occupying three or four pointing directions, they would often dart in a straight line toward the spot to which the current had by then carried the scent cloud. Both trout and tuna are good swimmers; they live in transparent waters. Their habitats differ, in that fast currents, turbulences, and eddies play a far greater role in the environment of the trout than in that of the tuna. Patterns of olfactory food search are different for each, but are well suited for locating the approximate position of a prey, leaving the final approach and the ingestion to eye-dominated reflexes.

Other fish, such as many of the species in the large order of catfish (Siluri-formes) live in murky waters and make little or no use of vision in feeding or

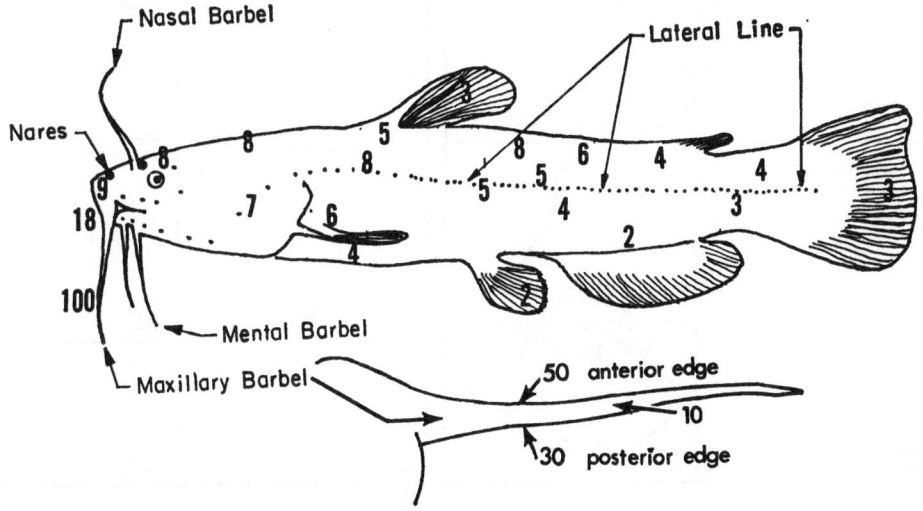

FIG. 2.   Distribution of taste buds on a bullhead catfish (*Ictalurus natalis*). The taste buds were counted under dissection microscope on various square millimeter plots. Note the con-centration of taste buds on barbels and head and the fairly even distribution on the flanks which is somewhat denser dorsally than ventrally.

social interactions. Mechanical sense receptors, including those of the lateral line and electrical perception (Dijkgraaf, 1968), as well as the senses of smell and taste, direct their behavior. In some, taste appears to have become the dominant, and a true distance sense; they have taste buds all over their bodies, especially concentrated on their barbels, reaching several hundred thousand in number on the surface of a single fish (Fig. 2), thereby enabling them to detect and subsequently locate with precision a source of scent that emanates from at least 30 fish lengths away (the limit of the observation tank) (Bardach et al., 1967b). Bullheads (*Ictalurus*) show no loss of the ability to detect and locate scents from afar even when surgically deprived of their sense of smell. At the first taste of a food stimulus the maxillary barbels are spread, and the head is moved slowly to and fro. Then the fish usually starts swimming towards the release point of the scent, accelerating as it goes. During the approach, bullheads hold their barbels erect with the tips of them forming the oval base of a "triangulation cone" that has its apex in the taste buds of the caudal peduncle. Under favorable conditions, that is, in still water where stimulus intensity would diminish concentrically as one proceeds outward from the stimulus source, blinded anosmic bullheads locate the source quickly, swimming toward it in an almost straight path. In a current with eddies, deviations from the straight path, sometimes reminiscent of the figure-8 search described for the tuna, albeit far slower, are the rule. The pattern of searching

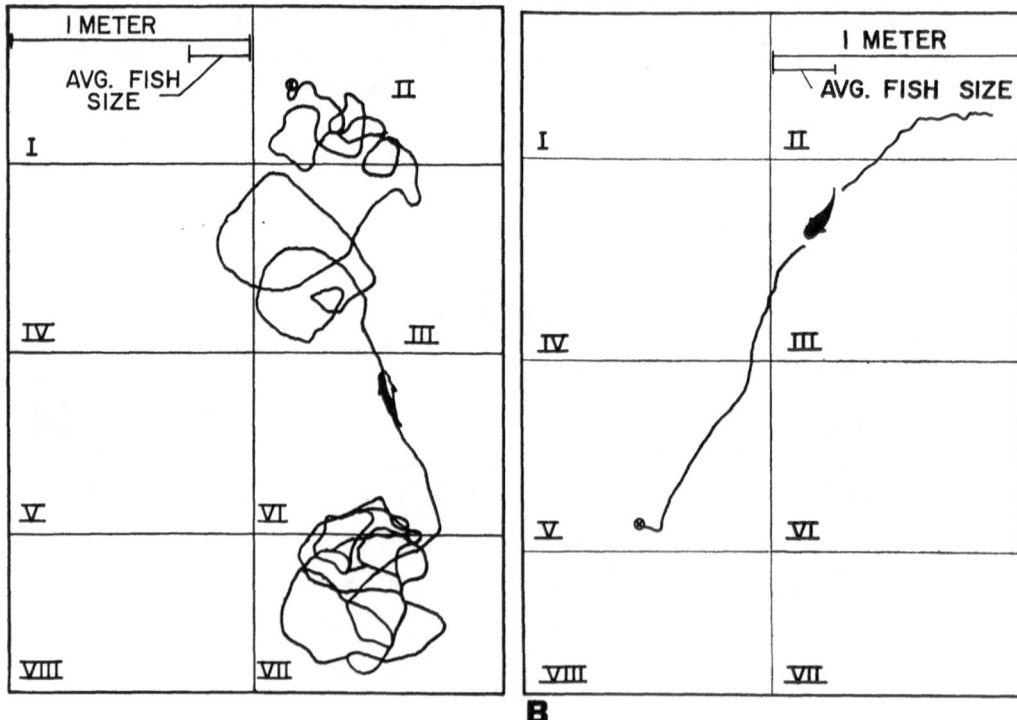

B

FIG. 3. Swimming tracks of blinded bullheads searching for the source of a stimulus. 3A. Fish deprived of taste buds on its right flank. 3B. Fish with taste buds bilaterally intact. (From Bardach, et al. 1967b. *Science*, 155(3767):1276-1278.)

and the arrangement of the taste buds on the body of the fish make it likely that concentrations of chemicals fore and aft, right and left, are compared simultaneously. This interpretation is supported by the behavior of bullheads unilaterally deprived of their taste sensors; such animals circle towards the intact side in seeking food (Fig. 3).

Aside from reacting to stream odors and food scents or tastes, fish show sexual and social reactions to close-up, water-borne pheromones. Pheromones are involved in the formation of pairs of the blind goby, *Typhlogobius californiensis*. After the pair is formed, each mate attacks only members of their own sex which enter their burrow. Parent blind gobies distinguish their young from other broods (Mac-Ginitie, 1939).

In another gobiid, *Bathygobius soporator*, mating occurs after a gravid female releases a pheromone. This courtship pheromone initiates vigorous courting movements in territorial males (Tavolga, 1956). Yet another bond, that between parents and their broods, is maintained by pheromones from the young in several species of cichlids (Kühme, 1964; Künzer, 1964; and Myrberg, 1966).

Recently it has been shown that chemical communication is essential for the social lives of yellow bullheads, *Ictalurus natalis* (Todd et al., 1967, and Todd, 1968). The social organization of the bullhead is hierarchic and territorial, and their communities are based upon their ability to recognize each other as individuals by means of pheromones. There is some evidence indicating that changes in status are also transmitted by chemical cues. Future research will probably show that pheromones contribute to the social behavior of a wide variety of fish and that many species have evolved complex chemosensory communication systems.

The evidence that has led various authors to describe the existence of chemical communication in fish will be discussed in a following section.

TESTS OF CHEMOSENSORY ACUITY OF FISH

The olfactory capacity of fish is both wide and acute and thus well suited to function in social behavior where chemical recognition of individuals would require correct responses to slight differences in the scents as they are given off by different individuals within a species. This capacity is the basis of the chemosensory social behavior mentioned above to be described later in this article in more detail.

The development of olfactory and gustatory senses differs from species to species and, more significantly, among the higher taxa. Smell is probably best developed in eels (Anguillidae, Muraenidae, and Congridae), and in a number of catfish (Siluriformes); it is also good in certain minnows and in tunas, salmon, and trout (see Anatomy of Smell and Taste Organs in Fish, p. 213). Little is known about the olfactory endowment of deep-sea fish, although some of them are described as having a well-developed nose (Marshall, 1954). Their ecology makes these fish likely candidates for using smell both in food search and in whatever social behavior they may be shown to possess.

Signal-detection experiments have shown that proper motivation will enhance the number of correct responses at low-stimulus intensities, barely above noise levels. Simple conditioned reflex experiments assure the experimenter that he can achieve good motivation in test animals. Whole animal threshhold data so gathered are likely to represent, for any sensory modality, the limits beyond which detection of a stimulus by any one species is no longer possible.

The European eel, *Anguilla anguilla* holds the record, to date, of detecting a scent with which it associated access to, and rest in a tube. The experimenter, H. Teichmann (1962), took advantage of the animals' strong tendency to seek shelter in a burrow. When they detected the test scent they were rewarded by being permitted access to an artificial burrow. Small eels detected in these experiments a concentration of pure β-phenylethyl alcohol at a concentration of 1.77×10^3 molecules/cm^3. The space in the nasal cavity of the small eels employed by Teichmann is barely larger than 1 mm^3. Thus no more than a few molecules of the stimulus substance could have been in contact with the receptive sites. Fish then may exceed in their olfactory performance that of "aerial macro-osmatic animals such as dogs or certain insects" (Teichmann, 1962).

Gustatory acuity has been tested in a number of fish, most thoroughly in the minnow *Phoxinus laevis* (Glaser, 1965), a fish well endowed with taste buds even on head and body, but not exhibiting the profusion of gustatory sensors found in some of the catfish. Table 1 illustrates the minnow's performance in taste tests.

Although Glaser (1965) asserts that an increased number of taste buds, within limits, does not improve taste acuity, it does enable the fish to locate food more efficiently once it has been detected. However, Humbach (1960) performed tests suggesting that blind cavefish (*Astyanax*), with many more taste buds than the minnow, have a taste sense that is 300 times as acute for bitter and between 2,000 and 4,000 times as acute for salty, acid, and sweet substances. Although

TABLE 1

Comparison of taste thresholds of man* with those of the minnow *Phoxinus*†

SUBSTANCE	CAPABILITIES (MOLES/LITER)		SUPERIORITY OF THE FISH OVER MAN
	MAN	MINNOW	
Raffinose	—	1/245760	
Sucrose	1/91	1/81920	900 fold
Lactose	1/16	1/2560	160 fold
Glucose	1/13	1/20480	1575 fold
Galactose	1/9	1/5120	569 fold
Fructose	1/24	1/61440	2560 fold
Arabinose	1/13	1/15360	1182 fold
Saccharin	1/9091	1/1536000	169 fold
Quinine hydrochloride	1/1030928	1/24576000	24 fold
Sodium chloride	1/100	1/20480	205 fold
Acetic acid	1/1250	1/204800	164 fold

* After von Skramlik. *In* Glaser, 1965. Z. *Vergl. Physiol.*, 53:1-25.
† After Glaser. 1965. Z. *Vergl. Physiol.*, 53:1-25.

the tests for taste acuity in fish have, so far, been restricted to the conventional four taste sensations known to be possessed by man, fish can also taste a variety of substances which are probably not perceived in terms of these sensations (Bardach et al., 1967a). Complex organic materials, such as would be contained in the emantions of prey, were also tasted, e.g., the scent of silkworm pupae in the case of Japanese carp varieties (Konishi and Zotterman, 1963). Well-controlled conditioning experiments with fish, testing the taste acuity for such compounds, would be of interest, as would be attempts to establish behavioral taste thresh-holds with the extremely taste-sensitive catfish. It is our opinion, albeit so far based only on behavior observations, that taste acuity in certain fish may well be greater than that described for the minnow by Glaser (1965). But even in the gustatorily best-endowed fish species, taste is not likely to rival smell in either acuity or discriminatory capacity, although in some groups with external taste buds, such as the catfish and cod family, it is a far better instrument of short-range location than olfaction. Fore and aft and lateral differences in concentration of taste sub-stances can be detected simultaneously because of the distribution of taste buds on the body and on fins or barbels (Bardach and Atema, in press; also see Anatomy of Smell and Taste Organs in Fish, below). As some of the species so endowed, e.g., the bullheads, rely on chemical communication in their social behavior, we suspect that taste also plays a role in the interaction between individuals, especially in en-counters involving body contact, but we have not yet been able to marshall evidence for the role of taste in fish's social behavior.

Anatomy of Smell and Taste Organs in Fish

The salient difference between the fish nose and the nose of higher vertebrates is that it is separate from the respiratory channel. Water containing olfactory stimulants is moved over the olfactory epithelium in various ways; lampreys have a single, well-developed median nasal apparatus with a hydraulic system that operates in synchrony with breathing (Kleerekoper and van Erkel, 1960). They must search for a scent source by comparing stimulus intensity on subsequent stations along their sinuous swimming paths. Sharks and rays have paired nares lying at varying distances from the oral opening. The nasal pits are oval- to lozenge-shaped and lie across the snout, or point obliquely towards the mouth. They are covered by skin flaps and have grooves fashioned so that the water flows through them from a lateral entrance to a median exit. The water may also be drawn in, to a greater or lesser degree, by the respiratory stream, depending on how close the nares lie to the mouth. As some sharks reach considerable size and others, such as the hammer-head (Sphyrna) with its broad wedge shaped head, have their nares far apart, they may be able to compare simultaneously the concentration of oncoming stimuli to the right and the left of them. Elasmobranch noses have numerous lamellae, a sign of high olfactory acuity, though rigidly controlled tests of their chemosensory capacities have yet to be done (Tester, 1963).

The nasal development of teleost fish varies with their ecology, ranging from the almost nonsmelling stickleback (Gasterosteus) of shallow clear waters to, for

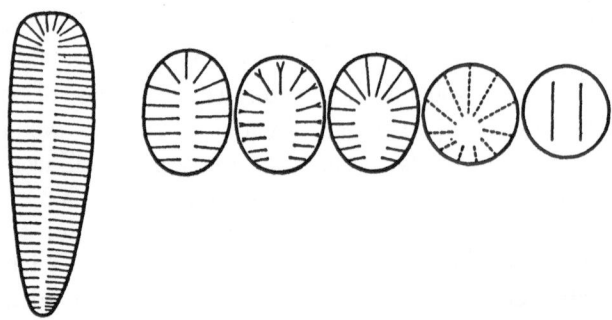

FIG. 4. Arrangement and number of olfactory folds in the nares of certain fishes. From left to right they are: stickleback (*Gasterosteus aculeatus*), pike (*Esox lucius*), rainbow trout (*Salmo gairdneri*), perch (*Perca fluviatilis*), the minnow Elritze (*Phoxinus phoxinus*), eel (*Anguilla anguilla*). (After Teichmann. 1954. Z. Morphol. u. Oekol. Tiere., 43:208.)

instance, the eels (Anguillidae), the catfish (Siluriformes), the morays (Muraenidae), and some deep-sea fish in which smell is probably the species' most important sense with respect to functioning prominently in communication. The number of nasal folds increases accordingly from very few in the stickleback to nearly a hundred in the adult eel (Fig. 4). The arrangement of the olfactory cells on the folds differs with species or families (Holl, 1965), and within a species folds also increase with increasing age. Electron microscopy of the olfactory cells of the minnow *Phoxinus* (Bannister, 1965) revealed, aside from water-propelling ciliated supporting cells, true olfactory cells with several kinds of apical processes. Some apical processes are ciliated, others have club-shaped structures sticking into the mucus, and others resemble, at their apices, the microvilli of supporting cells in taste buds. It is not known if the histologic differences are concomitant with functional ones.

For smell to function effectively, water must be renewed and moved over the olfactory cells; fish create a water stream in and out of their noses by various structural modifications. In some (e.g., goldfish) there is a dam of skin and connective tissue between the incurrent and excurrent naris deflecting water through the nose as the fish swims. In others, a pouch-like nasal sac acts as a hydraulic system when certain bones of the jaw are moved in breathing or snapping (Hasler, 1957). While cilia of the supporting cells on the olfactory lamellae, together with the other devices just mentioned, help propel the water, they are the main flushing agents in the noses of bullheads and eels—fish that show high olfactory acuity (Fig. 5). Their mode of beating creates small vortices over the lamellae in which the stimulus-carrying water reaches all portions of the epithelium. The fish also have the ability to increase the ciliary beat in each naris independently so as to sniff differentially with either naris (Chen and Bardach, in preparation).

Taste in catfish (Siluriformes), codfish (Gadidae), minnows (Cyprinidae), loaches (Cobitidae), and, presumably, in certain deep-sea forms has not only retained its function of oral testing of foods, but has also become a true distance sense (see Reactions to Nearby Stimuli, above). Fish taste by means of typical vertebrate taste buds, but they are also capable of sensing certain chemicals with free nerve endings and/or single epidermal cells, which, in some of their microstructural aspects, resemble taste cells in taste buds and were therefore presumed to be chemosensitive

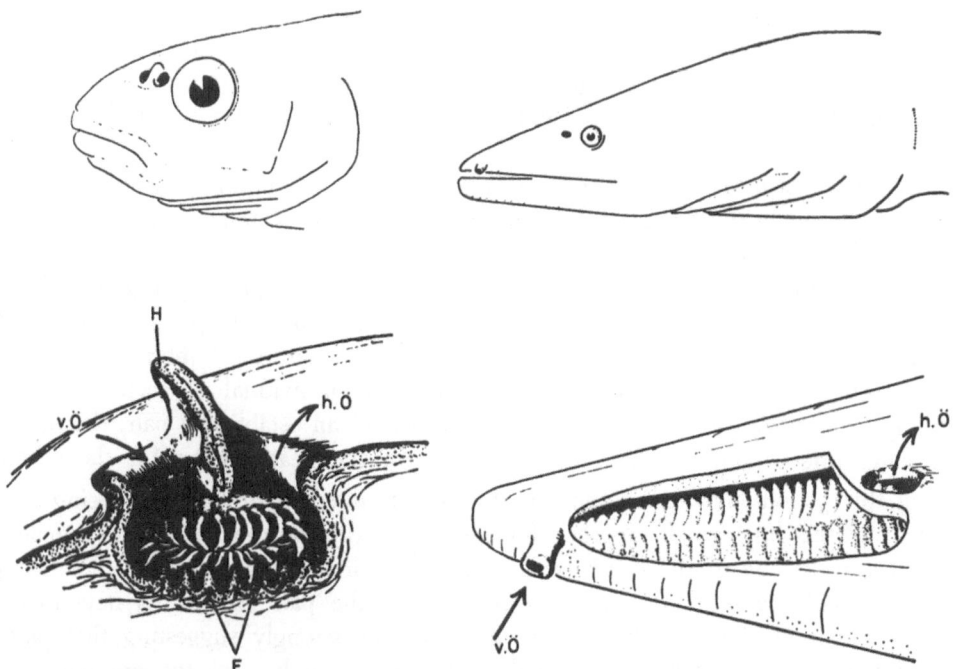

FIG. 5. Side view of the nasal chambers of 2 species of fish; left—the minnow Elritze (*Phoxinus phoxinus*) and right the eel (*Anguilla anguilla*). (After Teichmann. 1962. Die Umschau in Wissenschaft und Technik, Frankfurt a.M. 62(19):589.)

(Whitear, 1965). The groups of fish which use taste as a distance sense rely on a distribution of taste buds on their appendages and on the surface of their bodies innervated by a branch of cranial nerve VII, the ramus recurrens, which joins in the formation of the ramus lateralis accessorious (Freihofer, 1963).

Thousands of taste buds are also found in the mouth and on the gill structures of fish (Pfaffenzeller, 1923), an anatomic peculiarity presumably taken advantage of by mouth-breeders which are believed to recognize their young by taste (Myrberg, 1966). While taste has not been proved to play a role in other aspects of social behavior, the fact that fish have a better developed gustatory sense than other vertebrates, with prominent taste centers in the central nervous system, suggests that it may in the future be found to play a role, together with smell, in the detection of chemical signals.

EVIDENCE FOR INTRASPECIFIC CHEMICAL SIGNALS (PHEROMONES) IN FISH COMMUNICATION

Sexual Pheromones

Pheromones in pair formation and maintenance were well documented by MacGinitie (1939) in his studies of the blind goby, *Typhlogobius californiensis*. The blind goby is intertidal and lives commensally in pairs in burrows of the shrimp, *Callianassa affinis*. The burrows are constructed entirely by the shrimps, and the adult gobies inhabit their deepest regions. The eyes of newly hatched blind gobies are perfectly developed, but after taking up residence in the burrows of the shrimp, their retinas deteriorate with the result that in most adults the eyes are indiscernible.

No stimulus other than forced ejection will induce a blind goby to leave its burrow. The majority of these fish pair for life and share a burrow for as long as 12 years. Pair formation takes place within the first six months of life before the onset of the breeding season. Pairing may not be an active process of selection by the female or the male but the result of any member of the opposite sex being tolerated in the burrow. The extremely aggressive nature of blind gobies to members of their own sex prevents more than one individual of each sex from residing in a burrow. If a male invades the burrow of an established pair, the males will enter into a conflict which usually results in the death of one or the other. During this period, the female remains indifferent unless there are eggs being incubated in the burrow. If another female enters the burrow, the two females fight, and again one of the combatants is killed or driven out. If an invader does manage to kill the resident, the other member of the pair accepts its new mate without any indication that it is aware of a change, strongly suggesting that pair formation in gobies is based on mutual tolerance of members of the opposite sex and is not necessarily based on recognition of a specific mate.

As a test of the concept that in the depths of the burrows sex could only be determined by pheromones, MacGinitie (1939) placed individual gobies in cellophane bags into an occupied burrow. Each bag was punctured with a pin enabling the water surrounding the enclosed fish to flow outward to the pair in the burrow. When the odor reached the pair, the partner of the same sex as the fish in the bag immediately began an attack.

Another territorial gobiid, at least during its breeding season, is *Bathygobius soporator* of which Tavolga (1956) discovered that pheromones from the females elicit courtship behavior in the male. The males spend most of their time in shells or crevices which serve as the center or focal points of their territories. When other males enter, they are chased or fought. In the presence of a gravid or pregravid female, the coloration of the male changes and it approaches the female with fanning and gasping movements. If she is ready to spawn, she follows the male to the nest where she deposits the eggs on the inner surfaces. After spawning the male guards and fans the eggs until they hatch four or five days later.

When a gravid female was placed in a 50 ml beaker of sea water for two minutes, and the water was the pipetted into the tank with a resident male, the male promptly began courtship. As little as 0.5 ml of this water was effective in initiating the response. Thus gravid females produced a pheromone which elicited courtship behavioral patterns and the courtship color phase in the majority of territorial males. The courting movements were performed while swimming randomly throughout the tank. Relatively nonspecific visual and tactile stimuli were found to be essential for orienting the courtship behavior of the male.

A single introduction of the pheromone enabled the courtship to continue for approximately an hour. The pheromone was emitted from the urinogenital region and was found to be present in the ovarian fluids and the eggs. Tavolga suspected that the pheromone originated in the interior of the ovary. Freshly extruded eggs were effective in stimulating courtship for 15 minutes.

Males deprived of their sense of smell did not respond to any amounts of the courtship pheromone, and fish with plugged noses regained their sensitivity when the plugs were removed or fell out.

Jaski (1939, 1947), in another report of a pheromone involved in the reproductive behavior of fish, claimed to have discovered a substance which he called copulin, secreted by the male guppy, *Lebistes reticulatus*. This "hormone" [sic] supposedly triggered the ovipositor growth in the females and synchronized their estrus cycles. In the absence of copulin, individual cycles occurred. Baerends et al. (1955), Breder (1951), Clark and Aronson (1951), Stolk (1950), and van Koersveld (1949) have failed in their attempts to duplicate his experiments, and consequently have rejected Jaski's conclusions.

Among other authors suggesting that pheromones play a role in the courtship behavior of fish were: Leiner, 1930 (in the stickleback, slime secreted from the male stimulates the male's own courtship); Roule, 1931 (secretions from the male lamprey may attract the female; courtship behavior in the male shad, *Alosa alosa* may be initiated by a substance from the female); Eggert, 1931 (in *Blennius pavo* there are cutaneous anal glands in the male which may act to attract the female); Belding, 1934 (secretions of the salmon female attract the male); Breder, 1935 (pheromones are implicated in sexual discrimination in the bullhead, *Ictalurus nebulosus*); Seitz, 1940 (a male *Haplochromis multicolor* frequently mouths the urinogenital region of an intruder; this act may assist in chemical discrimination of the sexes); and Nelson, 1964 (the caudal gland in male Glandulocaudine fish probably produces a pheromone assisting to bring pairs together while mating).

The roles of pheromones involved in the sexual relations of fish have been little studied, but it seems likely that a variety of distinct signals are involved in this behavior. For numerous phylogenetically diverse groups of fish, following the reproductive season, the bond which ties parents to their young is facilitated by pheromones.

It was briefly noted above that Leiner (1930) observed that slime from a male stickleback stimulated courtship behavior in another male of the same species. A similar phenomenon has been analyzed in detail by Losey (1969). He discovered that at least three species of blennies, *Hypsoblennius* spp., produced a pheromone which facilitated male courtship behavior.

Water from tanks containing blennies at various stages of courtship was used as a test stimulus. In general, water from tanks containing mating blennies or pairs in intense courtship elicited a significant sexual response and attraction to the source of the stimulus, while water from tanks containing less-activity courting pairs or isolated individuals of either sex did not. The strongest reactions occurred when the source of the stimulus was conspecific, but ripe male *H. robustus* also responded positively to water from tanks containing mating pairs of *H. gentilis*. Egg-guarding males, which would still court and mate in the presence of females, were not attracted to the stimulus, nor were females.

Losey postulates that the pheromone facilitates courtship by attracting ripe nonparental males to a courting male, thus increasing sexual contacts which might lead to pair formation. He does not state whether a male normally courts one or several females at a time. If a male normally courts only one female at a time, it is difficult to see how attraction of more males would facilitate pair formation.

Equally questionable is Losey's assertion that the source of the pheromone is the courting male. It may be, but since courtship behavior was induced by introduction of a ripe female which was not chemically isolated from the male it is also possible that the pheromone was released by the female under the stimulus of intense courtship.

Parent–Young Relations

Parent fish which care for their broods must have evolved some means of distinguishing them from the young of other fish and from a wide variety of animals of similar size and shape which they normally eat. The dilemma here implied was vividly illustrated by Lorenz (1952) when he described the behavior of a jewel fish, *Hemichromis bimaculatus,* as it attempted to feed and care for its young at the same time.

> As I approached the container I saw most of the young were already in the nesting hollow, over which the mother was hovering. She refused to come for the food when I threw pieces of earthworm into the tank. The father, however, who, in great excitement, was dashing back and forth searching for truants, allowed himself to be diverted from his duty by a nice hind-end of earthworm. He swam up and seized the worm, but owing to its size, was unable to swallow it. As he was in the act of chewing the mouthful, he saw a baby fish swimming by itself across the tank; he started as though stung, raced after and took it into his already filled mouth . . . The fish stood stock still with "full cheeks," but did not chew. For many seconds the father jewel fish stood riveted . . . Then he solved the conflict in a way in which one was bound to feel admiration: he spat out the whole contents of his mouth: The worm fell to the bottom, and the little jewel fish did the same. The father turned resolutely to the worm and ate it up, without haste, but all the time with one eye on the child which "obediently" lay on the bottom beneath him. When he had finished, he inhaled the baby and carried it home to its mother.*

Breder (1939) noticed that parent brown bullheads, *Ictalurus nebulosus,* distinguished the presence of strange young in their nest. He suspected that the discrimination was visual, although chemical cues were probably involved. Parent blind gobies, *Typhlogobius californiensis,* will not eat their own young, at least not for many days after hatching, but blind gobies other than the parents will readily eat the new hatched larvae of other parents (MacGinitie, 1939). This observation strongly implicates pheromones in brood recognition by the parents for this species.

Chemical stimuli emitted by young jewel fish, *Hemichromis bimaculatus,* influence parental behavior (Kühme, 1963). When water from the tank in which the young resided was allowed to flow into the parents' tank, the parents oriented to

* From Lorenz. 1952. *Amer. Mid. Naturalist,* 21:489.

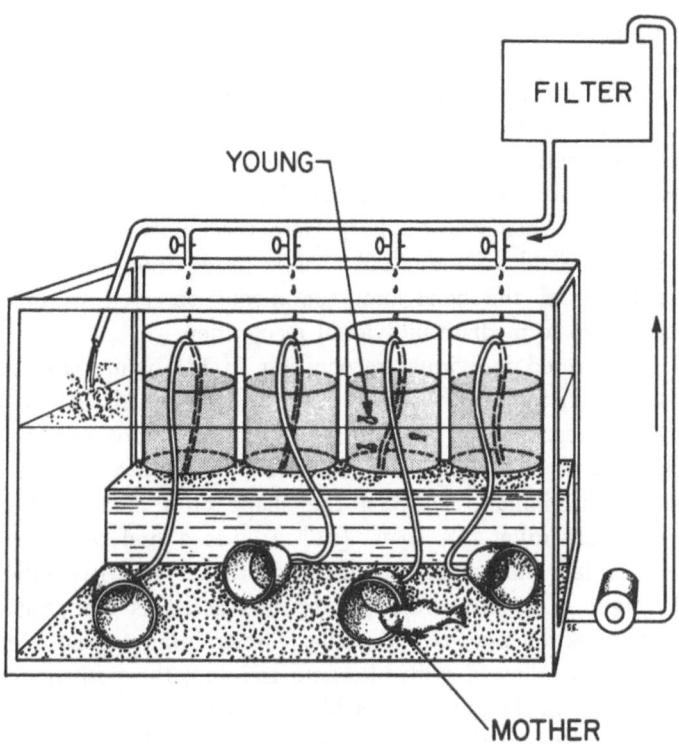

FIG. 6. Experimental arrangement for testing pheromones from young cichlids. (After Kühme. 1963. *Naturwissenschaften*, 51:120.)

the point of inflow and exhibited fanning and other parental behavior actions (Fig. 6). The attachment to the continuous inflow from the tank containing the young continued until the time when the young would normally separate from their parents. Kühme was able to manipulate the time at which parental behavior ceased by replacing young fry with older fry and vice versa, shortening or extending the behavior, respectively. The brooding period of one female was extended three weeks when pheromones from newly hatched broods were added to her tank. The parental bond of the jewel fish terminated as the production of pheromones from the larval fish ceased.

Parent jewel fish are capable of distinguishing their broods from others of comparable ages. When tank water from several broods was simultaneously introduced the parents selected the point of inflow of the tank water from their own young (Kühme, 1963). The dwarf cichlid, *Nannacara anomala* (Künzer, 1964) and *Cichlasoma nigrofasciatum* (Myrberg, 1966) are also capable of recognizing their own broods. In the latter fish, the females were observed to bite at the outlets carrying water from tanks occupied by other species. Since pheromones are important in regulating the parental behavior of these highly visual cichlids it is very probable that a wide variety of fishes which care for their young also make use of chemical cues emitted by the young.

Nonsexual Social Behavior and
Individual Recognition

Many fish exhibit sophisticated social behavior outside the reproductive season. Communities and coteries may be common among fish that establish territories and hierarchies, and where behavioral interactions are based on a wide variety of agonistic displays. This behavior is often based on individual recognition (Todd et al., 1967). Cooperative behavior resulting from the ability to discern individuals has been recently described (Todd, 1968). The recognition of individuals may be the basis for much of the behavioral complexity observed in fish. Chemical signals have been shown to mediate individual recognition.

Göz (1941) discovered individual recognition in fish while studying a single individual of a species of minow, *Phoxinus*. This fish he called *Phoxinus* 34. Since his findings may be questioned in the future, a detailed description of his unusual research is indicated. A colleague of Göz threw out *Phoxinus* 34 because it would not learn to discriminate between the shape of a circle and ellipse. Göz noticed that it was very quiet but "gay" [sic]. It was a preferential right turner. Number 34 was blinded and trained to discriminate between two other *Phoxinus* males of equal length. The tank water of the one acted as the positive stimulus and that of the other as the negative stimulus. The training regime began with 20 positive trials. This was followed with alternate training, the positive reward being food and the negative punishment being a blow with a glass rod. After the twenty-fourth negative trial, Göz beat the fish for 30 seconds as it fled frantically about the test tank. Twenty minutes later, on the next negative trial, the fish went to the feeding place, then turned and fled to a hiding place. Four and one-half hours later it fled the positive stimulus, but after 12 seconds returned to the feeding place and fed. In the next 20 minutes *Phoxinus* 34 scored correctly in four negative and four positive trials. After 15 further correct choices, *Phoxinus* 34 made several errors. This fish continued to score correctly in the majority of over 100 further trials. Later it was successfully used in 18 further discrimination experiments involving nine species of fish of different sizes and sexes. Truly, *Phoxinus* 34 was a phenomenon, and it is possible that its lack of visual acuity may initially have had some bearing on its performance as an olfactory discriminator, particularly since this species normally schools.

Individual recognition by pheromones and its importance in the social behavior of fish was first described for the yellow bullhead, *Ictalurus natalis* (Todd et al., 1967). This study combined a variety of methods for decoding the chemosensory information being transmitted between individual bullheads. Discrimination experiments using chemosensory cues only provided the basic information. This was followed by an analysis of various combinations of agonistic interactions between pairs of bullheads. Finally, communities of normal and anosmic bullheads were described. It will be apparent in the following sections that a wide variety of techniques are necessary for evaluating the information carried by the pheromones.

The initial determination of discrimination between individuals by bullheads by means of chemical cues alone was a conditioning experiment. Immature fish of approximately 20 cm in length were kept singly in 19-liters capacity aquaria for several weeks before the experiments. A clay flower pot in the rear of each tank served as a shelter. Silver-wire electrodes from an inductorium were mounted on the front of the tanks. The fish were divided into test and donor groups. Fifty milliliters of a donor's tank water was the discriminative stimulus. Water from the tanks of donors was gently poured over the bubble train of the porous "airstone" in the test aquarium. Each test fish was trained to a nonrandom, scrambled schedule (see Todd, et al., 1967, for sequences) by a food reward, or electric shock, to discriminate between two donor fish. The donors were moved from aquarium to aquarium to avoid discrimination on the basis of odors which might be indigenous to the tank and unrelated to the donor fish.

As fish learned to discriminate between two donors, their responses varied appropriately. They rose rapidly to the surface at the front of the aquarium and gulped, as if in search of food, when the positive odor was introduced, but fled to the shelter where they were safe from electric shock when presented with the negative stimulus. For the first few trials, a powerful shock was used. The high voltage propelled the animal out of the electric field to the area of the pot. After a few trials, the fish seldom attempted to locate the source of the negative odor. Subsequently, a mild shock was sufficient for the training.

TABLE 2

Odor discrimination between individual bullheads

NUMBER OF FISH TESTED	TOTAL NUMBER OF TESTS	NUMBER OF CORRECT RESPONSES	PERCENT CORRECT RESPONSES
10	937	895	95.52

Table 2 shows the results of discrimination tests by yellow bullheads. The test animals were trained to discriminate between the tank waters of the two individuals of a donor pair. Three donor pairs of immature yellow bullheads provided the stimuli throughout the study. Each test animal was trained to a single donor pair. The fish learned to discriminate between water from the positive and negative donors in about 26 trials, (range 19 to 32; mean 26.2 trials). Out of 937 tests, only 42 (or 4.5 percent) incorrect choices were made. All fish showed retention of the learned discrimination for at least 3 weeks without retraining.

To test whether the gustatory or the olfactory sense was responsible for the observed discrimination, the olfactory epithelia of three animals from the discrimination experiments, and of two fish without previous training (naive fish), were destroyed by cold cautery (Table 3). Stress induced by the cautery was short-lived: the animals began to feed after severals hours. After a week, the three trained animals were given 408 trials in which they compiled 197 correct scores (48.3 percent), as shown in Table 3. Training was attempted with the two naive fish, but they did not discriminate after 32 trials. In an additional 114 further trials they scored

43.9 percent correct scores (Table 3). After the training period, gross microscopic examination showed that the olfactory epithelium had been destroyed, and that no noticeable regeneration had taken place.

As these tests had shown that bullheads can use smell to discriminate between two individuals, we attempted to locate the source of the substance that provides the fish with their individual odor. In this quest, slime was wiped from the dorsal surface of each donor bullhead with cheesecloth and placed in a 50 ml vial filled with clean, aged tap water. The remainder of the experiment was similar to the preceding discrimination tests.

TABLE 3

Odor discrimination of bullheads with destroyed olfactory epithelia

	NUMBER OF FISH TESTED	TOTAL NUMBER OF TESTS	NUMBER OF CORRECT RESPONSES	PERCENT CORRECT RESPONSES
Fish from prior test*	3	408	197	48.3
Naive fish	2	114	50	43.9
Total	5	522	247	47.3

* See table 2.

In 32 trials, one of seven test fish made two errors to a negative stimulus. In the remaining 30 correct performances, the responses were less intense to the neutral water mixed with donor slime than to the donor water itself. This lessened intensity suggested that other substances, such as urine, might contribute to individual odor.

Bullheads observed in large aquaria of up to 8,000 liters behaved differently towards strangers than to tank mates of either inferior or superior status. In fact, the fish appeared to recognize each individual in the community and respond to them in a manner determined from their previous interactions. A stranger usually produced a uniform behavior in the tankmates—overt aggression.

An experiment was devised to test for differences in conflict behavior between pairs of strangers and pairs of acquaintances. In each experiment, two fish in a 200-liter aquarium were physically and visually separated by a barrier of frosted glass. Since there was an interchange of water between compartments, they were not chemically isolated. Figure 7 illustrates the interactions or behavioral "pathways" exhibited first by a bullhead (fish 1) and a stranger, then a dominant acquaintance and finally an acquaintance of comparable status.

Figure 7 is representative of a number of observed behavioral interactions of normal and blinded bullheads. In Figure 7A, the attack against the stranger was intense and vicious. The combatants engaged in biting, quivering, and mouth fighting. These three behavioral actions have only been observed together in conflicts between strangers. The set of sequences in Figure 7B illustrates the behavior of the same bullhead with a fish who had dominated it in a previous interaction some days earlier. Here, lower-keyed behavioral attributes such as displaying, circling, and nipping occurred, although the predominant behavior of the previously observed

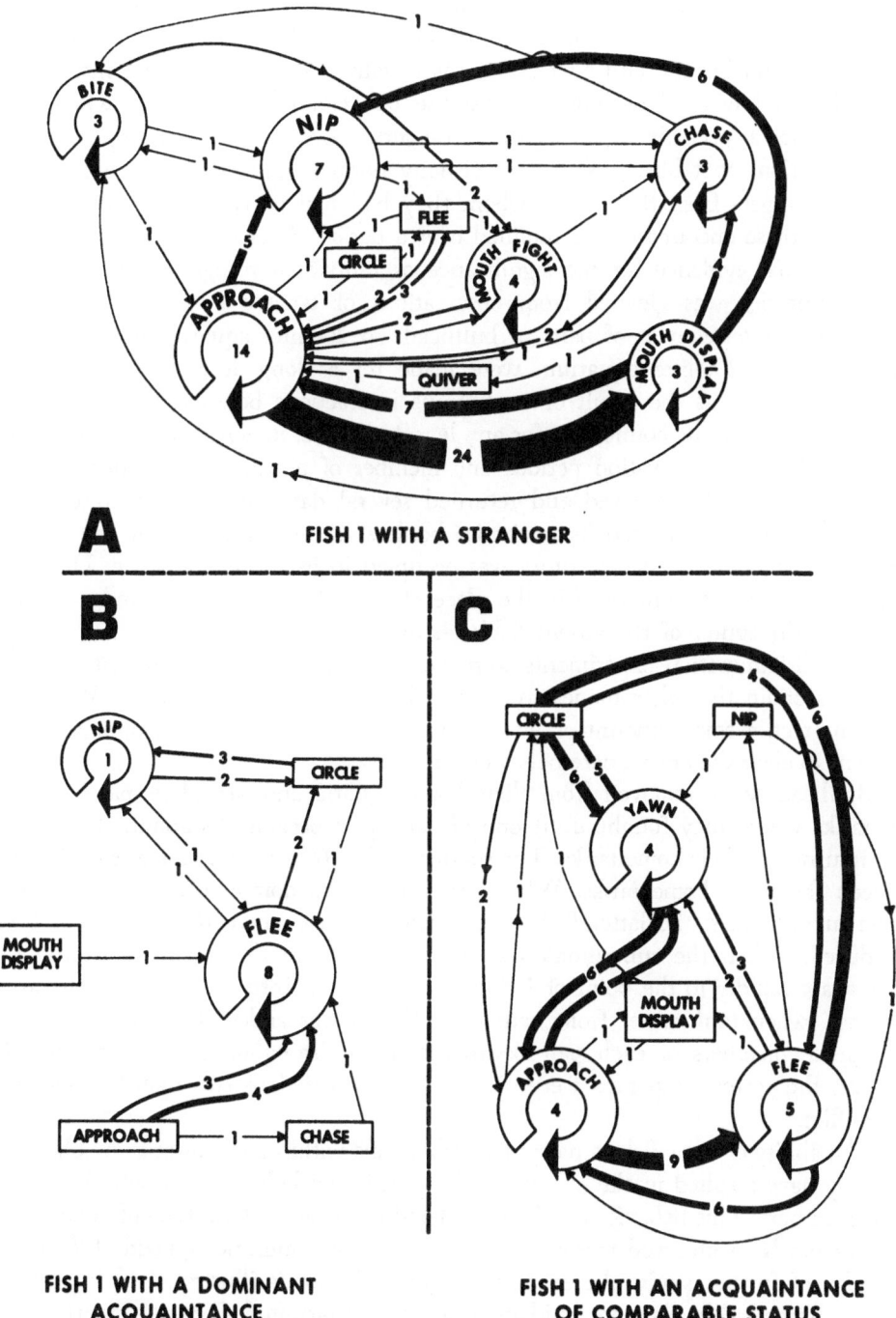

A

FISH 1 WITH A STRANGER

B

FISH 1 WITH A DOMINANT
ACQUAINTANCE

C

FISH 1 WITH AN ACQUAINTANCE
OF COMPARABLE STATUS

FIG. 7. Differences in the agonistic behavior patterns between bullheads that are strangers and acquaintances of comparable and dominant status. The numbers indicate the frequencies of occurrence of individual behavior units, as does, roughly, the thickness of the lines.

bullhead was avoidance of the fish of higher status. The third interaction (Fig. 7C) took place with an acquaintance of comparable status. There was a balance between approaching and fleeing, and aggression was ritualized in the sense that interactions occurred at a low level of intensity and the behavioral characteristics which lead to injury, such as occur in fights between strangers, were absent. It was apparent from our observations that blind, as well as visually normal bullheads, were able to distinguish strangers from those with whom they had had previous encounters. Their memory of these encounters was retained over a period of several weeks.

Additional evidence for the significance of individual recognition in maintaining communities was gleaned from observations of territorial behavior. Repeated recordings of the behavior of pairs of bullheads in 200-liter aquaria and small communities in a 1,200-liter aquarium were made for as long as five months. Stable borders developed as the result of continuous interactions between tankmates. If a fish failed to defend its boundaries for any length of time, its territory was encroached upon. During the observation periods one member of a pair, or a resident from a community would be removed and returned several days later. On its return, the inhabitants usually respected its territorial borders as they had been at the time of removal, even if the vacated territory was occupied in its owner's absence. However, when a stranger was introduced in the place of a resident, it was attacked even if it swam into the center of the absent fish's territory.

The discrimination experiments were not the only means by which it was possible to ascertain that pheromones were crucial in individual recognition. After bullheads interact, a small amount of water from the tank of one of the protagonists is sufficient to elicit the appropriate behavioral response.

Bullheads were removed from their home aquaria and placed in pairs in 200-liter tanks where they fought until one of each pair became dominant. They were then returned to their home tanks. The following day, 600 ml of water was exchanged between the pair's home tanks. When water from the dominant's tank was added to the subordinate's, the latter fled, or avoided the area where the water had been introduced. When the subordinate's water was added to the dominant's tank, the latter swam rapidly to the point of inflow and in several instances attacked the area. However, when tank water from neutral bullheads was added, both dominant and subordinate members of each pair began searching throughout their territories. In one test, the correct responses were observed for a period of two months following the conflict.

The ability of some fish to recognize individuals within a community by chemical cues may have resulted in the evolution of sophisticated behavior mutually beneficial to the members. This behavior can be considered cooperative. One type of cooperative behavior has been observed many times in bullhead communities (Todd, 1968) and is worthy of further study. An example of this behavior is illustrated in Figure 8A and 8B. Three bullheads were resident in a large aquarium for several months. The dominant fish was the largest and the fish at the bottom of the hierarchy was the smallest. Each animal resided in a territory around a clay tile. However, when water from the tanks of previous antagonists was added to the community, the subordinates immediately left their territories, fled to the clay tile owned by the dominant and came to rest on the dominant's back (Fig. 8A). The three fish would remain quiet

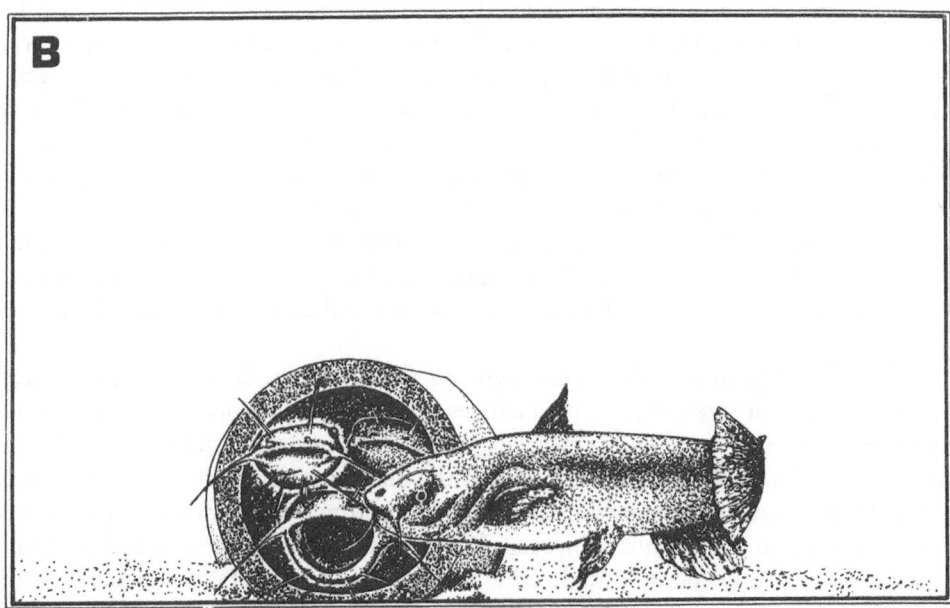

FIG. 8.   Cooperative behavior among bullheads. 8A. The subordinates enter the dominant's shelter shortly after a stranger was placed in the tank. 8B. The dominant begins its attack on the stranger while the subordinates remain in the shelter until the stranger is defeated. (Redrawn from flash photographs.)

for about 15 minutes and, if there were no further disturbances, the subordinates would be viciously evicted by the "boss."

When a stranger was placed in the community, the subordinates would dart to the tile as described above and remain on the back of the dominant fish even as it began to attack the stranger (Fig. 8A). During the fight, the subordinate remained in the dominant's shelter. After the defeat or flight of the stranger, the dominant would return to its tile and after a period evict its subordinate tankmates.

On several occasions it was observed that if a stranger tried to enter the tile of the dominant at the same time as the residents, the dominant was able to discriminate between them. The stranger was attacked while the residents were permitted to enter. Protection of subordinates within a bullhead community may be mutually advantageous. The inferior fish are provided with protection, and the dominant animal has the advantage of neighbors unlikely to threaten its dominance and territory.

Although the odor that gives an individual its identity appears the most significant information that is communicated, other information such as that indicating sex and age may be transmitted by pheromones. Some evidence indicates that a change in an individual's status perhaps occasioned by stress can be communicated to other bullheads by pheromones. We tested the effects of stress on individual recognition, as McKim (1966) had reported stress products in fish urine. We used the discrimination technique described previously and then observed conflicts induced in groups of several fish to shed light on the social significance of status changes.

First, the effects of stress on individual recognition were tested: six donor fish (i.e., the fish whose ambient water provided the stimulus to be discriminated) were subjected to mild electric shocks (15 volts at source) at regular intervals for 3 hours. After this moderate stress, donors were consistently recognized in discrimination trials by the test fish. To introduce a variation in stress, donors were removed from their tanks and made to fight with larger neutral fish. They lost the fights and, as a result, were altered chemically in such a manner that the test fish did not discriminate between them but approached the stimulus on every trial. This experiment did not indicate whether the defeated donor animal was no longer recognized as a specific individual, or if a change in status after fighting was responsible for the changed reaction on test fish.

The next experiment shed some light on this question. Behavioral observations had suggested that stress and status changes were noticed by other members of the community. Several pairs, each consisting of a dominant and a submissive bullhead, were made to share 200-liter aquaria. When the dominant fish were removed, isolated overnight, and returned, the submissive tankmates did not attack them, as they would have attacked strangers. When, however, the dominant fish were returned after having experienced losing encounters with other bullheads, the originally submissive tankmates immediately attacked them. The subordinates did not attack the returned tankmates with mouth fighting, biting, and quivering as if they were strangers, but as acquaintances of inferior status, with nipping and displaying.

The behavior suggested that information on both individual identity and loss in status was communicated chemically to the other fish which shared the tank. It is unlikely that the loser's behavior communicated any information to the subordinate on its return, since it was quite common for a fish to remain still on its return, whether it had been defeated or not.

Further experimentation is needed on the subject of status in relation to individuality in catfish behavior.

Social Behavior of Bullheads Deprived of Their Sense of Smell

An important experiment in any study on the role of pheromones in the lives of an animal is to see how the animal performs when deprived of the ability to sense these substances. If there are fundamental behavioral changes, then the existence and significance of pheromones can be ascertained.

A single attempt was made to establish a community of noseless bullheads in an 800-liter tank. Four bullheads had their sense of smell eliminated by cautery only in the month of December and were placed in separate aquaria for a month before introduction into the large tank. All were of similar size and were placed in the new tank together. Within the first three hours, five mouth fights (the type of fighting utilized by strangers), lasting up to 20 minutes each, were observed. Normally, mouth fights last for only a few seconds. One of the fights involved three fish. Two of them clamped their jaws together in the conventional mouth fighting posture, while the third seized the jaw of another at right angles to it. This behavior had not been observed before. As the day continued, each fish defeated the others and was defeated in turn. No dominance, hierarchies, or territories were established. On the following day the fish were badly damaged. They were still very aggressive although the frequency of conflicts had diminished. On the third day, the fish were still badly damaged from conflicts and no dominance was observed. On the fourth day, the animals were quiet, but after feeding they began to attack one another. Several brief mouth fights were observed. For the first time, there were flight components in the aggressive behavior of most of the animals. One of the individuals was more aggressive than the other three. On the fifth day, mouth fights were observed and the fish seemed badly cut. On the seventh day, round-robin aggression was recorded but one of the animals avoided most of the conflicts. Between the eighth and sixteenth days, the amount of fighting diminished. By the seventeenth day the wounded skin of the animals was healing, and for the first time, a hierarchy was established. Three of the animals had territories within the "caves" of cement blocks where they spent most of their time. An examination of the olfactory epithelia revealed that it had at least partially regenerated in all of the fish. It is possible that the restoration of the sense of smell coincided with the final establishment of a community. After "nose" regeneration, the behavior of the bullheads was found to be similar to that of normal communities of similar size.

The behavior of bullheads deprived of their sense of smell and placed singly into the resident communities of a 1,200-liter aquarium shed some light on the ability of an anosmic individual adapting to a normal "social situation." The first experiment involved one of the permanent members of the community which was removed and had its olfactory epithelium eliminated. Before its removal from the community it was the smallest and lowest member of the community in the hierarchy, with a small hiding place in the right rear corner of the tank. After cautery it was reintroduced and immediately began to attack all three residents in their shelters. Within a half hour it had been attacked and beaten by its three former tankmates. Then it began to flee, but within another half hour, it renewed its attacks on the residents and was defeated a second time. Then the "noseless" fish was removed. The following day it was reintroduced and immediately attacked the residents as before. A vicious bite to its head from number 2 resulted in the operate's defeat. The "noseless" animal moved throughout the tank and did not return to its old territory where it would have been safe from attack and was removed after one hour and fifteen minutes.

Anosmic strangers introduced into the community attacked the residents and were defeated. When they were reintroduced, they repeated their attacks. "Noseless" bullheads did not seek out the areas in a community where they were immune from attack, nor did they rapidly integrate themselves into a community. Tests allowing the anosmic bullheads to remain in the community for several days, in order to determine if over a longer period they would establish territories and adjust, have not been carried out.

Normal animals of similar size introduced into the same community learned after a brief period of conflict to avoid the territories of the residents and thereby to minimize conflict. In several instances they were able to establish small territories. Furthermore, when a normal animal was removed and reintroduced it was cognizant of the social structure, and avoided the residents and returned to the part of the tank where it had been somewhat immune from harassment the time before. Olfactory detection of pheromones seems to be prominent in the creation and maintenance of stable bullhead communities, at least under laboratory conditions.

Our observations have not been duplicated in nature; thus it is possible that the behavior is atypical. It is not known what percentage of wild adult bullheads establish territories when not breeding, or if they recognize each other as individuals in nature. However, it is likely that the behavior we have observed is duplicated in essence in lakes and ponds.

McLarney (1968) observed catfish of the genus *Rhamdia* in aquaria and found that they recognized each other as individuals by chemosensory means. Their social behavior was very similar to that of bullheads. He subsequently observed *Rhamdia* in their natural environment, a pool in a stream in Costa Rica. There also they appeared to recognize each other as individuals and behaved as they did in aquaria.

Schooling Behavior

Many fish school, particularly those which are utilized for food by man. Keenleyside (1955) defined a school of fish as "an aggregation formed when one fish reacts to one or more other fish by staying near them." The chief factor common to all schools is a definite mutual attraction between individuals. Although sight is the primary sense involved in the schooling behavior of fish (Morrow, 1948), pheromones have been shown to be important in the aggregating and dispersing of schools, the latter as a pheromone-induced response to predation. The cohesion of fish schools will be discussed first.

Wrede (1932) was one of the first to investigate chemical communication by fish in a manner that had bearing on the phenomenon of schooling. She used a three-compartment maze within an aquarium. Before each test one *Phoxinus laevis* was placed briefly in one of the compartments in order to give it a fish odor which the other two compartments did not have. Blinded *Phoxinus* swam to the chamber that had the odor of its species mate more often than to adjoining compartments. Further, the odor emitted from the chamber had a quieting effect on the test animals. Wrede suggested that these chemical cues, which she found to be in the slime, may aid in keeping schools of *Phoxinus* together.

AGGREGATING PHEROMONES. Nocturnal changes in orientation of some schooling species are by no means random (Keenleyside, 1955). Some schools may be maintained during darkness, and many deep-water fish may use senses other than vision to coordinate their schooling activities. As evidence, Keenleyside cited the fact that on dark nights some species are caught in dense aggregations at considerable depths.

The rudd, *Scardinius erythrophthalmus*, school during the day and remain off the bottom and active at night (Keenleyside, 1955). In order to test if some mechanism maintains the nocturnal aggregations, two small aquaria, one of which had one of its glass ends replaced with a perforated plastic sheet, were placed inside a large aquarium. The tank with the perforated end served as the test tank, while the other small aquarium functioned as a control tank. Experimental animals were blinded and acclimatized for 15 hours prior to the tests. The behavior of each test animal in the outer aquarium was recorded, with particular regard to its position in the tank, for 20 minutes before the experiment when eight rudd were placed in each of the two smaller aquaria. Then the movements of the test animals in the outer aquarium were recorded for an hour. The olfactory epithelia of the test rudd were then destroyed and the trials repeated. Before the tests, the animals moved at random through the large tank, but, after the rudd schools were placed in the smaller aquaria, the blinded test fish showed a clear preference for the area where water with the scent of the rudd school diffused into the tank. The fish slowed their rate of swimming and remained in the diffusion area. After their sense of smell was eliminated, they showed no preference for either end of the large aquarium, thereby demonstrating that it was not mechanical or gustatory stimuli from the school but olfactory information which attracted the test fish. That the response was specific to other rudd and not fish odors in general was demonstrated when the eight rudd in each of the smaller tanks were replaced by brown bullheads, *Ictalurus nebulosus*. The rudd did not show a preference for either end of the tank.

The roach, *Rutilus rutilus*, also emits a pheromone which attracts other members of the species (Hemmings, 1966). Young of the jewelfish, *Hemichromis bimaculatus*, which only school as young, orient to other young of their species by means of chemical stimuli (Kühme, 1964).

ALARM PHEROMONES. Some fish are capable of signalling the presence of danger to other members of their species and pheromones communicate this information for a wide variety of the Cypriniformes and Siluriformes.

Von Frisch (1938, 1941) discovered the alarm pheromone and reaction in fish, and it has been due to his efforts and the research of his students that more is known about this aspect of fish behavior than any other chemically induced behavior. Von Frisch (1941) defined an alarm substance as a substance which communicates the presence of danger, provided it is produced by members of the same species. This definition emphasized a response to danger and tended to direct, and possibly limit, the scope of the research on pheromones in fish (see discussion).

Von Frisch (1941) first noticed the alarm reaction in a minnow, *Phoxinus laevis*, the skin of which had been cut before it was returned to its school. Immediately the school dispersed. Subsequently, in a field study to determine the validity of the observations, a small feeding station was established on the edge of a pond and the minnows of the area were conditioned to feeding at the station. With the minnows conditioned, the alarm substance (prepared after a standardized technique: von Frisch, 1938) was introduced through the feeding tube five minutes after feeding. A flight response followed within 30 to 60 seconds. The fish at the feeding tray fled a short distance in an erratic manner, then grouped and retreated. The effect of the alarm substance on the school was long-lived, and in some instances several days passed before the fish would return to feed.

The fright reaction exhibited by *Phoxinus* was not stereotyped; thus von Frisch (1941) found it necessary to arbitrarily establish seven categories or intensities of response at comparable pheromone concentrations. At one extreme, the reaction involved the sudden flight to a hiding place, emergence, rapid swimming around the tank, and avoidance of the feeding area for a considerable length of time. At the opposite end of the spectrum, no reaction to the fright substance was observed. After 438 tests with schools of *Phoxinus*, von Frisch concluded that the variability appeared to be in part due to sex, season, fish from different waters, and nutritional state. The potency of the alarm pheromone diminished when the fish were starved, and after sixteen weeks its effectiveness had dropped to 25 percent of its normal level. Further, the minimum concentrations of the alarm pheromones required to elicit a behavioral response varied. Some fish responded to 1:500 dilution of the fright substance while others remained indifferent to the undiluted extract. Aquarium hatched *Phoxinus* responded to a dilution of 1:50,000. (Schutz, 1956; Pfeiffer, 1963).

Responsiveness to the alarm pheromones is variable both through circadian changes and as a result the experimental procedures themselves (Thinnes and Vandenbussche, 1966; Hemmings, 1966). The intensity of reaction of schools of *Rasbora heteromorpha* over 24 periods was recorded under normal daylight, artificial illumination at night, normal night conditions, and in complete artificial darkness during the daylight period, thereby separating the influences of circadian rhythms from light intensities. The schools varied considerably in their responses and the alarm substance produced its strongest reaction during the day even if the laboratory was in darkness. In the absence of the alarm substance, nighttime illumination did not increase the cohesion of the school significantly as compared to normal night dispersion. The readiness of *Rasbora* to school, in response to external stimuli, fluctuated diurnally (Thinnes and Vandenbussche, 1966). Long term diminution of responsiveness to alarm substances by *Phoxinus laevis* have been reported (Pfeiffer, 1963).

The problem of determining the causes of variability in responsiveness in a pheromone which supposedly acts to warn fish schools of a predator is further complicated by the predator also offering visual and mechanical cues. Two aquaria were placed side by side and their interiors were arranged as mirror images of each other so that two schools of *Phoxinus* were chemically but not visually isolated. When the pheromone was introduced into one of the tanks the chemically induced fright reaction in the first tank visually caused the same reaction in the school in the next tank (Schutz, 1956).

The alarm reaction varies between different types of fish although the majority of species studied react similarly to *Phoxinus:* upon sensing the pheromone they disperse, then regroup and flee (Pfeiffer, 1963). But the tench, *Tinca vulgaris,* and the crucian carp, *Carassius carassius,* swim with their heads on the bottom and their bodies at an angle of 60 degrees to the substrate, resulting in its disturbance. The murky waters serve to conceal the fish. Upon sensing the alarm pheromone, some bottom fish such as *Gobio fluviatilis* immediately cease movement, thus possibly avoiding their enemies. Hatchet fish, *Carnegiella strigata,* which are normally surface swimmers, dive and form dense schools (Pfeiffer, 1963). Other species such as *Esomus lineatus* rise to the water surface, crowd together, and leap out of the water (Schutz, 1956).

Alarm pheromones and reactions are widespread throughout the Ostariophysian fish, Cypriniformes and also Siluriformes that inhabit the marine as well as fresh-water environments. In some of the most recently evolved characins the alarm pheromones are absent, although they are Ostariophysi; it has been suggested that the ability to produce the pheromone has been secondarily lost in this group (Pfeiffer, 1963). Within the order Cypriniformes, the alarm pheromone from one species will elicit an alarm reaction in other species. The strongest reactions occurred between closely related species, even when they were widely separated geographically. The interfamilial responses were only slight (Schutz, 1956), which suggests that the chemical structure of the pheromone probably varies from species to species, being least similar in the most widely separated groups.

Steven (1959) reported the presence of the alarm pheromone and reaction in a non-Ostariophysian fish, the atherinid, *Hepsetia stipes* (*Atherinomorus stripes*). Skinner et al. (1962) described an alarm reaction in another atherinid, the top smelt, *Atherinops affinis.* Rosenblatt and Losey (1967) failed to find the pheromone, or reaction, in the top smelt and suggested that Skinner et al. described an alarm reaction caused by experimentally induced stress, possibly resulting from mechanical disturbances.

The Source and Nature of the Alarm Pheromone. The source of the alarm pheromone in the Ostariophysian fish appears to be similar throughout the group. Superficial injury of the skin showed that the epidermis contained the alarm substance (von Frisch, 1941). There is a particular type of club cell in the epidermis which is unique to those species which produce alarm pheromones. According to Pfeiffer (1960) these club cells only release their contents to the body surface upon injury. The barbel epidermis of the carp and some catfish contains no club cells or only a few very small ones, while the body epidermis of these species is abundantly supplied with these cells. The fright reaction was elicited by body skin, but not by barbel skin (Pfeiffer, 1963). Very little work has been done in ascertaining the chemistry of alarm pheromones. Boiling the substance for 5 minutes did not affect its potency, but boiling continued for 10 minutes reduced its effectiveness to about one-fifth of its normal level (von Frisch, 1941). It has been suggested that the alarm substance is a purine or pterin-like substance (Hüttel, 1941). These substances are concentrated only in xanthophores which are most numerous on the dorsal surface. Since ventral skin is an equally effective source of the alarm pheromone, Hüttel's suggestion is most likely not correct (Pfeiffer, 1963).

The alarm substance for *Phoxinus* is not volatile although extremely soluble in water (Schutz, 1956). Despite this fact the pheromone is detected by the fish's sense of smell. When the olfactory nerve was separated, *Phoxinus* was not able to detect its presence.

INTERSPECIFIC CHEMICAL COMMUNICATION

Fish interact most with their speciesmates, but also interact with other fish as well as invertebrates in their environment in relations that may be commensal, symbiotic, or of a predator-prey nature. Some of these interactions appear to involve various degrees of chemical communication, but few of them have been examined critically.

Among symbiotic relationships of fish with invertebrates involving species-typical behavior and suspected chemosensory components are those of *Fierasfer* with sea cucumbers, of *Aeoliscus* with sea urchins, perhaps that of the goby *Clevelandia ios* with the burrowing worm of mudflats, *Urechis*, (also called the fat innkeeper because it has several commensals), of *Nomeus* and the Portuguese man of war, and of the clownfish, *Amphiprion* with the giant sea anemone *Stoichactis*. Only the last of these relationshps has been investigated in any detail (Davenport, 1955).

One to several clownfish on a reef form a bond with an anemone; they are often breding pairs. The association—in which the fish has "the freedom of the anemone," hiding between its tentacles and darting in and out of them with impunity—is established gradually, and relies on progressively stronger and longer contact between the fish and the tentacles of the host. Nematocysts are not discharged, nor are feeding reactions displayed to the resident fish, to an isolated piece of its outer skin, or to its eggs, but they are displayed to its flesh without the skin. An intact alien fish species or its skin is stung and then ingested as the clownfish itself would be by another species of anemone. The skin mucus of the fish contains a heat-labile-active principle that is "discriminated" by the tentacles and raises the threshold in them to mechanically induce discharge nematocysts (Davenport and Norris, 1958).

It is not known whether the inhibition of a feeding reaction (clinging of tentacles to prey and transport of the latter to the mouth) caused by the mucus of *Amphyprion* is directly effective on the nervous system of the anemone or whether the inhibition occurs because no nematocysts are discharged. The rate of decay of the inhibitory compound or compounds in the mucus and the chemical nature of these compounds are also not known.

Specific scents are involved not only in commensal relations of fish, but prey is often recognized by its odor, and the scent of a predatory species modifies the behavior patterns of prey fishes. Compounds in the slime of the predator act on the olfactory channel of the prey in a manner comparable to the recognition of individuals by means of smell among the members of a specific community.

Motivated by the observation that freshly caught specimens of different species of fish have distinctive odors perceivable by man, Göz (1941) established that these odors are also discriminated by the fish themselves. The minnow *Phoxinus laevis* was trained to distinguish between waters in which several other species of fish had been kept. Some were in the minnow family; others were distantly related or unrelated to the test animals. *Phoxinus* could even learn to select the scent of a certain species from a mixture of scents of several species.

When blinded or isolated *Phoxinus* smell their most important predator, the pike (*Esox*), they may flee, occasionally attempting to jump; but most frequently they spread their fins and slowly glide to the bottom where they remain quite motionless, supporting themselves on the rims of their paired fins. The reaction resembles that exhibited to the fright substance contained in the minnow's own skin. When the fish is deprived of its sense of smell the reaction does not occur. As the pike locates prey predominantly through its lateral line sense, a behavior pattern that minimizes water movement, such as the one described above, would assume adaptive value.

When the minnows school, as is their habit, they are far less likely to react to pike odor in the manner indicated unless the predator is near them and has wounded a member of the school. Then the fright substance reinforces the effect of the predator smell. In experiments where pike odor was applied, after fright substance was introduced and subsequently flushed, the minnows again sank to the bottom and remained motionless.

Scents of other predators elicit similar reactions varying in intensity according to the degree to which *Phoxinus* is preyed upon by them. Comparable interactions based in part on chemical communication were observed by George (1960) between *Gambusia* and one of its predators, another member of the genus *Esox*. The experiments do not permit conclusions on the existence of a genetically determined behavioral response to predator smells. They do illustrate, however, as in the case of the reaction to the fish's own fright substance, that the responses are variable and easily modified.

A chemically triggered reaction of prey to potential predators that is more fixed exists in young jewelfish, *Hemichromis bimaculatus*, which are attracted to the odor of their own species, but avoid those of other species (Kühme, 1964). Baby jewel fish are heavily preyed upon and instead of identifying their potential predators, they avoid all other species of fish.

Another reaction to scents of potential predator species was observed in brown and yellow bullheads: these fish guard their schools of milling young. As the fry grow up, they disperse gradually and the parents relax their vigilance. Intrusion of predators such as bass, sunfish, and perch into the territory of parent and half-grown young elicits resumption of the guarding behavior of the adult bullheads. The behavior change can be elicited even when the predators are visually isolated from the parent (Atema, 1967, personal communication). The few examples cited here strongly suggest that other such chemosensory interactions between predator and prey species will be found to have resulted in species-specific adaptive behavior of the prey and perhaps also to having modified the behavior of the predator itself.

DISCUSSION

The study of chemical communication in fish has barely begun, and it is impossible to predict to what extent pheromones will be found to be important in their lives. The studies of MacGinitie (1939) and Tavolga (1956), who dealt with pair formation and maintenance, and of Kühme (1964), Künzer (1964), and Myrberg (1966), who concerned themselves with parental behavior, show that pheromones may have played a significant role in speciation and evolution. They have not yet been proved, however, to act as isolating mechanisms between fish, although many closely related species live sympatrically in habitats where visual mechanisms of sexual segregation would be of little value.

Pair formation and species segregation rely on behavior patterns that keep two or many fishes together; the slime of fish also has an important aggregating function which probably assists in maintaining schools of a species (Wrede, 1932; Keenleyside, 1955; Kühme, 1964; and Hemmings, 1966). Von Frisch (1938, 1941) and his students, Schutz (1956) and Pfeiffer (1963), demonstrated the presence of an alarm pheromone which was released by injured members of schools in a large number of Ostariophysian fish species and caused dispersal. They considered this substance solely as an adaptation to escape predation. The great variability of responsiveness to the pheromone among individuals from the same schools and between different schools of the same species, observed by them, casts some doubt on the adaptive value of a pheromone that functions only to warn a school of the presence of a predator in their midst.

An attractive hypothesis presents itself, suggesting that the alarm substance and the aggregating pheromones are in some way related. They may act in a manner similar to some insect pheromones which initiate aggregation at one concentration and cause flight and dispersion at another concentration (Bossert and Wilson, 1963). School aggregation and dispersion in response to chemical signals have both been described from Ostariophysian species. Their club cells, which may contain the alarm substance (Pfeiffer, 1963), do not connect directly to the outside, and injury is necessary for the release of the pheromone into the environment. The same substance, having seeped through the cell wall, may diffuse slowly outward and mix with the slime at a low concentration to give it its aggregating qualities. It is likely that the relationship between aggregating and alarm substances will prove to be complex and that other compounds are involved in one or the other reaction as well. We postulate, nevertheless, that the precursor to the alarm substance originally had a communicating function that was very likely to have elicited aggregation.

To deal with this question further, field as well as laboratory experiments, in which the behavioral responses of schools are analyzed for adaptiveness, will be necessary. The inability of schooling species to behave normally when confined to aquaria, no matter how large, will remain a serious drawback in the study of communication in fish schools. Semi-natural, penned-off lagoons or bays with water conditions permitting total area photography from above could provide a suitable work area for the study of fish schools. All or some members would be surgically deprived of their sense of smell prior to the initial observations.

Whatever new approaches or methods may be devised in the analysis of fish behavior involved in communication between fishes, experiments would become simpler and more precise if the chemical nature of the pheromone was known. Though slime may be amenable to a number of analytical treatments, these are complicated by the small size of the club cells and their diffuse nature; it will be difficult to secure, in relatively pure form, an adequate amount of their specific contents. A method for slime analysis could be adapted from the experiments of Hoese and Hoese (1967). In a quest for the nature of the substance which elicits feeding reactions in the oyster-feeding, naked goby (*Gobiosoma bosci*), they dialyzed oyster extracts and chromatographed them on Whatman paper to obtain spot chromatograms. These were cut into two strips, one of which was developed with the appropriate reagents and the other immersed in the aquarium where the gobies were observed for preference(s) for certain spots on the paper. While the tests gave insight into the chemical nature of attractants in certain specific foods of fish, they also illustrated that the method might be adapted for research on aggregating substances in fish. Compounds producing feeding reactions had as their main common characteristics 2 to 5 carbon atoms and an amino group, as well as one or two carboxyl groups, in a majority of cases. A sulfhydryl group, or its methyl substitute, or a methylated nitrogen atom were also components in the effective ingredients. All molecules that could be identified were simple and short, straight-chained with only one or two functional groups. Feeding tests using chromatographically separated fractions and liquid oyster extract alike indicated the presence of some other, unidentified effective substances with higher RF values than those of the several chromatographically identified amino acids. The above findings are of interest here because the one dispersing substance known to act on fish aggregations is serine from mammalian skin rinses, which salmon avoid when confronted with on their upriver migrations (Idler et al., 1956). Serine certainly is a compound like the ones described above and it can be smelled by a number of fish species, aside from salmon (Bardach et al., 1967a). The disparate items of information on water-borne stimuli eliciting certain behavior patterns in fish, and the necessity that the effective diffusible organic molecules be small, could at least serve as preliminary directives in a search for the nature of schooling and dispersing pheromones.

The social behavior of bullheads, mediated in a large part by compounds in their slime, was complex and sophisticated, involved discrimination between at least two but probably more individuals, and was variable with the status of individuals and the conditions of their encounters. We believe that bullheads and other species that exhibit such recognition have progressed beyond social releasers

(Lorenz, 1935; and Tinbergen, 1951) in the evolution of their socialization. Such social releasers are sign stimuli that trigger the behavior suitable for a particular social interaction. The interactions are specific: male for female, rival for rival, parent for offspring, schoolmate for schoolmate, and so on. Inherent in the concept of social releasers is the fact that little learning is involved.

Many social releasers are visual, such as in the stickleback, *Gasterosteus aculetatus*, a fish which is almost devoid of a sense of smell. Its behavior at the same time is rigid and dependent on sign stimuli (Tinbergen, 1951). Its red belly is the primary stimulus for attack; male sticklebacks will direct their attacks even toward a passing red van. The males court females with swollen bellies and females will follow red models in a tank. Tinbergen noted that in the stickleback even the complicated patterns of activities were dependent upon simple stimuli.

Intermediate between the olfactorily insensitive stickleback with its stereotyped behavior, and the yellow bullhead with its dependence upon smell, are the Glandulocaudine fish who derive their names from the fact that they have large glands on their tails. The exact communicating function of the glandular substance has not been determined, although its presence only in the males suggested that it serves as an isolating mechanism. Glandulocaudine fish are capable of recognizing individuals during social interactions (Nelson, 1964). *Corynopoma* males when offered a choice of females to court showed preferences based on prior interactions and not on size or aggressiveness. The preferences in one species were so consistent that one can well suspect the existence among them in nature of a true pair bond (Nelson, 1964). He did not investigate the senses which mediated this complex interaction; however, it seems likely that pheromones play a major role in it.

The yellow bullheads have many of the attributes of behavior usually ascribed only to animals much higher on the phylogenetic scale (Atema et al., 1969). They have an extremely well developed sense of smell essential to their social behavior, lending support to the suggestion by Maier and Schneirla (1935) that there exists in vertebrates a relationship between the sense of smell and behavioral complexity. Preliminary observations suggest that bullheads respond to bullhead models only on rare occasions, and no sign stimuli have been found to have significance for them, although these may exist during the mating season when stereotypy would tend to be adaptive.

Behavioral complexity in vertebrates is a correlate of nerve paths that permit increasing interactions between the components of the central nervous system. Whatever the roles of the fish forebrain may be, it certainly relays olfactory impulses to other centers. We have begun experimentation with bullheads whose forebrains were removed; they would attack a bullhead model, a hand waved in front of them, or even someone walking by their tank. As mentioned above, normal bullheads rarely respond to sign stimuli. The behavior of the forebrainless bullhead appears truly stereotyped and "primitive" in the first three postoperational months. It is noteworthy, though, that normal behavior gradually returns thereafter, without regeneration of the telencephalon (Atema, 1969).

Many fish probably have evolved pheromones as a means whereby individuality and status can be communicated irrespective of water transparency or time of day and, thus, they have taken a step to the subsequent evolution of complex social behavior. There are only a few indications as yet to support this hypothesis of a relationship between chemical communication and behavioral sophistication, but fish, with their great variation in brain structure and development, as well as behavior, are promising subjects for the study of the role of chemical communication in the evolution of behavior.

REFERENCES

Atema, J. 1967. Personal communication.

———— 1969. The chemical senses in feeding and social behavior of the catfish (*Ictalurus natalis*). Ph.D. Thesis, University of Michigan, Ann Arbor.

———— Todd, J. H. and J. E. Bardach. 1969. Olfaction and behavioral sophistication in fish. *In* Pfaffman, K., ed. Proceedings of the Third International Symposium on Olfaction and Taste. Rockefeller University, N.Y. pp. 241-251.

Baerends, G. P., R. Brouwer, and H. T. Waterbolk. 1955. Ethological studies on *Lebistes reticulatus* (Peters). *Behavior*, 8:249.

Bannister, L. H. 1965. The fine structure of the olfactory surface of teleostean fishes. *Quart. J. Micr. Sci.*, 104(4):333-342.

Bardach, J. E. 1967. Chemical senses and food intake in the lower vertebrates. *In* Kare, M. R., and O. Maller, eds. The Chemical Senses and Nutrition. Baltimore, The Johns Hopkins Press, pp. 19-33.

———— M. Fujiya, and A Holl. 1967a. Investigations of external chemoreceptors of fishes. *In* Hayashi, T., ed. Proceedings of the Second International Symposium on Olfaction and Taste. New York, Pergamon Press, Inc., pp. 647-665.

———— and R. Gooding. In preparation.

Bardach, J. E. and Atema, J. The sense of taste in fishes. *In* Beidler, L., ed. Handbook of Sensory Physiology, Vol. IV. Chemical senses. New York, Springer Verlag. (In press.)

———— J. H. Todd, and R. Crickmer. 1967b. Orientation by taste in fish of the genus *Ictalurus*. *Science*, 155(3767):1276-1278.

Belding, D. L. 1934. The spawning habits of the Atlantic Salmon. *Trans. Amer. Fish. Soc.*, 64:1-211.

Bossert, W. H., and E. O. Wilson. 1963. Analysis of olfactory communication among animals. *J. Theor. Biol.*, 5:443-469.

Breder, C. M. 1935. The reproductive habits of the common catfish, *Ameiurus nebulosus* (Le Sueur), with a discussion of their significance in ontogeny and phylogeny. *Zoologica*, 19:143-179.

———— 1939. Variations in the nesting habits of *Ameiurus nebulosus* (Le Sueur). *Zoologica*, 24:367-380.

———— 1951. *Quoted in* Clark and Aronson. Sexual behavior in the guppy, *Lebistes reticulatus* (Peters). *Zoologica*, 36:49-66.

Chen, C., and J. E. Bardach. In preparation.

Clark, E., and L. R. Aronson. 1951. Sexual behavior in the guppy, *Lebistes reticulatus* (Peters). *Zoologica*, 36:49-66.

Davenport, D. 1955. Specificity and behavior in symbioses. *Quart. Rev. Biol.*, 30:29-46.

——— and K. S. Norris. 1958. Observations on the symbiosis of the sea anemone *Stoichactus* and the pomacentrid fish, *Amphiprion percula*. *Biol. Bull.*, 115:397-410.

Dijkgraaf, S. 1968. Electroreception in the catfish, *Ameiurus nebulosus*. *Experientia*, 24(2):127-188.

Eggert, B. 1931. Die Geschlechtsorgane der Gobiiformes und Blenniformes. *Z. Wiss. Zool.*, 193:249-558.

Fagerlund, U. M., J. R. McBride, M. Smith, and N. Tomlinson. 1963. Olfactory perception in migrating salmon. III. Stimulants for adult sockeye salmon (*Oncorhynchus nerka*) in home stream waters. *J. Fish. Res. Board Can.*, 20:1457-1463.

Freihofer, W. C. 1963. Patterns of the Ramus Lateralis Accessorius and their systematic significance in teleostean fishes. *Stanford Ichthyol. Bull.*, 8(2):80-189.

George, C. J. W. 1960. Contributions to the natural history of the mosquito fish. Harvard University Doctoral Dissertation (unpublished).

Gerking, S. D. 1959. The restricted movement of fish populations. *Biol. Rev.*, 34:221-242.

Glaser, D. 1965. Intersuchungen über die absoluten Geschmackschwellen von Fischen. *Z. Vergl. Physiol.*, 53:1-25.

Göz, H. 1941. Ueber den Art und Individualgeruch bei Fischen. *Z. Vergl. Physiol.*, 29:1-45.

Gunning, G. E. 1959. The sensory basis for homing in the longear sunfish, *Lepomis megalotis* (Rafinesque). *Invest. Indiana Lakes, Streams*, 5:103-130.

Hara, T. J., K. Ueda, and A. Gorbman. 1965. Electroencephalograph studies of homing salmon. *Science*, 149(3686):884-885.

Hasler, A. D. 1957. Olfactory and gustatory senses of fish. *In* Brown, M., ed. The Physiology of Fishes. New York, Academic Press, Inc., Vol. 2, pp. 187-207.

——— 1966. Underwater Guideposts, Madison, University of Wisconsin Press.

——— and W. J. Wisby. 1951. Discrimination of stream odors by fishes and its relation to parent stream behavior. *Amer. Natur.*, 85:223-238.

Hemmings, C. C. 1966. Olfaction and vision in fish schooling. *J. Exp. Biol.*, 45:449-464.

Herrick, C. J. 1907. The tactile centers on the spinal cord and brain of the Searobin, *Prionotus carolinus*. *J. Comp. Neurol.*, 17:307-325.

Hoese, H. D., and D. Hoese. 1967. Studies on the biology of the feeding reaction in *Gobiosoma bosci*. *Tulane Studies in Zoology*, 14(2):55-62.

Holl, A. 1965. Vergleichende morphologische und histologische Untersuchungen am Geruchsorgan der Knochenfische., *Z. Morph. Okol. Tiere.*, 54:707-782.

Humbach, J. 1960. Geruch und Geschmack bei den augenlosen Höhlenfischen *Anoptychthys jordani*, Hubbs und Innes und *Anoptychthys hubbsi* Alvarez. *Naturwissenschaften*, 23:557.

Hüttel, R. (1941). *In* Pfeiffer, W. 1963. Alarm Substances. *Experientia*, 9:1-11.

Idler, D. R., U. M. Fagerlund, and H. Mayo. 1956. Olfactory perception in migrating salmon I. 1-Serine, a salmon repellent in mammalian skin. *J. Gen. Physiol.*, 39:889-892.

Jaski, C. J. 1939. Ein Oestruszyklus bei *Lebistes reticulatus* (Peters). *Proc. Kon. Ned. Akad. Wet.*, 42:201-207.

——— 1947. *In* Bretscheider and Duvvene de Wit. eds. Sexual Endocrinology of Non-Mammalian Vertebrates. New York, Elsevier, pp. 129-131.

Keenleyside, M. H. S. 1955. Some aspects in the schooling behavior of fish. *Behavior*, 8:183-248.

Kleerekoper, H. 1967. Some effects of olfactory stimulation on locomotor patterns in fish. *In* Hayashi, T., ed. Olfaction and Taste, II. Proceedings of the 2nd International Symposium, Tokyo. Oxford, Pergamon Press, Inc., pp. 625-645.

——— and G. A. van Erkel. 1960. The olfactory apparatus of *Petromyzon marinus*. *Can. J. Zool.*, 38:209-223.

Konishi, J., and Y. Zotterman. 1963. Taste functions in fish. *In* Zotterman, Y., ed. Olfation and Taste, I. Proceedings of the 1st International Symposium, Wenner Gren Center, Stockholm. New York, The Macmillan Company, pp. 215-233.

Kühme, W. 1964. Chemisch ausgeloeste Brutpflege und Schwarmreaktionen bei *Hemichromis bimaculatus* (Pisces). *Z. Tierpsychol.*, 20:688-704.

——— 1964. Eine chemisch ausgeloeste Schwarmreaktion bein jungen Cichliden (Pisces). *Naturwissenschaften*, 51:120-121.

Künzer, P. 1964. Weitere Versuche zur Auslösung der Nachfolgereaktion bei Jungfischen von *Nannacara anomala* (Cichlidea). *Naturwissenschaften*, 51:419-420.

Losey. 1969. Sexual pheromone in some fishes of the genus *Hypsoblennius*, Gill. *Science*, 163(3863):181-183.

Leiner, M. 1930. Fortsetzung der ökolgischen Studien an *Gasterosteus aculeatus*. *Z. Morphol. Okol. Tiere*, 16:499-522.

Lorenz, K. 1935. Der Kumpan in der Umwelt des Vogels. *J. Ornithol.* 83:137-213, 289-413.

——— 1952. King Solomon's Ring. New York, T. Y. Crowell Company.

MacGinitie, G. E. 1939. The natural history of the blind goby, *Typhlogobius californiensis* (Steindachner). *Amer. Mid. Naturalist*, 21:489-505.

MacKim, J. M. 1966. Stress hormone metabolites and their fluctuations in the urine of the Rainbow Trout (*Salmo Gairdneri*). Ph.D. Thesis, University of Michigan, Ann Arbor.

Maier, N. R. F., and T. C. Schneirla. 1935. Principles of Animal Psychology: New York, McGraw-Hill Book Company.

Marshall, N. B. 1954. Aspects of Deep Sea Biology. New York, Philosophical Library.

McLarney, W. O. 1968. Personal communication.

Morrow, J. E. 1948. Schooling behavior in fishes. *Quart. Rev. Biol.*, 23:27-38.

Myrberg, A. A. 1966. Parental recognition of young in Cichlid fishes. *Anim. Behav.*, 14:565-571.

Nelson, K. 1964. Behavior and morphology in the Glandulocaudine fishes (Ostariophysi, Characidae). *Univ. Calif. Publ. Zool.*, 75 (2):59-152.

Pfaffenzeller, J. 1923. Über Sinnesorgane am Reusenapparat verschiedener Fische. *Zool. Jahrb. Abt. Anat. Ontog. T.*, 44:615-626.

Pfeiffer, W. 1960. Über die Schreckreaktion bei Fischchen und die Herkunft des Schreckstoffes. *Z. Vergl. Physiol.*, 43:578-614.

——— 1963. Alarm substances. *Experientia*, 19:1-11.

Rosenblatt, R. H., and G. S. Losey, 1967. Alarm reaction of the Top Smelt, *Atherinops affinis*: Reexamination. *Science*, 138:671-672.

Roule, L. 1931. Les Poissons et le Monde Vicant des Eaux. Paris, Delagrave.

Schutz, F. 1956. Vergleichende Untersuchungen über die Schreckreaktion bei Fischen und deren Verbreitung. *Z. Vergl. Physiol.*, 38:84-135.

Seitz, A. 1940. Die Paarbildung bei einigen Cichliden 1. *Z. Tierpsychol.*, 4:40-84.

Skinner, W. A., R. D. Mathews, and R. M. Parkhurst. 1962. Alarm reaction of the Top Smelt, *Atherinops affinis* (Ayres). *Science*, 138:681-682.

Steven, D. M. 1959. Studies in the shoaling behavior of fish, 1. *J. Exp. Biol.*, 36:261-280.

Stolk, A. 1950. The ovarian occlusion apparatus in the viviparous cyprinodonts, *Lebistes reticulatus* (Peters) and *Xiphorus helleri* Heckel. *Proc. Kon. Ned. Akad. Wet.*, 53:1-33.

Tavolga, W. N. 1956. Visual, chemical and sound stimuli in the sex discriminatory behavior of the gobiid fish, *Bathygobius soporator*. *Zoologica.*, 41:49-64.

Teichmann, H. 1962. Die Chemorezeption der Fische. *Ergeb. Biol.*, 25:177-205.

——— 1962. Was leistet der Geruchssinn bei Fischen? *Die Umschau in Wissenschaft und Technik.* Frankfurt a. M. 62(19):588-591.

——— 1954. Vergleichende Untersuchungen an der Nase der Fische. *Z. Morphol. u. Oekol. Tiere.*, 43:171-212.

Tester, A. L. 1963. Olfaction, gustation and the common chemical sense in sharks. *In* Gilbert, P. W., ed. Sharks and Survival. Boston, D. C. Heath.

Thinnes, G., and E. Vandenbussche. 1966. The effects of the alarm substances on the schooling behavior of *Rasbora heteromorpha* Dunker, in day and night conditions. *Anim. Behav.*, 14:296-302.

Tinbergen, N. 1951. The Study of Instinct. London, Oxford Univ. Press.

Todd, J. H. 1968. The social behavior of the Yellow Bullhead, *Ictalurus natalis*. Ph.D. Thesis, University of Michigan, Ann Arbor.

——— Atema, J., and J. E. Bardach. 1967. Chemical communication in the social behavior of a fish, the Yellow Bullhead, *Ictalurus natalis*. *Science*, 158:672-673.

van Koersveld, J. 1949. *Cited in* Baerends, G. P., et al. 1955. Ethological studies on *Lebistes reticulatus* (Peters). *Behavior*, 8:249.

von Frisch, K. 1938. Zur Physiologie des Fischschwarmes. *Naturwissenschaften*, 26:601-606.

——— 1941. Über einen Schreckstoff der Fischhaut und seine biologische Bedeutung. *Z. Vergl. Physiol.*, 29:46-145.

Von Skramlik, C. 1948. *In* Glaser, D. Intersuchungen über die absoluten Geschmackschwellen von Fischen. *Z. Vergl. Physiol.*, 53:1-25.

Walker, T. J., and A. D. Hasler. 1949. Detection and discrimination of odors of aquatic plants by the bluntnose Minnow (*Hyborhynchus notatus*). *Physiol. Zool.*, 22:45-63.

Whitear, M. 1965. Presumed sensory cells in fish epidermis. *Nature*, 208(5011): 763-764.

Wrede, W. L. 1932. Versuche über den Artduft der Elritzen. *Z. Vergl. Physiol.*, 17:510-519.

9

Chemical Perception in Reptiles

GORDON M. BURGHARDT

Department of Psychology
The University of Tennessee
Knoxville, Tennessee 37916

INTRODUCTION 241
REPTILIAN EVOLUTION AND MORPHOLOGY OF THE CHEMICAL SENSES 242
THE ROLE OF THE CHEMICAL SENSES IN REPTILIAN BEHAVIOR 245
CHEMICAL PERCEPTION IN NEWBORN SNAKES 277
CONCLUSION 303
ACKNOWLEDGMENTS 304
REFERENCES 304

INTRODUCTION

It is an unfortunate fact that while turtles, snakes, and other reptiles are familiar to almost everyone, they are little known scientifically compared to the other groups of vertebrates and many groups of invertebrates. Reptilian sensory abilities and behavior, especially, have been neglected, although that situation is changing.

Reptiles have invaded almost all habitats, and extinct forms have lived in many areas unoccupied by extant forms. Great structural divergence accompanied such ecological differences; consequently, reptiles have widely varying behavior patterns, sense organs, and, therefore, sensory abilities. While there exists no general systematic account of reptile behavior, some sources for an introduction to the richness and diversity of reptilian structure, ecology, and behavior are: Bellairs (1960), Carr (1966), Ditmars (1936), Goin and Goin (1962), Mertens (1960), Oliver (1955), Pope (1955), and Schmidt and Inger (1957).

This chapter will be concerned primarily with the chemosensory aspects of reptilian behavior, but similar reviews of other modalities are urgently needed. The anatomy of the chemical sense organs will be discussed briefly for orientation.

Physiological work will be omitted as it will be more appropriately covered in a later volume of this series. The main portion of the chapter begins with a brief treatment of the methods used in the study of chemical senses in relation to behavior. Reptile behavior has been divided into somewhat arbitrary functional categories (e.g., courtship, feeding) and within each of these, the results for the major reptilian groups are treated separately. Since feeding behavior has been studied most, it will constitute the major part of the review. While not meant to cover all reports which implicate the chemical senses, the plan was to discuss most of the experimental work with emphasis on both methodology and findings. A few references were encountered which could not be obtained. It is hoped that this review may be valuable in suggesting techniques and problem areas for future studies of reptile behavior.

REPTILIAN EVOLUTION AND MORPHOLOGY OF THE CHEMICAL SENSES

From the late Permian to the Cretaceous periods (about a 190-million-year span), reptiles were the dominant terrestrial form of life. Successfully invading all available ecological niches, they showed a remarkable structural diversity. The present reptile fauna is sparse by comparison but, even so, a wide variety still exists. Because of the diversity in the structure, habitat, and evolutionary history of reptiles, it is not very useful to describe a "typical" reptile, especially with regard to the structure and role of the chemical sense organs.

In reptiles, as in most vertebrates, gustation and olfaction are the most important chemoreception processes. The gustatory channel seems to be considerably less important in most forms and, consequently, has been the least studied. For some groups, the olfactory apparatus itself can be divided into two separate sense organs: Jacobson's organ (also referred to as the vomeronasal organ), and the

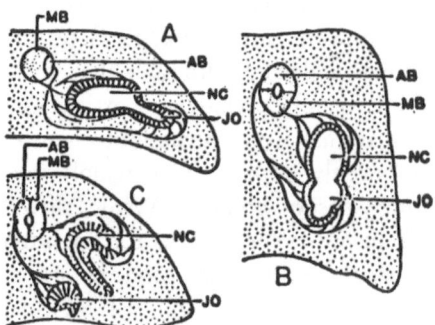

FIG. 1. Diagrams of the nasal areas of various forms to show the innervation of the nose. In order to show the nasal cavities in transverse section, the olfactory bulbs are figured as lying dorsomedial to the center of the nasal cavity; actually, they are well posterior to this position. The distinctness of the two main subdivisions of the olfactory nerve is considerably exaggerated, especially in A. A, a urodele; B, a turtle; and C, a snake. Abbreviations: AB, accessory olfactory bulb; JO, Jacobson's organ or its probable homolog; MB, main olfactory bulb; and NC, nasal cavity. The solid lines represent the branches of the olfactory nerve. (*From* Parsons. 1959a. *Evolution*, 13:177.)

typical olfactory organ associated with the nose or snout. The term *olfaction*, henceforth, will be used only in relation to the latter organ. In those amphibians (Fig. 1A) and mammals, as well as some reptiles (Fig. 1B), which possess Jacobson's organ, both the olfactory and Jacobson's organs occur together, and they are difficult to distinguish anatomically and functionally. Consequently, Jacobson's organ is rarely mentioned except in detailed anatomical reports.

In general, the nose consists of paired membranous sacs which open anteriorly at the external nares or nostrils and posteriorly into the palate via the internal nares. The nasal apparatus has, in addition to olfaction, the important function of the "preparation of the inspired air for contact with the delicate lung surfaces, involving moistening, cleansing, and thermoregulation . . ." (Stebbins, 1948, p. 184). Thus, the nasal apparatus will vary in reptiles not only in response to chemosensory needs, but also in response to climate and habitat on an even more fundamental level. For instance, Stebbins (1943) has shown that in lizards of the sand-dwelling genus *Uma*, modifications of the nasal apparatus have occurred to prevent inspiration of minute particles and to allow respiration beneath fine sand.

Considerable early literature exists on the detailed anatomy of the nasal cavity and associated structures in reptiles. Nevertheless, many species still have not been examined. Although a detailed review of this literature is not feasible here, valuable recent reviews and bibliographies, upon which the following summary is based, are given by Parsons (1959a, 1959b, 1967).

Chelonia

The nasal cavities of most turtles are relatively simple (Fig. 1B). Olfactory epithelium lines most of the olfactory region. In most turtles, sulci of the intermediate region are lined with vomeronasal epithelium. Turtles possess a large accessory olfactory bulb, which contains most of the fibers leading to the vomeronasal epithelium plus some of those leading to the olfactory region. These fibers constitute the medial trunk of the olfactory nerve. Fibers from the lateral and dorsal walls of the olfactory area form the lateral trunk of the olfactory nerve which enters the main bulb. There is a controversy as to exactly what constitutes Jacobson's organ in turtles.

Crocodilia

Crocodilians possess a complex nasal cavity. Jacobson's organ appears completely vestigial in these species, and it quickly disappears in the developing embryo. In addition, there is no vomeronasal epithelium and, correspondingly, no accessory olfactory bulb.

Rhynchocephalia

Sphenodon punctatum—the lizard-like tuatara from New Zealand—is the only living member of this order. The group is closely related to the Eosuchia, from whence the Squamata probably arose (Bellairs, 1960). The nasal anatomy of

Sphenodon punctatum is quite similar to that of lizards, but little is known behaviorally, and we will not consider this species again.

Squamata

The lizards and snakes constitute the most widespread and diversified group of reptiles extant today. The anatomy of the nasal structures in lizards and snakes has been comparatively well studied. In these organisms, Jacobson's organ has become completely separated from the nasal cavity. It is a paired structure, lying below the nasal cavity, opening by separate passages into the palate in the anterior part of the mouth (Fig. 1C). It is a roughly spherical structure whose ventral side is normally invaginated to form a large mushroom-shaped body. A narrow duct leads to the palate and enters the oral cavity anterior to the choana. Vomeronasal epithelium is limited to Jacobson's organ, and olfactory epithelium is restricted to the nasal cavity. All families of snakes have a well-developed Jacobson's organ. However, much more variation is found in lizards. In some lizards, almost all the chemosensory areas are poorly developed and may even be completely absent. Most squamates have a very large accessory olfactory bulb which is roughly medial to the main

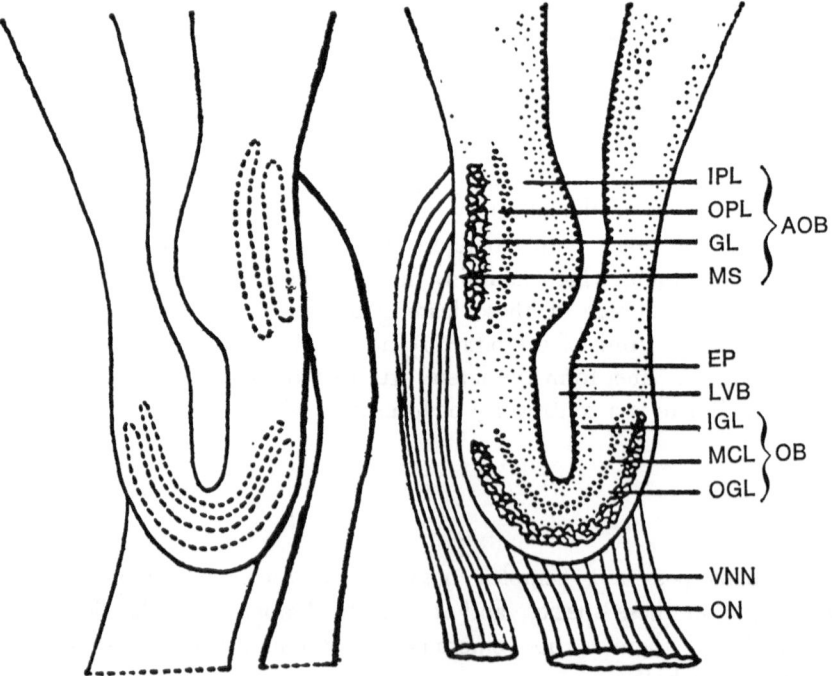

FIG. 2. Diagram of the olfactory and accessory olfactory bulbs in a reptile. A frontal section is shown to illustrate the various cell layers. AOB, accessory olfactory bulb; EP, ependyma; GL, glomerular layer; IGL, inner granular layer; IPL, inner plexiform layer; LVB, lateral ventricle of the brain; MCL, mitral cell layer; MS, myelospongium; OB, olfactory bulf; OGL, outer granular layer; ON, olfactory nerve; OPL, outer plexiform layer; VNN, vomeronassal nerve. (*From* Parsons. 1959b. *Bull. Mus. Comp. Zool.*, Harvard Univ., 120:237.)

olfactory bulb. This bulb receives the nerve fibers from Jacobson's organ via the vomeronasal nerve, which is quite distinct from the rest of the olfactory nerve (Fig. 2). The relative development of the accessory and main olfactory bulbs is closely correlated with the relative amounts of vomeronasal and olfactory epithelium, respectively. In the blacksnake (*Coluber c. constrictor*) the accessory bulb is even larger than the olfactory bulb (Carey, 1967).

Recently, Burns (1969) has discovered oral papillae in the upper and lower jaws of various snakes, most prominently in sea snakes. That these are chemosensory is indicated by the presence of taste bud-like structures innervated by the fifth cranial nerve (trigeminal).

The frequent tongue extensions in snakes and many lizards have led to much speculation—some of it going back many centuries—concerning function. The tongue in most higher vertebrates has a clear chemosensory function, that of taste. It is only to be expected then that the role of gustation should be attributed to the tongue. Indeed, Aristotle thought that the snake tongue is forked into two tips "and has a fine and hair-like extremity because of their great liking for dainty food. For by this arrangement they derive a twofold pleasure from savours, their gustatory sensation being as it were doubled." (1912, 660^b 10). In point of fact, the serpent tongue does not possess taste buds (Oliver, 1955; Klauber, 1956; Payne, 1945).

Klauber (1956) has discussed in some detail the various functions advanced for the tongue. Some of the more imaginative functions advanced were that the tongue served to keep the serpent's nose clean, to lick and thereby to moisten prey before swallowing, to serve as an attractant or decoy for prey, to facilitate cooling via evaporation, to lap up liquids, to serve as a feeler, or even to function as a poisonous stinger.

The view here is that the tongue does play a role in chemoreception, but not by any of the mechanisms suggested above. The detailing of this role and the experimental evidence for it constitutes an important subplot in the story that follows.

THE ROLE OF THE CHEMICAL SENSES IN REPTILIAN BEHAVIOR

All attempts to demonstrate whether a particular behavior is dependent upon a chemical sense employ one of two methods. The first and most common method is to exclude systematically use of the various sense organs themselves. This may be done by destroying the sense organ (e.g., by cauterizing Jacobson's organ) or by temporarily making it functionless (e.g., by covering the eyes or by plugging the nostrils). A much less used technique of eliminating the function of a sensory receptor is the severing of the nerves which innervate the structure. These changes are irreversible.

A less common method used to investigate the role of the various senses is to eliminate systematically sensory cues from a normally effective stimulus. Examples include testing the animal in the dark to rule out visual cues or placing the stimulus in a sealed glass container to eliminate chemical cues. Experiments involving both methods of sensory restriction simultaneously are of great value. If a particular behavior seems especially dependent upon a certain aspect of the usual stimulus object, the cue is referred to as a *sign stimulus*.

The implication of the chemical senses in a given behavior is certainly not in and of itself very satisfying. It is necessary to know how the chemical senses function in regulating reptilian behavior. For most behaviors, however, so little is known about the significance of chemoreception that in some areas speculation or informal observations constitute the total information available. Some readers may feel that trivial results utilizing crude methods are discussed in too great detail. The only defense is that a science must begin somewhere and a knowledge of the methods and results of the past can serve to inspire future studies.

Orientation

Studies on the role of the chemical senses in general orientation to gross environmental situations (which exclude orientation to prey, predators, and species members) are limited mainly to turtles. The orientation of hatching turtles to bodies of water after digging out of the nest has received much experimental attention in both the field and laboratory (see Ortleb and Sexton, 1964, for a thorough bibliography of studies up to that time). Odors have been considered in a few studies, but no observations implicate the chemical senses at this time.

Noble and Breslau (1938) showed that hatchling snapping turtles (*Chleydra serpentina*) did not orient to water from the river toward which the turtles were migrating when collected. Their method was to place the turtles in the center of a rectangular tank with the water in a dish at one end. The trials were performed in the dark with groups of hatchlings. The nonsignificant preference for the river water was attributed to a humidity gradient since controls with other types of water at the opposite end were not performed. Although there is as yet no evidence, water scents from the area where young green turtles are hatched may play a role in the remarkable transoceanic migrations of these animals (Carr, 1965), similar to that shown in salmon (Hasler, 1966).

Ortleb and Sexton (1964) investigated the factors underlying orientation in painted turtles (*Chrysemys picta*), an aquatic fresh-water species. They used a large water-filled Y maze, in which the stem was 6 feet long and each arm 4 feet long; the width and depth was 10 inches. Water could be introduced into each stem. Using a two-choice technique, they investigated the turning of the animals towards and away from arms varying in the following parameters: temperature, current, illumination, water color, and odor. The last parameter involved a choice between simulated pond water and tap water. No choice was evident. These workers avoided the two-choice control problem neglected by Noble and Breslau.

Although there is no experimental or even anecdotal evidence implicating the chemical senses in the orientation and homing of turtles, it would not be surprising if such evidence is soon found. Critically needed are homing experiments in which the senses are systematically eliminated, as has been done with newts (Grant et al., 1968).

Studies on homing in lizards have typically involved diurnal visually-dominated forms, such as inguanids. The sensory basis of such behavior has not yet been studied, although vision will undoubtedly prove to be of major importance. Studies on the more secretive fossorial forms, such as skinks and teiids, would be interesting.

There has been little study of homing behavior in snakes. Some evidence for homing in the water snake *Natrix s. sipedon* has been found by Fraker (personal communication) who feels that the chemical senses are involved. If these findings are substantiated and the role of chemoreception is supported, it would be interesting to know whether airborne or substrate terrestrial cues are more important to this semiaquatic species or whether aquatic cues predominate. Autumnal mass denning migrations in some snakes are other, perhaps similar, phenomena where chemical cues may be important.

Intraspecific Aggregation Behavior

Some species of serpents, if disturbed or exposed to unfavorable conditions, will come together and form a compact intertwined mass. Both young and adults engage in this behavior regardless of season, so it is clearly not tied to either sexual behavior or hibernation. Gravid females, however, do not aggregate. Aggregation is especially prominent in the North American brown snake (*Storeria dekayi*). Using this and other species, Noble and Clausen (1936) performed a series of fascinating experiments on the factors involved in this behavior. The present discussion is limited to their experiments on sensory control, which were influenced by the earlier German work on feeding behavior discussed later in this chapter.

In determining the roles of the various senses, Noble and Clausen tested for the aggregation response in a very simple manner. The test situation was a small cage containing a layer of slightly dampened gravel. The stimulus was to shake the cage suddenly or merely to tap one of the sides. This was continued for three minutes and the distribution of the snakes was noted. If more than 90 percent of the snakes aggregated, the response was considered positive; if less than 20 percent of the snakes aggregated, it was scored negative. Usually, several groups were tested under the given conditions and repeated testing of the same group was employed.

The senses were eliminated in the following ways (Table 1): snakes were blinded (B) by covering the eyes with either adhesive tape and india ink or with a mixture of collodion and ink; smell was eliminated (N) by plugging the nostrils with cotton smeared with Vaseline and covering the plug with collodion; Jacobson's organ was presumably eliminated by severing the mediating tongue caudal to the fork (T) or by cauterizing the organ itself (J).

The snakes were tested in both an illuminated and a dark room. Table 1 gives the results. Noble and Clausen concluded that vision is most important, that olfaction is less so, and that Jacobson's organ is virtually not involved at all. However, some comments are in order. The authors reported only results for the tongueless (T) animals, but nowhere do they present the results for the cauterized (J) group. A more complete presentation of the results, particularly a percent aggregation score rather than a simple positive-negative score, might be useful. The table oversimplifies the results, since the authors qualify some of the results with a

TABLE 1

The influence of sense organs on aggregation in *Storeria dekayi**

SERIES†	ROOM‡	NUMBER OF SNAKES		TRIALS	REACTION
		EXPERIMENTAL§	CONTROL‖		
B	L	15	0	3	Positive
B	L	15	10	3	Positive
B	D	15	0	4	Positive
B	D	15	10	3	Positive
N	L	14	0	3	Positive
N	L	14	10	3	Positive
N	D	14	0	8	Negative
N	D	14	10	3	Negative
T	L	12	0	4	Positive
T	L	12	10	5	Positive
T	D	12	0	4	Positive
T	D	12	10	4	Positive
BN	L	14	0	5	Negative
BN	L	14	10	5	Negative
BN	D	14	0	4	Negative
BN	D	14	10	4	Negative
BT	L	12	0	4	Positive
BT	L	12	10	3	Positive
BT	D	12	0	2	Positive
BT	D	12	10	3	Positive
NT	L	11	0	3	Positive
NT	L	11	10	5	Positive
NT	D	11	0	5	(4) Negative
NT	D	11	10	5	Negative
BNT	L	11	0	4	Negative
BNT	L	11	10	4	Negative
BNT	D	11	0	3	Negative
BNT	D	11	10	4	Negative

* From Noble and Clausen. 1936. Ecol. Monogr., 6:271.
† Designates types of experimental animal used, i.e., - B, blindfolded; N, nose-stopped; T, tongue-less; BN, blindfolded and nose-stopped; BT, blindfolded and tongueless; NT, nose-stopped and tongueless; BNT, blindfolded, nose-stopped and tongueless.
‡ L - Daylight (in front of window); D - Darkroom.
§ Animals from which sensory structures were removed.
‖ Untreated animals.

"longer latency" note. As with much of the data of Noble and his colleagues, discussed below, some statistical treatment was clearly needed.

Further experiments showed that *Storeria* will not aggregate with narcotized individuals of the same species or with specimens infiltrated with paraffin or with colored wax models. For normal *Storeria*, movement appears necessary since freshly killed specimens are ineffective. Tongueless snakes reacted like normal ones, but 66 percent of the plugged-nostril snakes, frequently in the dark, aggregated with models and narcotized individuals. These results indicate that olfaction plays an important role in the discrimination of the "proper" stimulus. The tongue has only a minor role, if any at all, according to the authors.

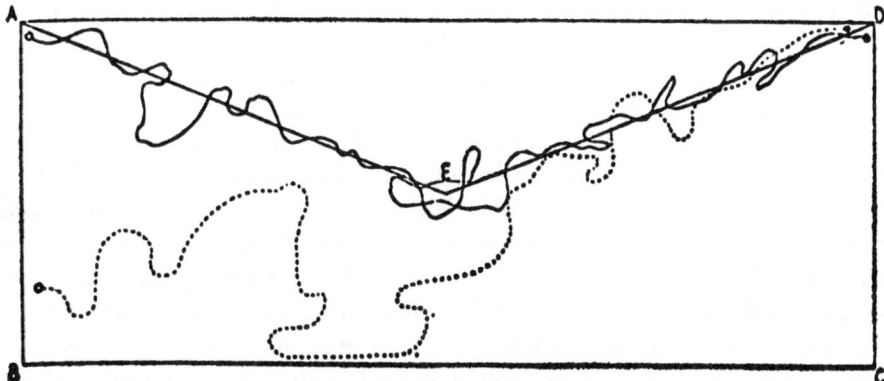

Fig. 3. The trail of two male *S. dekayi* starting at AB and ending at D. Trail AED was made by rubbing body integument over the course. (*From* Nobel and Clausen. 1936. *Ecol. Monogr.*, 6:297.)

The role of chemical stimuli was further shown by dyeing individuals different colors and assaying the effect of this on aggregation. There was none. If *Storeria dekayi* and *Thamnophis sauritus* individuals were mixed, the two species formed separate aggregations, but in all cases where the nostrils were plugged in both species, a certain number of mixed aggregations occurred. Tongue removal had no effect.

What, then, is the source of the olfactory stimulus? Covering the cloaca eliminated the cloacal secretion but it did not eliminate aggregation. Covering the body of the snake with vaseline, however, did abolish the response, which suggests the chemical comes from the integument. Noble and Clausen made use of a trailing technique which involved the following of a trail made by rubbing the cloacal region or the integument on gravel, concrete, or sandstone. Figure 3 shows some typical results. Olfaction appeared to be the critical modality. Blind, and blind and tongueless subjects trailed normally. Again, only the integument was found effective.

Dundee and Miller (1968) have recently returned to the study of aggregation in snakes. Using ringneck snakes (*Diadophis punctatus arnyi*), it was found that if a group are placed in a circular, sand-filled arena containing only ten identical inverted 12-inch discs in a circular arrangement, the snakes would most frequently be found under one or only a few discs. Further, they would choose pans under which snakes had previously aggregated, suggesting that a chemical cue was involved. This was supported by 1) rotating the arena so that external visual factors were varied, and 2) transferring the "conditioned" sand to under unpreferred discs. Results of both tests support the importance of "habitat conditioning" via chemical (excretory?) factors. The relative strength of the aggregation tendency varied with substrate dampness. (See also discussion on p. 276.)

Alarm Reaction

Frequently, snakes release a rather obnoxious-smelling fluid from the postanal or cloacal glands. Speculation concerning its function has been limited to a consideration of either a defensive role directed against a potential predator, or a role in attracting other individuals, as in sexual behavior or in aggregation.

Recently, Brisbin (1968) has reported an observation which he feels is evidence that this secretion is used as an alarm substance. He used the King snake (*Lampropeltus getulus*). A discussion of his evidence indicates how desperate we are for knowledge about reptile behavior. In this experiment, one female was obtained as a young hatchling, probably less than two weeks old. It was raised in isolation from other snakes for more than three years. The snake was periodically measured but she never released any scent from her postanal gland, which many snakes do normally upon being disturbed. This individual was quite phlegmatic. One day, however, a newly captured, adult, male king snake of a different subspecies was brought into the laboratory and routinely measured on the laboratory work table. This individual did release a copious amount of secretion on the table top. After he was weighed and measured, the table was thoroughly rinsed with water, but it still retained (to the human nose) a musky odor. Shortly thereafter, the hand-raised king snake was brought into the room and placed on the table to be measured. The snake began to move rapidly back and forth on the table top, displaying rapid tail vibration as if it was extremely disturbed, and then it began to emit musk, probably for the first time in its life. Was this a coincidence or a causal situation? Brisbin argued that it was possible that the postanal musk had an alarm function. While the reporting of the single case has an important function to play in scientific research, some questions must be asked. Why, for instance, was the experiment not repeated? What would happen if this or other snakes were placed in the vicinity of a snake which was releasing musk or disturbed to the point of doing so? Since so little ethogram information is available on snakes, should it be concluded that an alarm response was observed? No importance was placed on the fact one snake was a male and the other a female. Nevertheless, if the observation can be replicated, it does indicate that the inexperienced king snake can recognize an alarm substance or, at least, the scents emanating from another snake. Determining whether this is species-characteristic depends upon further research.

Territorial Markings

Glandular secretions which serve to designate an individual "home range" or "territory" have not yet been demonstrated in any reptile. The most promising research in this area is concerned with the femoral glands in lizards. Many lizards have pores on the ventral side of the hind thighs. These epidermal follicular glands have been described in many lizards. An excellent recent study is that by Cole (1966b) on *Crotophytus collaris*. In general, males have larger glands than females, the glands are most active during the breeding season, and either castration or administration of sex hormones can alter their size (Cole, 1966a). Although several studies are underway (Cole, 1966a), the nature of the secretion has not been determined for even one species.

The function of these glands is unknown, but Cole (1966a) lists several suggestions that have been advanced. An early hypothesis was that the secretion served to fasten the male to the female during copulation. Cole suggested that close observation of copulatory behavior eliminated this notion. The secretion is often neither strong enough nor sticky enough to have this function. Noble and Bradley (1933) suggested that in some species the accumulated secretions serve as a tactile stimulus to enable the male to quiet the female during mating. Although this cannot be ruled out, Cole notes that males of some species possessing femoral glands do not rub the females in the manner described by Noble and Bradley, and some species lacking the glands do rub the females in this manner.

However, two hypotheses do suggest a chemoreceptive function. The secretion may be deposited in the home range of the male either facilitating the pairing of the sexes or serving to define a territory, presumably to ward off other males. Both could be operative, but it is not even known if the secretion is odoriferous. Hathaway (1964) has made some suggestive observations on *Crotophytus*: it may be that the organs are vestigial in all lizards and have no function, or that the above suggestions are completely erroneous.

Courtship

CHELONIA. Behavior in courtship, especially species and sex recognition, often relies on chemical cues. In reptiles, however, actual evidence is hard to find. Auffenberg (1965) performed a particularly neat series of observations on two partially sympatric species of South American tortoises (*Geochelone*). During the breeding season, he found two phases involved in sex and species discrimination. Males would challenge all turtle-like objects with species-characteristic, horizontal head movements. Reciprocal, identical head movements by the challenged turtle served to identify it as a conspecific male. This was entirely visual. If a challenged tortoise did not respond, the male challenger would walk to the posterior part of the shell and smell the cloacal area. If the second turtle was a conspecific, sexually mature female, the male mounted immediately. This, then, was the second phase which involved, presumably, an olfactory cue (Fig. 4). A more detailed analysis has not been carried out. Auffenberg did note that females of the other species were mounted only when no conspecific female was available. He also attempted to smear cloacal exudations of sexually active females of either species on other species. This, however, "only infrequently resulted in mounting attempts" [p. 339]. A male *G. carbonaria* did try to mount a skeletonized shell of the same species when cloacal exudations were rubbed on its posterior portion. Finally, Auffenberg noted a male trying to mount and to copulate with a head of lettuce, which a female had climbed over (with some difficulty) a few moments previously.

Auffenberg saw the second (olfactory) stage as the more primitive, and he saw the first (visual) stage as deriving from a basic motor pattern associated with olfaction. In this, he followed Eglis (1962), who classified olfactory "sniffing" movements in tortoises and found them to be of taxonomic value.

CROCODILIA. McIlhenny (1935), as well as others, has described the pungent musk emitted by bull alligators (*Alligator mississippiensis*) during the mating season. These bulls also roar. Both stimuli are perceived by humans at considerable distances and are thought to function in attracting females.

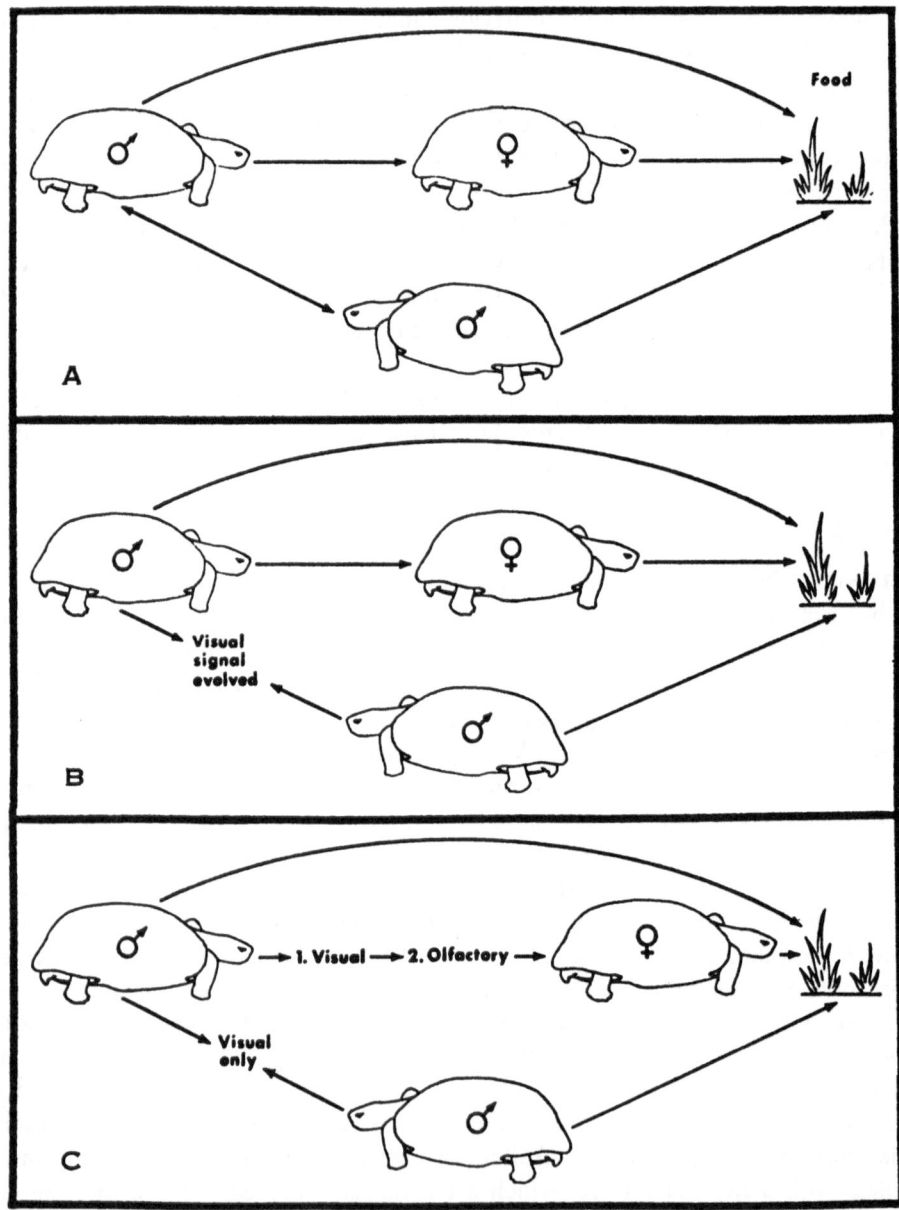

FIG. 4. Hypothetical stages (top to bottom) in the evolution of sex and species discrimination in two South American tortoises, *Geochelone carbonaria* and *G. denticulata*. A. Hypothetical primitive recognition pattern probably involved only olfactory signals. B. In more advanced recognition patterns, visual signals were probably derived from simple olfactory types. C. In present species, sex recognition involves a two-stage pattern for identification of females, and a one-stage pattern for identification of males. (*From* Auffenberg. 1965. *Copeia*, p. 341.)

SAURIA. In the nocturnal gecko, *Colenyx variegatus*, Greenburg (1943) concluded that sex recognition by males can be mediated by chemical clues. He anesthetized both sexes, broke off and interchanged tails, introduced them to males, and found courtship responses mainly to female parts. Noble and Bradley (1933) observed courtship in a large number of lizards, but elimination of the sense organs disturbed the males so much that no conclusions could be reached.

Carpenter (1962) described "cloacal rubbing" by male racerunners (*Cnemidophorus sexlineatus*) as an indication of impending sexual activity. The behavior involved rubbing the cloacal area on the ground in the process of wagging the entire pelvic region. The male, using its forelegs, moved forward slowly. Cloacal rubbing frequently was associated with a female, but it also was observed when a male "appeared to have picked up a scent of some kind" [p. 147]. Carpenter does not discuss the possible role of this behavior in depositing a scent which influences another lizard. Instead, he felt it was probably an autoerotic mechanism triggered by the presence of a female.

SERPENTES. Using mainly *Thamnophis sirtalis* and *T. butleri*, Noble (1937) performed a series of pioneering experiments on the sensory control of sex discrimination. As with most snakes, these species have virtually no visual sexual dimorphism. Noble could not get interspecific matings; even staining females brilliant colors would not confuse the male snakes. Blindfolded males performed well if they were in the immediate vicinity of a female, but they did not seek out or find females as well as did normals.

Noble then tested the effectiveness of trails made on glass plates in eliciting following by males. Is the cloacal region of snakes as important as it apparently is in turtles? As in the species mentioned above, the cloacal secretion of garter snakes is particularly powerful and vile smelling to humans. Noble found, however, that cloacal secretions of females never attracted males, but that males would respond with a "chin pressing" response to any portion of a female's body they chanced upon. The integument was devoid of glands, according to Noble, and had little odor; yet, this was what a male would respond to and follow. The only effective trail was that made by a female in estrus. However, Noble and Clausen (1936) did find that *T. sirtalis* would follow trails of its own species, regardless of sex, in the fall when the "aggregation drive" predominated. As discussed above, *Storeria* did not show sex discrimination in trailing. However, these experiments were performed in autumn, too. Since *Storeria* is closely related to *Thamnophis*, it would be surprising if they did not also show sex discrimination in trailing during the breeding season.

Would an airborne odor be as effective as a trail? Noble placed actively courted females of either *T. sirtalis* or *T. butleri* in cloth bags or paper-covered cloth bowls. No male responded, which seemed to indicate that a trail was essential for attracting a male. The necessity of a visual stimulus was ruled out on the basis that a blindfolded male made vigorous efforts to come in contact with uncovered females.

Washing a female with soap, alcohol, or ether did not affect the responses, which meant that an accumulated substance soluble in these chemicals was not involved. Could the response be blocked by covering the back of the female with rather potent substances? Quinine sulphate, picric acid, ammonium hydroxide, thymol, and sodium bicarbonate had no effect, but sodium chloride and vaseline were effective in blocking the response. The vaseline probably blocked the response by covering the source of attraction, while the other chemicals merely competed with it. Nonetheless, it was remarkable that a male was able to detect a small amount of odoriferous material even in the presence of very sour or bitter substances. At least, that was Noble's view. It must be pointed out that these substances are sour or bitter to humans, but not necessarily to snakes. More important, some of the foregoing compounds have considerably less "odor" than "taste."

What is the important chemosensory organ involved in following an estrus female's trail? When the nostrils of T. sirtalis and T. butleri males were plugged in the usual fashion, active courtship ceased; when the plugs were removed, the animals resumed courting. (There was one exception: one freshly captured male T. sirtalis.) Also, when portions of the tongue were removed, courtship ceased and rarely resumed. Noble concluded that in courtship both olfaction and Jacobson's organ were involved and that either organ may function alone, although the obstruction of one organ usually was sufficient to disrupt courtship.

Maternal Behavior

Maternal behavior is a rarity among reptiles. Care and protection or incubation of the eggs is known for some snakes, notably pythons (python) and cobras (Naja), as well as lizards, in particular skinks (Eumeces) and glass lizards (Ophisaurus). Extended care after hatching was reported only in the Crocodilia (see Mclhenny, 1935).

Suggestive observations on possible chemosensory functions in such maternal behavior have been made almost exclusively on Eumeces. Noble and Mason (1933) observed maternal behavior in E. laticeps and E. faciatus. The mother spent a great deal of time coiled around the eggs, usually in or under decaying wood. Whenever the female returned, she would touch one or more of the eggs with her tongue or thrust her snout between them. Was she testing them? A female would also retrieve her eggs if they were scattered about. She would retrieve eggs of the other Eumeces species observed, but she would not retrieve eggs from other lizard genera, such as Sceloporus or Ophisaurus.

Was odor the cue? Noble and Mason performed some experiments based only on extremely small samples (usually one or two). A paraffin egg model was rejected. If a female's own eggs were shellacked, they usually were not brooded. Experiments of the sensory-elimination type, which Noble was to exploit later so fully, were inconclusive. If the nostrils were plugged with cotton and grease, the females forced the plugs out. Cutting the tips of the tongue in a brooding female led to the cessation of brooding, but the blindfolded females did find and brood their eggs. The authors concluded that Jacobson's organ was most important. While this evidence was suggestive, it should stimulate more competent experiments.

Noble and Kumpf (1936) retracted the above conclusion when Noble later engaged in research which showed that Jacobson's organ was not very important compared to olfaction. If the nostrils were plugged in a brooding female E. laticeps, she still found the eggs. The same result occurred even when she was additionally blindfolded. When the tongue was removed, she also found her eggs. Presumably olfaction was operative; whether the animal was also blindfolded, a crucial point if meaningful comparisons between the two chemosensory functions are to be made, was not clear. Curiously, the authors concluded that Jacobson's organ was accessory and that olfaction was of the most importance.

Evans (1959) extended the early work on maternal behavior with his observations on a brooding *Eumeces obsoletus*. He noted actual rotation of the eggs by the female. He verified this observation by marking the eggs and noting their positions every day. Evans carried his observations further than had previous studies—until the eggs hatched. Evans noted that the mother seemed to help and to assist in the hatching process, "smelling and nosing" [p. 106] empty shells, rubbing her chin on hatchlings, and licking the cloacal vent of the young even up to 10 days after hatching.

Predator Recognition

One particularly suggestive behavior pattern and its incipient analysis will be discussed in this section. Cowles (1938) described a peculiar response of rattlesnakes (*Crotalus*) to California king snakes (*Lampropeltis getulus californiae*). This response was in marked contrast to the typical defensive reaction which consists of a raised head with an S-shaped loop of the anterior part of the body, accompanied by tail rattling. In this radical variant, a broad loop or bend in the body was formed and raised above the ground, while the head was held on the substrate. Almost one third of the body was elevated above the loop. When the king snake (a famous predator of rattlers immune to their poison) approached, the rattlesnake literally threw this body loop at the king snake. If the predator continued the attack, the rattlesnake retreated and, if escape was impossible, hid its head under several loops of the body (an ignominious attitude for this species!).

Bogert (1941) attempted to anaylze the sensory cues involved in eliciting this unusual response. A king snake was placed in a large glass jar for several hours. The snake was removed and a sidewinder (*Crotalus cerastes*) was introduced within five minutes. The defensive posture was obtained. Rattlesnakes blinded with adhesive tape also responded to the king snake. If the entire trunk of the tongue was severed, however, the response was abolished. This indicated to Bogert that recognition was mediated by the tongue-Jacobson's-organ system (see feeding studies described below). He then attempted to localize the source of the warning odor on the body of the king snake. He rubbed a stick on a part of the king snake's body and presented the stick to the rattlesnake. A response was given to sticks rubbed on the dorsal surface but not to sticks rubbed on the cloacal gland secretion. Sticks rubbed on the ventral surface were ineffective according to Bogert, but he admits there were some procedural failings during these tests, a generalization which could be extended to the whole series of experiments.

Bogert also found that other snake-eating snakes would elicit the response, and that the eastern race of one species of king snake was effective with western rattlesnakes. This point eliminates individually acquired recognition as a factor, unless the races do not differ in the crucial scent.

Unpublished observations by Inger, who analyzed the response of rattlesnakes into 10 separately scored units, demonstrated that the response to the same species of king snake did not differ statistically between species of rattlesnakes. Neither did the different king snake species vary in effectiveness. Even the relative size of the king snake had no noticeable effect on the responses of the rattlesnakes.

Snakes are preyed upon by more than other snakes. Do they recognize other predators? Cowles (1938) reported that a rattlesnake rattled to the odor of skunk oil and went into the defensive posture similar to that elicited by king snakes (see also Cowles and Phelan, 1958).

In conclusion, it would seem that this is a very promising area for future research. [The use of chemicals by blind snakes to repel both dangerous insects (army ants) and snake-eating (ophiophagus) snakes is discussed on page 276.]

Feeding Behavior

CHELONIA. There have been few studies on the chemosensory aspects of food finding in turtles. Turtles, however, have been the most frequently used reptiles in learning and in visual-discrimination research.

The first study on smell in turtle feeding was that of Honigmann (1921). He performed some informal experiments that merit mention on a variety of European freshwater species. Food was placed in sealed glass vials or linen bags. Empty vials or bags, filled with equal amounts of sand or stones, were used as controls. In this way chemical and visual stimuli could be tested separately. The stimuli were tested both on land and in water. The results were not presented in any systematic or tabular fashion; only brief protocols were given. The control vials were ignored, while many individuals attacked the bags but not the glass; therefore, there is some indication that chemical cues could be more important than visual ones. However, it should be remembered that movement, generally a potent cue for reptiles, was not one of the visual cues involved. It was also of interest that some turtles learned to inhibit their biting of the fish-filled bag. This showed that the turtle could modify its behavior toward olfactory cues. Results were not clearly presented for the terrestrial versus water tests. In water, chemical cues may be important in responses to different food items (Burghardt and Hess, 1966).

Noble and Breslau (1938) performed somewhat similar tests with young, but probably not ingestively naive, snapping turtles (*Chelydra serpentina*) and painted turtles (*Chrysemys picta*). They found that turtles would attack food in a sealed, glass vial placed in the water. They also wrapped food in muslin and used pebbles wrapped in muslin as a control. They suspended these in water and found no significant difference in the number of turtles orienting to each. They tested their animals in the dark, however, and so they were probably testing localization rather than discrimination. They ran one trial in light with musk turtles (*Kinosternon odoratum*), again with no response. They concluded that the sight, rather than the odor, of food was important. That a different species was used in the inadequate light situation may be important. More important for the reconciliation of their results with Honigmann's (of which Noble and Breslau seemed unaware) may be the fact that Honigmann used adult turtles that had been fed the food used in the olfactory tests for many months, perhaps even years, prior to the experiments.

With a terrestrial turtle, the eastern box tortoise (*Terrapene carolina*), Allard

(1949) found no clear differential results for fish or stones wrapped in burlap bags. Unfortunately, fish are not a normal food item for the species although fish are eaten in captivity. Allard did note that box turtles almost always touch food with their nose before feeding.

Poliakov (1930) used classical conditioned-reflex procedures (with electric shock) to condition turtles to discriminate amyl acetate, camphor, clove oil, and turpentine. After 300 trials, a discrimination was established.

The most formal experiments were reported by Boycott and Guillery (1962). These authors, who cited almost every learning study on turtles, seemed ignorant of the work of both Honigmann and Noble and Breslau. The species they used was the red-eared terrapin, *Pseudemys scripta elegans*. These authors were not interested in investigating a preference already present but in training the turtles to discriminate odors. They claimed to find no such preferences and, hence, no "innate olfactory responses" [p. 568]. The authors do not explain how one could discover innate responses using animals as old as three years and obtained from commercial dealers. The apparatus they used was a square tank filled with water. Three or four pieces of chopped lamb heart were placed at one end of the tank and a few milliliters of tap water or olfactory stimulus were allowed to diffuse through the tank. If water was introduced, the turtle was allowed to feed; if the olfactory stimulus was present, it received a shock. Two trials to each condition were run and then the tank was washed out thoroughly. The turtles did learn to discriminate amyl acetate, but lost it after destruction of more than half the olfactory fibers. Discriminations were also made to vanillin and eucalyptus oil. Some animals, however, were able to discriminate as well a number of weeks postoperatively as they did preoperatively. The authors give several possible explanations: the taste buds took over the amyl acetate discrimination; the general chemical receptors took over the discrimination; the lesions were incomplete or the olfactory nerve regenerated.

Gustation has been much less studied. Honigmann (1921) described a large male *Kinosternon cruentatum* who, for seven years, ate virtually nothing but mealworms. Honigmann rolled a thin strip of fish around a mealworm and gave it to the turtle. The turtle took the combination into his mouth, worked off the fish, spat it out, and then ate the mealworm. A similar result was obtained with meat and fish rolled together. Honigmann also reported drastic and sudden changes in the food preferences of his turtles (see also Burghardt, 1967c; Burghardt and Hess, 1966).

My own experience with taste in turtles (unpublished observations) was quite frustrating. I wanted to find something that could be mixed with chopped horsemeat to make it aversive to young diamondback terrapins (*Malaclemys terrapin*). Mixing quinine powder into meat until it was white did not deter the turtles. Neither did other substances—which I anthropomorphically decided they would not like—repel them such as red cayenne pepper, lemon juice, and vinegar. Similar results were obtained with snapping turtles, but a threshold could eventually be reached. It should be noted that these animals had been fed on the meat for many months and that it was one of their most preferred foods.

SQUAMATA. The majority of the findings implicating the chemical senses in the control of reptile behavior have involved snakes, and the behavior most frequently studied has been feeding. All snakes are carnivorous, a condition which requires that they actively hunt for food. Many anecdotal observations by zookeepers and others attest to the importance of chemical cues in the feeding behavior of captive snakes (e.g., Kauffeld, 1953; Lowe, 1943). Many fascinating and ingenious techniques have been developed to lure snakes into accepting substitute prey. This is frequently necessary since snakes may have highly specialized diets.

Snake-feeding research before 1935. The earliest systematic experimental study on chemical senses in snakes was performed by Baumann and reported in a series of papers (1927, 1928, 1929). These papers, quite influential on later work, were the first to describe the development of a technique, the trailing method, which still is used in snake-behavior research. It has also been exploited in the study of many other animals, such as ants, as reported elsewhere in this volume.

Reported in 1927, the first study, as well as most of his subsequent work, utilized the European asp viper (*Vipera aspis*). Baumann's test situation was a one-meter-square wooden box covered with sand or gravel in which were located four smaller boxes, one in each corner (Fig. 5). There was a hole in each of these smaller boxes. Baumann was studying the striking and the locating of prey, usually mice. Normally the feeding behavior in the viper was as follows: it struck and injected venom in a living mouse. The mouse ran off while the poison acted. The viper followed the route taken by the mouse, located the now deceased victim, and swallowed it. What is the sensory control of this behavior? In contrast to a viper which has not struck

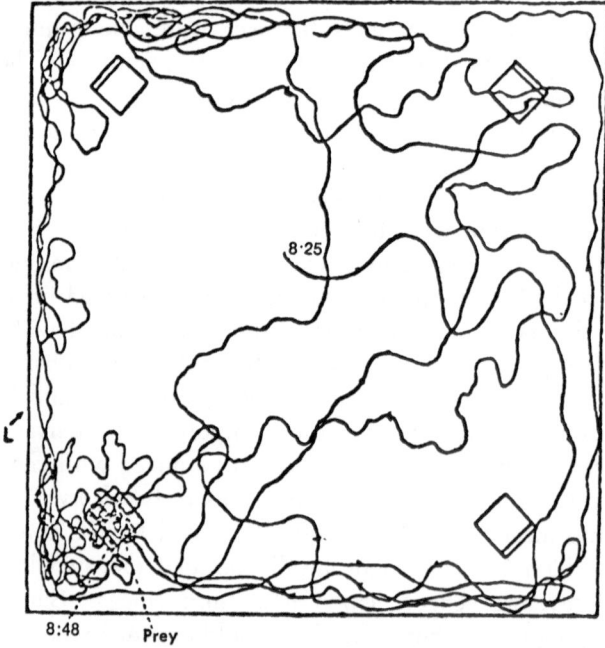

FIG. 5.   Baumann's 1 meter square test arena for vipers. The prey animal (mouse) is placed in one of the 4 small boxes. In the present instance the viper struck the mouse and was placed in the arena and the mouse in the box. "Searching" movements were released in the snake, and he localized the prey and pulled it out of the box in 23 minutes. Airborne chemical cues seemed most important. (*Adapted from* Baumann, 1927. Rev. Suisse Zool., 34:177.)

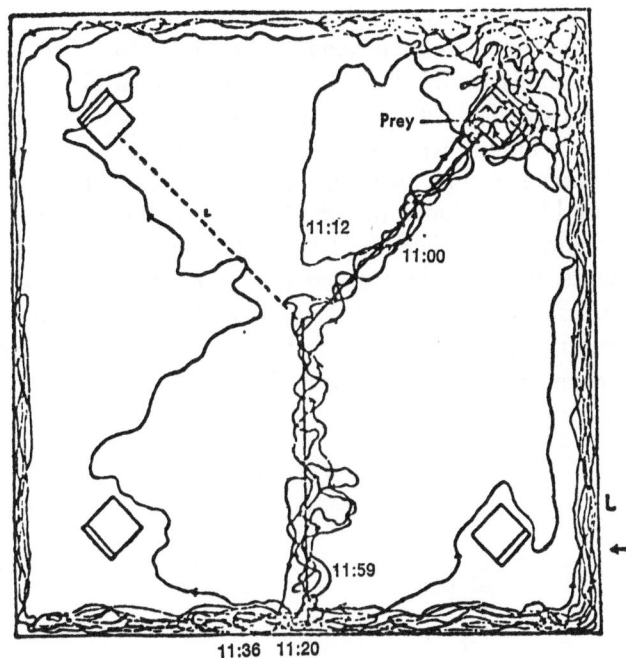

FIG. 6.   Same situation as in Figure 5, except the viper is provided with a trail of a mouse killed by viper poison (continuous line) and a trail of a living mouse (broken line). "Searching" movements are directed to the dead mouse trail as well as toward the box. (*Adapted from* Baumann, 1927. *Rev. Suisse Zol.*, 34:181.)

at a mouse, vipers which have just struck performed "searching" type movements until they located the trail. This behavior seems identical to *appetitive behavior* as used by ethologists (Craig, 1918). By placing dead prey in one of the four boxes, Baumann was able to eliminate sight, touch, taste, and auditory stimuli as being responsible for the finding of prey. This left, as he argued, only the olfactory type of senses. Several groups of animals were tested on the various conditions: some struck at a living or a dead mouse before being permitted to search for the prey; some did not bite a mouse at all; others bit a living mouse and were placed in the large box with trails of dead or living mice. In Figure 5 is a representative result where the snake bit, but without a trail. In Figure 6 is represented a test with a trail. Baumann showed that vipers could find prey on the basis of airborne chemical cues alone, but that trails, if present, were used. He concluded that they could perceive tracks made by both dead and living mice, but that they could discriminate between the two, since the trail of a dead mouse was much more effective than the trail of a living mouse. In addition, he found that the trail of a living mouse appeared to disturb the searching movements normally seen after biting a mouse. This, it may be added, would seem to be adaptive since under natural conditions it would indicate that a viper's strike had not been effective.

In his next paper Baumann (1928), essentially repeated these results using larger numbers of animals. In addition, he made trails (1) of living mice, (2) of poisoned mice, and (3) of mice killed with a quick blow on the head. He found that the former trail was clearly distinguished from the latter two trails, and in addition, he found indications that trails from (2) were more effective than from (3). Also, a strike was necessary to obtain good trailing behavior. The manually killed mouse-trail (3) was followed only if the mouse had been dead for some time. The exact duration was not specified. He also showed that the viper could discriminate between a trail made by a bitten, living mouse from a trail made by an unbitten, living mouse. Baumann claimed to find no evidence of learning as measured by the time taken to find food, and he concluded that the behavior of the viper was instinctive.

In 1929, Baumann republished the preceding results in expanded form, along with new experiments, in a lengthy monograph concerning the feeding behavior of the viper. His conclusions concerning odors and trails were similar to those in his earlier reports. He divided the behavior of the viper into several components with each under a characteristic stimulus control. I have organized his results into the following diagram.

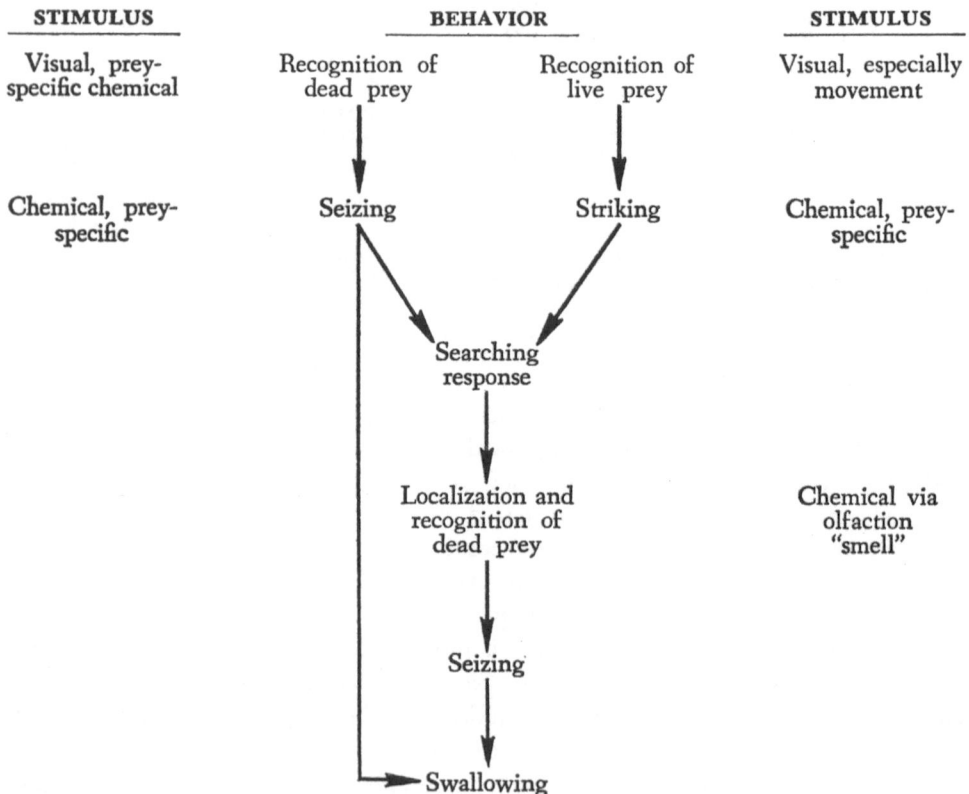

STIMULUS	BEHAVIOR		STIMULUS
Visual, prey-specific chemical	Recognition of dead prey	Recognition of live prey	Visual, especially movement
Chemical, prey-specific	Seizing	Striking	Chemical, prey-specific
	Searching response		
	Localization and recognition of dead prey		Chemical via olfaction "smell"
	Seizing		
	Swallowing		

Note that dead prey normally did not elicit the poison injecting strike.

A paper by Gettkandt (1931) reached quite different conclusions. Using von Uexküll's "functional circle" approach (a translated paper on this topic is available in Schiller, 1957) and realizing the possible artificiality of laboratory experimentation, he set out to study the sensory control of feeding in the coachwhip (*Zamenis flagelliformis*) and the grass snake (*Natrix natrix*). He used a well-designed test chamber to record the movements of the subjects. The glass and wire covering was evenly divided into squares to facilitate the recording of the animals' movements in a manner similar to the psychologist's open-field apparatus. His results led him to believe that optical stimuli from prey were most important, especially "regularity" of movement. Shape, size (within limits), and color were of little importance. He spent considerable time discussing the possible role of the tongue as a touch receptor but ruled this out since he noticed tongue contact prior to feeding less than 10 percent of the time. What did the tongue do? He suggested that it is important in drinking (which can be seriously questioned by observing a serpent drink) and that its darting during feeding is solely to whirl odoriferous air to the nostrils. Why should Jacobson's organ be brought in as a chemical sense when the more obvious olfactory organ had not been ruled out? His tongueless *N. natrix* readily ate frogs. That smell or taste was of little importance he tried to demonstrate with the following evidence. Mice covered with oil of cloves or essence of mocha were eaten by the coachwhip. The hidden assumption here is that such substances would effectively overpower any chemical stimuli in mice used by snakes. The fact that the snake did not show any obvious aversion to sodium chloride, magnesium sulfate, quinine, or grape sugar solutions was adduced as evidence against taste being present to any significant extent in snakes. A clear aversion to 5 percent acetic acid was found, however, but whether smell or taste was involved was not clear to Gettkandt. Indeed, the snake quickly learned to avoid the dish containing acetic acid. Objects without scent were struck at, but this is really not conclusive until it is shown that (1) a defensive strike was not involved, and (2) conditioning of a chemical cue to visual cues was not involved in a feeding strike to "odorless" objects. Baumann, as we have seen, clearly implicated chemical cues as most important in recognition, although not in the original strike. To this, Gettkandt called for replication. In addition, he rightly pointed out that species differences are to be expected. Before Gettkandt's paper was even published, however, a replication was in press. But it was an independent replication of Gettkandt's work, and, as it turned out, a critical one which led to future workers' ignoring almost completely Gettkandt's contribution.

These replications were made by Wiedemann (1931, 1932). He provided interesting, provocative, and even humorous descriptions of serpent behavior, but he did not perform systematic experiments as Baumann or Gettkandt had done nor did he describe his procedures, methods, and results nearly as completely. However, a number of important points wre made. In his first paper, he was concerned with a variety of species of European snakes, but most of his observations were made on the European grass snake (*Natrix natrix*), also used by Gettkandt.

The main part of the paper was devoted to the significance of the sense organs in the feeding behavior of these snakes. He considered two main modalities: visual and chemical. Concerning the visual, he found that in *N. Natrix* visual stimuli were very important in arresting the attention of the snake as well as in arousing "searching" behavior. Movement was of great importance while form or color, especially of prey, were of little importance. Wiedemann even thought that they might not even be able to discern forms very well. To show that form perception was not important to feeding behavior, Wiedemann presented the snakes with a carefully prepared model of a frog made from plaster. They took no notice of it. However, a freshly killed but motionless frog was immediately siezed, indicating that the actual ingestive response was mediated by chemical cues. This was shown also by the observation that movement of an object could elicit an open-mouth strike in this species, but that the object was immediately released if it was unpalatable. The ingestive reflex was released only by chemical stimuli. Wiedemann records many fascinating examples of snakes swallowing inappropriate objects which had been rubbed with amphibians or fish—foods that the animal would readily eat. For instance, beef would be eaten if it were rubbed with an Axolotl or left lying in the amphibian's aquarium. He could get snakes to swallow rubber sponges, cigars, and even his finger if they were rubbed with normal prey. Since some of the objects themselves had strong odors, at least to humans, he felt that the snake could perceive and respond appropriately to highly specific chemical stimuli. Note that Gettkandt reached opposite conclusions with similar results from similar experiments. Blinded and nostril-covered animals were capable of responding in a similar fashion. Therefore, Wiedemann concluded that Jacobson's organ was the chemical sense most involved and that the tongue was the mediator. Using several concentrated "taste" substances, he found little or no aversion or attraction on the part of the snakes. Solutions of salt, sugar, and quinine were drunk with no obvious aversion. Rubber sponges rubbed on prey and then sprinkled with the above substances were also swallowed. Strong vinegar did release a slight response in the snakes, but, of course, it has a strong odor. A weaker solution which has lost its pungent odor to a human, was acceptable to the snake. He concluded that strong-tasting substances (to humans) without odor were undetected. Therefore, the "four taste qualities" were not found in snakes. Again, the clear similarity to Gettkandt's study is apparent. Obviously, a more detailed study is required before any firm conclusions can be reached. Wiedemann made a brief addendum to his paper noting Gettkandt.

Wiedemann's next paper (1932) was concerned exclusively with the viper, *Vipera berus*. He repeated Baumann's trailing experiments on *Vipera aspis* and obtained similar results. They were so similar, in fact, that Wiedemann did not even bother to report his data. It is interesting to speculate what he would have done in the 1931 paper if he had been aware of Gettkandt's results earlier. In addition, he found that trails made with lizards or frogs were also effective with this species of viper. From field studies, he determined that lizards and frogs, as well as young mice, constituted the main part of the diet. Wiedemann found that the snake would not find prey without a trail unless he came within 3 or 4 centimeters of it. This would suggest that the success of Baumann's vipers, when trails were not present, was due to their fortuitously passing close to the box containing the prey. This would be much less likely outside the confines of a small arena. He also found that the viper could find lizards buried in shallow sand. An incidental observation, which assumes great interest in light of modern biological research, was his statement that the animals were more successful in the early morning hours or in late afternoon. A number of factors could be working here. Are the chemical senses more sensitive at certain parts of the day-night cycle? How cyclic is the hunger "drive" itself?

Wiedemann determined that the poison itself was an important factor in helping the viper locate its prey. He found that rubbing prey with poison enhanced its chance of being found and eaten. He showed that trailing was accompanied by constant tongue flicking, again implicating Jacobson's organ. He repeated with vipers his early demonstrations with *Natrix* on the swallowing of various artificial objects coated with the proper chemical stimuli, in this case, frog slime, lizard blood, etc.

The next major advance in the study of the chemosensory control of feeding behavior in snakes was the work of Kahmann (1932, 1934), who was influenced by Wiedemann. Kahmann (1932) considered the previous literature concerning Jacobson's organ from the morphological viewpoint. He was particularly enamored of the theories of Broman (1920), who postulated two functions for Jacobson's organ: smelling objects inside the mouth, and scenting external stimuli with the help of the tongue. It might be added that human nasal olfaction seems to serve both of these functions. Kahmann then proceeded to a behavioral analysis of the role of Jacobson's organ and his paper makes clear his contempt for Gettkandt's conclusions. Although Baumann had clearly shown the importance of chemical reception in prey-finding and Wiedemann had also made important and interesting observations, Kahmann felt that they had not made a clear distinction between the role of nasal olfaction and Jacobson's organ, as Gettkandt's position made clear. The closest attempt was Wiedemann's (1931) with his blinded and nostril-covered snakes.

Although Kahmann (1932) studied several species of snakes and lizards, most of his work is concerned with *Natrix*, especially *N. natrix*, one of the species used by Gettkandt. His experiments usually were repeated frequently enough so that we can have confidence in that aspect of his results. In the first series of experiments, Kahmann used normal, adult snakes and tested them in a large arena similar to that used by Baumann. He performed two series of experiments with normal snakes, one with and one without the use of tracks laid by the prey. As usual, he recorded the movement of the animal. He found that a hungry snake placed in a box where frog skin had been rubbed in two of the four corners would spend most of its time exploring in those corners. If he used an actual prey object, either nonmoving or dead, he found that the snake eventually would find the prey and explore frequently in that area.

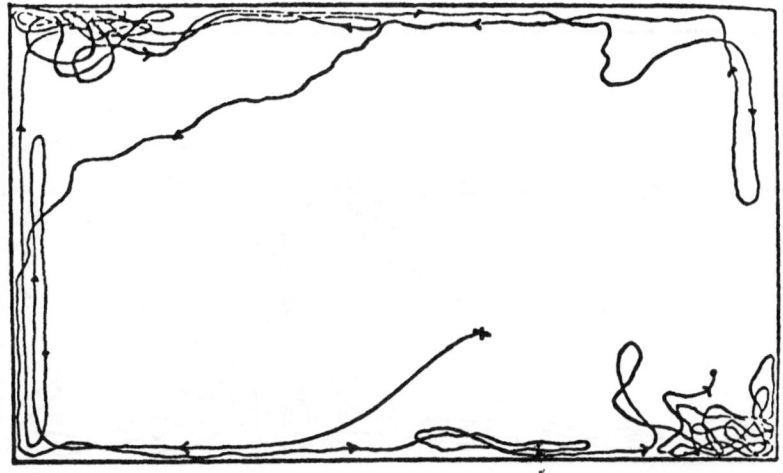

FIG. 7. A hungry European grass snake (*Natrix natrix*) spends more time in the two corners of a test arena rubbed with frog secretions. (*From* Kahmann. 1932. *Zool. Jb. Abt. Allg. Zool. Physiol.,* 51:202.)

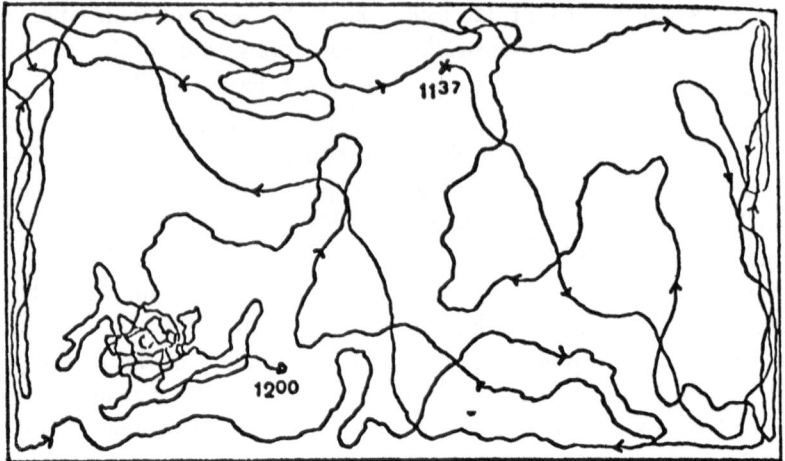

FIG. 8. A N. *natrix* has struck a living frog which is killed and placed in the arena. In contrast to Figure 7, wide ranging appetitive movements are evident. Compare Figure 5. (*From* Kahmann. 1932. *Zool. Jb. Abt. Allg. Zool. Physiol.*, 51:204.)

Confirming Wiedemann, he found that active exploring and interest was aroused only when a snake came by chance into the area where the scent or prey had been placed. He also had the snake attempt to find the prey it had already bitten. A hungry snake attacked a frog, which was then taken from the snake, killed, and placed in the experimental box. The *Natrix* was then introduced and its movements recorded. The movements were wide ranging as compared with the snake that had struck previously at the prey (see Figs. 7 and 8). He repeated this last experiment using tracks to guide the snake to the place where the prey was laid in the experimental arena. An example of this is seen in Figure 9. It was quite clear that the animal discovered and followed (or at least devoted more attention) to the area where the track had been laid. He concluded these tracking experiments with the following observations. First,

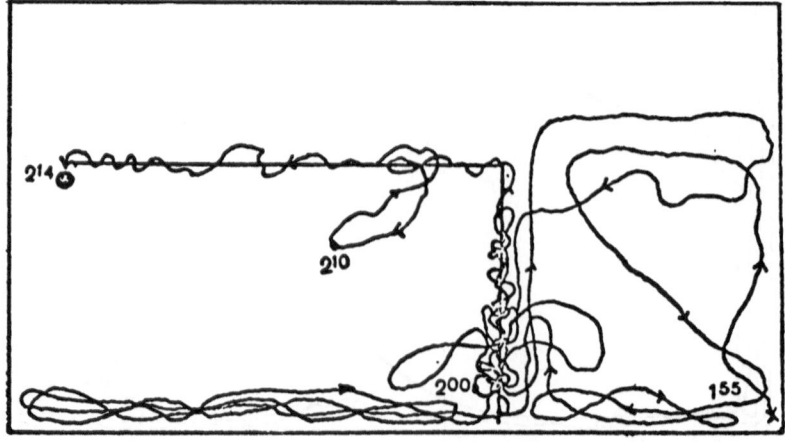

FIG. 9. Same situation as in Figure 8, except a prey trail is provided. (*From* Kahmann. 1932. *Zool. Jb. Abt. Allg. Zool. Physiol.*, 54:206.)

when the snake discovered the track, tongue-flick movements became very short and quick and the head movements changed. Secondly, the snake's head and tongue orientation was close to the floor where the track was. However, it must be noted, even in Kahmann's results, that a snake did frequently lose the trail. This may have been due to extraneous vibrations or other disturbances. Snakes were also able to follow tracks which did not lead to prey, a rather nice paradigm of the human search for the pot of gold at the end of the rainbow.

Most of the findings up to now have not been new, but Kahmann determined the relative importance of the nose and Jacobson's organ in this behavior. His first experiment involved blocking the nose with paraffin. The eyes were covered, but the animals were still usually successful in finding prey. Kahmann tried to block the internal nares but with unsatisfactory results. It can be concluded that blinded and nostril-covered snakes, although able to find the prey in the absence of a track and to follow the track very clearly for a short distance, were not as efficient as normal snakes. Whether this was due to the role of these other organs or whether it was caused by the experimental manipulations upon the subjects is not clear. Kahmann then attempted to eliminate Jacobson's organ. He did this by cauterization with a hot needle using histological techniques to confirm the destruction of the organ. The results were clear in that trail-following of any type appeared to be abolished (Fig. 10). However, both operated and normal snakes were able to orient to quick movements of prey or other objects and appetitive movements were activated. Kahmann concluded that Jacobson's organ was an important, perhaps the most important, sense organ of many snakes.

Kahmann's last series of experiments concerned the role of the tongue in the operation of Jacobson's organ. By cutting off just the tip of the tongue, he was able to show that tracking ability was reduced, if not abolished. Kahmann concluded that

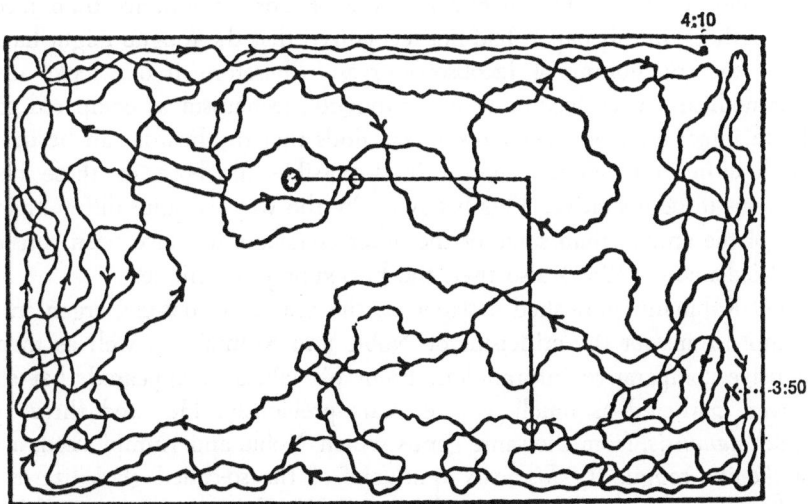

FIG. 10. As in Figure 9, except Jacobson's organ cauterized. The search movements are released, but the trail is not recognized nor the prey found. (*From* Kahmann. 1932. *Zool. Jb. Abt. Allg. Zool. Physiol.*, 54:222.)

the tips of the tongue enter into the openings of Jacobson's organ to stimulate it with the substances the tongue had picked up. The next question was how the tongue received the smell substances. In trailing or prey recognition, must the tongue come in contact with the substrate or prey? Kahmann concluded that it frequently did. Exceptions occurred if the snake had been in a region of strong odors, in which case the air itself might have contained enough suspended odoriferous particles to stimulate Jacobson's organ. In 1934 Kahmann studied various other species of snakes and showed that many of these species also were able to follow prey and that this ability was abolished by cutting the tongue. He also made some observations on the comparative development of Jacobson's organ in various snakes.

Lizards. The early German work with snakes was marked by a certain unity of approach and methods. Before going into the more recent and nonteutonic studies on serpents, a brief digression on lizards is in order. Kahmann, in the studies just discussed, made a few observation on lizards, concluding that lizards with long, bifid, snake-like tongues used Jacobson's organ in the same way as did snakes and hence it was as important to them. Unfortunately, as Noble and Kumpf (1936) pointed out, very little experimental evidence was presented to support this conclusion concerning lizards. These authors performed a number of studies with lizards. They used *Ameiva exsul*, a teiid lizard with a long, bifid tongue. In cages containing three to five lizards, they would hide under gravel various baits, such as cotton saturated with beef juice and egg, tomato juice and cod liver oil, or mealworms. These represented foods which the lizards had eaten for several weeks. The lizards were able to find the baits. The authors then performed various operations on the animals to eliminate certain sense organs. The animals were either blinded, made anosmic, had their tongues cut off (which would eliminate gustation according to the authors), or Jacobson's organ was cauterized. Eleven groups of animals with eleven combinations of operations were utilized. Although they concluded that no sense organ was of unique importance, they did conclude that olfaction was of more importance than Jacobson's organ. The authors, in this paper, had not recognized that the tongue might have been important in the functioning of Jacobson's organ. Therefore, if one looks at results for animals with an intact Jacobson's organ-tongue mechanism as compared to those with only an intact olfactory mechanism, one finds that the lizards can be quite successful with either one alone. Hence, it is possible for both of these chemical sensory systems to be operative. This behavior, by the way, is quite different in what is required of the animal than some of the other behaviors we have been considering, although Wiedemann (1932) also used the buried-prey technique.

In 1939 Kahmann published a paper on the role of Jacobson's organ in lizards in an attempt to answer the criticisms of Noble and Kumpf was well as to provide more extensive comparative information. From his tables, it appeared that some of the data were gathered as much as five years previously. He used three species: *Ameiva saurimanensis,* from the same genus which Noble and Kumpf used, a species with a snake-like tongue; *Ophisaurus apus,* one of the so-called glass lizards which has a rather thick, fleshy tongue; and *Acanthodactylus scutellatus,* a fringe-toed lacertid lizard which also has a rather thick tongue.

Kahmann started with four groups of 10 *Ameiva* each. In the first experiment, these animals, after being blinded, had to find a piece of food. All 40 animals were given two trials each. For the 80 trials, the food was found by 69 of them, (86 percent). In the next phase of the experiment, Kahmann cauterized Jacobson's organ in one group of 10 animals and removed the tongue in another group of 10 animals. He then had these animals find buried food. The eyes also were covered in these animals and each one was run four times for a total of 40 tests under each condition. Overall he found that there was 37½ percent success with Jacobson's organ cauterized and 45 percent success with the tongue removed. Kahmann concluded that, in a certain proportion of the time, nasal olfaction alone was capable of localizing prey and releasing the ingestion reflex. The higher proportion of tongueless animals finding prey could be interpreted on the basis of the lizard being able to stimulate Jacobson's organ to a certain extent even without the tongue. This possibility is discussed later under the work of Wilde (1938). Kahmann removed nasal olfaction in another group of 10 animals each, which were already blinded. He tested these animals as he did the blind-only group. Prey were lying free in the cage. He found that 66 to 67.5 percent (the range is due to an inconsistency in Table 3, p. 677) of the time the prey was found. He concluded that Jacobson's organ alone was involved in this recognition and localization of prey. Why he used a different testing procedure with the Jacobson's organ and tongue-eliminated groups was not clear and made comparison difficult, although the control and nose-blocked lizards seemed to have had the easier task. Without giving data, he discussed the lack of ability of Ameiva to follow trails. Less extensive experiments with the other two species yielded different results. In both cases, the removal of Jacobson's organ did not noticeably decrease the success of the animal, but removal of nasal olfaction had a greater effect. Even here, however, the blinded and olfactory-deprived animal found prey 40 to 90 percent of the time in O. apus and 50 to 70 percent of the time in A. scutellatus. Some trailing ability was seen in Ophisaurus.

From these various papers by Kahmann, his views, especially as expressed at the end of his last paper (1939), can be summarized as follows: Jacobson's organ functions both as a distance sensor and trailing receptor—via the tongue mechanism— as well as a chemical sensor for objects within the mouth of the organism. In most, if not all snakes, both functions are very important, a conclusion attested to by the experiments performed by him and others. The highly developed tongue with the forked tip was inserted into the organ in the transfer of the chemically active substances. In snakes, olfaction played a very small role as compared to Jacobson's organ. In lizards, from such families as Varanidae and Teiidae, Jacobson's organ functioned both as a distance and mouth sensing organ, although not quite as efficiently as in snakes. In these and other lizards with well-developed tongue mechanisms, Jacobson's organ played a predominant role, although olfaction was more important than in snakes. As the tongue becomes more fleshy and less pointed, olfaction increases in importance and the distance-sensing and trailing function of the tongue-Jacobson's-organ system decreases. However, Kahmann noted that in most lizards which did not have well-developed forked tongues, Jacobson's organ was still highly developed. This was probably because the organ was still important for testing objects already in the mouth, especially for determining the palatability of foods. An interesting exception was the chameleon where Jacobson's organ is small or even absent. These animals have a well-developed tongue, but it is used to capture prey, apparently having no

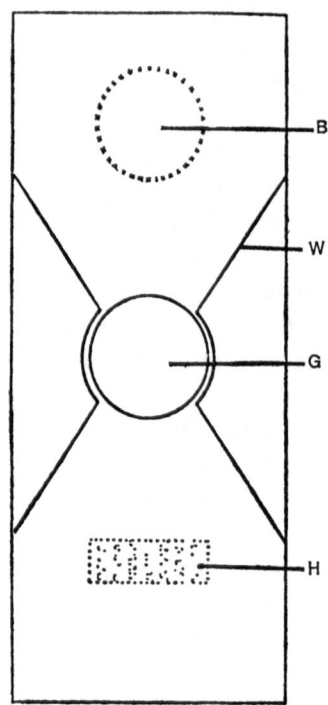

FIG. 11. Apparatus used in studies of taste discrimination in lizards. B, light bulb; G, drinking dish; H, electric heater; W, wall. (*From* Rensch and Eisenkraut. 1927. *Z. Vergl. Physiol.*, 5:609. Courtesy of Springer-Verlag, Berlin.)

distance sensory function. Kahmann also enumerated the interesting inverse relationship between the development of the eye and the prominence and importance of Jacobson's organ.

In only one study (Rensch and Eisentraut, 1927) has gustation in lizards been investigated. Figure 11 illustrates their apparatus. The dish, containing the various taste solutions or water, was placed in the center of a narrow passageway. Through the use of light bulbs and electric heaters, a constant dry heat was produced in the terrarium which made the animals active and thirsty. In their activity, they would pass from section to section of the tank via the narrow passageway, continually passing close to the dish and with frequent opportunities to drink. One may immediately question, of course, the suitability of this technique since the animals were in a more agitated state than normal. Rensch and Eisentraut studied three species of *Lacerta* and a quite unrelated lizard, *Anguis fragilis*, the European slow worm. Most of their results were obtained with *Lacerta vivipara*. After the animals became accustomed to drinking tap water from the dish in the center of the experimental tank, they introduced various taste stimuli into the water and observed the reaction of the animal. The most clear indication of discrimination was an avoidance or "disgust" reaction involving vigorous shaking of the head and cessation of drinking. They used the four classic taste qualities in selecting their stimuli. Common salt (sodium chloride) in concentrated solutions was aversive to all lizards. Five percent and weaker solutions resulted in different reactions among the animals, but even a one percent solution led to reduced drinking in two of the species. The control for this type of observation was to reintroduce tap water after the tests with the salt solutions and observe whether more "hearty drinking" occurred. One-half of one percent sodium chloride solution seemed to lead to reduced drinking. The sour stimulus was prepared with tartaric acid. This, too, was considered to be an unpleasant stimulus for the lizards. Some aversiveness was seen to solutions as low as 0.25 percent. Two bitter stimuli were used. The first was aloe, a bitter substance prepared from leaves of certain members of the lily family and used as a laxative; this was ingested in suspension. Quinine, which is an even more bitter substance, resulted in clear aversive responses from some species. However, it appeared that lizards were considerably less sensitive to bitter than are humans. For sweet stimuli, concentrated sugar solutions (sucrose?) were used. The authors found that more liquid was ingested than normally would be the case with tap water and they felt that this was a positive stimulus for the animals. The authors concluded that, except for bitter, lizard taste sensations were similar to humans, if somewhat less sensitive.

It is probably somewhat ungracious to strongly criticize a study which was not only the first, but still a rather lonely study of tastes in reptiles. However, some comments are in order to encourage replications and elaborations of this type of study. First, it appears that the presentation of the stimuli did not follow any systematic order. Second, no large variations in concentration were given; therefore, threshold determinations were, at most, suggestive. Third, the meaningfulness of structuring a taste study in lizards around the four classic "taste" sensations is questionable. These are becoming increasingly less useful for human taste studies and their extrapolation to a different species is most tenuous. Fourth, no consideration was taken of the possibility that olfactory stimuli were also operative. This criticism also applies to many mammalian studies of taste. It has been found that rats can discriminate between two concentrations of a "taste" substance on the basis of olfactory stimuli (Miller and Erickson, 1966).

It should also be considered whether sensations and thresholds were, or even could be, measured in any meaningful fashion by this technique. Motivation, of course, was not controlled in any precise manner. In addition, it must be emphasized that because an animal drank a solution, it does not mean that it could not discriminate it from pure water. It may be that it was not aversive enough to the animal to outweigh its need or "desire" for water. Hence, we could determine only thresholds for "aversiveness" as measured by certain behaviors, and nothing about the sensations or the ability to perceive in general.

The informal observations by Wiedemann (1931) and Gettkandt (1931) were in marked contrast, since they found little discrimination, although the methods were quite dissimilar. Noble (1937) also found that bitter substances did not hinder male recognition of female *Thamnophis* integument. Interestingly, salt had an effect which showed that serpent taste-sensitivity may vary qualitatively in the same manner as in lizards. Obviously, some good, controlled studies on all types of reptiles are needed.

Snake-feeding research after 1935. Noble and his associates in the 1930's performed several important studies on aggregation, courtship, and trailing in snakes which were discussed previously. They repeatedly concluded that olfaction was the most important of the chemical senses.

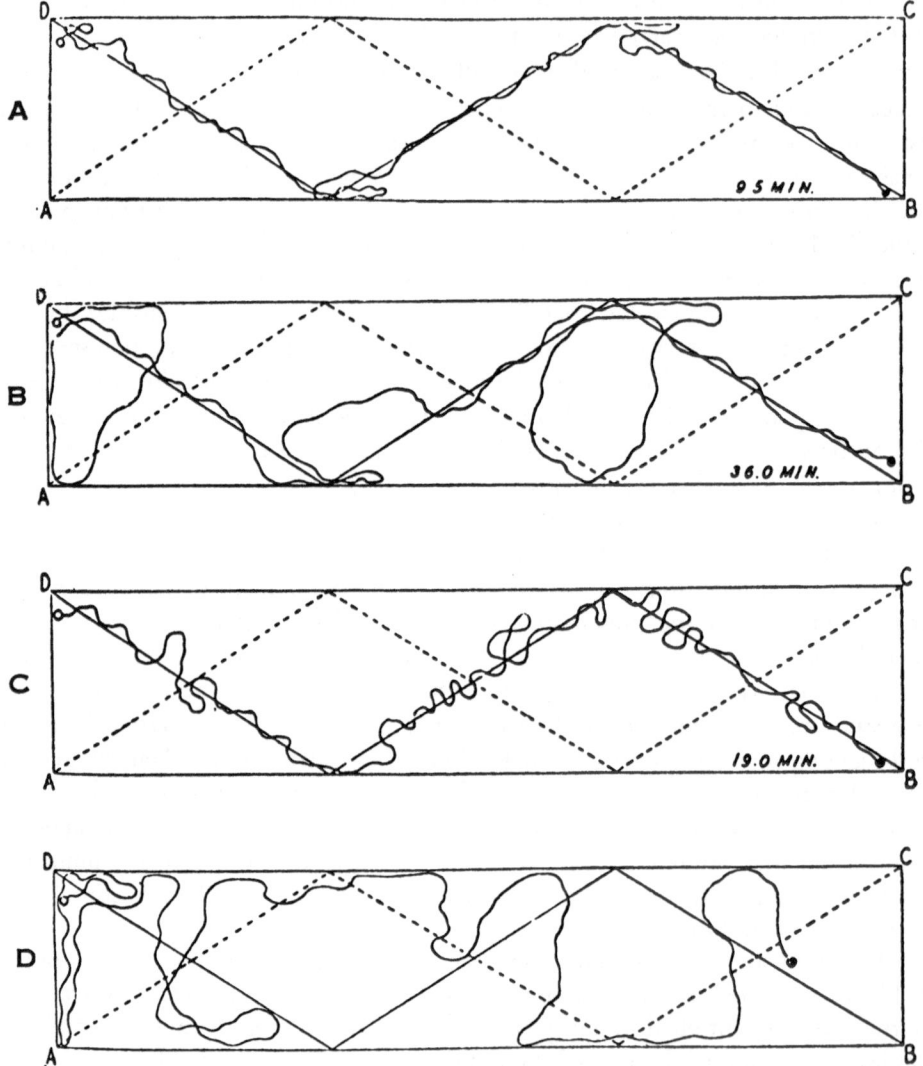

FIG. 12. Diagram of soapstone surfaces. Continuous line in the area, the food-scented trail; broken line, the distilled water trail. A, the trail of normal S. dekayi beginning at D and ending at B, with the time required to cover the trail; B, the trail of an S. dekayi with olfaction eliminated; C, the trail of an S. dekayi with Jacobson's organ incapacitated; D, the trail of an S. dekayi with both Jacobson's organs and the olfactory organs incapacitated. No time reaction is listed here, since the trail was not followed from D to B and the time involved here was not significant. Compare with Kahmann, Figures 9 and 10. (*From* Noble and Clausen. 1936. *Ecol. Monogr.*, 6:301.)

Noble and Clausen (1936) tested brown snakes (*Storeria dekayi*) and garter snakes (*Thamnophis sirtalis*) on food trails placed on a soapstone surface. The trail was prepared from the liquid resulting from crushing earthworms and adding water. The method was similar to the aggregation-trailing experiments discussed earlier. By covering the eyes, covering the nostrils, cutting the tongue, or cauterizing Jacobson's organ in various combinations, they were able to show (Fig. 12) that when the nostrils were covered, trailing ability was abolished or the latency greatly increased. Cutting the tongue and cauterizing Jacobson's organ had some effect in both species.

The conclusions reached by Noble and Clausen were that the olfactory organs alone are an adequate sensory mechanism for food trailing in these species, and the tongue and Jacobson's organ in combination make the trailing possible, but either structure alone is inadequate.

You will note the discrepancy with Kahmann. Although Kahmann used a different genus (*Natrix*), it is closely related to the genera (*Thamnophis* and *Storeria*) used by Noble and his associates. Noble, as has been mentioned, was under the impression that the tongue was the seat of gustatory sensations. But the tongue, as noted above, possesses no taste buds. It is interesting to note that although a clear contradiction of Kahmann's *data* was found, Noble and Clausen (1936) stated (p. 312) that they "found very little to support the views of Kahmann (1932)."

Some of the studies by Noble and his colleagues have been criticized for being "elaborate mutilation experiments" devised merely to disprove the findings of the earlier German workers (Evans, 1959, p. 108). A study by Wilde (1938) has been referred to as disproving Noble and as proving the importance of Jacobson's organ. Few people noted that Wilde worked with what he called the *feeding reaction*, which is referred to here as the *prey-attack response* to differentiate it from other aspects of feeding behavior, such as trailing, which have been more commonly studied.

Wilde clearly stated that his study was an attempt to find a response more dependent upon Jacobson's organ than on the olfactory organ and to eliminate this organ in a less injurious way than by cauterization. Testing adult eastern garter snakes (*Thamnophis sirtalis*), he utilized the prey-attack response given to a water extract of nightcrawlers absorbed onto cotton attached to the end of a glass rod. He found that 63 percent of the time snakes would attack if only the tongue touched the swab; they attacked 95 percent of the time if the lips were allowed to touch the swab. Control swabs dipped in distilled water or sucrose solutions elicited little tongue flicking. This implied that visual and tactile stimulation alone were inadequate. Wilde then performed several operations. Severing the olfactory tract eliminated use of both olfactory and Jacobson's organs and abolished the attack response. Cutting a very large part of the diffuse olfactory nerves did not abolish the attack response, but severing the vomeronasal nerve to Jacobson's organ did abolish the response.

Wilde also tested Kahmann's theory that the bifid tongue must fit into the paired openings of Jacobson's organ for stimulation to be effective. He progressively cut off more and more of the tongues of his snakes. He removed portions of the tips, leaving the stumps either too short or too thick to enter the lumens of Jacobson's organ. Nevertheless, the contact of the swab by the tongue elicited the attack response. He then removed all of the tongue beyond the origin of the fork and still obtained a response, although the frequency was reduced. Wilde suggested that the tongue stump had difficulty touching the swab. If the entire tongue was removed, the possibility of tongue contact was eliminated. But even here the response could be obtained if the swab touched the snake's lips. Therefore, while the tongue aided in transporting substances to the organ, it was not itself critical. Wilde also found that his snakes with severed vomeronasal nerves would not eat the scalded worms he used as food. This contradicted the results of Noble and Clausen (1936) who found that their snakes with cauterized Jacobson's organ would feed normally—although the conditions were unspecified. Wilde kept his snakes isolated in separate cages. Perhaps Noble and Clausen's animals were kept in a group and fed live prey. Factors such as social facilitation, prey movement, and conditioning may have been involved.

This work by Wilde, in short, demonstrated a clear cut response to chemical stimulation mainly controlled by the tongue-Jacobson's-organ system. Furthermore, he used a simple, yet efficient testing technique—one that could be modified in our work with newborn snakes almost 30 years later. That his work has been somewhat ignored is shown by a recent major text in animal physiology which states that the tips of the tongue are inserted into Jacobson's organ (Gordon et al., 1968, p. 405).

Thurow (1949, unpubl.) has attempted an overall analysis of feeding behavior in adult *T. radix* and its sensory control. On the bases of his observations, as well as previous work, he arrived at the following conclusions which I have abstracted.

BEHAVIOR	STIMULI
Pointing (orientation to prey)	tactile, vibrational, chemical, movement, contrast, sharp boundaries
Approach	movement, contrast, sharp boundaries
Identification	movement, contrast, boundaries, chemical
Strike	chemical
Swallowing	tactile, chemical

Thurow favors Jacobson's organ over olfaction as the more important chemical sense. He feels that the above are the necessary elements and that others, such as "searching," can be dispensed with. It would be of great interest to analyze further these links in terms of motivational state, previous experience, type of food, and so on.

Another technique, used by Fox (1952), consisted of passing air over prey such as frogs or slugs and observing the snakes' behavior. In the garter snakes (*Thamnophis*) observed by Fox, the animals became more active and oriented to various prey odors. No details of the experiment or of the results are given. Clearly, a substrate trail was not involved here. Cowles and Phelan (1958) exploited this technique with rattlesnakes. They passed air through aqueous solutions of odors and introduced them into cages. It might be added that research such as this must be done very carefully in order to eliminate all extraneous cues, especially the visual, vibrational, and chemical ones. If the odors were effective, the snakes would uncoil and orient or locomote to the source. Decomposed milk was found to be particularly attractive to the animals and putrid rats were preferred over freshly killed mice; this suggested that rattlesnakes may scavenge more than was previously thought. However, too much food decomposition was repellent. One of the authors' main goals was to find a repellent substance to ward off rattlesnakes. You will recall that 20 years earlier Cowles found the defensive response to skunk odor. Using thio-alcohol *n*-butyl mercaptan, which was similar to skunk musk, they found no fear response unless visual or tactual stimuli were added—a very interesting finding.

They then proceeded to record heart-rate changes rather than overt behavior, and found increases to king snake odor, mercaptan, and human odor (produced by passing air through a used undershirt). They recorded the largest change, however, to the sight of humans. Taping the mouth shut to prevent tongue-flicking did not eliminate the heart-rate change, but effective odors were associated with tongue-flicking. The authors concluded that olfaction was a high-sensitivity organ with low discrimination which served as a trigger for the tongue-sampling and the ensuing, more explicit, analysis by Jacobson's organ. Hainer (1964) reviewed the literature and concluded the opposite: that Jacobson's organ achieves greatest sensitivity at the price of discriminability with olfaction going the reverse route. A third alternative, based on our work, is presented on p. 274.

In Cowles' laboratory, Dullemeijer (1961) continued work on olfaction in rattlesnakes. He used the trailing technique, but with the modification of pitting a visual and a chemical stimulus against each other. The visual stimulus was a mouse in a jar and the chemical stimulus was a dead mouse in a perforated box. Unless the chemical cue was very strong, such as a long-dead mouse, the visual cue was responded to most often. However, movement was the essential visual stimulus. After the strike, the rattlesnake does not pay attention to living, free mice, but proceeds to follow the trail of the envenomed one. Covering the nostrils did not affect the behavior, but cutting the tongue tips did, which is further evidence for the importance of Jacobson's organ.

At the present time Naulleau (1964, 1966) is an active worker in this area. He has re-examined and extended Baumann's early work on *Vipera aspis* and analyzed the behavior more thoroughly. His methods were basically those of Baumann; he used trailing experiments and measured the success of the animals in finding food, although he improved the method by using much larger arenas and several different substrates. He eliminated the organs and found that Baumann's and Kahmann's conclusions were basically correct.

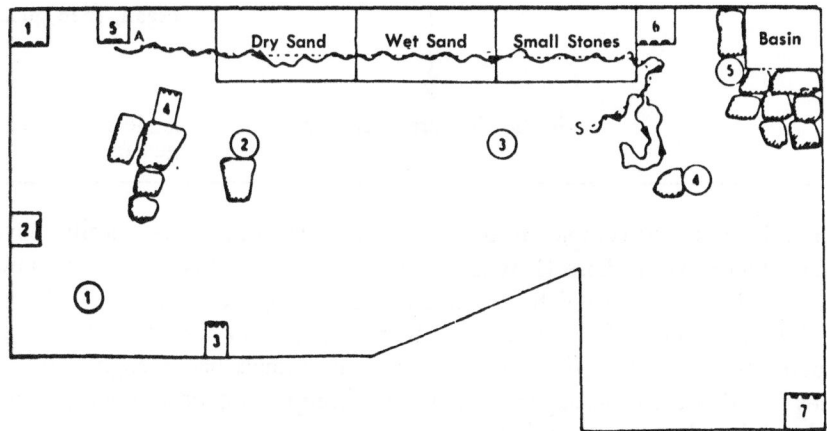

FIG. 13. A larger, more naturalistic arena for testing trailing in snakes. This figure illustrates a normal viper (*Vipera aspis*) following a mouse trail (broken line) over differing substrates to locate prey. (*Adapted from* Naulleau. 1966. Thesé, p. 87.)

Naulleau, on the basis of his experiments, characterized feeding behavior as follows (compare with Baumann, p. 260).

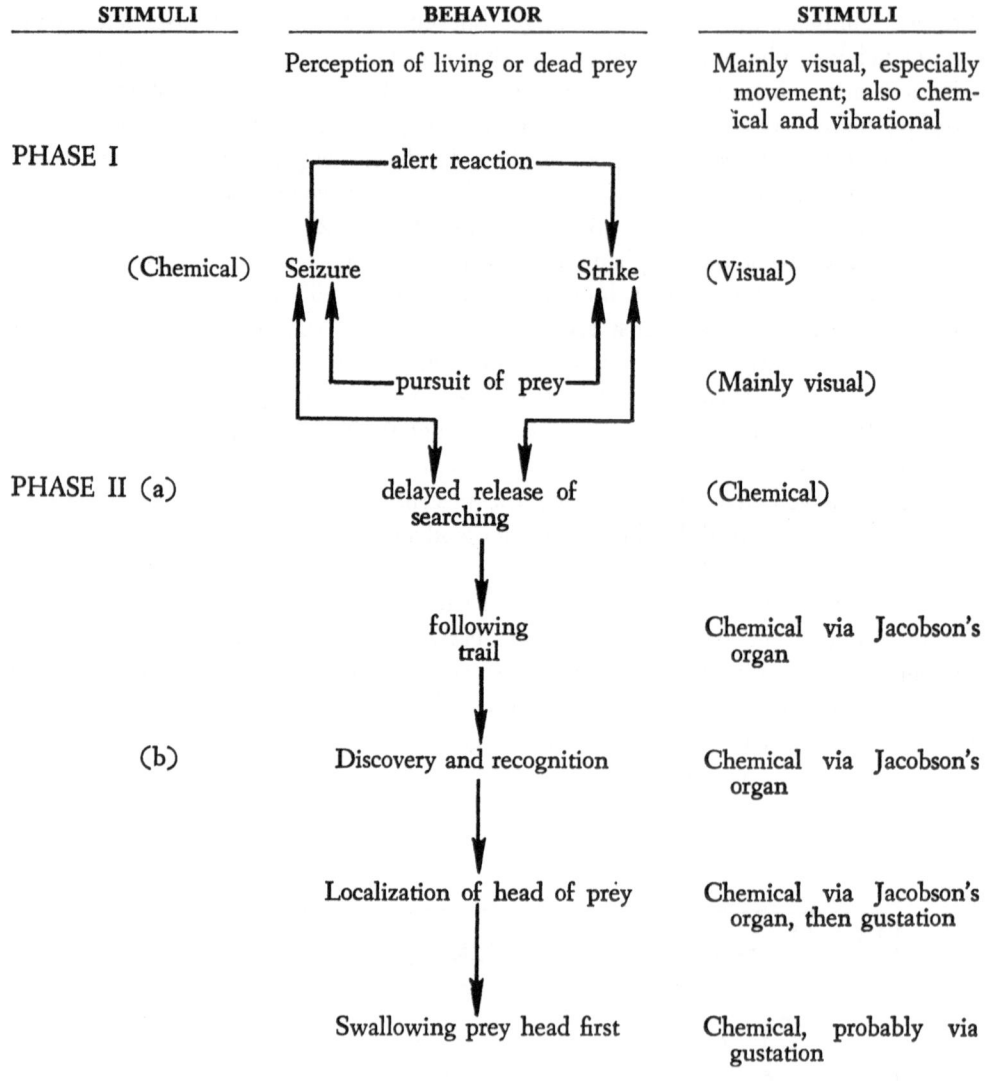

STIMULI	BEHAVIOR	STIMULI
	Perception of living or dead prey	Mainly visual, especially movement; also chemical and vibrational
PHASE I	alert reaction	
(Chemical) Seizure		Strike (Visual)
	pursuit of prey	(Mainly visual)
PHASE II (a)	delayed release of searching	(Chemical)
	following trail	Chemical via Jacobson's organ
(b)	Discovery and recognition	Chemical via Jacobson's organ
	Localization of head of prey	Chemical via Jacobson's organ, then gustation
	Swallowing prey head first	Chemical, probably via gustation

Phase I was almost completely controlled by visual stimuli, specifically movement. Our main interest is in phase II, where chemoreception seemed most important. The "search" reaction was released by the seizing or striking behavior or, if phase I was bypassed, by the "odor" of the dead prey. In both parts (a) and (b), blinded animals performed quite normally (Fig. 13). The normal animals had a slight advantage in that they could use the visual stimuli of the disappearing prey; hence, they could orient themselves better and find the trail sooner.

Blocking the nostrils had no effect upon the unfolding and completion of phase II (Fig. 14). However, some respiratory impairment was noticed. Cutting the tongue tips of the viper did not eliminate the search response, but it did destroy trailing

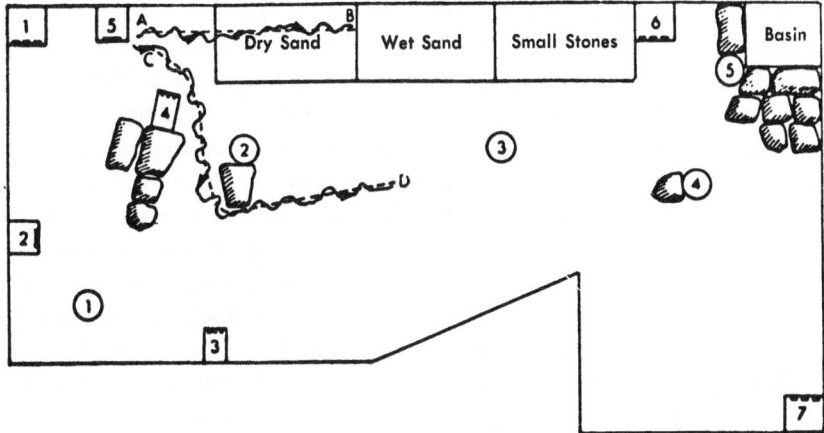

FIG. 14.   Same situation as in Figure 13, except the viper's nostrils are blocked. No difference is apparent from the performance of the normal snake. A-B: trail made by unbitten mouse. C-D: trail made by dragging the same mouse over the substrate after being bitten by the snake. (*Adapted from* Naulleau. 1966. Thesé, p. 87.)

ability (Fig. 15). If part (a) was bypassed, the snake was able to localize the head of the prey and to swallow it normally. Naulleau concluded (1966), as did Kahmann (1932), that nasal olfaction was of little, if any, importance in feeding.

Jacobson's organ in conjunction with the tongue is crucial in trailing, but what about releasing the search reaction and localizing the prey's head? Could Jacobson's organ be functioning here without the tongue? Naulleau attempted to eliminate Jacobson's organ by cauterization, but he had little success. The vipers were disturbed greatly, would not eat properly, and succumbed to infection. The one surviving subject responded as the severed-tongue snakes did. The results were not conclusive, as Naulleau himself recognized, and the organ may not have been completely destroyed (no histological examination was referred to). However, Naulleau assumed that Jacobson's organ had been functionally eliminated, leaving him in a quandary as

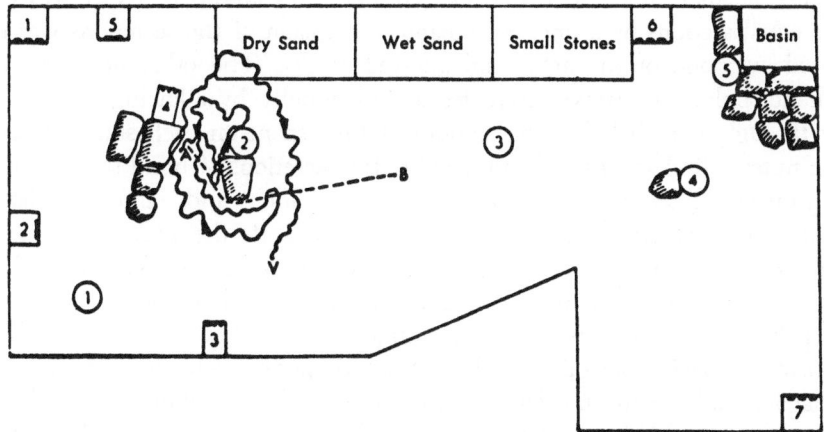

FIG. 15.   Same situation as in Figure 13, except the viper's tongue has been severed. No trailing or prey localization is evident. A-B: trail of mouse killed by the snake a day earlier. Compare Figure 9. (*Adapted from* Naulleau. 1966. Thesé, p. 91.)

to what chemical (or other) senses were involved in the behaviors undisturbed by his treatments. He considered tactile cues of little importance and this left, by default, gustation, which Naulleau concluded was the mediating sense. However, in the absence of definite studies and knowledge concerning the location of taste buds (which Naulleau admitted must be in the anterior regions of the mouth to play the role he attributed to them), a decision was premature. Since then, sensory oral papillae have been found in some snakes, which may resolve matters (Burns, 1969). It would be helpful to have experiments run with animals that were nostril-covered, tongueless, and had Jacobson's organ cauterized, to evaluate the possibility advocated by Noble and Clausen (1936) in relation to trailing: that both Jacobson's organ and the tongue play a role and can function alone; but that if both are eliminated the ability is abolished. An additional interesting finding of Naulleau was that the viper has a very limited ability to determine the direction of a trail.

The chemistry of the effective substances is of great interest. For instance, the skin of a mouse is more attractive than the skinned carcass. Naulleau has not, however, been successful in extracting any relevant substances from mouse skin. These ongoing studies by Naulleau are important, nonetheless, and should finally enable us to understand the complete feeding behavior of at least one serpent.

Recently Watkins et al. (1967) have added another interesting example of the mechanisms by which snakes use the chemical senses in food finding and in defensive behavior. They clearly showed that blind snakes (*Leptotypholps dulcis*) followed pheromone trails of two species of army ants. Field observations supported the notion that the blind snakes followed pheromone trails of raiding columns to ant nests where they fed on the army ant blood. The ants did not attack these blind snakes, but snakes manually introduced into raiding columns are attacked. Blind snakes so attacked, coiled and released an anal gland secretion which they smeared on their bodies, repulsing the ants (Gehlbach et al., 1968).

A later report analyzes the phenomenon further (Watkins et al., 1969). Anal gland secretion was collected by exposing snakes to army ants or by surgically removing the glands. Secretions were also extracted from army ants (*Neivamyrmex nigrescens*). A small arena with the floor covered with absorbent paper was divided into two equal sections by a line. An ethanol suspension of the secretions was evenly distributed over one of the areas and allowed to dry. The other area was treated with ethanol only. The snakes were tested individually by allowing them to enter the box through a small hole. The amount of time spent on each side was recorded for 30 minutes. Blind snakes were attracted to the secretions of blind snakes and army ants. Sympatric insectivorous (e.g., *Sonora*, *Tantilla*) and ophiophagous (e.g., *Diadophis*, *Lampropeltis*) snakes were repelled by both the blind snake and the army ant secretions. The authors categorized the snake-army ant system as follows: (1) The army ant used trail pheromones to coordinate its activities. (2) This pheromone repelled certain predators, but (3) attracted others (*L. dulcis*). (4) The anal secretion of *L. dulcis* (the blind snake) attracted species members but (5) repelled ophiophagous snakes and (6) deterred army-ant attacks. In addition, (7) the blind-snake–army ant association was mutually repellent (summation?) against predators (on either blind snake or ant) and competitors (insectivorous snakes). Clearly, this fascinating phenomenon showed the interrelatedness of the functional categories into which this survey has been divided.

This review of snake feeding behavior has shown the importance of the chemical senses and some of the disagreements concerning the importance of Jacobson's organ. Since nearly every study, except those of Noble's group and Gettkandt, indicated that Jacobson's organ was more important than nasal olfaction, especially in trailing, we can conclude that Jacobson's organ is indeed important. But, as in many areas of science, we must avoid too dogmatic a stance on this issue as couched in an either/or framework. Clearly, further work is needed, especially to incorporate recent advances in the study of animal behavior.

CHEMICAL PERCEPTION IN
NEWBORN SNAKES

Our research on chemical perception in snakes has stemmed from an interest in the behavior of newborn animals and a fascination with the ethological sign stimulus concept. Using feeding behavior, or more precisely the prey-attack response of Wilde (1938), I wanted to determine the role that the chemical senses could play, and also to determine how "innate" and "learned" processes interacted in determining the behavior at later stages. The first question asked was whether newborn snakes without feeding experience would respond to chemical cues from species-characteristic prey.

The first study (Burghardt, 1966) involved a litter of 20 newborn, unfed, eastern garter snakes (*Thamnophis sirtalis*). Garter snakes, which were used most often in these studies, are viviparous and can have as many as 60 or more young in one litter, although 15 to 30 is the typical number. These garter snakes eat earthworms and fish among other prey when in captivity as well as in the field. Would newborn garter snakes respond to chemical stimuli from worms and fish with prey-attack behavior? How would they respond to extracts from prey rarely, if ever, eaten?

The 20 newborn, unfed snakes were placed in small glass tanks, one individual per tank. The tanks were separated with partitions so that each individual was visually isolated from its neighbor. Water extracts were made of the skin-surface substances of prey, either horsemeat, meal worms (*Tenebrio molitor* larvae), minnows (*Notropis atherinoides acutus*), or redworms (*Eisenia foetida*). They were washed in tap water, dried, and placed in 60°C distilled water for one minute in the proportion of 20 cc per 3 grams of prey. They were stirred during this period and then removed. The remaining liquid was filtered and the clear, usually colorless, supernatant liquid was stored in the refrigerator until used.

The extracts were presented to the snakes as follows: a cotton swab was dipped into the extract, the excess removed, and the swab slowly introduced into the snake's cage. The swab was brought to within 2 cm of the snake's snout. If no attack was made within 20 seconds, the swab was moved closer until the lips were touched. Naive newborn snakes dramatically attacked the cotton swab that had been dipped in either worm or fish extract. All 20 attacked the worm extract and 13 of the 20 attacked the

FIG. 16.   A naive garter snake attacking a cotton swab dipped in an earthworm extract.

fish extract. They did not, however, attack swabs dipped in the insect extract, the meat extract, or the distilled water control. The attack response, given any effective stimulus, comprised the following: the snake increased its rate of tongue flicking and lunged forward at the swab with its jaws open at about a 45-degree angle (see Fig. 16). This response appeared identical with that toward live redworms and small fish which were fed to the snakes several days later. Although these young snakes normally refused to eat a piece of the horsemeat, they readily attacked and ate the horsemeat if a drop or two of worm extract was placed on it. This is congruent with the observations on adult snakes by Weidemann (1931) and Ramsey (1947), with *Natrix* and *Tropidoclonion*, respectively. Therefore, we can say that the inexperienced snake can recognize, on the basis of chemical stimuli alone, what it "should" attack as potential prey.

In order to study in detail a phenomenon such as this, it should be possible to test individuals repeatedly. The next question asked, therefore, was whether this stimulus-response connection was reasonably permanent. Two snakes, which had not been involved in the first tests, yet were from the same litter and previously unfed, were presented with the worm extract at five-minute intervals for 20 consecutive trials. The control swab was presented before the first test, after the tenth and after the last trial. Figure 17 shows the results of this experiment. Both subjects responded on every trial and there was no gradual lengthening of the response latency, although there was some variation. Subsequently, I have shown that a naive snake can be tested 40 consecutive times at five-minute intervals and will attack each time (Burghardt, 1967b). However, some individuals did cease attacking after considerably fewer tests. This may have been partially due to a function of motivational state, inter-trial interval, or extract strength. In later litters it was found that sometimes individuals never attack and also refuse to eat, perhaps because of the artificial nature of the

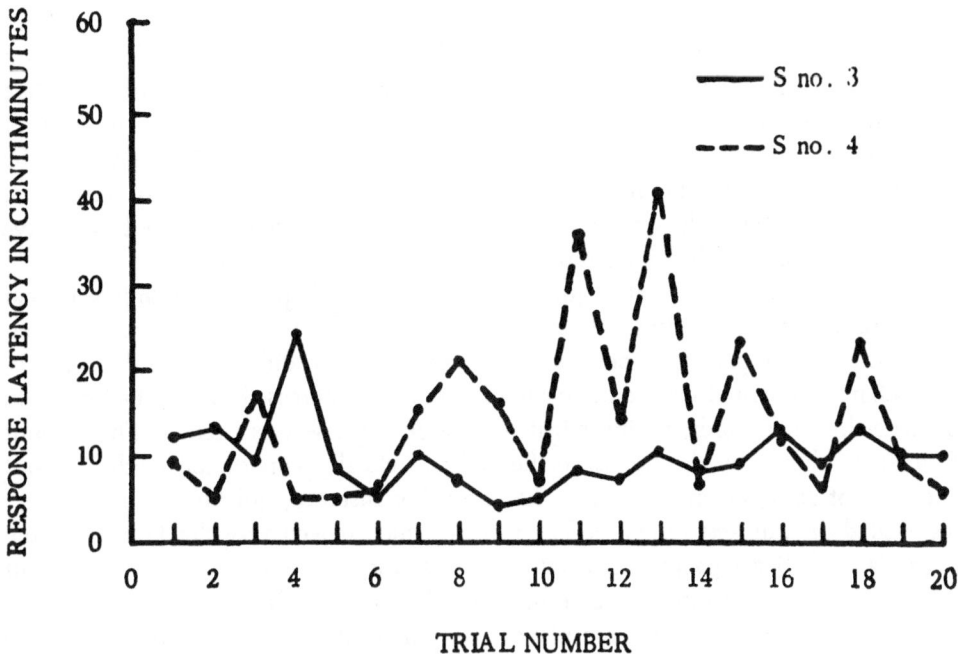

FIG. 17.   Latency of prey attack response to swabs dipped in redworm extract and presented at 5-minute intervals. Inexperienced *Thamnophis sirtalis* were used. (*From* Burghardt, 1966. *Psychomic Sci.*, 4:38.)

environment. These preliminary results convinced me that the chemically elicited responses of newborn snakes were amenable to investigation from a number of approaches. Using this and other litters of newborn young colubrid snakes (family Colubridae), we (colleagues, students, and assistants) have attempted to test some of the parameters involved. This is very much a progress report, then, and details of many of the following experiments have been or will be reported elsewhere (Burghardt, 1966, 1967a, 1968, 1969; Burghardt and Hess, 1968; Sheffield et al., 1968). The testing and extracting techniques have remained fairly constant, although some modifications and improvements have been made, such as the use of frozen extracts.

The Role of the Various Senses

In these experiments both methods discussed at the beginning of this chapter were used, that is, both the elimination of sensory information from the stimulus and the elimination of the senses of the organism itself. Obviously, the first experiment told us something about the stimulus control of the response. With the 20 individuals used in the first experiment, we proceeded to investigate the responses to visual aspects of prey animals from which the snakes could obtain no chemical information. In other words, the animals could see the normal visual characteristics of prey which they would attack, but they received no chemical cues from them. This was accomplished by presenting the prey animals in small, sealed, glass vials which were introduced into the home cage of each individual snake. These experiments were performed on the fifth day after birth and the snakes had not yet eaten. Ten snakes were presented with a vial containing water and three lively guppies. The control vial used for this situation contained water but no guppies. Other subjects were presented with vials containing three redworms or five mealworms. The control vials for these animals were empty. Both the control and test vials remained in the tanks for a period of one minute. Any behavior, including latency of responses, directed toward the vials was recorded. The 10 subjects not presented with live guppies were tested with dead guppies in water the following day.

The results here were also quite clear. The most important finding was that in no case did the visual stimulus of living prey elicit the attack response. However, there was a definite pattern of behavior often present which differed from the controls. This behavior consisted of orientation toward the vial, increased tongue flicking, and approach. In other words, an investigative type of response was elicited. The amount of "interest" shown by the snakes to the vial was clearly proportional to the amount of movement going on in the vial at that time. For instance, quickly swimming guppies elicited the quickest response. The worms, which moved very slowly, elicited the longest latency, if a response occurred. The mealworms, intermediate in the amount of movement, were also intermediate in their effectiveness. Of course, the three prey species also differed in external morphology and mode of locomotion. That the crucial factor was movement, however, seems substantiated by the response being directed to that part of the vial where there was movement. The ten subjects given dead guppies showed no response. It appeared that these visually elicited responses soon habituated since reintroduction of the vial shortly afterwards led to little or no response. We are now in a position to say that visual stimuli alone are not sufficient to elicit the prey attack response, but that they do play a role in orientation.

In the cotton swab presentation of chemical stimuli, there was a visual factor which undoubtedly aided in directing the response, although not in its elicitation. However, possibly visual information of some kind was necessary for an attack response to be made when the proper chemical cues were present. Another problem, of course, was concerned with which of the chemical senses were involved in the response to chemical cues. This, on the basis of the previously discussed research, resolved itself to the question of whether nasal olfaction or Jacobson's organ was involved. The work previously discussed by Wilde, who severed the olfactory or the vomeronasal nerves, showed that with the adult eastern garter snake (*T. sirtalis*) Jacobson's organ was crucial to the actual attack of prey. Was Jacobson's organ also the critical chemical sense in naive newborn snakes? In the following experiment, 20 ingestively naive newborn plains garter snakes (*T. radix*) were used which have similar habits to *T. sirtalis*. The young had vision and/or olfaction eliminated in various combinations by covering the eyes and nostrils with an opaque collodion mixture: 5 animals were made anosmic and blind, 5 animals were made anosmic only, 5 animals were blinded only, and 5 animals were left normal and untreated. The collodion mixture could be removed easily after it had dried with no apparent harm to the snakes. Snakes do not have movable eyelids, but have instead transparent scales over the eye. This made the procedure for blinding the snake much easier than with many other animals.

The stimuli were presented in the standard manner on cotton swabs. The sub-

TABLE 2

Effects of Sensory Elimination on Response to Worm Extract in 20 inexperienced *Thamnophis r. radix*

GROUP	N	MODALITY ELIMINATED	TOTAL ATTACKS	MEAN ATTACK LATENCY (SEC.)	MEAN TONGUE FLICK ATTACK SCORE	MEAN EXTRACT MINUS CONTROL SCORE
1	5	Olfaction	6	34.1	64.1	53.0
2	5	Vision	4	19.0	61.7	53.3
3	5	Olfaction and Vision	4	16.7	58.1	47.6
4	5	Normal	4	26.1	50.7	43.6

jects were tested first on a distilled-water control swab and then with a nightcrawler extract (*Lumbricus terrestris*). Then the control and nightcrawler swabs were presented to the subjects again. Table 2 gives the results of this experiment. Note that snakes attacked the nightcrawler swab regardless of what senses were eliminated. This clearly showed that normal snakes did not do better than the other three groups. In addition to attack frequency, note that a tongue-flick attack score was utilized in presenting these results. An extract, when not actually eliciting a prey-attack response, often would elicit a large number of tongue flicks over and above that elicited by distilled water. It appeared that the frequency of tongue-flicking was correlated with the intensity of arousal by (or "interest" in) the swab. Previous experiments indicated that the prey-attack response in garter snakes was mediated by the tongue-Jacobson's-organ system (Wilde, 1938) and the present attack data also supported this view since olfaction did not seem important to the newborn snakes. Therefore, Jacobson's organ seemed the most likely candidate. Tongue flick data, therefore, could be considered together with attack data in assessing the relative "releasing values" of various extracts. A scoring system, somewhat arbitrary, was devised and used to score the extract given to each snake. This scoring system was based on two assumptions. First, that an actual attack was a more definitive feeding response than any number of tongue flicks and, secondly, that a more potent stimulus led to an attack with a shorter latency than would a weaker stimulus. The base unit of this system was the maximum number of tongue flicks given by any individual of the experimental group tested to any of the test stimuli. The maximum was almost invariably given to a swab which contained an effective extract. A snake which did not attack was given a score identical with the base unit for that group of snakes, plus one point or fraction for every second or fraction less than one minute that it responded. The score for an attacking subject could then be represented by the following formula where the response latency was measured in seconds.

$$\text{Score} = \text{base unit} + (60 - \text{response latency})$$

For instance, if a snake gave 65 flicks to the nightcrawler extract and that was the highest number of tongue flicks given by any individual without attacking, then 65 would be the base unit. A snake which attacked at the end of 20 seconds would then get a score of 65 plus 60 minus 20 (which equals 105).

Using this system, it can be seen from Table 2 that the tongue-flick attack score results were comparable to the attack results alone. In demonstrating that vision was not necessary for prey-attack behavior, this experiment agreed with the results of the previous study in which the chemical cues were eliminated. It further demonstrated that vision was not necessary even as a supportive factor. Visual aspects of prey, therefore, were less important for snakes than for lizards, especially Iguanids (Burghardt, 1964). Of course, it should be remembered that the generality of this finding is limited. Although olfaction did not appear necessary if the source of the chemical stimuli was close to the snake, this does not prove that olfaction is not involved in feeding behavior.

Up to this point, we have shown the following about the prey-attack response in the newborn garter snake: it was possible to elicit responses with worm or fish extracts in the inexperienced snake. Visual stimuli alone were neither sufficient in themselves nor necessary for the response to occur. A response could be elicited repeatedly, without habituations, from naive snakes, and it was highly probable that the tongue-Jacobson's-organ system was the crucial sensory system involved.

The Comparative Perspective

We are now able to pose further questions. Would, for instance, newborn garter snakes respond to extracts of other prey normally eaten by the species? Would they do this without ever experiencing such prey before? Would different species respond in a similar fashion? Would they respond to the same extracts or different extracts, and how does their responsiveness correlate with their species-characteristic feeding habits? A number of such comparative studies have been done (see Burghardt, 1967a, 1967b; 1968; 1969). We began by looking at various species in the garter snake genus *Thamnophis*. Snakes of this large and widespread genus have differing food habits (Smith, 1961; Wright and Wright, 1957).

In these and subsequent comparative experiments, the following procedures were employed: all the newborn snakes were tested before any feeding experience whatsoever and before any previous extract-testing experience. Snakes were usually less than one week old. They were tested only once on each of the various extracts, and normally 12 extracts were employed. All the animals were isolated in individual tanks before they were tested, and they remained there for at least one day before being tested. In most instances, the water control was presented only once, as were the other extracts, but in some litters the water control was presented six times. Important differences were noted with the various number of control tests employed (see Burghardt, 1969). All extracts were prepared in an identical manner using the standard ratio of 3 g of prey per 20 cm^3 of distilled water. The results in general showed a surprising congruence between the responses of the inexperienced newborn young and known feeding habits of the species.

Let us first consider the common eastern garter snake, *Thamnophis sirtalis*. Figure 18 shows the results of the comparative tests. In these and similar figures, whenever

THAMNOPHIS SIRTALIS SIRTALIS

N=19

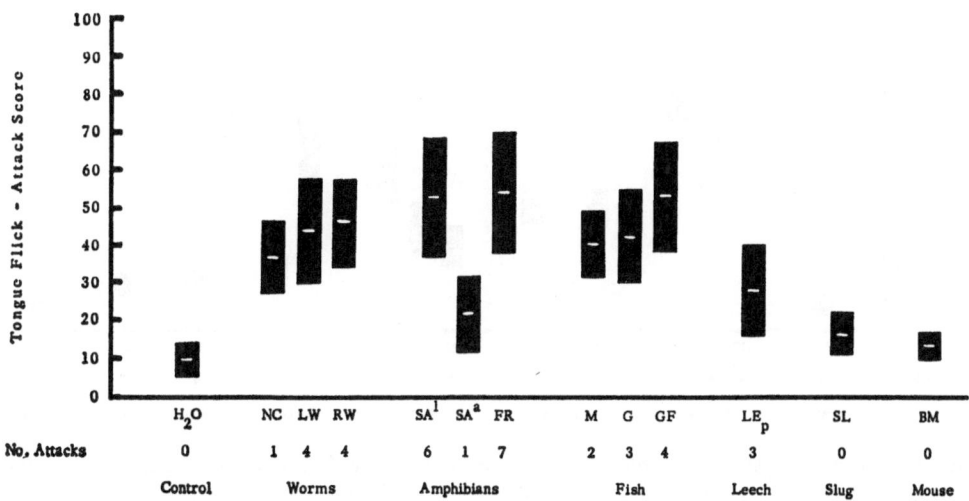

FIG. 18. Extract response profile of *Thamnophis s. sirtalis*. NC-nightcrawler; LW-leaf-worm; RW-redworm; SA¹-larval *Ambystoma* salamander; SAᵃ-metamorphosed *Ambystoma* salamander; FR-cricket frog; M-minnow; G-guppy; GF-gold fish; LEₚ-turtle leech; SL-slug; BM-baby mouse. (*From* Burghardt. 1969. *Behaviour*, 33:86.)

bars for two extracts do not overlap, the difference in the mean prey-attack–tongue-flick scores are significant at least at the .05 level.

All the test extracts attacked yielded a significantly higher score than the water control with the exception of the adult-salamander extract. It was interesting that the larval salamander extract was significantly more potent than the metamorphosed adult extract from the same species. This relationship has subsequently been found in a number of species which include salamanders in their diet. From data about the normal feeding habits of this species (Smith, 1961), we knew that earthworms, amphibians, fish, and leeches were included in the normal diet and that these animals were readily eaten by captive adults. It was clear, then, that naive young could recognize, by chemical means alone, a large number of the food items utilized by this species in nature.

Figure 19 shows the results for a litter of the eastern plains garter snake (*T. radix*). This species showed a similar profile in that worms, fish, amphibians, and the turtle leech were effective. A second leech was used, the giant leech, but this other leech elicited no attacks from this litter, although the tongue flick scores were significantly above the control level. The adults would not eat the giant leeches either; however, they readily were eaten by *T. sirtalis* adults, and other experiments with *sirtalis* young showed that they would attack extracts from these leeches. Perhaps it is at the level of the response to various leeches that we can distinguish the profiles of the two species. The normal feeding habits of *radix* are similar to *sirtalis* (Smith, 1961). Note also the similarity to the *sirtalis* in the metamorphosed versus larval

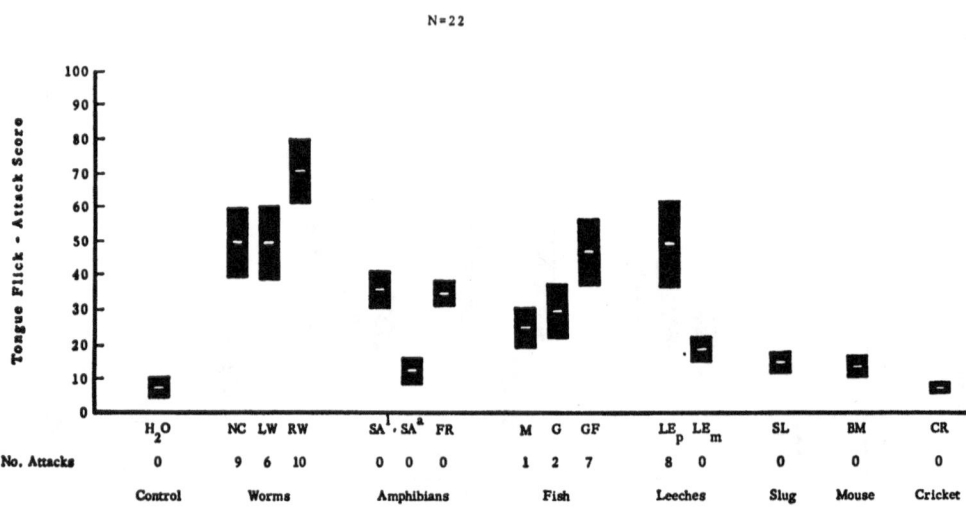

THAMNOPHIS RADIX RADIX

N=22

FIG. 19.   Extract response profile of *Thamnophis r. radix*. Symbols as in Figure 18 plus LE_m-giant leech; CR-cricket. (*From* Burghardt. 1969. *Behaviour*, 33:96.)

salamander results. Metamorphosis is accompanied by great physiological and morphological changes which undoubtedly influence the chemical stimuli emitted. We have attempted to determine when during the metamorphosis the change in *T. radix* preference occurs. With the help of Thomas Uzzell and Robert Storez, *Ambystoma laterale* salamanders were reared from eggs. When they reached about 2 cm in length, several were used in preparing an extract every few weeks. These extracts were kept frozen. An extract was also prepared from a fully metamorphosed yearling, resulting in 8 extracts in all. These extracts were tested on 8 naive newborn *T. radix*, using completely balanced orders. All extracts *except* the yearling extract elicited attacks, with no difference between the larvae extracts. Indeed, almost identical numbers of attacks were released with similar tongue flick-attack scores (ranging from 77.4 to 91.6). There was no evidence of decrease in effectiveness with increasing development of the larvae, although all four limbs had appeared, gills had disappeared, and crawling out of the water occurred. Since we were unsuccessful in completing metamorphosis in this group, we can only conclude that the preference change occurs fairly rapidly during later stages, probably when pigmentation and other skin changes are more pronounced.

Now consider a species of *Thamnophis* with a quite different type of extract-response profile, this was the aquatic garter snake *T. elegans aquaticus* from California (name revised by Stebbins in 1966 to *T. couchi aquaticus*). According to Stebbins (1954) and Fox (1952), this species eats mainly fish and amphibians. From the extracts presented to the newly born young of this species, it was clear that only fish and amphibians were effective (Fig. 20). Extracts from these organisms gave much higher scores than the control or other extracts. Unlike the preceding species, the worm extracts were little different from the controls. Since this species does discriminate between fish and worms in that it attacks the former and not the latter, it is reasonable to assume that the effective substances responsible for prey attack to fish are

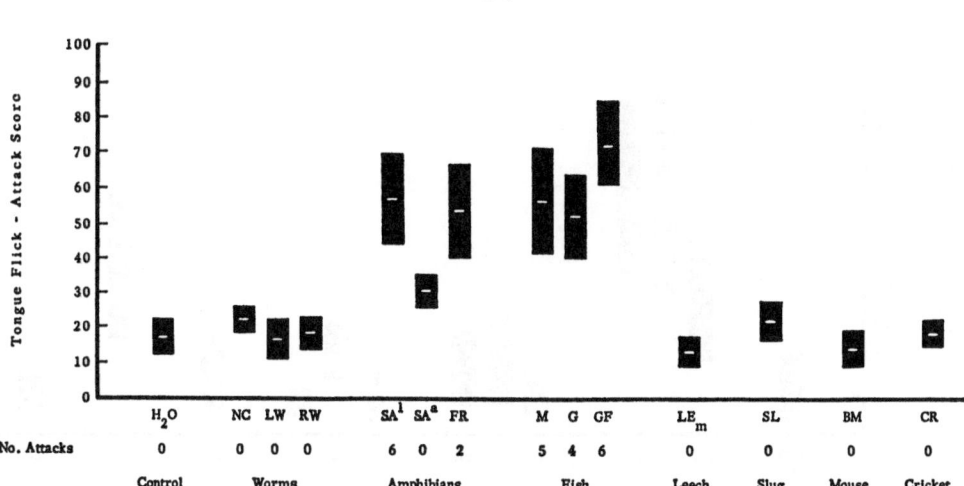

FIG. 20. Extract response profile of *Thamnophis elegans aquaticus*. (*From* Burghardt, 1969. *Behaviour*, 33:99.)

different from those involved in the prey attack to worms, even in snakes responding to both objects. Whether there is a "fish" or "worm" factor that is similar for all fish or all worms, or large portions of them, cannot be determined on the basis of this data. This question will be discussed again later.

Other species of *Thamnophis* have also been tested (Burghardt, 1969), all of which have been found to attack or to respond with increased tongue-flicking to certain extracts. The extracts to which they responded were highly correlated with preferred feeding habits. It would seem clear that natural selection has acted on the chemical perceptual selectivity of snakes.

In Butler's garter snake (*T. butleri*), the situation was a little more complex for fish, and amphibians constituted little, if any, part of the species' normal diet (Carpenter, 1952). Yet specimens readily ate fish in captivity and newborn young responded significantly to fish and amphibian extracts. It was, therefore, apparent that the normal feeding habits and ecology of a species were not sufficient to explain the response to chemical cues in newborn young. In this case, Butler's garter snake may have retained the perceptual side of a releasing mechanism lacking selective advantage in its present mode of life. Of course, retention by inexperienced snakes of the potential to respond to chemical cues from fish would have an adaptive advantage should fish become a necessary or more available food source. Therefore, in relation to the extracts used, Butler's garter snake seemed to possess more innate perceptual responsivity than did the aquatic garter snake, which did not respond to any of the worm extracts. Butler's garter snake has been considered to have evolved from the plains garter snake, which not only innately responds to extracts from fish and amphibians but also normally eats them (Schmidt, 1938).

What about snakes of other genera? A number of species have been tested and newborn young have been found to have similar perceptual discrimination preferences based on chemical cues. Let us look first at the genus *Storeria*, small snakes

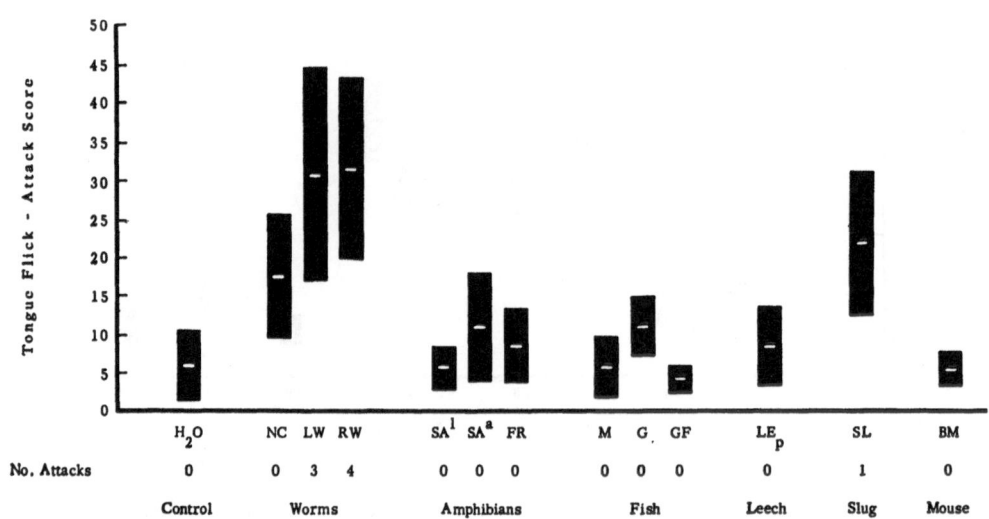

FIG. 21.   Extract response profile of *Storeria dekayi wrightorum.*

related to the genus *Thamnophis.* The midland brown snake, *S. dekayi wrightorum,* is a species which feeds mainly upon earthworms and slugs. A litter of young from this species were tested with a series of extracts, and they responded most to the worm and slug extracts, as seen in Figure 21. Similar results were found for another member of this genus with similar feeding habits, the red-bellied snake, *Storeria o. occipito-maculata.*

Let us now look at another genus, *Natrix,* the water snakes. Three sympatric species of *Natrix* young were tested on a series of extracts. One of these species, the common banded water snake, *N. s. sipedon,* responded mainly to fish and amphibian extracts, as was to be expected from what we know of their feeding habits (Smith, 1961). The other two species, Graham's water snake (*N. grahami*) and the queen snake (*N. septemvittata*), rarely feed on fish and amphibians, relying more on crayfish, especially newly molted ones (Smith, 1961). The young of these two species responded significantly above the control level only to crayfish extracts and more so to extracts prepared from newly molted soft-shelled crayfish. These results are illustrated in Figure 22 and are presented more fully in Burghardt (1968).

Turning to a snake which normally feeds upon warm-blooded prey, the fox snake (*Elaphe v. vulpina*), we see (Fig. 23) that young responded more to the baby mouse extract than to other extracts presented and that this was the only extract to which they responded significantly more often than to the water control. Unfortunately, the fox snake was defensive immediately upon hatching (this species is oviparous) and the introduction of swabs in the normal manner frequently elicited an attack which was not a prey-attack but a defensive attack. Rapid vibration of the tail which could, under proper circumstances and environmental conditions, sound very much like a rattlesnake, also accompanied these attacks. The fact that these attacks

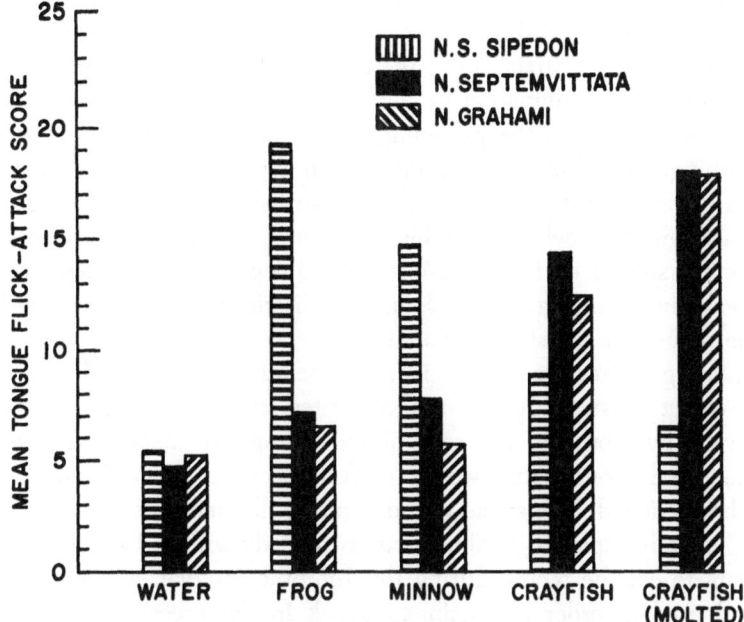

FIG. 22. Extract response profiles of selected extracts presented to 3 sympatric species of North American water snakes (Natrix). (*From* Burghardt. 1968. *Copeia*, 736.)

ELAPHE VULPINA VULPINA

N=10

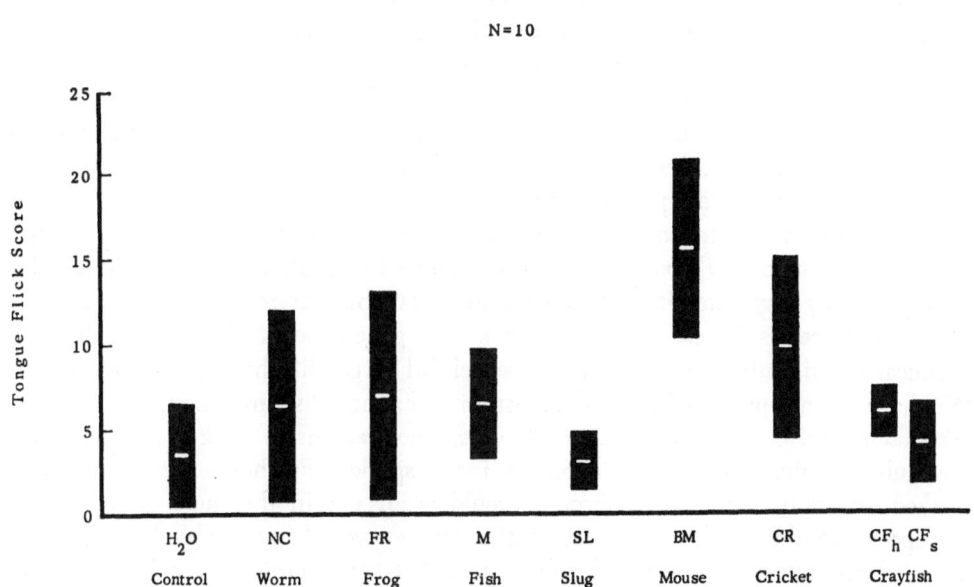

FIG. 23. Extract response profile of *Elaphe v. vulpina*. Symbols as in Figure 18 plus CF_h-normal hardshelled crayfish; CF_s-molted crayfish.

occurred hindered the use of attack scores. Only tongue-flick scores could be used, and these only when the snake did not attack. Therefore, attacks given to certain extracts necessitated that extract being presented again at a later time. Baby mouse extracts did lead to a discrimination in behavior. The result was modest and not too impressive in relationship to the preceding results (Fig. 23). However, it is worthwhile mentioning, since Naulleau, working with vipers (see above), has been unsuccessful in extracting from the skin of mice any substances which were effective in eliciting feeding-type behavior in his subjects (personal communication). Also, using a different species of *Elaphe*, Morris and Loop (1969) showed prey extracts were more effective than water, but found no discrimination between rodents and unacceptable prey such as fish and worms.

Rodent-eating snakes are usually more deliberative and respond more slowly to actual food objects than do garter snakes, water snakes, and many others. Even placing a live baby mouse in the cage of the young fox snakes did not elicit the almost immediate responses that worms or fish placed in a newborn garter snake cage would have elicited. The snake would approach the mouse cautiously, flicking its tongue and generally exploring various parts of the body with it. In no case did the fox snakes attempt to eat the mouse during the 1-minute interval, which was the maximum interval used in order to obtain an attack in the preceding species. In fact, after 15 minutes none of the snakes ate the mouse, but overnight each did eat its mouse. Therefore, it seemed clear a different type of methodology was needed for working with these animals. Using a 2-minute interval, more concentrated extracts, and repeated testing did lead to reliably increased responding to mouse extracts in corn snakes, *Elaphe g. guttata* (Burghardt and Abeshaheen, in preparation).

Looking at another oviparous species, the western smooth green snake (*Opheodrys vernalis blanchardi*), we see clear-cut results. This species, in contrast to the preceding, is insectivorous, also eating spiders and other small arthropods. A litter of naive young of this species were tested on the same extracts, at the same age, and on the same date as the plains garter snake. The results, however, were different, as seen in Figure 24. Here only the cricket extract was effective. Subsequent experiments, using other prey which this species might be thought to eat, has enlarged the list of effective extracts considerably. Extracts which have elicited attacks in naive young have included various caterpillars and small moths.

It is clear that the answers to the questions posed at the beginning of this section are affirmative. *Thamnophis sirtalis* responded with attack behavior to the wide variety of prey organisms it has been found to eat. Similar relations were found in many other species. These results should not be generalized to all snakes, although chemical identification would probably be helpful in the life history of most species. We have been unsuccessful in demonstrating extract discriminations in *Coluber*, *Heterodon*, and *Lampropeltis* young. However, more extensive work is needed before the null hypothesis should be accepted for a species. Furthermore, we saw that marked qualitative behavior differences could occur within the same genus, where one species would attack worms and another species would not. It should be possible by this technique to discover and collect *extract-response profiles* of a large number of species. Unfortunately, some problems are encountered in obtaining the prerequisite naive young snakes. Why should newborn young have to be used? Results

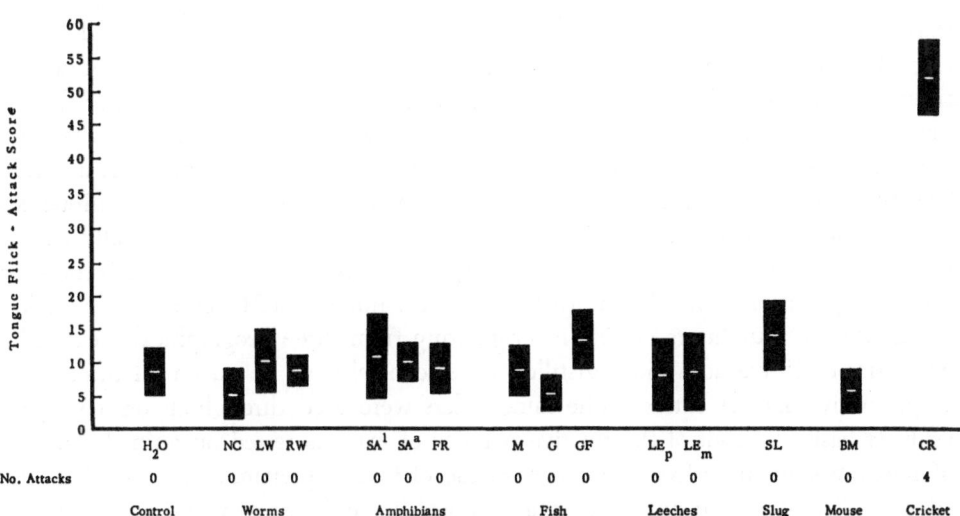

FIG. 24.   Extract response profile of *Opheodrys vernalis blanchardi*.

discussed below show that it is imperative that inexperienced young be used in constructing these extract profiles. But given that stipulation, there is at least the theoretical possibility that the use of these extract-response profiles might help to unravel some of the systematic and evolutionary questions surrounding many common serpents. In other words, I am proposing that perceptual preferences can be used to determine the relationships between species in a similar manner to the way in which behavior patterns have been of help in the study of animal evolution (e.g., Dilger, 1960; Lorenz, 1941).

Up to now we have shown that species differences do exist in the type of extract attacked or responded to by newborn, previously unfed snakes. Was it not possible to show some differences within the same species of newborn snakes? Different populations from various geographical areas have been exposed to different habitats with different prey availability for many generations. It would certainly be reasonable to inquire whether snakes from varied ecological habitats showed similar or different responses. Perhaps all snakes of a given species have the same inherited abilities which are then altered by individual experience with various food objects. On the other hand, some "canalization" of the innate behavior to specific habitat requirements could be expected to be found. Our tests with *Thamnophis sirtalis* suggested this possibility. I tested newborn *T. sirtalis* litters from Illinois and Indiana (Burghardt, 1969). Using similar series of extracts, I found that there were some differences between litters in the relative releasing values of effective extracts. Two litters from the same area were quite similar to each other as compared to litters of *T. sirtalis* from other areas. However, I was reluctant to conclude that geographical variation had been shown in the newborn. There were several reasons for this. One is that, in contrast to the previous examples of species differences, here the differences within *T. sirtalis* were not in the type of extract attacked so much as its relative effectiveness. Another factor was that these litters were tested under conditions that could have allowed the entrance of certain contaminating variables. The litters were of varying sizes. Hence, order, although systematically balanced as far as possible, was not completely eliminated as a factor. Some extracts were presented earlier than others, and vice versa, and, as shall be shown later, the position of the extract in relation to a control can affect its effectiveness. In addition, the extracts were prepared at different times because the litters were born on different dates over a period of many weeks. Fresh extracts were always used in these early studies and differences in the preparation of the extract could not be avoided. Therefore, we specifically set out to investigate the question of innate intraspecific variation in the effectiveness of chemical-sign stimuli (Burghardt, 1970). Identical numbers (21) of newborn snakes were used from four litters of *sirtalis*, originating from three geographical areas. The same series of 15 extracts and a distilled water control were tested on all litters, one test per individual per extract. The same orders were used throughout on a subject-to-subject basis. To control the relative potency of extracts tested on animals born at different times, we used frozen extracts instead of freshly prepared extracts. Since the earlier studies, it had been found that extracts could be frozen and would retain their effectiveness for much longer periods than refrigerated extracts. Therefore, samples from the same batches prepared weeks before any of the snakes were born could be used on naive young born weeks apart. All the snakes were tested on exactly the same days after birth. The snakes were from the Chicago area in Illinois, from Iowa, and from northern Wisconsin. Two litters were tested from the Chicago area since there was the possibility that differences between litters from different areas, if found, represented normal interlitter variability.

In Figure 25 are the data for all four litters. Only the attack frequency measure

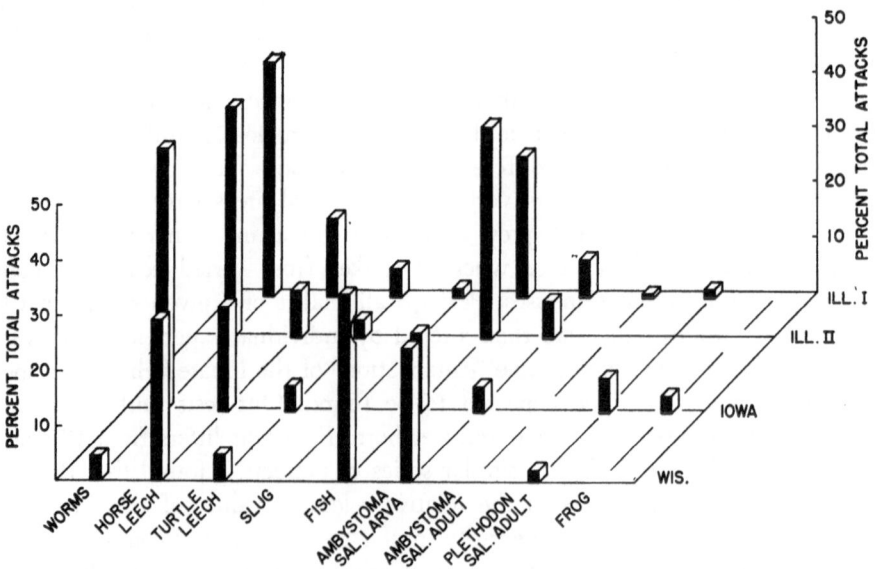

FIG. 25. Extract-attack profiles of 4 litters of *Thamnophis sirtalis* from three different geographic areas. (*From* Burghardt. 1970. *Behaviour*, 36:249.)

is given. The three scores for worms and fish were grouped. The differences were quite clear. The litter from Wisconsin responded at a very low level to worms and at a very high level to fish. The Iowa litter responded at a high level to worms and at a rather markedly lower level to fish. The two litters from Chicago, in contrast, responded at a high level to both fish and worms. There were other differences also, as seen in the responses to larval salamanders, the two species of leeches, and the Plethodontid salamander. There was a high relation between the two Chicago litters ($r_s = .90$) and significantly lower correlation with and between the other litters. Use of the tongue-flick data did not alter these conclusions.

These results confirmed the suspicions aroused by the earlier studies. Since these data were gathered, Dix (1968) has shown that newborn garter and water snakes from various geographical areas showed differences in food preferences. Therefore, the work based only on chemical cue preferences was collaborated by work on actual feeding behavior in newborn and adult snakes. We have also found evidence of individual differences in chemical preferences with a litter. For instance, some snakes consistently respond better to a worm extract than to that of a fish, and others do the reverse. Hence, *perceptual polymorphism* exists which has ecological and evolutionary implications.

Stimulus and Response Deprivation

Since the newborn animal seemed to have precise chemical recognition abilities, which were adaptive in the context of its normal life history, it was reasonable to consider how stable these perceptual abilities were in the absence of normal experience. While there was no way in which Jacobson's organ could be eliminated reversibly, it was possible to restrict the experience of the animals to specific and behaviorally crucial stimuli. An important question concerning the prey-attack response and the "innate perceptual schema" or "releasing mechanism" was how dependent the behavior pattern was upon the performance of the response and the perception of the sign stimulus. In other words, did "instinctive regression" occur if the animal was deprived of certain experiences (Thorpe, 1956)? It is known that some sensory abilities, like physical structures, can atrophy if not used or exercised (Hubel and Wiesel, 1963).

Adult snakes maintained in captivity readily ate food which they probably had never encountered in nature. Perhaps experience with any type of prey releasing the innate prey-attack response was sufficient to keep activated the response to all prey objects to which the young snake would respond at birth. In order to assess the role of experience in the chemical elicitation of prey attack and the fixed action pattern of the attack itself in the newborn snake, an experiment was performed (Burghardt and Hess, 1968). The rational was as follows: if the animal was raised from birth on an artificial diet which it would not eat by itself and was tested on its responses to extracts at a time when it was much older than it could possibly have been in nature without feeding, would the prey-attack response and the "releasing mechanisms" still be present, altered in some way, or absent? In addition, would there be any change in the response to the artificial food itself? Experiments began with a litter of 46 Chicagoland garter snakes (*T. sirtalis semifasciata*). The young were,

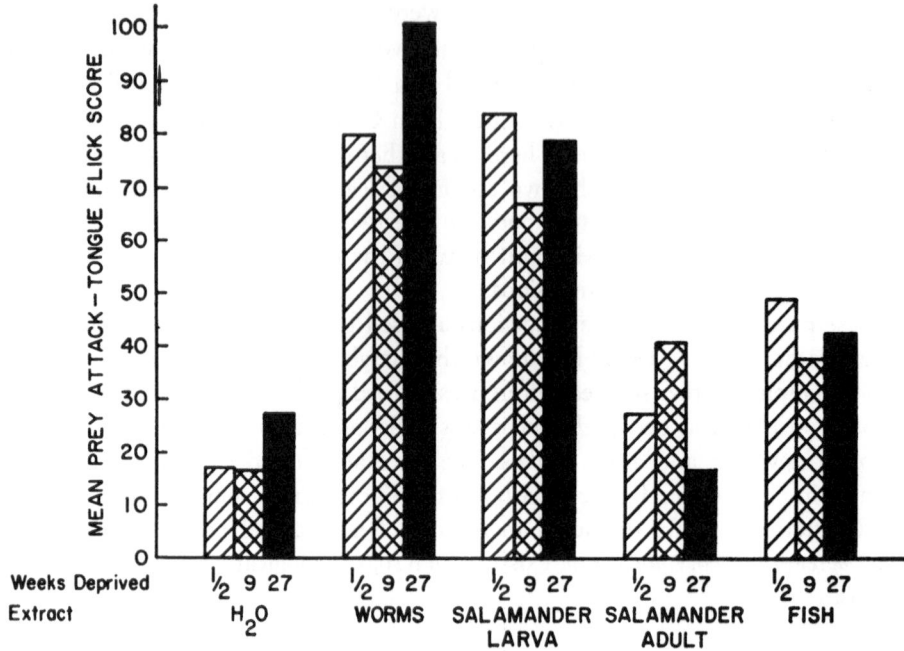

FIG. 26. Comparison of 2 and 6 month stimulus- and response-deprived *Thamnophis sirtalis* with littermates tested shortly after birth. No significant differences within extracts are evident.

as usual, previously unfed. Fifteen were chosen at random to undergo testing shortly after birth on twelve different prey extracts and a water control. This was part of the comparative testing described earlier. The rest of the snakes were raised in isolation from the prey animals used for extracts. They were kept approximately for either two months or six months on an artificial diet consisting mainly of strained liver baby food which was forcibly injected into the snakes' stomachs at periodic intervals. The snakes surviving this drastic regimen were tested at either two or six months of age with a series of extracts similar to those on which their litter mates had been tested shortly after birth. A brief summary of the results is seen in Figure 26. The important point is that the snakes tested at either two or six months of age showed an attack response toward test swabs indistinguishable from that of the newborn snakes. This was true across all the classes of prey objects effective in the newborn snakes. Rank correlations between the extracts common to the various deprived and nondeprived earlier tested groups were all significant, at the .01 level. No increased interest was shown in the extract from the liver, nor would they eat the liver when it was placed in front of them. Therefore, we can conclude that changes due to deprivation did not occur in the prey-attack response nor in the "innate schema" within six months after hatching and that the rank ordering of the releasing values of the various stimuli remained quite constant. Further, no learned association between the odor of the forced food and the life-sustaining food reinforcement was demonstrated.

It would seem evident, then, that we were dealing with a phenomenon which fulfilled all the requirements for an innate sign stimulus or, by inference, for the existence of an innate releasing mechanism (IRM) as discussed at length by the early ethologists (Schiller, 1957). Of course, no releasing mechanism (or triggering device, or differential responsivity, to use somewhat equivalent terms) could be totally impervious to experience. This was also true of the present behavior, as will be shown later. What we could say from the present experiment was that the relative preference for food-related chemical stimuli was stable and did not need reinforcement during ontogeny.

Concentration of Chemical

An important factor in the prey-attack response to chemical stimuli was the concentration of chemical needed to elicit the response. The corollary was how did the effectiveness of an extract vary with its concentration? Some experiments have been done attempting to look at this problem. Since we do not yet know what the effective chemicals were, we had no absolute measure of various concentrations. We could only measure concentration in relation to the standard extract as prepared in the manner described earlier. In the first experiment (Burghardt and Hess, 1968), 20 newborn *Thamnophis sirtalis* were utilized. They were previously unfed. Standard extract (3 g/20 cm^3 water) prepared from an earthworm was diluted to either one-half or to one-fourth of the original strength with distilled water. Another extract was prepared by using double the normal weight of prey per given amount of water.

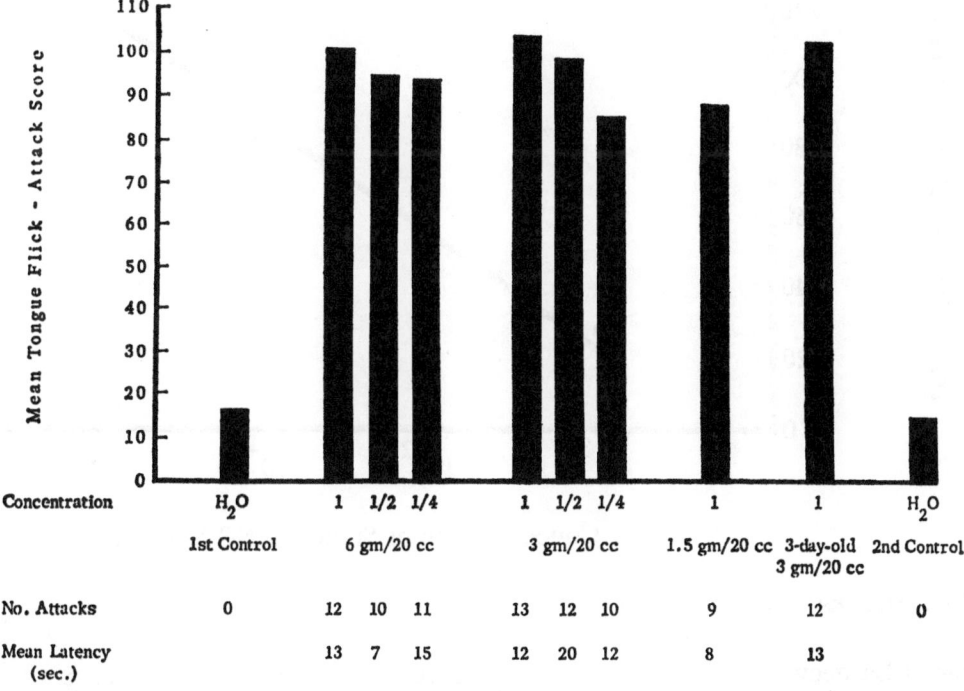

Fig. 27. Mean scores by 20 newborn *T. sirtalis* to extracts prepared with three different weight to water ratios and in various concentrations. A standard extract kept refrigerated for 3 days was also tested. Each extract was tested only once. Distilled water was tested twice.

This extract also was diluted to one-half or to one-fourth of the original strength. Another extract was prepared using only 1.5 g of worms to the standard amount of distilled water. In this experiment, the newborn young responded to all these extracts at nearly the same level (Fig. 27). The diluted extracts were somewhat less effective than the more concentrated extracts. Nonetheless, there was remarkably little variation; so little variation, in fact, that it seemed reasonable to conclude that minor variations involved in preparing an extract were of little consequence. However, it was also reasonable to assume that diluting the extract enough would effect its releasing value. Therefore, another experiment was performed in which the extract was presented in successive dilutions by a factor of ten. In this experiment, young Butler's garter snakes (T. butleri) were used. Standard nightcrawler extract was made and then diluted to either 10 percent, 1 percent, or 0.1 percent of its original strength. Twelve animals were tested once with each extract and a water control. Figure 28 shows the results with a line drawn indicating the logarithmic relationship between concentration and relative effectiveness of the various strengths of extracts. Even a one percent solution was effective in that one snake gave an attack to it. This clearly illustrated the sensitivity of the snakes to meaningful chemical stimuli. Until the active substances are identified, however, we cannot express this relation-

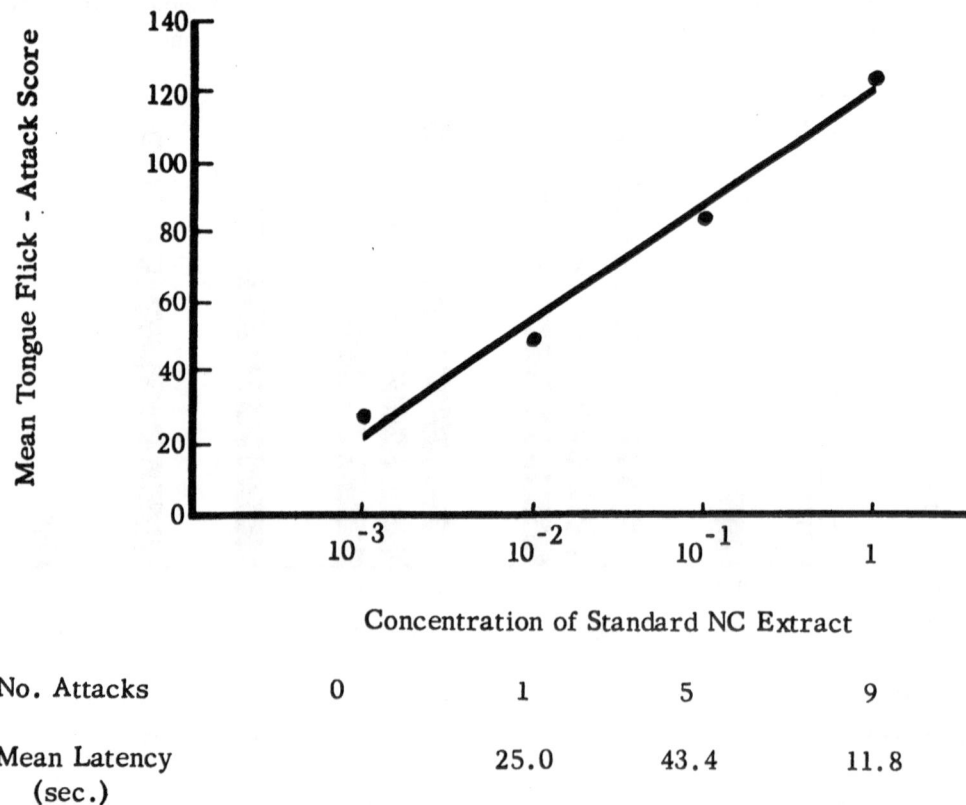

No. Attacks	0	1	5	9
Mean Latency (sec.)		25.0	43.4	11.8

FIG. 28. Responses of 12 *Thamnophis butleri* to varying concentrations of nightcrawler extracts. (*From* Burghardt and Hess. 1968. *J. Comp. Physiol. Psychol.*, 66:292.)

ship in precise chemical or quantitative terms as can be done with known odoriferous substances (see Chap. 12; Engen, 1964), although more elaborate psychophysical studies are underway.

Motivation

If the responses elicited by the extracts normally used in these studies were functionally equivalent to feeding responses, then the effectiveness of extracts should vary with the "hunger" of the animal, as measured for example by time since the last feeding. In order to investigate this in newborn young, 12 previously unfed *Thamnophis sirtalis* were tested. On the first day of this experiment, the snakes were tested once each on a redworm extract prepared in the standard manner as well as on 10 percent and 1 percent dilutions and distilled water. The snakes were fed a piece of redworm about 0.4 g and then retested on all four extracts about 90 minutes later. Approximately the same time each day for the following five days, each subject was tested again on the four extracts. The results are seen in Figure 29. It was clear

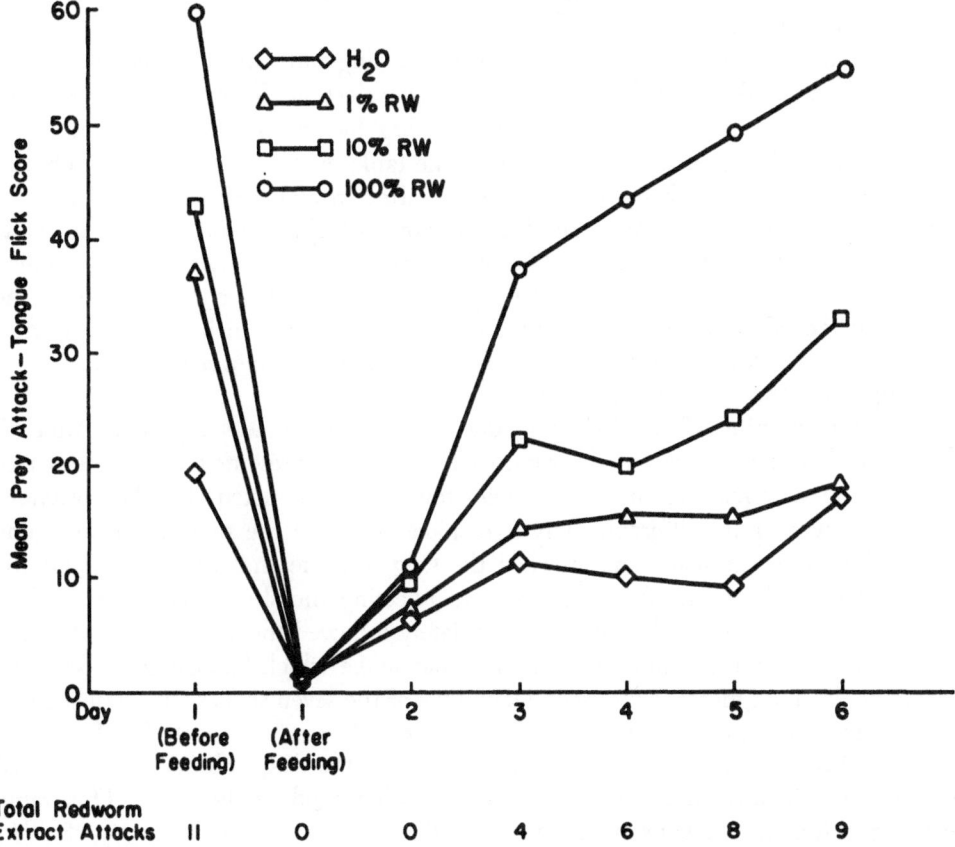

FIG. 29. Recovery of responsivity to redworm extract after 12 *Thamnophis sirtalis* were each fed 0.4 g of redworm. A 30-second test was used. The snakes had not been fed previously.

that responsivity to the extracts fell off drastically after feeding and then recovered over the following days. At the end of the sixth day, responsivity had almost reached the pre-feeding level. Note also that tongue flicking to the control water swab varied in the same direction as the worm extracts, although at a much lower level. It could be stated quite confidently that the level of responsivity to prey extracts was correlated with the deprivation level. This ability to measure "motivation" without feeding may be quite valuable for future experimentation.

Modifying the "Innate Schema"

We will now turn to some studies performed in an attempt to modify the stimulus-response relations shown by the newborn snakes.

An experiment raising some questions developed from one of the comparative series of tests on *Thamnophis sirtalis* (Burghardt, 1969). In one litter, one of the naive snakes attacked a swab dipped in a slug extract. Since this species was never reported as eating slugs, and the adults we had in captivity would not eat slugs, we decided to offer the young live slugs. One of these was placed in the tank of each of the 12 previously unfed snakes and allowed to remain for one minute unless it was eaten or in the snake's mouth.

On the first day, seven snakes attacked the slug but only one actually ate it. For six of the snakes, attacks were followed by rejection. This rejection took two forms: either the snake actively rejected the slug by backing away from it or by rubbing his head against the glass bottom of the tank; or it ceased the usual behavior pattern of ingestion which flowed on smoothly after an attack and remained motionless with its jaws over the slug. In this situation, the slug would proceed to crawl out of the snake's mouth to safety. The one snake who actually ate the slug seemed to do so by "mistake," a behavior that was neither as intense nor as qualitatively the same as when taking a food these animals are known to eat, such as fish or worms. It seemed to swallow the slug in spite of its own rejection movements. Subsequent tests supported this conclusion.

On the morning of the following day, this procedure was repeated. After all the snakes had been tested on the slug, each snake was presented with a dead redworm in the same manner and the behavior towards it was noted also. The reactions to the slug were quite different. Only two snakes attacked the prey-object and none ate it. One of the animals who attacked the slug was the snake who ate it on the preceding day. However, after attacking and rejecting the slug twice he ignored it.

The test with the dead redworm immediately followed the slug test. Eight of the snakes ate the worm in 1 minute, including that snake which had eaten the slug the preceding day but ignored it the second day. Five of the seven snakes who attacked and rejected the slug on either day ate the redworm. It can be concluded from this experiment that the snakes rapidly learned not to eat the slugs. Hunger was removed as a variable because the majority would eat other food (redworms). This rapid learning to inhibit a response to certain stimuli has been noted in a variety of other animals, and it usually involves visual cues. A well-known example is that of the toad and the bee (Brower et al., 1960).

The first indication that experience with extracts alone could modify behavior resulted from the comparative studies which employed several distilled water controls (Burghardt, 1969). Although control swabs never elicited an attack, analysis of the tongue-flick responses showed that a control presented after an attack-releasing extract elicited significantly more tongue flicks than a control following an ineffective extract. This was true for several species of *Thamnophis*. Conditioning appeared the most likely explanation, although other interpretations could not be ruled out. Further work on this question is planned, but we decided first to modify responsivity to extracts effective in the newborn young.

The first experiment to successfully modify the stimuli eliciting the prey-attack response in newborn garter snakes (*T. sirtalis semifasciata*) was carried out in collaboration with Jannon Fuchs (Fuchs and Burghardt, in prep.). We have found generally with this subspecies that scores for worms as a class are clustered together and fish scores are clustered together. If feeding experiences altered responsivity to chemical cues from that prey, would the alteration generalize to chemical cues of related prey not experienced by the snakes? Would the first feeding cause a more lasting or more dramatic change than later feedings, such as has been found in turtles (Burghardt, 1967; Burghardt and Hess, 1966). By the use of various concentrations of extracts, we wanted to find out if the threshold concentration would vary. Could we, by testing extracts during the course of a feeding period, chart the changes that occurred in relative preference?

Twenty-two previously unfed young from the same litter were studied. Extracts were prepared from two species of fish (guppies and minnows) and two species of earthworms (nightcrawlers and redworms). One percent and 10 percent solutions of these extracts also were used. They were kept frozen throughout the period except when being tested. In this way, extracts from the same batch were utilized. Each group of snakes was fed a given amount of either freshly killed guppies or redworms every fourth day for a total of eight feedings. Then the foods were switched, group I being offered fish rather than worms and group II being fed worms rather than fish. This procedure also continued for eight feedings.

The subjects were tested twice on each of the three concentrations of the four extracts plus distilled water before any feeding experiences, after the feeding experiences with food A, and after the feeding experiences with food B. In addition, on the day before every other feeding, beginning before the first, each snake was tested once on distilled water, once with the redworm extract, and once with the guppy extract. These were the standard undiluted extracts. In Figure 30, we can see that with the test of redworm and guppy extracts after every other feeding the two groups did diverge and, by the end of the eighth feeding, the guppy-fed group and the redworm-fed group were considerably different in their responses to guppy or redworm extracts. When the foods were reversed, the situation rapidly changed. However, even at the end of the eighth feeding on the second food, there was still a slight residual effect in favor of the first food eaten. Therefore, the short series of extracts tested every other feeding indicated progressive changes in the direction of the food being fed and, therefore, could be used to map the development of preferences.

In addition, we could say quite clearly that feeding experiences which the animal itself initiated and carried through were effective in altering relative releasing "effectiveness" in contrast to the situation described earlier involving the forced-fed food.

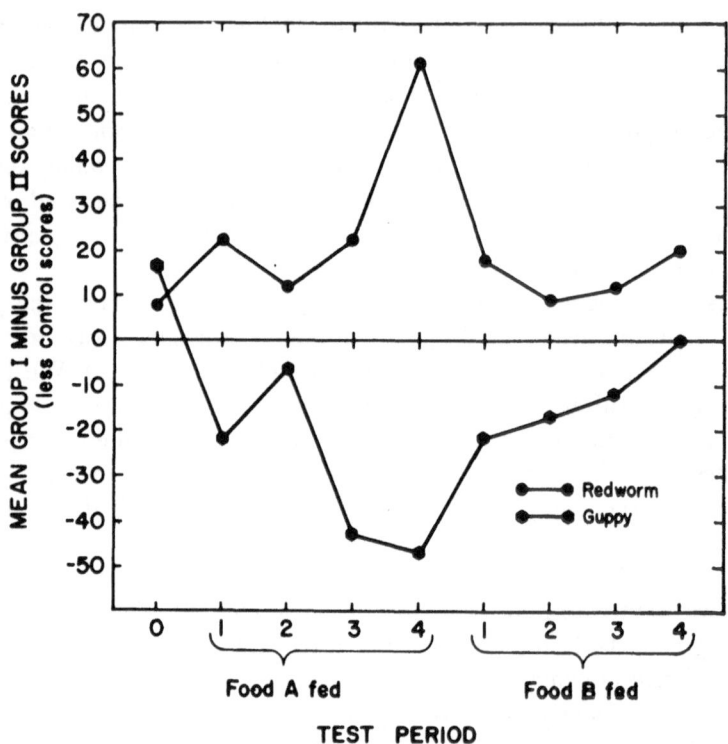

FIG. 30. Changing responsivity of snakes fed guppies for 8 feedings and then redworms (Group I) and snakes fed redworms and then guppies (Group II). The snakes were tested before any food was given at all (0) and after every 2 feedings. Note the divergence in favor of the first-fed food extract and the return to normal levels when the foods were changed.

All subjects ate either the guppy or the redworm at the beginning of the experiment but, when the foods given the subjects were switched, some individuals would not eat the new food. Subjects who did not accept the new food within a few days after the switch could not be tested at intervals along with other subjects. Therefore, the figure was biased in favor of individuals who readily switched and were less likely to be "imprinted" on the first food eaten. One subject, as a matter of fact, starved to death rather than switch his feeding preference.

The other important finding from this study concerned the responses to the nightcrawler and minnow extracts, the inexperienced foods. These data, as seen in Figure 31, showed quite clearly that little generalization occurred to extracts from related organisms. It was seen also that this difference between fed and unfed foods persisted even into the period when the first-experienced food was not being fed. Therefore, the snake discriminated between various species of worms and fish, animals from the same class of prey objects. This could not be shown from the testing of the inexperienced snakes. Indeed, to test chemical discrimination ability of serpents, this technique might be very useful. It is possible that one could add a known odor to a readily eaten food source and, after feeding the animal for several weeks, test it on

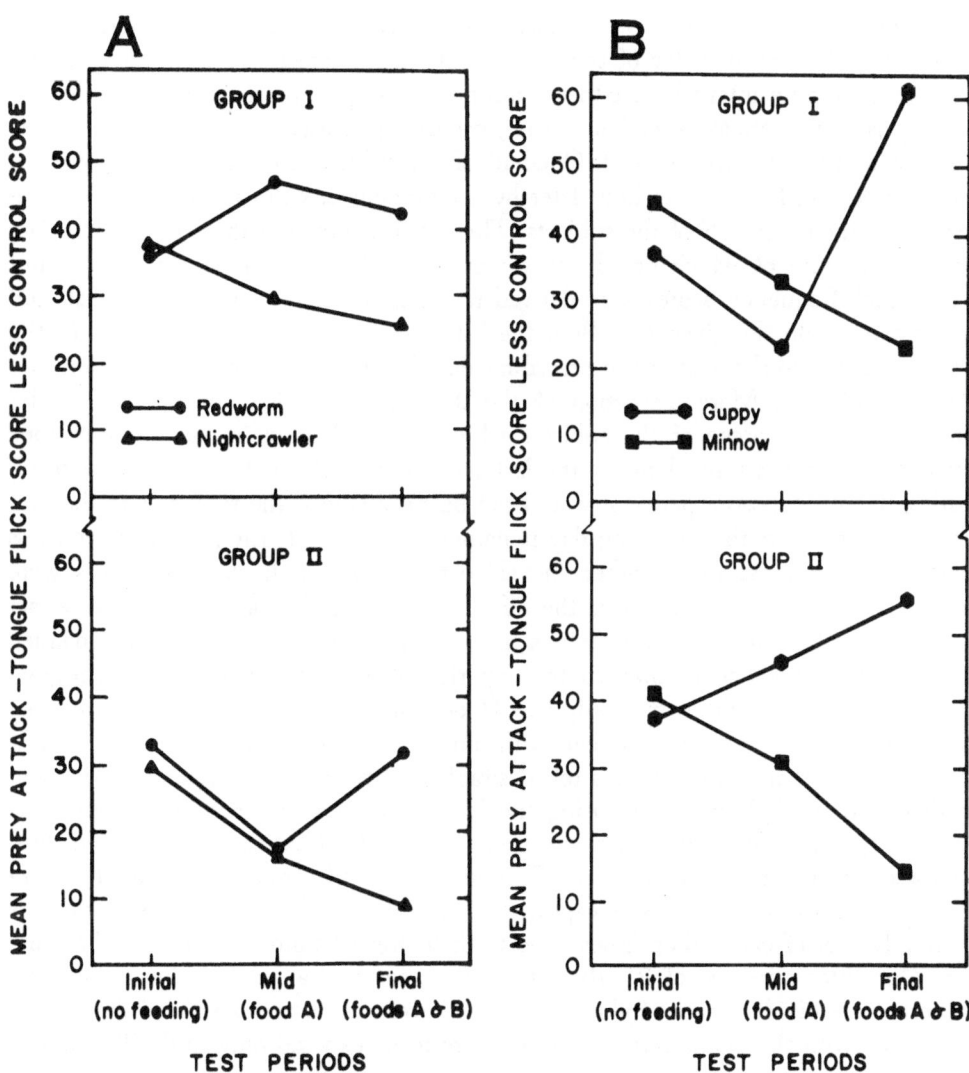

FIG. 31. Discrimination between fish and worm species is shown by selectively feeding snakes one species of worm or fish. Group I was fed redworms, then guppies, and vice versa for Group II.

extracts from normal and treated foods in an attempt to determine whether the animal could discriminate them.

In the above experiment, there are two factors which cannot be separated but which may either singly or together be responsible for the modification of chemical-cue preference. It is impossible to separate out the role of food reinforcement from that due to mere exposure to the relevant chemical stimuli. The following experiment represents an attempt to modify the chemical preferences of inexperienced snakes through pre-exposure to the chemical cues alone and hence remove any possibilities of "primary drive reduction." All the experiments were performed with newborn naive *Thamnophis sirtalis.*

Nightcrawler and minnow extracts were used with 24 snakes divided into two groups. On the first day of the experiment, each subject received four presentations at 60-minute intervals of either a minnow or nightcrawler extract. On the following day, approximately 24 hours after, each subject was tested once on both the minnow and nightcrawler extract. One-half of the subjects from each group received fish extract first, followed an hour later by a nightcrawler extract. The order was reversed for the other half of the subjects. The experimenter ran these tests blind, i.e., he was unaware of which one he was presenting. Based upon attack latency and tongue-flick frequency, scores were derived for each subject's response to the fish and worm extracts on the given day. Both the fish and worm scores were higher for the groups exposed to the extract the preceding day; however, these differences were not highly significant. More inspection of the data revealed some interactions on the pre-exposure day. Some of the subjects which attacked early in the series of four presentations stopped attacking by the last presentation. If habituation had occurred, during this pre-exposure period it would not be surprising if the responses to the pre-exposed extracts on the following day remained at a low intensity, perhaps even less than to the response to the other "novel" extract. Indeed, if the habituators were eliminated from the analysis then the effect in favor of the pre-experienced extract was even greater. Similarly, those which habituated to pre-exposure extracts responded more to the novel extract than to the experienced extract. Therefore, our original division of the possibilities into only two (food reinforcement, and simple exposure) was insufficient. Exposure coupled with an unhabituated, consummatory attack-response appears necessary. It was felt desirable to repeat the experiment, with some modifications on a larger scale, using the hypotheses that non-habituating attackers would respond more to the experienced extract, and habituators would respond less. In these experiments, two litters of *Thamnophis sirtalis* from Massachusetts were utilized. The gravid females were captured on the same small mound in Massachusetts within 10 feet of each other. They gave birth in the laboratory on exactly the same day. One litter consisted of 21, the other of 31, for a total of 52 subjects. The 52 subjects were divided in the following way: 21 were to be pre-exposed with the fish extract, 21 with the worm extract, and 10 were to be exposed only to distilled water. This testing period was on the seventh day after birth. The subjects, of course, had not been fed. They were exposed to the extract only three times instead of four as in the preliminary experiment. The three pre-exposure trials were at 20-minute intervals for each subject. On the following day, the subjects were tested on both the fish and the worm extract, the order being balanced as in the preceding studies.

From the trials on day 1, the snakes receiving the worm and fish extracts were divided into two groups, habituators and nonhabituators, utilizing the following operational criterion: if the snake attacked on the final trial of the three he was considered a nonhabituator, if he did not attack on the final trial he was considered an habituator. The control animals tested with distilled water did not, of course, attack, and only emitted several tongue flicks.

The results on the following day when both extracts were presented to all snakes are shown in Figure 32. No differences between the control, worm-, or fish-exposed groups are seen for the fish or worm extracts. However, for both extract-exposed groups, the nonhabituators responded on the average more to the experienced extracts,

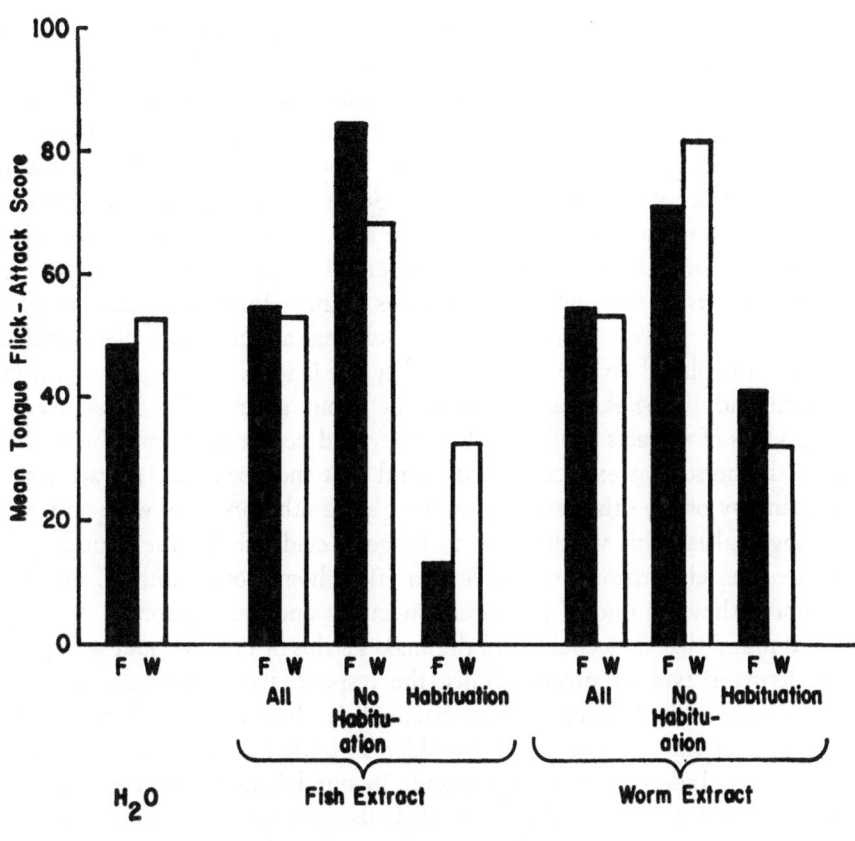

FIG. 32. Effects of prior experience (day 1) with either water, fish extract, or worm extract on responses to fish and worm extracts on following day. Relative effectiveness of the two extracts was influenced by responses on preceding day (see text).

and the habituators showed the reverse effect. Statistical analysis confirmed these results.

It would appear, then, that performance of the act of prey-attack toward a chemical prey-extract can bias the response of the snake in favor of that extract at a later date, if the attack response itself does not habituate. This indicated that the performance of the consummatory responses themselves acted as reinforcers in this situation. This is similar to the proposed role of locomotor effort in classical imprinting (Hess, 1959). While the effect is not very large in relation to either the congenital differences, or differences produced by actual feeding, it is real and illustrates another of the many factors which enter into the dynamic and complex nature of what may at first seem as the simple response of the lowly serpent to a Q-tip.

Up to now we have been dealing with modifying the "extract-response profile" in so far as to raise or to lower the effectiveness of extracts to which the newborn animal normally responds. Was it possible somehow to "build in" responsivity to a new extract which initially had no inherent attractiveness to the organism? The earlier experiment with the force-fed liver showed that technique could not demonstrate such a process. Definitive experiments on this point have not yet been completed. However, some interesting suggestive results are available. Newborn garter snakes, *Thamnophis sirtalis*, would not respond to horsemeat extract, nor did they eat pieces of horsemeat. Dropping some worm extracts on horsemeat, as noted earlier, would lead to it being swallowed. What would happen if this process was continued and then the extract was slowly eliminated. Some authors (Snediger, 1963) have suggested that, by an equivalent process, garter snakes could be made to eat chopped meat voluntarily and readily. To test this potential ability, newborn Butler's garter snakes, *Thamnophis butleri*, who responded as young almost the same as newborn *sirtalis* (Burghardt, 1969), were fed exclusively on a horsemeat diet. Initially, a worm extract was placed on the meat until they ate it readily; then the worm extract was removed and the snake fed on meat for about six months. They were then tested on a series of extracts similar to those they had been tested on at birth with the addition of the horsemeat extract. It was found that the horsemeat extract was more effective than any of the other extracts, although the other extracts were still effective in the young snakes. This would seem to be good evidence for the "building in" of a new preference. Unfortunately, people had taken home some adult *T. butleri* from the laboratory (they are one of the most innocuous and best tempered of the garter snakes) and found that they could feed the snakes strips of beef without going through any such decoying type of process. Since the responsivity of newborn *Thamnophis butleri* to horsemeat had never been ascertained, it obviously would be incorrect to make too much of the responsivity of the older *T. butleri* to horse meat extracts after eating horsemeat. Indeed, recent experiments in our laboratory showed that a small proportion of naive *T. sirtalis* will eat small balls of horsemeat without the addition of any "decoy" extract. However, the experiments again illustrate the fact that eating experiences can accentuate a preference. Whether these preferences must be present to begin with is still an unanswered question.

Chemical Analysis of the Effective Substances

What is the chemical nature of the substances eliciting prey-attack in newborn snakes? A preliminary study on garter snakes has been reported by Sheffield et al. (1968). These studies involved analysis of the nightcrawler extract. An extract prepared at 60°C was most effective. It was shown that the newborn snake must actually touch his tongue tips to the swab or food item before attacking it. This helped to explain why the numerous attempts reported in that paper to get a volatile fraction were unsuccessful. In general, the results showed that the attack eliciting substances were nonvolatile, stable under a wide variety of conditions, and consisted of large molecules (molecular weight over 5,000), as determined by gel-filtration techniques. Mucus from worms was not effective. It seems reasonable to suggest that Jacobson's organ is most sensitive to low volatile compounds, while olfaction is most effective with the more volatile chemicals. Clearly the problem of chemical analysis needs much more attention and will not turn out to be very simple. However, the chemical analysis will be of relevance even to the comparative studies. Whether newborn young of different species attacking the same prey extract are responding to the same chemicals or different ones has evolutionary implications.

Additional Questions

The newborn snake, then, is capable of species-characteristic discriminations involving chemical stimuli. Although feeding has been the only behavior studied until now, other behaviors might be examined profitably, such as predator recognition, alarm substances, and aggregation. Further studies of the psychophysical parameters, habituation, airborne odors, the role of the tongue, and the effects of food and extract experience would be useful. Genetic studies are also needed. The role of prenatal and maternal experience, if any, should be evaluated before the preferences of the newborn are ascribed completely to a genetic innate perceptual schema (or releasing mechanism). Since this has rarely been done with any animal the chemical release of prey-attack in snakes assumes theoretical importance. Many of the above studies are already underway in our laboratory.

CONCLUSION

A summary of all the foregoing findings is not necessary. The work reported—tentative, speculative, and spotty as it is—certainly indicates that the chemical senses are of great importance in the lives of reptiles, particularly snakes and many lizards.

Reptiles, especially snakes, would seem to have many advantages for chemosensory research. They certainly rely heavily on the chemical senses and, in contrast to other groups, including olfactory oriented mammals, many of their responses seem almost completely controlled by chemical cues. They are much easier to work with than most workers might think. The highly precocial nature of the young allow experiences to be carefully controlled. Cleaning and disagreeable olfactory stimuli are both minimal.

There are some disadvantages, of course. Their reproductive biology is little known and controlled breeding is more or less nonexistent. Reptilian diseases are little studied. On the other hand, balanced diets and ultraviolet light seem to eliminate most health problems. Mites brought in by wild stock are the biggest problem, but a number of commercial products are available which virtually eliminate losses. The relatively long life span makes genetic and breeding studies more difficult than with the usual laboratory animals.

In future research, a most critical area will be the physiological mechanisms involved in the specificity and stimulus control shown by snakes, especially via Jacobson's organ. Peripheral encoding is certainly of importance, but the modifiability we have demonstrated shows that it cannot be the whole story: central processes are also involved. A good start has been made on olfactory recording studies in reptiles, especially turtles (e.g., Moulton and Tucker, 1964). The work on Jacobson's organ has been reviewed by Tucker (in preparation).

In addition to the molecular approach, I think we also are in substantial need of an ecologically based attempt to put the whole organism in its olfactory perspective. The world is full of as much chemosensory variety as it is of visual variety. A constellation of potential odors and tastes confront the organism with every move it makes. Some species are more sensitive than others; some individuals are more sensitive than others of the same species. Similarly, differences may be qualitative rather than quantitative. Now that we are at long last recognizing the manifold ways in which odors stimulate and control animal and human behavior, we must go out into the real world and attempt to create an ecological chemistry parallel to Gibson's ecological optics (Gibson, 1966). Experienced woodsmen can describe the forest they are in quite accurately when blindfolded; salmon, it appears, can recognize the chemical gestalt of the water in which they were hatched (Hasler, 1966); and snakes possess similar abilities without learning at all. These are the things we must not lose sight of in our search for mechanisms.

ACKNOWLEDGMENTS

Many persons helped to make the studies reported here possible. The research was also supported by National Institute of Mental Health grants MH-776 awarded to E. H. Hess and MH-13775 and MH-15707 awarded to the author. L. S. Burghardt contributed her editorial talents and J. Spurgeon spent many hours in typing and retyping drafts of this chapter.

REFERENCES

Allard, H. A. 1949. The eastern box turtle and its behavior. *J. Tenn. Acad. Sci.*, 24:146-152.

Aristotle. De partibus animalium. Translated by W. Ogle. *In* Smith, J. A., and Ross, W. D., eds. 1912. The Works of Aristotle, Vol. 5. New York, Oxford University Press, Inc.

Auffenberg, W. 1965. Sex and species discrimination in two sympatric South American tortoises. *Copeia*, 335-342.

Baumann, F. 1927. Experimente über den Geruchssinn der Viper. *Rev. Suisse Zool.*, 34:173-184.

—— 1928. Über die Bedeutung des bisses und des Geruchssinnes für den Nahrungserwerb der Viper. *Rev. Suisse Zool.*, 35:233-239.

—— 1929. Experimente über den Geruchssinn und der beuterwerb der Viper (*Vipera aspis* L.). *Z. Vergl. Physiol.*, 10:36-119.

Bellairs, A. d'A. 1960. Reptiles: Life History, Evolution, and Structure. New York, Harper & Brothers.

Bogart, C. M. 1941. Sensory cues used by rattlesnakes in their recognition of ophidian enemies. *Ann. N.Y. Acad. Sci.*, 41:329-343.

Boycott, B. B., and Guillery, R. W. 1962. Olfactory and visual learning in the red-eared terrapin, *Pseudemys scripta elegans* (Wied.). *J. Exp. Biol.*, 39:567-577.

Brisbin, I. L., Jr. 1968. Evidence for the use of postanal musk as an alarm device in the king snake, *Lampropeltis getulus*. *Herpetologica*, 24:169-170.

Broman, I. 1920. Das Organon vomero-nasale Jacobsoni - ein Wassergeruchsorgan. *Anat. Hefte,* Abt. 1, 58:137-197.

Brower, L. P., Brower, J. vz., and Westcott, P. W. 1960. The reactions of toads (*Bufo terrestris*) to bumblebees (*Bombus americanorum*) and their robberfly mimics (*Mallophora bomboides*). *Amer. Natur.,* 94:343-355.

Burghardt, G. M. 1964. Effects of prey size and movement on the feeding behavior of the lizards *Anolis carolinensis* and *Eumeces fasciatus. Copeia,* 576-578.

———— 1966. Stimulus control of the prey attack response in naive garter snakes. *Psychonomic Sci.,* 4:37-38.

———— 1967a. Chemical-cue preferences of inexperienced snakes: Comparative aspects. *Science,* 157:718-721.

———— 1967b. Chemical perception in newborn snakes. *Psychology Today,* 1(4): 50-59.

———— 1967c. The primacy effect of the first feeding experience in the snapping turtle. *Psychonomic Sci.,* 7:383-384.

———— 1968. Chemical preference studies on newborn snakes of three sympatric species of *Natrix. Copeia,* 732-737.

———— 1969. Comparative prey attack studies in newborn snakes of the genus *Thamnophis. Behaviour,* 33:77-144.

———— 1970. Intraspecific geological variation in chemical food cue preferences of new born garter snakes (*Thamnophis sirtalis*). *Behaviour,* 36:246-257.

———— and Hess, E. H. 1966. Food imprinting in the snapping turtle, *Chelydra serpentina. Science,* 151:108-109.

———— and Hess, E. H. 1968. Factors influencing the chemical release of prey attack in newborn snakes. *J. Comp. Physiol. Psychol.,* 66:289-295.

Burns, B. 1969. Oral sensory papillae in sea snakes. *Copeia,* 617-619.

Carey, J. H. 1967. The nuclear pattern of the telencephalon of the blacksnake, *Coluber constrictor constrictor. In* Hassler, R. and Stephen, H., eds. Evolution of the Forebrain. New York, Plenum Publishing Corporation, pp. 73-80.

Carpenter, C. C. 1952. Comparative ecology of the common garter snake (*Thamnophis s. sirtalis*), the ribbon snake (*Thamnophis s. sauritus*), and Butler's garter snake (*Thamnophis butleri*) in mixed populations. *Ecol. Monogr.,* 22:235-258.

———— 1962. Patterns of behavior in two Oklahoma lizards. *Amer. Midland Natur.,* 67:132-151.

Carr, A. 1965. The navigation of the green turtle. *Sci. Amer.,* 212 (5):78-86.

———— 1966. Reptiles. New York, Time-Life Books. A division of Time Inc.

Cowles, R. B. 1938. Unusual defense postures assumed by rattlesnakes. *Copeia,* 13-16.

———— and Phelan, R. L. 1958. Olfaction in rattlesnakes. *Copeia,* 77-83.

Cole, C. J. 1966a. Femoral glands in lizards: A review. *Herpetologica,* 22:199-206.

———— 1966b. Femoral glands of the lizard, *Crotophythus collaris. J. Morph.,* 118: 119-136.

Craig, W. 1918. Appetites and aversions as constituents of instinct. *Biol. Bull.,* 34: 91-107.

Dilger, W. 1960. The comparative ethology of the African parrot genus *Agapornis. Z. Tierpsychol.,* 17:649-685.

Ditmars, R. L. 1936. The Reptiles of North America. Garden City, N. Y., Doubleday & Company, Inc.

Dix, M. W. 1968. Snake food preference: Innate intraspecific geographic variation. *Science*, 159:1478-1479.

Dullemeijer, P. 1961. Some remarks on the feeding behavior of rattlesnakes *Koninkl. Nederl. Acad. Van Wetenschappen, Série C*, 64:383-396.

Dundee, H. A., and Miller, M. C., III. 1968. Aggregative behavior and habitat conditioning by the ringneck snake, *Diadophis punctatus arnyi*. *Tulane Studies in Zool. and Bot.*, 15:41-58.

Eglis, A. 1962. Tortoise behavior: A taxonomic adjunct. *Herpetologica*, 18:1-8.

Engen, T. 1964. Psychological scaling of odor intensity and quality. *Ann. N. Y. Acad. Sci.*, 116 (2):504-516.

Evans, L. T. 1959. A motion picture study of maternal behavior of the lizard, *Eumeces obsoletus* Baird and Baird. *Copeia*, 103-110.

Fox, W. 1952. Notes on the feeding habits of Pacific coast garter snakes. *Herpetologica*, 8:4-8.

Fuchs, J., and Burghardt, G. M. Effects of early feeding experience on the responses of garter snakes to food chemicals. (in preparation.)

Gehlbach, F. R., Watkins II, J. F., and Reno, H. W. 1968. Blind snake defensive behavior elicited by ant attacks. *Bioscience*, 18:784-785.

Gettkandt, A. 1931. Die Analyse des Functionkrieses der Nahrung bei der Kutscherpeitschenschlange *Zamenis flagelliformis* L. nebst Ergän zungsversuchen bei der Ringelnatter *Tropidonotus natrix* L. *Z. Vergl. Physiol.*, 14:1-39.

Gibson, J. J. 1966. The Senses Considered as Perceptual Systems. Boston, Houghton Mifflin Company.

Goin, C. J., and Goin, O. B. 1962. Introduction to Herpetology. San Francisco, W. H. Freeman and Co. Publishers.

Gordon, M. S. 1968. Animal Function: Principles and Adaptions. New York, The Macmillan Company.

Grant, D., Anderson, O., and Twitty, V. 1968. Homing orientation by olfaction in newts (*Taricha rivularis*). *Science*, 160:1354-1356.

Greenburg, B. 1943. Social behavior of the western banded gecko, *Coleonyx variegatus* Baird. *Physiol. Zool.*, 16:110-122.

Hainer, R. M. 1964. Some suggested critical experiments in olfactory theory. *Ann. N.Y. Acad. Sci.*, 116(2):477-481.

Hasler, A. D. 1966. Underwater Guideposts: Homing of Salmon. Madison, University of Wisconsin Press.

Hathaway, L. M. 1964. Suggested function of the femoral glands in *Crotophytus collaris collaris* (Say). *Bull. Ecol. Soc. Amer.*, 45(3):117 (abstract).

Hess, E. H. 1959. Imprinting. *Science*, 130:133-141.

Honigmann, H. 1921. Zur biologie der Schildkröten. *Biol. Zentrlbl*, 41:241-250.

Hubel, D. K., and Wiesel, T. N. 1963. Receptive fields of cells in striate cortex of very young, visually inexperienced kittens. *J. Neurophysiol.* 26:994-1002.

Inger, R. F. Some reactions of rattlesnakes (genera *Crotalus* and *Sistrurus*) to king snakes (genus *Lampropeltis*). (unpublished manuscript.)

Kahmann, H. 1932. Sinnesphysiologische Studien an Reptilien—I. Experimentalle Untersuchungen uber das Jacobsonische Organ der Eidechsen und Schlangen. *Zool. Jb. Abt. Allg. Zool. Physiol.*, 51:173-238.

———— 1934. Zur Chemorezeption der Schlangen (ein Nachtag.). *Zool. Anz.*, 107: 249-263.

———— 1939. Über das Jacobsonsche Organ der Echsen. *Z. Vergl. Physiol.*, 26:669-695.

Kauffeld, C. F. 1953. Methods of feeding captive snakes. *Herpetologica*, 9:129-131.

Klauber, L. M. 1956. Rattlesnakes. Berkeley and Los Angeles, University of California Press. 2 vol.

Lorenz, K. Z. 1941. Vergleichende Bewegungsstudien an Anatiden. *J. Ornithol.*, 89:194-294.

Lowe, C. H. 1943. An improved method of snake feeding. *Copeia*, 58.

McIlhenny, E. A. 1935. The alligator's life history. Boston, Christopher Publ. House.

Mertens, R. 1960. The world of amphibians and reptiles. New York, McGraw-Hill Book Company.

Miller, S. D., and Erickson, R. P. 1966. The odor of taste solutions. *Physiol. Behav.*, 1:145-146.

Morris, D. D., and Loop, M. S. 1969. Stimulus control of prey attack in naive rat snakes: a species duplication. *Psychonomic Sci.*, 15:141-142.

Moulton, D. G., and Tucker, D. 1964. Electrophysiology of the olfactory system. *Ann. N.Y. Acad. Sci.*, 116(2):380-428.

Naulleau, G. 1964. Premieres observations sur le comportement de chasse et de capture chez les viperes et les couleuvres. *La Terre et la. Vie*, 1:54-76.

———— 1966. La biologie et le comportement prédateur de *Vipera aspis* au laboratoire et dans la nature. Thesé, Paris, P. Fanlac.

Noble, G. K. 1937. The sense organs involved in the courtship of Storeria, Thamnophis, and other snakes. *Bull. Amer. Mus. Nat. Hist.*, 73:673-725.

———— and Bradley, H. T. 1933. The mating behavior of lizards; Its bearing on the theory of sexual selection. *Ann. N. Y. Acad. Sci.*, 35:25-100.

———— and Breslau, A. M. 1938. The senses involved in the migration of young fresh-water turtles after hatching. *J. Comp. Psychol.*, 25:175-193.

———— and Clausen, H. J. 1936. The aggregation behavior of *Storeria dekayi* and other snakes with especial reference to the sense organs involved. *Ecol. Monogr.*, 6:269-316.

———— and Kumpf, K. F. 1936. The function of Jacobson's organ in lizards. *J. Genet. Psychol.*, 48:371-382.

———— and Mason, E. R. 1933. Experiments on the brooding habits of the lizards *Eumeces* and *Ophisaurus*. *Amer. Mus. Novitates*, 619(1):1-29.

Oliver, J. A. 1955. The Natural History of North American Amphibians and Reptiles. Princeton, D. van Nostrand.

Ortleb, E. P., and Sexton, O. J. 1964. Orientation of the painted turtle *Chrysemys picta. Amer. Midland Naturalist.*, 71:320-334.

Parsons, T. S. 1959a. Nasal anatomy and the phylogeny of reptiles. *Evolution*, 13:175-187.

———— 1959b. Studies on the comparative embryology of the reptilian nose. *Bull. Mus. Comp. Zool.*, Harvard Coll., 120:101-277.

———— 1967. Evolution of the nasal structure in the lower tetrapods. *Amer. Zool.*, 7:397-413.

Payne, A. 1945. The sense of smell in snakes. *J. Bombay Nat. Hist. Soc.*, 45(4):507-515.

Poliakov, K. L. Cited in Boycott and Guillery. 1962. Olfactory and visual learning in the red-eared terrapin, *Pheudemys scripta elegans* (Wied.), *J. Exp. Biol.*, 39:567-577.

Pope, C. H. 1965. The Reptile World. New York, Alfred A. Knopf, Inc.

Ramsey, L. W. 1947. Feeding behavior of *Tropidoclonion lineatum. Herpetologica*, 4:15-18.

Rensch, B., and Eisentraut, M. 1927. Experimentelle Untersuchungen über den Geschmackssimn der Reptilien Z. *Vergl. Physiol.*, 5:607-612.

Romer, A. S. 1966. Vertebrate Palentology. Chicago, University of Chicago Press.

Schiller, C. H. (ed.). 1957. Instinctive behavior. New York, International Universities Press.

Schmidt, K. P. 1938. Herpetological evidence for the post-glacial eastward extension of the steppe in North America. *Ecology*, 19:396-407.

———— and Inger, R. F. 1957. Living Reptiles of the World. Garden City, N. Y., Doubleday & Company, Inc.

Sheffield, L. P., Law, J. M., and Burghardt, G. M. 1968. On the nature of chemical food sign stimuli for newborn snakes. *Comm. Behav. Biol.*, 2:7-12.

Smith, P. W. 1961. The amphibians and reptiles of Illinois. *Ill. Nat. Hist. Surv. Bull.* 28:1-298.

Snedigar, R. 1963. Our Small Native Animals: Their Habits and Care. New York, Dover Publications, Inc.

Stebbins, R. C. 1943. Some aspects of the biology of the iguanid genus *Uma. Ecol. Monogr.*, 14:311-322.

———— 1948. Nasal structure in lizards with reference to olfaction and conditioning of the inspired air. *Amer. J. Anat.* 83:183-221.

———— 1954. Amphibians and Reptiles of Western North America. New York, McGraw-Hill Book Company.

———— 1966. A Field Guide to Western Reptiles and Amphibians. Boston, Houghton Mifflin Company.

Thorpe, W. H. 1956. Learning and Instinct in Animals. Cambridge, Harvard University Press.

Thurow, G. Feeling behavior in garter snakes. (unpublished manuscript.)

Tucker, D. Jacobson Organ and the Trigeminal Responses. *In* Handbook of Sensory Physiology, Vol. 4, Chemical senses. New York, Springer-Verlag New York Inc. (in preparation.)

Watkins, II, J. F. Gehlbach, F. R., and Kroll, J. C. 1969. Attractant-repellent secretions in the intra- and interspecific retations of blind snakes (*Leptotyphlops dulcis*) and army ants (*Neivamyrmex nigrescens*). *Ecology*, 50:1098-1104.

———— Gehlbach, F. R. and Baldridge, R. S. 1967. Ability of the blind snake, *Liptotyphlops dulcis*, to follow pheromone trails of army ants, *Neivamyrmex nigrescens* and N. *opacithorax. Southwestern Naturalist*, 12:455-462.

Wiedemann, E. 1931. Zur biologie der Nahrungs-aufnahme eurapäischer Schlangen. *Zool. Jb. Abt. Syst.*, 61:621-636.

———— 1932. Zur biologie der Nahrungs-aufnahme der Kreuzotter, *Vipera berus* L. *Zool. Anz.*, 97:278-286.

Wilde, W. S. 1938. The role of Jacobson's organ in the feeding reaction of the common garter snake, *Thamnophis sirtalis sirtalis* (Linn.). *J. Exp. Zool.* 77:445-465.

Wright, H. W., and Wright, A. A. 1957. Handbook of Snakes of the United States and Canada. 2 vols. New York, Comstock.

10

The Role of Pheromones
in Mammalian Reproduction

W. K. WHITTEN and F. H. BRONSON

The Jackson Laboratory
Bar Harbor, Maine 04609

INTRODUCTION 309
MOUSE PRIMER PHEROMONES 313
PRIMING PHEROMONES IN OTHER MAMMALS 317
ECOLOGIC SIGNIFICANCE 318
PRIMER PHEROMONES AS EXPERIMENTAL TOOLS 321
ACKNOWLEDGMENTS 322
REFERENCES 322

INTRODUCTION

Communication among members of a mammalian species probably depends upon a variety of signals directed to each of several receptor organs from which information is relayed to higher centers. The same amount and type of information could be conveyed by different signals within each modality or even by different receptors and pathways. For example, a male may recognize a receptive female by odor, vocalization, visual cues, or by absence of aggressive response to his advances. The information conveyed through the sense of smell may itself be complex and could be derived from any or all of the following sources: skin glands, vaginal secretion, urine, feces, or expired air, or may be acquired from the environment. This complexity makes understanding difficult, and man as the observer is handicapped because either many of the signals appear to be outside the range of his comparable receptor or the receptor itself, like the vomeronasal organ, may be vestigial.

Our concept of the role of chemical signals in mammalian reproduction has developed largely from observations of domesticated animals in nonexperimental situations. It also depends on generalizations based on our own olfactory responses as well as on comparative studies of the structure of olfactory systems. Some human responses have been quantified by olfactometry but others are vague, like the sensation which inspired the name Exaltone for a musk-like lactone. Our response to other substances may be a revulsion. This may have inhibited experimentation with compounds like valeric and caproic acids which could be important chemical signals attractive for some animals.

Reproductive pheromones in mammals have already been reviewed by Parkes and Bruce (1961), Bruce and Parkes (1965), Bruce (1966), Whitten (1966), and Bronson (1968), so that only sufficient background will be given here to make this review comprehensible. It is no longer necessary to define the term pheromone, but some qualifications suggested by entomologists should be examined before adopting this term for use in reference to the more complex behavior of mammals. We would like to see *primer pheromone* used to describe a substance acting on the endocrine system, probably through the central nervous system, to produce a response which may take some time to develop. One of us (Bronson, 1968) has already pointed out that the term *releaser pheromone* which is used to describe a substance eliciting an immediate behavioral response of the stimulus-response type should not be generalized to mammals. This term denotes both an innateness and a degree of fixedness of response that are probably not applicable to many complex and modifiable mammalian behavior patterns. *Signalling pheromone* was suggested in this context since the term implies only that information is being transferred and does not designate the nature of the response. The other of us (Whitten), while in general agreement, considers that the factor in the urine of male cats which evokes the rolling and rubbing and also the estrous crouch and treading of estrous females (Todd, 1963; Michael and Keverne, 1968) may prove to be a typical releaser pheromone. He also considers that "signalling" is a redundant qualification but has no better term to offer. The need for these generalizations may not persist when more of these substances are identified and can be referred to by chemical names or by the characteristic behavioral or physiologic responses they induce.

In the remainder of this review we propose to consider some of the general aspects of pheromones in mammalian reproduction and to describe in some detail recent results of primer pheromones in mice. We will conclude with a discussion of the biologic implications of these substances.

The Receptors

We would like to be able to start by identifying the receptors for the known reproductive pheromones of mammals but unfortunately we have very little definite information. The receptors most suitably adapted to respond to such pheromones appear to be those of the peripheral olfactory system. The little experimental evidence available indicates that these are in fact the receptors concerned for the pheromones so far established. Nevertheless, one must interpret with caution results from such crude procedures as removal of the olfactory bulbs (Whitten, 1966), and even suspect the technique of packing the olfactory area with iodoform-impregnated gauze to induce anosmia (Michael and Keverne, 1968). An animal treated in this way may be so concerned with the discomfort that he is not attracted to females. Even if it is established that a pheromone acts through the olfactory system, further work would be required to locate the receptors themselves. Are they in the terminal or the trigeminal nerves, in the septal organ, the vomeronasal organ, or in the olfactory neuroepithelium proper? If in the latter, are they restricted to any selected area? The observation by Planel (1953) that section of the vomeronasal nerves of male guinea pigs did not prevent orientation towards hidden females indicates that this structure does not contain all the receptors for sex attractions.

There is no good evidence that mammalian pheromones can act through taste; nevertheless the frequency with which animals lick or nuzzle each other and sample urine as it is voided warrants investigation (Fraser, 1968). It should be noted that olfactory receptors might be involved even in these situations because dyes placed in the mouth have been observed to move up the philtrum to the nostrils (Whitten, unpubl.).

Finally, there is the possibility that primer pheromones may stimulate internal receptors after absorption by inhalation, ingestion, or even from the vagina after copulation. It has been suggested that prostoglandins may induce uterine contractions and thus influence sperm transport (Eliasson, 1959).

The Pheromones

The importance of pheromones for mammalian reproduction has been clearly demonstrated by controlled behavioral and physiologic studies; for earlier references see Bronson (1968), and Michael and Keverne (1968).

Two primer pheromones have been demonstrated in the urine of male mice. One of these induces estrus while the other causes pregnancy block, and both will be discussed fully in a later section. No mammalian pheromone has been identified chemically. However, the brief unconfirmed report by Lissak (1962) that valeric acid induced estrus-like behavior accompanied by localized activity in the anterior hypothalamus of cats suggests that this substance may be a pheromone.

Many odorous substances have been identified from the urine and secretions of scent glands of mammals (Lederer, 1950; Prelog et al., 1944; Kingston, 1964) but few have been subjected to critical behavioral or physiologic studies. The crude secretions of the scent glands of commercial importance contain many substances, some of which are determined by diet while others may reflect the endocrine status of the animal. Thus, the characteristic odor of castoreum, for humans, is not attributable to a single entity but to a blend of the numerous ingredients. This complexity may confer individuality on the secretion from a single animal and so facilitate recognition and appraisal. It is perhaps significant that some of the important constituents from scent glands (e.g., civitone, muscone) are used as fixatives to enhance the effect of more volatile ingredients of perfumes. This suggests that these compounds may function as natural fixatives for the more subtle individual odors, determined by genetic and environmental factors or the endocrine status of the animal, as well as providing the basic scent characteristic of the species.

Wilson and Bossert (1963) have established some tentative characteristics of volatile insect pheromones. They concluded that the compounds most suitable for such functions will contain between 5 and 20 carbon atoms and have molecular weights between 80 and 300. To date all the known insect pheromones comply with this prediction and there is reason to expect that these characteristics will apply to mammalian volatile pheromones.

The Context and the Response

An examination of the context of odors in which pheromones function may help to understand the role that pheromones perform. Odors provide the major sensory input for many mammals and it may be difficult for an observer to distinguish between the response to a pheromone and the response to a trivial odor that has been previously associated with sexual reward. Therefore much careful work, with animals raised in isolation and with adequate control of environmental odors, will be required to solve these problems and to understand the role that imprinting plays in these phenomena (Mainardi, et al., 1965).

Individual recognition has been demonstrated by Bowers and Alexander (1967) to depend, in mice, on olfactory clues. Both male and female mice were able to distinguish between individual males of the same inbred strain. This discrimination must therefore be of an extremely high order because differences due to genetic, endocrine, or environmental factors would be minimal in these animals. This ability compares with that of dogs which, as Kalmus (1955) has shown, can distinguish between identical twins. It is possible that the dogs used cues derived from food and cosmetics.

Ropartz (1967) considers that the odorant responsible for individual recognition is produced by the limb pads. Whitten (1966) suggested that such an "identifier" may provide the context of "strangeness" in which the pheromone from the male induces pregnancy block. A strange male would possess scents common to the species and he would carry odors characteristic of the stock from which he came. He would also be devoid of odors peculiar to the stock into which he was introduced. Recognition of strangeness may result from the detection of unfamiliar odors or failure to observe familiar ones. Animals may, of course, be adapted to their own odors including those common to the species and readily detect the unfamiliar odors of the stranger. Recognition of individuals within a society may follow a similar pattern, but odors which reflect or even determine status within the society may also contribute. Status within a hierarchy has been shown to be androgen-dependent, and some androgen metabolites have distinct odors. It is, therefore, tempting to postulate that the concentration of such substances in the urine may reflect the gonadal activity of the animal and consequently its status. Secretions by androgen-dependent scent glands should also be proportional to the secretion of testosterone and function similarly.

Animals probably locate one another largely by chemical signals. These may be general air-borne odors or they may be left incidentally on a trail or deposited deliberately on marked objects. (See Chapters 11 and 13.) The coming together of two receptive animals may follow responses to several cues: signals from trails, from marked territory, and from substances which cause sexual arousal. For example, substances in the urine or vaginal secretions of estrous females which excite males may be deposited erratically and may have limited range and durability and thus provide little directional information. They may, however, arouse the male so that he searches more diligently for the trail which provides persistent and adequate directional information.

Signoret and du Mesnil du Buisson (1961) reported that odors from boars produced arousal of the central nervous system of sows. It has been suggested that such arousal may be the first critical component typical of the responses to all mammalian pheromones even those which are considered to act through the endocrine system (Harris, personal communication).

MOUSE PRIMER PHEROMONES

Most of our knowledge of mammalian primer pheromones has been derived from studies with house mice. Many people have failed to find evidence for these substances in the other common laboratory rodent, the rat, but they may occur in wild strains or when the food intake is restricted (Cooper and Hayes, 1967). Fortunately, most of the observations with mice have been confirmed with deermice (*Peromyscus maniculatus bairdii*) so we are confident that primer pheromones are not limited to the highly domesticated laboratory mouse.

Three main effects have been described: suppression of estrous cycles in all-female groups, induction of estrus by males, and the block to pregnancy produced by a strange male. Each of these will be briefly described, and the most recent advances in our knowledge of the properties, quantitation and production of the respective pheromones will be given.

Inhibition of Cycles in Grouped Females

Andervont (1944) first described a reduction in ovarian activity when female mice were housed in groups. He became aware of this effect because the development of hormone dependent tumors was retarded by grouping the female hosts. He followed the estrous cycles by vaginal smears and observed that they occurred earlier, were more frequent, and continued longer in segregated mice. Lee and Boot (1955, 1956) and Dewar (1959) confirmed this observation but found that the prolonged cycles were pseudopregnancies. Working with another strain and away from the odors of males, Whitten (1959) observed anestrus in a proportion of animals when they were kept in large groups. These effects of grouping are somewhat similar and the differences may be due to environmental or genetic factors.

The evidence that these effects are produced by pheremones is far from proven, but the phenomenon is described because of that possibility and because the suppression of estrus induced by all-female grouping may modify the response to the primer pheromone from males. Lee and Boot (1956) and Mody (1963) showed that removal of the olfactory bulbs prevented the occurrence of pseudopregnancy in grouped mice and assumed that it was produced by an "olfactory stimulus" conducted to the hypothalmus. There is, however, another possible explanation for this observation. Dewar (1959) and Whitten (unpubl.) have observed male-like mounting in all-female groups and the former author has plausibly suggested that it is this activity which induces the pseudopregnancy. Removal of the olfactory bulbs from all of the group could conceivably block the cue which induces other members of the group to mount the female in estrus. Removal of olfactory bulbs itself abolished the estrous cycle in the strain studied by Whitten (1956) so no inference can be made about the pathway for this effect. However, when precautions were taken to exclude visual, tactile, and auditory signals the anestrus continued, in Whitten's studies, which suggests that chemical communication was important.

QUANTITATION AND PROPERTIES. It has not been possible to quantitate suppression of cycles or to demonstrate the presence of a pheromone in the urine. An indirect approach to quantitation of this phenomenon would be to correlate the number per cage with the mean length of the estrous cycle. However, no such correlation was observed with SJL/J mice.

ENDOCRINOLOGY. A possible source of an inhibitory pheromone could be substances which function in the negative feedback between ovary and pituitary, and which spill over into the environment. Even when suppression is maximal some few animals continue to exhibit cycles and frequently exhibit prolonged estrus. Such animals could be the source of the pheromone and its production estrogen dependent. However, injection of estradiol into a small proportion of a group did not result in total suppression of the cycles of the remainder (Chapman and Whitten, unpubl.).

It is unlikely that the suppression is due to stress produced by the high population density because there is no loss of weight and most animals can return to estrus within 72 hours and exhibit normal fertility. This is supported by the recent study by Bronson and Chapman (1968) which showed that the initially stressful response is to isolation rather than to grouping and that there are no real differences in the concentration of corticosterone in the plasma of isolated and grouped females after a period of time. No doubt strain differences occur in adrenal responses to grouping as well as in aggression and male-like mounting.

GENETICS. Pseudopregnancy in female groups has been observed in both inbred and random-bred strains of mice, but the anestrous response has not been found in inbred animals. The degree of estrous synchrony described by Land and McLaren (1967) in the outbred Q stocks and that which has been observed in some of our own heterozygous animals suggests that anestrus may depend on diversity within the group. Thus, if anestrus is dependent on chemical communication several compounds may be involved. The anestrous response appears to be dependent on population density and this would be less obvious when all members of the group smell alike.

Estrus Induction by Males

Whitten (1958) showed that estrous cycles were shorter and more regular when a male mouse was nearby. This shortening, particularly when combined with a release from suppression associated with all-female groups, results in synchronization of estrus on the third night after pairing. Marsden and Bronson (1964) proved that this effect was produced by a pheromone because the application of drops of male urine to the nostrils of female mice regularly over a 2-day period resulted in a comparable degree of estrous synchrony.

QUANTITATION. The outstanding problem for the study of this pheromone is quantitation. Bronson and Whitten (1968) have described a method in which urine is delivered for 3 days by an infusion pump into cages containing ten SJL/J female mice. Vaginal smears are evaluated on the third and fourth morning and the percentage of animals exhibiting estrus recorded. Females from this strain do not exhibit suppression when grouped, but in the absence of males they have a mean cycle length of between 7 and 8 days, while in the presence of a male it is 4 or 5 days. Thus, even with control material one would expect 30 percent of the experimental females to have cornified smears during the period of observation. A further 30 percent would be unable to respond during the test period because they had recently ovulated. These limits are observed in practice and even with three replicates the assay is severely restricted and time consuming. The assay would be improved if animals which show suppression could be obtained. Otherwise it will be necessary to characterize and verify another response. So far no associated behavioral response has been observed even in blinded animals, nor has it been possible to demonstrate a suitable electrical change in the olfactory tract.

PROPERTIES. Characterization of the pheromone is difficult because of the poor assay and because a prolonged period of exposure is necessary before a response can be detected. A significant shortening of the estrous cycle was observed only when the females had been exposed for 48 hours, though some indication was found after 24 hours. Such a long exposure is unexpected and indicates that adaptation which is common with odor perception may not occur. The need for a long exposure complicates estimation of the response threshold, fadeout times, and chemical stability.

All attempts to extract the pheromone from urine appear to have been unsuccessful, but solvent residues may have interfered with the assay. Likewise it has not been possible to trap the active substance from air passed over groups of males. The pheromone has been shown to be transmitted by movement of air over a 2 m gap, which indicates that it is volatile and probably acts through olfactory receptors (Whitten et al., 1968).

PRODUCTION OF THE PHEROMONE. The pheromone is present in urine collected directly from the bladder and is, therefore, not derived from any of the accessory sex glands or the preputial glands (Bronson and Whitten, 1968). It is absent from the urine of castrates but is produced by androgenized females. From this, it is concluded that it is a steroid metabolite or from some structure under endocrine control. Steroids are usually excreted as glucuronates which are odorless, but most male mice excrete sufficient *β-glucuronidase* to liberate the odorous compound from such complexes. Males from the C3H/J strain are deficient in this enzyme but nevertheless induce estrous synchrony which suggest that either the pheromone is not a steroid or that it is not excreted as a conjugate.

Bartke and Wolff (1966) failed to observe estrous synchrony when females were paired with males heterozygous for the lethal yellow allele. Unfortunately, these workers used bigamous mating and it appears that their results were confounded with reduced ability of the mice to produce copulatory plugs (Whitten, in preparation).

GENETICS OF THE RESPONSE. So far only one inbred strain of mice, BALB/cDg, has been found that does not respond to the estrus-inducing pheromone. It is perhaps significant that the females from this strain have short cycles which continue after removal of the olfactory bulbs. Surprisingly, these females respond to a strange male with pregnancy block.

Pregnancy Block by Strange Males

Bruce (1959) observed that if a recently mated female was removed from the stud male and placed with a strange male, the female returned to estrus usually in 4 days. In other words, the cycle proceeded as though the first mating had not occurred. It has since been shown that this block to pregnancy can be induced by pooled urine from the strange males so the existence of a pheromone has been established (Parkes and Bruce, 1962).

QUANTITATION. Many of the problems which have been troublesome with the assay for estrus-inducing pheromone are again apparent. Between 10 and 20 percent of control animals return to estrus and the maximum response in the test is rarely over 80 percent. The assay takes about a week and so far no dose-response lines have been established.

PROPERTIES. Bruce observed a block to pregnancy in some animals after a 12-hour exposure but again the maximum was reached only after 48 hours. Parkes and Bruce (1962) and Dominic (1966) showed that the stimulus needed to be renewed at least every 12 hours. However, Chipman et al., (1966) observed block with three exposures of 15 min/day for 4 days. These females may have become contaminated with the pheromone which returned with them to their cage.

Parkes and Bruce (1962) could not demonstrate the pheromone in urine stored without preservatives, but Dominic (1966) showed that it could be stored with antibiotics and antioxidant for 7 days. Unfortunately, no indication of the degree of fecal contamination is given nor of the method of collection—from metabolism cages or direct from the animal.

PRODUCTION. The production of the pheromone is androgen dependent; castration causes it to disappear from the urine and androgen treatment of male or female castrates causes its secretion (Bruce, 1965; Dominic, 1965).

Bruce (1963) noted that strain differences are related to the effectiveness of a male, and Marsden and Bronson (1965) were unable to find inbred mice which induced pregnancy block, but Chipman and Fox (1966) observed it in wild mice. Recently, however, two inbred strains have been found which exhibit pregnancy block. Our studies have shown that susceptibility and ability to induce pregnancy block both appear to be inherited as recessive characters (Chapman and Whitten, 1969).

PHARMACOLOGY AND ENDOCRINOLOGY OF THE RESPONSE. Dominic (1966) has shown that reserpine treatment prevents pregnancy block, and Brown-Grant (1966) suggests that avertin may interfere. Snyder and Taggert (1967) have provided evidence to support the contention that pregnancy block is mediated through the adrenal. Unfortunately, significant control groups are absent from these studies.

The Relationship Between the Pheromones

There is a great similarity between the substances which induce estrous synchrony and block implantation. Ovulation earlier than would otherwise occur is common to both responses and follows a similar time sequence. Both substances occur in the urine and are androgen dependent and labile. Parkes (1960) suggested that the active material was a spectrum of odors and that the blend was different in the stud and strange males. Whitten (1966) suggested one pheromone was responsible for both effects, but an additional factor or identifier was required to change the context. It seems possible that the pheromone responsible for estrous synchrony may be one of several related substances and that pregnancy block occurs if there is sufficient difference between the pheromones of the two males.

The Neuroendocrine Pathways

All of the evidence suggests that the hypothalamus and pituitary are involved in the responses. Barraclough and Cross (1963) and Scott and Pfaffmann (1967) have recorded spikes in the hypothalamus after olfactory stimulation. Dominic (1966) has observed that reserpine abolishes pregnancy block and so do ectopic pituitaries. Bruce and Parkes (1960) observed that the original pregnancies could be maintained by prolactin.

PRIMING PHEROMONES IN OTHER MAMMALS

The question of commonness of these phenomena among mammals is of obvious importance since it is on this basis that pheromones will be accorded greater or lesser import in the overall picture of mammalian reproduction. The three priming phenomena that have been discussed have been documented in only one of the 19 possible orders of mammals, the rodents. Physiologic information leading one to suspect their functional presence may be found for three additional orders, but the nature of this evidence is tenuous. Direct relationships between olfactory stimulation or ablation and ovarian function have been documented for rats, cats, guinea pigs, pigs, and rabbits (Takewaki, 1949; David et al., 1952; Magnotti, 1936; Signoret and Mauleon, 1962; Franck, 1966). Estrous cycles of two species, sheep and goats, are known to be synchronized by the presence of a male but the role of olfaction in this respect is not known (Schinckel, 1954; Hulet, 1962; Shelton, 1960). The CNS arousing effect of male urine in pigs is also worthy of note (Signoret and du Mesnil du Buisson, 1961).

Probably the most convincing argument that priming pheromones are more widespread than now documented rests on the fact that all three phenomena occur and seem to function identically in two widely divergent genera, *Mus* and *Peromyscus* (Bronson and Eleftheriou, 1963; Bronson and Marsden, 1964; Bronson and Dezell, 1968). These two genera belong to different families of rodents and, while one evolved in Asia (*Mus*), the evolution of the other has apparently been limited to North America. Thus, the similarity between priming functions in these two species could simply be an interesting case of parallel evolution but, considering the few species which have actually been investigated, it is more probable that they are only representative of a larger number of species and that future work with a wider spectrum of mammals will prove rewarding. Important in this aspect is the necessity for looking beyond our normal laboratory or domestic stocks of animals since these are end products of long, and often unintentional, selection for reproductive efficiency and, hence, probably selection against many pheromonal effects. The loss of pregnancy blocking capacity in many inbred strains of mice is a case in point (Marsden and Bronson, 1965). The findings that incidence of estrus is not altered by male exposure in laboratory rats unless gonadotropin release is first suppressed by starvation is another (Cooper and Hayes, 1967).

ECOLOGIC SIGNIFICANCE

At least three questions concerning ecologic significance are apparent: (1) how effectively do priming pheromones function under natural conditions?; (2) does the capacity to either secrete or respond to them confer selective advantages?; and (3) could pheromones act as regulators of population density?

Questions concerning the effectiveness of priming pheromones under natural conditions and potential advantages associated with them have an obvious answer: we would probably not be observing these phenomena in the laboratory today unless they had previously operated in the field and had conferred selective advantages. This is, of course, *post priori* reasoning but considering the host of factors that could operate both independently and interactively to influence reproduction in the field, we may never have more adequate proof. It is reasonable at this time, then, to ask whether the phenomena as we know them in the laboratory appear compatible with what is known about natural populations of the two genera for which adequate laboratory documentation of priming pheromones exist.

Wild stocks of the house mouse (*Mus musculus*) and the prairie deermouse (*Peromyscus maniculatus bairdii*) exist under conditions which are first of all highly variable and, secondly, only remotely similar to those under which laboratory stocks are maintained and studied. Like many rodents these two species are not particularly colonial. Once having established a home range, adults tend to limit their daily activities to relatively small areas, the centers of which tend to be spatially dispersed or even overdispersed, at least during the breeding season and with respect to the same sex. Such dispersed spatial distributions are maintained by territoriality, mutual avoidance, avoidance of a dominant animal by a subordinate, or other forms of nonterritorial aggression. The two species are to some degree nocturnal, and olfaction probably plays a prime role in intrapopulation communication. Granted a resounding lack of knowledge about the social and physical structure of natural populations there are still consistent ecologic differences between *Mus* and *Peromyscus*. Prairie deermice are usually found in open grassland densities of less than 20 per acre (Terman, 1966). *Mus*, on the other hand, may be found living under a variety of field conditions, commensal with humans, or in a combination of these two types of environments. Densities are usually more erratic and considerably higher; up to 300 per acre in grassland (DeLong, 1967), while several dozen per square yard (equivalent to several thousand per acre) may live in small chicken coops or English grain ricks (Southwick, 1958; Selander, personal communication). A prime difference between the two species, then, would seem to be an average higher rate of social interaction for house mice than for deermice.

Major characteristics that would seem necessary for efficient pheromone function in natural populations such as those described above are essentially those ensuring that an animal receives an adequate dose of pheromone: amount of the pheromone secreted, transportability of the active molecule, fade-out time, and sensitivity of the recipient. On the basis of laboratory studies, it does not seem too difficult to visualize good effectiveness for male-secreted, estrus-inducing pheromone during random daily activities in wild house mice because of their normally high population densities and particularly when they occur in relatively confined areas such as chicken coops or grain sheds where air flow is reduced. For example, Chipman and Fox (1966) found the incidence of estrus altered in wild *Mus* by simply housing males in the same poorly ventilated room with experimental females; and Bronson (unpubl.) found significant estrus-induction by dripping as little as 0.1 ml male urine into cages containing 10 wild *Mus* females. It is more difficult to visualize a good degree of effectiveness in outdoor populations of *Mus* and particularly so in the normally low density deermouse populations where the chance of females acquiring an adequate dose during random social interaction would seem quite low. It could be effective, however, if diestrous females shared a poorly ventilated burrow with a male or if the male anointed her with urine or some similar secretion. Deermice, nevertheless, have been shown to be even more sensitive to male urine than wild *Mus* in the laboratory (Bronson, unpubl.) and a final thought is that restriction of activity to runways as occurs in some rodents (e.g., many Microtines) would certainly enhance pheromonal effectiveness.

While the induction of estrus by male odor would appear to work most efficiently at high density, laboratory studies indicate that the block to implantation caused by a strange male might be most effective at a medium density. Blocking effectiveness is reduced by pre-exposure to many males or if the stud or other females are present along with the strange male (Bruce, 1963). In a naturalistic approach, Chipman et al. (1966) were able to induce blocks in the wild *Mus* with as little as three 15-minute exposures per day. These workers also found that males confined to maze-like, wire-bottomed runways effectively blocked pregnancies in wild *mus* females confined to similar criss-crossed runways below them. It is apparent then that this phenomenon could work in the field but would require a density that is both high enough to ensure opportunity to encounter strange males but low enough to avoid interference with the block by the presence of other animals. Selective advantages associated with the male-secreted pheromones seem rather obvious. The ability to induce estrus in a female even at the expense of another male's insemination must exert considerable selective pressure.

The potential for estrus-inducing pheromone to enhance the reproductive isolation of incipient species is obvious and Bruce has suggested that the strange male effect could benefit a population by perpetuating some degree of outbreeding.

It is not difficult to envisage a function for the pheromones produced by males which would confer a selective advantage, but the same cannot be said for the all-female priming phenomenon. All-female groups do occur during the breeding season but are apparently uncommon in most rodent species. Frank (1957) reports the development of "mother-families" among German Microtines at high density where males are excluded unless a female is in estrus. Possibly pertinent are reports of *Mus* females crowding together to avoid aggressive males and in large complex pens that such grouping was accompanied by decreased incidence of estrus (Crowcroft and Rowe, 1957). It should be noted that diestrus may be induced by socially stressful situations entirely separate from priming pheromone effects (e.g., Christian et al., 1965) and, considering the relative ease with which such suppression is broken by male odor in the laboratory, the pheromonal significance of the findings of Crowcroft and Rowe seems questionable. The evolutionary significance of all-female pheromonal effects also seems tenuous. It is difficult to visualize a selective advantage associated with mutual suppression of the estrous cycle by females unless this occurs within a framework of population selection as proposed by Wynn-Edwards (1962). A population which is slower breeding is less likely to outstrip its food supply. It should be noted, nevertheless, that one potentially important function for this phenomenon could be in the beginning of the breeding season as an aid to synchrony of reproductive efforts within the population. Whether or not this would confer selective advantages is problematic.

Since the effectiveness of the known priming pheromones is undoubtedly related to population density it is reasonable to ask whether or not these factors could also function as regulators of population density. Growth of populations of mammals, as of other species, may be limited either by many environmental factors or by a factor intrinsic to the population itself, i.e., a social factor. Large pen populations of both deermice and house mice, given a surfeit of all environmental requisites except space, will cease growth at characteristic levels because of a decrease in reproductive success and/or an increase in mortality (Terman, 1968; Christian et al., 1965). Theories to account for the social regulation of density include both behavioral-endocrinologic feedback systems with ACTH playing a prime role and changes in genetic constitution. The suggestion has been made that priming pheromones could also operate at high density to inhibit reproduction and hence bring about a cessation of population growth (Chipman and Fox, 1966). Two known priming phenomena could function in this manner; the induction of pregancy block by strange males and the suppression of estrus by all-female grouping. The latter effect, as discussed previously, is probably a rare occurrence in natural populations and is of questionable biological significance. Given appropriate behavioral circumstances, i.e., a high rate of movement on the part of adult males, it is conceivable that their priming pheromones could depress reproductive success at high density by continuously blocking implantation. This, however, does not seem to be a prime characteristic shown by either *Mus* or *Peromyscus* at high density at least in large pen population studies. In addition, the strange male block to pregnancy would seem to be a relatively inefficient mechanism for inhibiting reproduction in a population when compared to other intrinsically-induced depressions such as complete failure to produce young or mortality during the pre-weaning stages. This is not to say that there are not other yet undiscovered priming pheromones that could operate to limit density. One experiment with deermice has evaluated this possibility in the laboratory; Terman (1969) reported an increase in reproductive success rather than the expected inhibition when pairs were maintained on bedding that had been soiled by large pen populations. This cannot be considered a definitive experiment, however, because of the lack of knowledge about the lability of priming pheromones and possibly countering this experiment is a report by Petrusewicz (1957) showing that large pen populations of *Mus* recommenced growth when placed in new, clean cages. In summary, then, it seems unreasonable at this time to postulate these mechanisms as efficient regulators of population growth. Further knowledge from both the laboratory and field may change this picture. It would seem on the surface, however, that a prime function of these mechanisms would be to provide alternative stimulatory pathways leading to reproductive success rather than to failure given the overriding importance of successful reproduction in the maintenance of a species.

PRIMING PHEROMONES AS EXPERIMENTAL TOOLS

The final question of significance, the importance of priming phenomena to other areas of physiology or behavior, could have many answers. Priming pheromones, because of their shaping effects on behavior and their possible wide effects on endocrine organs, could prove of interest to a number of biologic areas. The one point we would like to make here, however, is their potential importance to neuroendocrine research. Environmental regulation of endocrine activities is largely brought about by way of the hypothalmus and a large endeavor has been mounted for understanding the relationships between higher brain areas and the hypothalamus and between the hypothalamus and the adenohypophysis. The role of environmental cues in models developed to understand these relationships is usually evaluated in terms of incident light or photoperiodic effects. This is reasonable given the effects of seasonal photoperiod on reproduction and, in particular, the diurnal rhythm in the release of the ovulating flush* of LH in the rat and mouse. Nevertheless, light lacks the specificity of response at the level of the hypothalamus to yield really definitive experiments on environmental regulation. For example, almost every endocrine organ shows some photoperiodic effects and the ovulating flush, at least in the rabbit, is composed of

* Flush, meaning fully supplied, fertile, or fecund.

many pituitary hormones (Desjardins et al., 1967). It is conceivable that should the effects of priming pheromones be relatively exclusively on gonadotropin and should isolation and identification of these substances occur, a tool of a higher degree of specificity would be available for studying neural pathways and integrative processes above and in the hypothalamus.

ACKNOWLEDGMENTS

Part of this work was supported by the United States Public Health Research Grants HD-00473 and HD-00767 from the National Institute of Child Health and Human Development, National Institutes of Health. The principles of laboratory animal care as promulgated by the Council of the American Physiological Society are observed in this Laboratory.

REFERENCES

Andervont, H. B. 1944. Influence of environment on mammary cancer. *J. Nat. Cancer Inst.*, 4:579-581.

Barraclough, C. A., and B. A. Cross. 1963. Unit activity in the hypothalamus of the cyclic female rat: effect of genital stimuli and progesterone. *J. Endocrinol.*, 26:339-359.

Bartke, A., and G. L. Wolff. 1966. Influence of the lethal yellow (A^y) gene on estrous synchrony in mice. *Science*, 153:79-80.

Bowers, J. M., and B. K. Alexander. 1967. Mice: individual recognition by olfactory clues. *Science*, 158:1208-1210.

Bronson, F. H. 1968. Pheromonal influences on mammalian reproduction. *In* Diamond, M., ed. Perspectives in reproduction and sexual behavior: a Memorial to Wm. C. Young. Indiana University Press.

―――― and B. E. Eleftheriou. 1963. Influence of strange males on implantation in the deermouse. *Gen. Comp. Endocrinol.*, 3:515-518.

―――― and V. Chapman. 1968. Adrenal-oestrous relationships in grouped or isolated female mice. *Nature*, 218:483-484.

―――― and Helen E. Dezell. 1968. Studies on the estrus-inducing (pheromonal) action of male deermouse urine. *Gen. Comp. Endocrinol.*, 10:339-343.

―――― and H. M. Marsden. 1964. Male-induced synchrony of estrus in deermice. *Gen. Comp. Endocrinol.*, 4:634-637.

―――― and W. K. Whitten. 1968. Oestrus-accelerating pheromones of mice. Assay, androgen dependency and presence in bladder urine. *J. Reprod. Fertil.*, 15:131-134.

Brown-Grant, K. 1966. The effect of anesthesia on endocrine responses to olfactory stimuli in female mice. *J. Reprod. Fertil.*, 12:177-181.

Bruce, H. M. 1959. An exteroceptive block to pregnancy in the mouse. *Nature*, 184:105.

―――― 1963. Olfactory block to pregnancy among grouped mice. *J. Reprod. Fertil.*, 6:451-460.

———— 1965. The effect of castration on the reproductive pheromones of male mice. *J. Reprod. Fertil.*, 10:141-143.

———— 1966. Smell as an exteroceptive factor. *J. Anim. Sci.*, 25 (Suppl.):83-89.

———— and A. S. Parkes. 1960. Hormonal factors in exteroceptive block to pregnancy in mice. *J. Endocrinol.*, 20:xxix.

———— and A. S. Parkes. 1965. Pheromones and their role in mammalian fertility. *In* Austin, C. K., and J. S. Perry, ed. Agents Affecting Fertility. Boston, Little Brown and Co.

Chapman, V. M., and W. K. Whitten. 1969. The occurrence and inheritance of pregnancy block in inbred mice. *Genetics*, 61:S9.

Chipman, R. K., and K. A. Fox. 1966. Oestrous synchronization and pregnancy blocking in wild house mice (*Mus musculus*). *J. Reprod. Fertil.*, 12:233-236.

———— J. A. Holt, and K. A. Fox. 1966. Pregnancy failure in laboratory mice after multiple short-term exposure to strange males. *Nature*, 210:653.

Christian, J. J., J. A. Wood, and D. E. Davis. 1965. The role of endocrines in the self-regulation of mammalian populations. Rec. Progr. Hormone Res. 14:501-578. New York, Academic Press.

Cooper, K. J., and N. B. Hayes. 1967. Modification of the oestrous cycles of the underfed rat associated with the presence of the male. *J. Reprod. Fertil.*, 14:317-324.

Crowcroft, P., and F. P. Rowe. 1957. The growth of confined colonies of the wild house mouse (*M. musculus* L.) living in confined colonies. *Proc. Zool. Soc. Lond.*, 192:359-370.

David, R., G. Theiry, M. Bonvallet, and P. Dell. 1952. Effets de la stimulation des bulbes olfactifs sur le cycle sexul de la chatte. *Comptes Rendus Soc. Biol. (Paris)*, 146:670-672.

Desjardins, C., K. T. Kirton, and H. D. Hass. 1967. Anterior pituitary levels of FSH, GH, ACTH and prolactin after mating in female rabbits. *Proc. Soc. Exp. Biol. Med.*, 126:23-26.

DeLong, K. T. 1967. Population ecology of feral house mice. *Ecology*, 48:611-634.

Dewar, A. D. 1959. Observations on pseudopregnancy in the mouse. *J. Endocrinol.*, 18:186-190.

Dominic, C. J. 1965. The origin of the pheromones causing pregnancy block in mice. *J. Reprod. Fertil.*, 10:469-472.

———— 1966. Block to pseudopregnancy in mice caused by exposure to male urine. *Experientia*, 22:534.

Eliasson, R. 1959. Studies on prostoglandins: occurrence formation and biological actions. *Acta Physiol. Scan.*, 46 (Suppl.) 158.

Franck, H. 1966. Effets de l'ablation des bulbes olfactifs sur la physiologie genitale chez la Lapine adult. *Comptes Rendus Soc. Biol.*, 160:863-865.

Frank, F. 1957. The causality of microtine cycles in Germany. *J. Wildl. Mgmt.*, 21:113-121.

Fraser, A. F. 1968. Reproduction Behaviour in Ungulates. London, Academic Press.

Hulet, C. V. 1966. Behavioral, social and psychological factors affecting mating time and breeding efficiency in sheep. *J. Anim. Sci.*, 25 (Suppl.) :5-20.

Kalmus, H. 1955. The discrimination by the nose of the dog of individual human odours and in particular of the odour of twins. *Brit. J. Anim. Behav.*, 3:25.

Kingston, B. H. 1964. The chemistry and olfactory properties of musk, civet, and castoreum. Proc. 2nd Int. Cong. Endocrinol., London.

Land, R. B., and A. McLaren. 1967. The response of female mice to repeated injections of human chorionic gonadotrophin. *J. Reprod. Fertil.*, 13:321-327.

Lederer, E. 1950. Odeurs et parfums de animaux. *Fortschr. Chem. Organ. Naturstoffe.*, 6:87-153.

van der Lee, S., and L. M. Boot. 1955. Spontaneous pseudopregnancy in mice. *Acta Physiol. Pharmacol. Neerl.*, 4:442-443.

――― and L. M. Boot. 1956. Spontaneous pseudopregnancy in mice II. *Acta Physiol. Pharmacol. Neerl.*, 5:213-214.

Lissak, L. 1962. Olfactory-induced sexual behavior in female cats. Symposium XIV, International Congress Series No. 47. *Excerpta Medica*, 1:653-656.

Magnotti, T. 1936. L'importanza dell'olfatto sullo sviluppo e funzione degli organi genitali. *Boll. Mal. Orecch. Gola Naso*, 54:281.

Mainardi, D., M. Marsan, and A. Pasquali. 1965. Causation of sexual preferences of the house mouse. The behavior of mice reared by parents whose odour was artificially altered. *Atti Soc. Ital. Sci. Nat. e Mus. Civico di Storea Nat. Milano*, 104:325-338.

Marsden, H. M., and F. H. Bronson. 1964. Estrous synchrony in mice: alteration by exposure to male urine. *Science*, 144:3625.

――― and F. H. Bronson. 1965. Strange male block to pregnancy: its absence in inbred mouse strains. *Nature*, 207:878.

Michael, R. P., and E. B. Keverne. 1968. Pheromones in the communication of sexual status in primates. *Nature*, 218:746-749.

Mody, J. K. 1963. Structural changes in the ovaries of IF mice due to age and various other states: demonstration of spontaneous pseudopregnancy in grouped virgins. *Anat. Rec.*, 145:439-447.

Parkes, A. S. 1960. The role of odorous substances in mammalian reproduction. *J. Reprod. Fertil.*, 3:312-314.

――― and H. M. Bruce. 1961. Olfactory stimuli in mammalian reproduction. *Science*, 134:1049-1054.

――― and H. M. Bruce. 1962. Pregnancy-block in female mice placed in boxes soiled by males. *J. Reprod. Fertil.*, 4:303-308.

Petrusewicz, K. 1957. Investigation of experimentally induced population growth. *Ecologia Polska, Series A*, 5:281-301.

Planel, H. 1953. Etude sur la physiologie de l'organ de Jacobson. *Archs. Anat. Histol. Embryol.*, 36:198-206.

Prelog, V., L. Ruzicka, and P. Wieland. 1944. Steroide und Sexualhormone. *Helv. chim. Acta*, 27:66-68

Ropartz, P. 1967. L'effect de Groupe chez les animaux. Colloques internationaux die centre National de la Recherche scientifique No. 173.

Schinckel, P. G. 1954. The effect of the presence of the ram on the ovarian activity of the ewe. *Aust. J. Agric. Res.*, 5:465-469.

Scott, J. W., and C. Pfaffmann. 1967. Olfactory input to the hypothalamus: electrophysiological evidence. *Science*, 158:1592-1594.

Shelton, M. 1960. Influence of the presence of a male goat on the initiation of oestrous cycling and ovulation of angora does. *J. Anim. Sci.*, 19:368-375.

Signoret, J. P., and P. Mauleon. 1962. Action de l'ablation des bulbes olfactifs sur les mecanismes de la reproduction chez la truie. *Ann. Biol. Anim. Biochem. Biophys.*, 2:167-174.

—— and F. du Mesnil du Buisson. 1961. Etude du comportement de la truie en oestrus. Fourth Int. Cong. Anim. Reprod. Artif. Insem. (The Hague).

Snyder, R. L., and N. E. Taggart. 1967. The effects of adrenalectomy on male induced pregnancy block in mice. *J. Reprod. Fertil.*, 14:451-456.

Southwick, C. H. 1958. Population characteristics of house mice living in English corn ricks: density relationships. *Proc. Zool. Soc. Lond. B.*, 131:163-175.

Takewaki, K. 1949. Occurrence of pseudopregnancy in rats placed in vapor of ammonia. *Proc. Japan Acad.*, 25:38-39.

Terman, C. R. 1966. Population fluctuations of *Peromyscus maniculatus* and other small mammals as revealed by the North American classes of small mammals. *Amer. Midland Natural.*, 76:419-426.

—— 1968. The dynamics of *Peromyscus* populations. *In* King, J. A., ed., Biology of *Peromyscus*. Lawrence, Kansas, American Society of Mammalogists.

—— 1969. The influence of pheromones produced in asymptotic populations on the reproductive function and maturation of isolated pairs of prairie deermice. *Anim. Behav.*, 17:104-108.

Todd, N. B. 1963. The catnip response. Doctoral dissertation, Harvard University. 57 p.

Whitten, W. K. 1956. The effect of removal of the olfactory bulbs on the gonads of mice. *J. Endocrinol.*, 14:160-163.

—— 1958. Modification of the oestrous cycle of the mouse by external stimuli associated with the male. Changes in the oestrous cycle determined by vaginal smears. *J. Endocrinol.*, 17:307-313.

—— 1959. Occurrence of anoestrus in mice caged in groups. *J. Endocrinol.*, 18:102-107.

—— 1966. Pheromones and mammalian reproduction. *In* McLaren, A., ed., Advances in Reproductive Physiology, Vol. 1. London, Logus and Academic Press, p. 155-177.

—— F. H. Bronson, and J. A. Greenstein. 1968. Estrus-inducing pheromone of male mice: transport by movement of air. *Science*, 161:584-585.

Wilson, E. O., and W. H. Bossert. 1963. Chemical communication among animals. *Recent Progr. Hormone Res.*, 19:673-710.

Wynne-Edwards, V. C. 1962. Animal dispersion in relation to social behaviour. Edinburgh and London, Oliver & Boyd.

11

The Role of Skin Glands
in Mammalian Communication

R. MYKYTOWYCZ

CSIRO Division of Wildlife Research
Canberra, Australia

INTRODUCTION 327
APPROACHES TO THE STUDY OF SCENT GLANDS IN MAMMALS 329
OCCURRENCE OF SCENT GLANDS IN MAMMALS 330
DISTRIBUTION OF THE SECRETIONS FROM SKIN GLANDS 332
HISTOLOGIC STUDIES OF SKIN GLANDS 335
TYPES OF MESSAGES CONVEYED BY ODOR FROM SKIN GLAND SECRETIONS 336
COMPARATIVE STUDIES OF SKIN GLANDS 347
SKIN GLANDS IN PRIMATES 348
CHEMICAL STUDIES OF SKIN GLANDS 350
CONCLUSION 352
REFERENCES 353

INTRODUCTION

The Need for Understanding
Communication in Mammals

The need for a better insight into the problem of animal communication has emerged over the last two decades or so because of the rapid development of ethologic studies. These have shown that social organization is an integral part of the biology of many species of animals and not only of those such as ants and bees, which have for long been recognized as "social" species. Sociality in animals has recently been reviewed by Eisenberg (1966).

It has become apparent that for the complete understanding of the life of a species, information on its method of communication is essential. Numerous reports dealing with various species of mammals have demonstrated that decreased reproductive efficiency and survival rate could be induced not only by disease, food shortage, predation, and climatic conditions, but also by a factor which various authors have termed "social interaction," "social stress," "interspecific conflict," or "social pressure" (Christian and Davis, 1964; Chitty, 1964; Myers, 1966; Nelson, 1967). These involve some communication between individuals.

Some investigators, interested in the control of economically important species, also hope that through a better understanding of the animal's "language" they may be able to design more efficient methods leading to the reduction, or even the complete elimination, of a pest species. Their expectations have undoubtedly been stimulated by recent successes in the use of pheromones in insect control (Jacobson and Beroza, 1964; Wilson 1965; Shorey and Gaston, 1967).

On the other hand, a better knowledge of animal communication may assist those interested in the preservation of species of wildlife or in the management of domestic animals (McBride et al., 1967). Others maintain that information on communication in mammals may provide the background for an understanding of all communicative behavior, including man's (Sebeok, 1965).

Systems of Communication in Mammals

The mechanisms of animal communication have recently been reviewed by Marler and Hamilton (1966), and the following is based largely on their views.

Animals use different systems to communicate with one another. The signals which one animal sends to another can be visual, acoustic, tactile, olfactory, or gustatory. Each system has advantages and disadvantages which depend on the physical environment and situation under which it is employed. Often a combination of the different systems is the only way to convey a message. These combinations enrich the animal's signalling repertoire. The functions of different systems overlap extensively, hence the loss of one sense may be compensated by the development of another.

One advantage of olfactory signals is their effectiveness over great distances. In this respect they are superior to tactile or visual communication which are most effective at close range. A unique property of olfactory signals is their persistence in time. This could be of particular advantage for some purposes, as olfactory communication can thus be effective even in the absence of the signalling animal. However, in some cases where a rapid change of signals is required, this durability may be a disadvantage. In such instances the use of acoustic signals is of special importance. Situations may arise when signalling may have to take place at the same time as other activities. Visual communication does not fulfill this requirement for which sound, and to a lesser extent olfactory stimuli, are more suited.

Sources of Odor in Mammals

There are different sources of odors in mammals. Urine and feces are obvious ones and it was Hediger (1944) who emphasized their communicative properties in mammals. Other sources of odor are the cutaneous glands which are the subject of this paper.

Some authors also refer to the odoriferous properties of saliva. Thus, Schultze-Westrum (1965) in a study of the role of olfaction in the intraspecific communication of the sugar glider, *Petaurus breviceps papuanus,* demonstrated that saliva could be active in olfactory communication. However, he could not say definitely what was the source of its odor.

Dutt et al. (1959) reported that the sebaceous glands situated in the preputial diverticulum are the source of the characteristic odor in boars. The surgical removal of the preputial glands from immature males was carried out. Later, at slaughter, no odor could be detected in the carcasses of boars from which the glands had been completely removed. Odor persisted, however, in the animals with partly removed glands.

It is thus possible that the respective smell of urine and feces, as well as saliva, may also be modified by odor from the skin glands. Lederer (1949) found that one third of the compounds isolated from castoreum—the secretion of the beaver's anal gland—was also present in the urine of mammals.

Approaches to the Study of Scent Glands

The existence of odor-producing skin glands did not escape the attention of early mammalogists. All that they did, however, was to record the existence of the glands and describe their macroscopic appearance. As only sporadic and limited information was then available on the behavior of animals, it was impossible to determine fully the role of glands which function for behavioral purposes.

As knowledge of patterns of mammalian behavior developed, speculations were made about the territorial, sexual, and identifying role of scent glands, but in most cases these were not substantiated by systematically collected observations or experimentally derived proofs.

Similarly, early observations on the histology of skin glands and their chemical composition did not take into consideration the behavioral function of the glands.

With recent advances in ethology, approaches to the study of skin glands are changing, and more attention is being paid to experimentation and to the behavior of animals.

Thus an experimental approach has been used by Thomson and Pears (1962) in the study of the secretions of the sternal and anal glands of the brush-tailed possum, *Trichosurus vulpecula.* Using caged animals, they tested the reaction of individuals to the scent derived from their own and from foreign species.

Schultze-Westrum (1965), working with an experimental population of caged sugar gliders of known age, sex, group membership, and social status, made quantitative measurements of their reactions to the scent of anal, sternal, frontal, and pouch glands, and of saliva, urine, as well as to the odors collected from other parts of the body. The marking activity of the Mongolian gerbil, *Meriones unguiculatus,* which uses the secretion from its ventral glands, has been studied under experimental conditions by Thiessen et al. (1968). Studies of chin and anal gland marking by the rabbit, *Oryctolagus cuniculus*, have been made by Mykytowycz (1965, 1966a, and unpublished).

In recent years there has been an effort to introduce chemical analysis into the study of mammalian scent glands. Müller and Lemperle (1964) used gas-liquid chromatography to analyze the secretions from the skin glands of the fox, chamois, red deer, and Alsation dog.

More recently, Müller-Schwarze (1967), while working with confined populations of mule deer, *Odocoileus hemionus*, subspecies *columbianus*, separated out by distillation and gas-liquid chromatography various fractions of the tarsal gland secretions and tested the reactions of other mule deer to these (see also Brownlee et al., 1969).

Hesterman and Mykytowycz (1968) applied psychometric methods to measure the relative intensities of odors from scent glands. Using panel techniques, which are commonly employed in industry for the standardization of aromatic products and in olfactory studies generally (Benjamin et al., 1965), the intensities of the odors from glands of rabbits and hares of different ages, sex, and reproductive stages were compared. It is now obvious that for the detailed study of organs functioning for behavioral purposes, as do odor-producing glands, quantitative information on the behavior of the species is essential to back morphologic, histologic, and chemical observations of the organs. The acquisition of precise behavioral data may not be practicable for many species living under natural conditions, but for some species, including the European wild rabbit in Australia, it has been found possible to get these behavioral data. Since frequent references will be made here to studies made by myself on the skin glands of this species, a brief outline follows of the the experimental conditions under which these studies were made. Experimental populations of wild rabbits were confined to paddocks within which their movements and activities above ground could be watched day and night. Underground positions of individual animals could also be checked when necessary. Each rabbit was distinctly and permanently marked to permit quick individual identification. The dates of birth and death of all rabbits living in the enclosure could be recorded. Animals could be collected at various intervals for weighing and other biologic examinations. Under these conditions it was possible to record the history of each animal and its territorial, reproductive, and social activities. Measurements were taken of the behavior patterns in relation to frequency of occurrence, time of day, and season, and scent glands were obtained for histologic and other examinations.

This work established that the wild rabbit is a strongly territorial animal which lives in sharply defined groups, each having its own hierarchy of linear order (Myers and Poole, 1959, 1961; Mykytowycz, 1958, 1959).

OCCURRENCE OF SCENT GLANDS
IN MAMMALS

In most publications, particularly recent ones, dealing with the general biology of mammals reference is made to the wide occurrence of scent glands, and their importance in mammalian communication (Bourlière, 1954; Wynne-Edwards, 1962; Corbet, 1966). According to Müller-Schwarze (1967a) scent glands have been reported from 15 of the 18 mammalian orders, and on the basis of their location on the animal's body there are 40 different types of glands. Schaffer's (1940) extensive monograph dealing mainly with the histology of skin glands is a most valuable source of information on their occurrence and on the early literature. Between the years 1910 to 1927, R. J. Pocock published a series of papers appearing mainly in the Proceedings of the Zoological Society of London, in which he described the morphology and occurrence of skin glands in a variety of species. More recently Fiedler (1964) reviewed the occurrence of skin glands in mammals and discussed their possible communicative role.

In many species skin glands occur in more than one region of the body. For instance, in Cervidae, glands are present in 10 different sites: retrocornual, supraorbital, antorbital, circumcaudal, infracaudal, preputial, metatarsal, tarsal, and interdigital on both hind and forelegs (Müller-Using and Schloeth, 1967). As most of them are bilateral, the source of skin gland odor is increased.

Lagomorphs have chin glands and paired anal and inguinal glands. In addition there are Harder's, infraorbital, and lachrymal glands, situated in the orbit, which are derived from the skin. Their functions are not clearly understood, although some authors surmise that they have a behavioral role. Thus Rue (1965) has reported that the male cottontail rabbit, *Sylvilagus floridanus*, uses scent glands in the corners of its eyes as an adjunct in the breeding season. It rubs these glands against twigs or blades of grass as a scent cue for other rabbits. Marsden and Holler (1964) reported similar marking behavior in cottontails.

There is great diversity in the complexity and external appearance of the glands. Some are scattered diffusely, but most appear as well-organized, sizable organs with muscular investiture, storage pouches, and external ducts or wide openings allowing extrusion of the glands.

In many species there are hair arrangements associated with the glands for combined visual and olfactory signals. In most cases these arrangements appear to be adapted to retain the odoriferous secretions. In the mule deer the tufts of hairs overlying the metatarsal glands are up to four and one-half inches in length (Linsdale and Tomich, 1953).

In many species the hair surrounding the glandular areas is conspicuous by being denser, coarser, and often colored with pigment. Eadie (1954) emphasized that pelage descriptions in American mole genera, especially *Scalopus*, have been confused in the past because stains from the secretions of skin glands have been often mistaken for permanent color patterns. On the chest of male red kangaroos, *Megaleia rufa*, during certain periods, the skin and fur show distinct signs of pigmentation from apocrine glands (Mykytowycz and Nay, 1964).

The wide occurrence of skin glands emphasizes the importance of these organs to mammals. They are best developed among nocturnal species—in which olfactory communication might be expected to play a more important role than other methods of communication—and among gregarious animals—which obviously possess a greater need for communication.

DISTRIBUTION OF THE SECRETIONS
FROM SKIN GLANDS

There are various ways in which the odorous secretions from skin glands are distributed, but they can be divided into two basic categories: passive and active (Schultze-Westrum, 1965). One can also differentiate between direct and indirect distribution of gland secretions.

Passive Marking

The mere presence of glands on the body may endow an animal with a characteristic odor of communicative value.

An example of passive marking is seen in the caribou, *Rangifer* (Quay, 1955), the Virginia deer, *Odocoileus virginianus* (Quay, 1959), and the mule deer. All these animals have interdigital glands from which secretions are deposited where ever they move. Linsdale and Tomich (1953) reported that mule deer commonly follow the trails or routes of other deer by occasionally sniffing at the ground. Dogs regularly track deer by ground scent, and there are some indications that this is also done by wild cats and coyotes. This type of passive marking is carried out also by man and by other animals possessing pedal glands. Talbot and Talbot (1963), referring to the function of the pedal glands of wildebeest, point out that the secretion could also be smelled by man in the absence of wildebeest, and indeed this was often used in tracking down these animals.

In passive application the secretion does not have to be deposited on objects in the environment; the scent may emanate directly from the source and be detected over some distance. Thus the scent from the tarsal glands of the mule deer is released directly into the air and may be detected by man at a distance of 150 feet (Linsdale and Tomich, 1953).

The passive marking of members of a social group by the occupation of common shelters also occurs. The young of marsupials, even after leaving the pouch, but while still suckling, stay in contact with the mother's scent. During suckling they pick up the secretion from the pouch glands. For the sugar glider, Schultze-Westrum (1965) has reported red markings on the head, neck, and shoulders of the young which they get by coming into contact with the mother's pouch secretion.

Active Marking

In active marking, the secretions are often applied directly in gland-to-object contact. Marking with the secretion of the submandibular gland by the wild rabbit (Mykytowycz, 1965) is carried out in this way. This animal rubs its chin on objects such as grass blades, the entrance to a burrow, the corner of a post, a stump, a stone, another rabbit, or even fecal pellets and food.

Similarly the pika, *Ochotona princeps*, applies the secretion from its apocrine cheek glands to various objects (Harvey and Rosenberg, 1960), and various species of deer apply the secretions from their antorbital and retrocornual glands. Obviously the scent of skin glands can be directly applied to objects with greater precision than can odors deposited in the urine or feces.

Secretions from glands situated on less flexible parts of the body than the head can also be used for direct marking. Frädrich (1967) has described marking by peccaries, *Tayassu tajacu*. This animal ruffles the hairs overlying its lumbar glandular area, simultaneously bending the hindlegs and, with rubbing movements, applies the milk-like secretion of the gland to grass, tree stumps, and other objects.

In the examples of direct application of secretion mentioned above, the forms of behavior are not particularly unusual, but in other cases seemingly unnatural, grotesque postures are adopted by the animals. Thus martens, *Martes martes*, and other Mustelids, when selecting elevated objects for marking with their anal glands, adopt a "handstand" posture (Goethe, 1964). Herpestinae use a similar stance in marking with their anal gland (Dücker, 1965), as do different species of Viverridae (Zannier, 1965).

The dragging of hindquarters with extended anal gland has been reported for many mammals including the marmot, *Marmota marmota* (Koenig, 1957; Kratochvil, 1964), the dormouse, *Glis glis* (Koenig, 1960), agouti, *Dasyprocta aguti aguti* (Roth-Kolar, 1957), the ground squirrel, *Citellus richardsonii* (Sleggs, 1926), and the echidna or spiny anteater, *Echidna aculeata* (Dobroruka, 1960).

In some cases the secretion is applied from a distance. This method is used especially by species which employ their secretion for purposes of defense. Bourlière (1954) reports that the skunk, *Mephitis mephitis*, can direct the secretion from its anal glands with great accuracy over a distance of four to five meters.

Not all animals apply the secretion from their glands directly. Some use different parts of their body to transfer the secretion. Frank (1956) described methods used by the mouse, *Arvicola terrestris*, of applying secretion from the lateral glands. This animal touches its flanks with the hindfeet, and the secretion is subsequently transferred to the ground during characteristic foot-drumming. It has also been suggested that, when grooming, lagomorphs and rodents pick up the secretions from the glands situated on the head—including those present in the eye orbit—and the scent is subsequently transferred to the ground by their feet.

In the mule deer, Müller-Schwarze (1967a) reports that the forehead is regularly rubbed against the inside of the animal's own hind leg thus picking up material from the tarsal and metatarsal glands. The head in turn may be rubbed against dry branches or bushes which are regularly checked olfactorily by all members of a group. The ringtail lemur, *Lemur catta*, frequently passes its tail from base to tip between its two forearms presumably anointing it with secretion from brachial glands (Hill, 1953).

Some species of animals use the secretions from skin glands in combination with saliva. Eibl-Eibesfeldt (1964) reporting on the marking procedures of the tenrec, *Echinops telfairi*, stated that two tenrecs marked their attendant by salivating on the spot to be marked and then scratching with one foot alternately on the salivated spot and on their own bodies. In this way the saliva seemed to transfer their own body odor to the marked spot which afterwards smelled noticeably of tenrec.

Secretion from the skin glands can also be used in combination with urine. Many canids mark stones, stumps, and other projecting objects with products of their preputial glands mixed with their urine (Bourlière, 1954). Mule deer regularly urinate on their tarsal glands while rubbing them together (Linsdale and Tomich, 1953). The reason for this behavior has not been explained, but it is possible that the scent from the tarsal glands is mixed with the urine which then serves as a medium of transportation.

Frequency of Distribution of Secretions from the Anal Glands

While direct marking with the secretion from the anal glands is employed by many mammals, others use their own feces as the vehicle for distribution of scent from the anal glands (Hediger, 1944; Ortmann, 1960). This is the case with the secretion from the anal glands of the rabbit, and probably other lagomorphs.

Until recently it was believed that the anal gland in rabbits functioned for the lubrication of hard fecal pellets, but recent observations have disproved this. Extirpation of these glands did not affect defecation (Takaki and Tagawa, 1961). Also, direct observations on the behavior of the animals and experimental data suggest the territorial function of the scent from these glands (Mykytowycz, 1966a; 1968).

It has been established that rabbits produce two types of hard fecal pellets: those produced during random defecation and those produced for marking purposes. Through experimental manipulation it is possible to collect these two types of pellets from the rabbit. When both types were presented for comparison to a panel of human assessors, the "marking" pellets were judged to be markedly the more odorous (Hesterman and Mykytowycz, 1968).

There have been few studies carried out with the specific intention of measuring the frequency of use of a scent gland secretion. In the European wild rabbit, the frequency of application of the secretion from the anal gland and its spatial distribution have been investigated. One of the methods has been to study the distribution of dung-hills within natural rabbit territories. These dung-hills are characteristic of areas occupied by populations of rabbits and are formed by the deposition of pellets at discrete sites. The densities, size, and position of dung-hills have been analyzed in relation to special features of the terrain, distance from burrows, position of other warrens and available feeding areas. The results (Mykytowycz and Gambale, 1969) suggest that the dung-hills are not distributed randomly. Apart from topographic features, which sometimes affect distribution, the dung-hills appear to be denser along the boundaries with adjacent colonies.

Under controlled laboratory conditions the production of fecal pellets by animals of known biologic background was measured after they were exposed to strange dung-hill material and fresh, uncontaminated grass sods, which were placed in their home runs. Similarly, data on "chinning" activities were obtained under the same conditions. These laboratory studies confirmed the conclusions drawn from the field obesrvations of natural and experimental populations. Production of fecal pellets and frequency of "chinning" on the dung-hills were highest in dominant males. Females were not so strongly attracted to dung-hills as males, and immature animals remained indifferent (Mykytowycz and Hesterman, 1970).

In the wild rabbit the application of the secretion is dependent on many variables related to the behavior of the species and to individual behavior. Thus, the frequency with which an animal uses the secretion from its skin glands varies according to the needs imposed by the position it occupies within the community; this position is determined by its age, sex, and social status.

Gustatory Perception of Skin Gland Secretions

Although it is generally assumed that the signals emanating from the skin glands are perceived by the nose, this is not the only way of detecting them: they can also be perceived by taste or, in lower animals, absorption through the epithelium of the respiratory tract.

Reiff (1956) referred to two types of odor perception in mammals: (1) nasal smelling during which airborne molecules are perceived from a distance; (2) gustatory smelling, in which molecules of odor taken into the mouth enter the turbinates and subsequently come into contact with the olfactory epithelium. Henkin (1967) suggested three different regions in which odor could be sensed by persons deprived of the olfactory bulbs: the area of the nasal cavity innervated by a branch of the trigeminal nerve; the upper part of the pharynx supplied by the pharyngeal plexus; the lower pharynx innervated by branches of the vagus nerve.

Many species of mammals employ licking in their communication. This could occur in the course of general grooming of another individual or be directed towards a specific source of secretion. Thus, Buechner and Schloeth (1965), in describing the ceremonial mating behavior in the Uganda kob, *Adenota kob thomasi,* state that the male nuzzles the inguinal region of the female by pushing his outstretched head between her straddled thighs. The male licks either the udder or the inguinal glands or both. The inguinal glands in both sexes are blind pouches in which a yellowish wax with a pungent odor accumulates. Sometimes the male uses the tip of his horns to poke and stroke the hips and inguinal region of the female. Licking of the secretion from mandibular, throat, and abdominal glands of the sexual partner is also a common precopulatory behavior of male kinkajou, *Potos flavus* (Poglayen-Neuwall, 1966). Consequently, it is possible that skin glands function for communication through taste more frequently than we realize.

Tasting by licking the urine of estrous females is a common form of sexual behavior displayed by giraffes, cattle, horses, rabbits, and probably many other mammals.

HISTOLOGIC STUDIES OF SKIN GLANDS

Odor-producing glands in mammals are mainly of two types: holocrine sebaceous and apocrine sudoriferous. Although differing in their gross anatomic features, the basic histology of these glands is the same in all mammals. Sebaceous glands are the more common of the two (Schaffer, 1940). Eccrine sudoriferous glands occur less commonly and are best developed in man and the higher primates. Other mammals usually have only a few of them restricted to the volar surface of the paws and digits. Some South American monkeys have eccrine glands on the ventral surface of the tail (Montagna, 1963).

Apocrine and sebaceous glands may occur separately as, for instance, the apocrine chin glands of the rabbit and hare, the cheek glands of the pika, and the sebaceous lateral gland of the hamster, *Mesocricetus auratus auratus*. Or they may occur in combination, as in the inguinal glands of rabbits and hares, or lateral glands in the short-tailed shrew, *Blarina brevicauda* (Pearson, 1946). The occurrence of apocrine glands in frequent combination with sebaceous glands has raised speculation that sebum in these cases may act as the base for a sudoriferous secretion (Schaffer, 1940; Dryden and Conaway, 1967).

In some species, the glands may be disseminated in the skin as are the apocrine glands of the brush-tailed possum (Bolliger and Hardy, 1944), and of the kangaroos, *Megaleia rufa*, *Macropus canguru*, and *Macropus robustus* (Mykytowycz and Nay, 1964). In others they occur as sharply defined organs.

Schaffer's (1940) voluminous publication reviews all the early literature on the histology of skin glands in mammals, while Ortmann (1960) deals specifically with anal glands in more than 100 species. The main interest of early histologists in these glands was in describing and recording their existence. This enriched the list of occurrences but added very little to our knowledge of their functions. The lack of a deeper interest in skin glands among experimental histologists in the past is easy to understand if one considers that the behavior of animals was not adequately understood. As a result, very little, apart from a few morphologic facts, is known about the sebaceous glands of mammals other than man, the rat, and the mouse (Montagna, 1963). The same is true of apocrine glands.

An attempt to relate the histology of the skin glands to the behavior of the animal has been made more recently during studies of the odor-producing glands of the wild rabbit in Australia. In a series of papers (Mykytowycz, 1965, 1966a, 1966b, 1966c), it was demonstrated that the size and secretory activities of various glands—the chin, anal, inguinal, and Harder's—were correlated with the age, sex, social status, and reproductive stage of the rabbits. It could clearly be seen that in a given population the histologic picture of the glands varied widely between animals of the same sex and age, reflecting the different degrees of importance of their glands to different individuals. Considering this wide variation in the histologic picture of skin glands in seemingly similar indvduals one should be cautious about generalizing from the results obtained from the histologic examination of limited numbers of samples, especially those unaccompanied by background biologic information.

TYPES OF MESSAGES CONVEYED BY
ODOR FROM SKIN GLAND SECRETIONS

Before discussing the communicative function of skin glands, it may be proper to emphasize that communication may not be their only role. For example—as happens in man—sebum may be required for the protection of hair and skin from moisture, and from fungal and bacterial activity. Sebum, like any oil, has a smoothing effect on skin (Kligman, 1963). Some glands, such as the sebaceous glands in the brush-tailed possum, are also believed to be of importance because of the light-absorbing properties of their secretion which facilitates heat control (Bolliger, 1944a). The secretion from skin glands may also possess venomous properties, as for instance the crural glands of the playtypus, *Ornithorhynchus anatinus* (Calaby, 1968).

Not all the secretions of skin glands can be detected by the human nose, e.g., the secretions from the submandibular glands of the rabbit, the cheek glands of the pika (Harvey and Rosenberg, 1960), or the sebaceous side glands of the musk shrew, *Suncus murinus*. As Dryden and Conaway (1967) have pointed out in connection with the side glands of the shrew, the absence of an odor detectable by humans does not necessarily mean that other mammals cannot detect that odor.

It is also conceivable that the action of bacterial decomposition may be necessary to produce a perceptible odor. Shelley (1956) demonstrated that in man the odorless secretion from axillary glands acquires a characteristic odor only under the influence of specific microorganisms.

With regard to the types of signals conveyed by animals through odors from skin glands, we still know so little that efforts to tabulate them are no more than intelligent guesses based on conjecture. Nevertheless, this is done below in order to introduce further discussion in this paper.

Schultze-Westrum (1965) gives a list of 15 situations in which mammals use odors as signals; this list could be extended and, no doubt, will be enlarged later as knowledge of animal behavior progresses.

Types of messages which could be conveyed by mammals by means of olfaction:

Intraspecific communication	*Interspecific communication*
Individual appraisal	Individual
Group membership appraisal	Species membership
Age apprasial	Prey
Social status appraisal	Predator
Sex appraisal	Warning
Reproductive stage indication	Defense
Trail marking	
Territory marking	
Identification with home range	
Warning	
Defense	
Alarm	
Submission	
Attention-seeking	
Greeting	
Encouraging approach	
Distress-signalling	
Pain indication	
Gregariousness	

In many cases in which the role of olfaction in animal communication has been studied, the exact source of the odor has not been specified. Mainardi et al. (1965) showed that it is possible to train mice to recognize the odor of commercial perfumes as signals. Female mice, reared by perfumed parents, in their adult life demonstrated a tendency to avoid normal males, preferring perfumed ones. Considering this type of evidence one can safely assume that in experiments in which the source of odor was not specified, the scent derived from skin glands would also be of importance.

For instance, in mice, according to Lane-Petter (1967), the characteristic mousy smell comes from preputial glands. Whitten (personal communication) noted this odor in urine drawn directly from the bladder. This does not exclude the possibility that the preputial glands may be the original source of scent—as in the boar, quoted above.

Godfrey (1958) investigated the olfactory mechanism of racial isolation in bank voles (*Clethrionomys* spp.) using the general body odors of female voles carried by an air current as the stimuli for a male in a two-point choice situation. Moore (1965), whose experiments were aimed at demonstrating olfactory discrimination as an isolating mechanism in *Peromyscus* spp., used as the criterion the attractiveness of cages in which different mice were previously kept. Carr and Caul (1962) and Carr et al. (1965), when testing the response of rats to sex odors, also used cages previously occupied by different groups of animals as the olfactory stimuli. Le Magnen (1952) tested responses of the white rat to sexual odors by confronting them with the whole body smell of other individuals in specially designed "discrimination boxes." De Maude (1940) demonstrated the effect of olfactory cues on the maze-learning of rats by using the linings from the animal's own cage.

Individual Appraisal

There is no doubt that many olfactory messages may be passed during an actual encounter between one animal and another. Among the first would be those concerning species and individual recognition.

Schloeth (1956) studied 814 encounters in 214 species of animals and reported that during these encounters olfactory cues were of the utmost importance. The sites of odor production as well as the places of odor deposition are examined regularly by animals meeting one another. Naso-nasal, naso-genital, naso-anal, and naso-glandular examinations take place most commonly.

Müller-Schwarze (1967a), working with mule deer, reported that when a new male was introduced into a group of two females and a male, the tarsal glands of both males were sniffed at by females 21 times in 205 minutes. When another male was introduced to the above four individuals, checking occurred 19 times in 23 minutes. However, the newcomer never checked any individual's tarsal glands. In undisturbed deer groups the tarsal gland of an individual is examined on an average of once an hour.

Results obtained during observations on the role of reproductive pheromones in mice clearly indicate that there are not only distinctions between the odors of different strains of mice, but that differences also exist within the same strain which allow female mice to recoginze individual males (Bruce, 1965).

Bowers and Alexander (1967) specifically investigated the ability of mice to recognize individuals by olfactory cues. Female mice discriminated between two male mice of the same inbred strain on the basis of olfactory cues. Mice could also discriminate by olfactory cues between two different species, *Mus musculus* and *Peromyscus maniculatus*, and between males and females.

Reports of casual observations on the use of scent glands in individual recognition are numerous, and evidence of this has also been gathered under experimental conditions for the sugar glider (Schultze-Westrum, 1965) and mule deer (Müller-Schwarze, 1967a).

Kulzer (1961) demonstrated experimentally the ability of female rousette bats, *Rousettus aegyptiacus*, to recognize their progeny by smell.

The ability of nestling rabbits to recognize their mother's odor has also been tested experimentally. Young rabbits were used at the age when they were still blind and deaf. They were confronted with swabs saturated with the anal or inguinal gland secretions of their mothers, strange adult females and males, and strange juveniles. Their reactions, in the form of avoidance or attraction, were measured. It was found that attraction demonstrated by following the swabs was commonly shown to the mother's anal secretion, while avoidance of this secretion was seldom recorded. Avoidance, often accompanied by facial distortion (grimacing), was, on the other hand, the most characteristic reaction to anal gland odors of strange rabbits. The differentiation between the mother's and strange animals' inguinal secretions was not so strong, thus suggesting a different function for the odors of the two glands (Mykytowycz, unpubl.).

Group Appraisal

The behavior of many species of insects and vertebrates shows their ability to differentiate between members of different social groups. The role played by olfaction in this recognition is generally accepted.

Apart from inferences made from the pregnancy-block work (see Messages Related to Reproduction, p. 341), there is other evidence to show that different strains of mice have distinctive odors. Thus in trapping mice from wild populations, Nichols (1944) has noted that apart from variations in some external features, mice from separate colonies can be distinguished from one another by differences in odor.

The role of smell as the factor commonly permitting the identification of social groups of house mice and brown rats is emphasized by Lorenz (1966), who refers to the original observations by Eibl-Eibesfeldt and Steiniger.

Parkes (1963) points out the differences in odors between strains of mice, quoting that professional perfumers are able to separate by smell the clothes on which different strains of mice have been kept for a few hours. The work of Bowers and Alexander (1967) on mice quoted earlier is also relevant to this problem.

The odor of a strange group is recognized and stimulates aggression in the sugar glider (Schultze-Westrum, 1965). In the course of the study of aggression in wild rabbits, females harrassed kittens which were members of the same colony but killed those from strange groups.

Age Appraisal

General observation leaves little doubt that there is age discrimination by olfactory means in many species of animals. This is most important in the operation of the inhibiting mechanism preventing adults of some species from seriously attacking young individuals. The behavior of dogs is an example (Lorenz, 1966).

Experimental evidence for the European wild rabbit has been obtained (Hesterman and Mykytowycz, 1968) to show that the skin glands may be used in age discrimination. Swabs of the secretions from the anal glands of rabbits of different ages were compared for intensity of odor by a panel of human assessors. Using statistical scaling techinques it was shown that the odor intensity of the swabs was directly related to the age of the rabbits from which they were obtained, the odor of adult individuals being the strongest.

Social Status Appraisal

Circumstantial evidence collected during the course of observations on many species of animals suggests that odor may assist in the maintenance of social status.

From observations on the sugar glider, Schultze-Westrum (1965) believes that scent alone could be responsible for the acquisition of social rank.

Examination of anal, inguinal, and submandibular glands of the wild rabbit in Australia revealed a statistically significant correlation between social status of individuals and the weights of their glands (Mykytowycz and Dudzinski, 1966). Histologic examinations also showed that similar correlations existed between the secretory activities of the glands and the social ranking of the animals (Mykytowycz, 1965, 1966a, 1966b). The skin glands are larger and more active in animals of high social rank which are more frequently involved in communication. Christian et al. (1965), in describing the results of a number of stress experiments involving mice, refer to the larger preputial glands in dominant individuals. It remains to be investigated if more frequent use alone determines the size of the glands. Most likely the hormonal predisposition is responsible for their size since the higher social status in the wild rabbit seems to be also correlated with higher sexual activity. (See below, The effect of Injection of Sex Hormones.) Observations that the weights of preputial glands of both brown and albino mice decrease approximately linearly as the population increases, supports the view of endocrine control of skin glands (Christian et al., 1965).

Observations under entirely natural and experimental conditions, indicated that dominant individuals applied secretions from the submandibular glands more frequently than subordinates. Indeed, the frequency of "chinning" under standardized conditions may indicate the relative ranking of two male rabbits prior to their actual physical contact (Mykytowycz, 1965).

Similarly, the frequency of the deposition of fecal pellets among animals exposed to situations stimulating marking, was the highest for adult dominant bucks (Mykytowycz and Hesterman, 1968).

Observations on the higher frequency of marking by sugar gliders of dominant status were made by Schultze-Westrum (1965). Subordinate individuals which marked infrequently assumed the marking behavior of the dominant animals after the latter were removed or died.

The analysis of the weights of olfactory bulbs of wild rabbits did not indicate any positive correlation with the size of the anal or inguinal glands, nor with social status of the animal. However, a statistically significant negative correlation was found in the female material. The largest olfactory bulbs occurred in the socially lowest individuals and the smallest ones in dominant does (Mykytowycz, 1968). It would be interesting to know whether this phenomenon is general or occurred just in the experimental population studied. This stimulates speculation on the possible connection between the acuity of the sense of smell and intrauterine mortality in rabbits. Intrauterine mortality is a constant phenomenon within all free-living rabbit populations and its incidence is much higher in subordinate females than in dominant ones.

Messages related to Reproduction

An extreme example of the importance of olfaction in reproduction is seen in the pregnancy-block phenomenon in mice (Parks and Bruce, 1961; Benjamin et al., 1965; Whitten, 1966; Whitten and Bronson, Chap. 10 of this volume). From work in this field it has been established that olfactory stimuli in mammals may produce effects on the estrous cycle and pregnancy by neurohumoral pathways acting slowly through the anterior pituitary body as is the case with light or other exteroceptive factors (Parkes, 1963).

The role played by olfaction in precopulatory behavior also attracted the attention and has been tested in a number of mammals including the rat, guinea pig, and pig. After experimental ablation of the olfactory bulbs, animals of both sexes carried out normal reproductive behavior, thus showing that olfactory stimuli may not be essential to assure successful copulation (Carr and Caul, 1962; Donavan and Kopriva, 1965; Sink, 1967). These observations suggest that a function as important as mating may not depend on a single sense but rather may be initiated by any one of a number of stimuli. For example, in swine the "mating stance," a characteristic posture of the female in estrus, normally released by pressure on the back, can also be induced by olfactory stimuli (Signoret and du Mesnil du Buisson, 1961).

Even in species in which olfaction is not essential for mating, there is ample evidence to show that under natural conditions it is of importance as a factor regulating the selection of sexual partners and so may affect such vital population phenomena as speciation or the prevention of inbreeding.

In studies related to the pregnancy-block phenomenon it has been demonstrated that reproductive pheromones are contained in the urine of mice (Dominic, 1964; Marsden and Bronson, 1964). It is, however, reasonable to assume, and indeed apparent from the behavior of the animals, that the role of the odoriferous glands may also be important in reproduction. The importance of skin glands in the reproduction of rodents has been emphasized by, among others, Mitchell (1967), working with the ventral gland of the Mongolian gerbil, and by Stanley and Powell (1941) with regard to the preputial gland in the white rat.

Evidence for the involvement of skin glands in reproduction is provided by characteristic behavior patterns involving smelling; licking and secretion from skin glands; sexual dimorphism in skin gland size; the effects of castration and of the injection of sex hormones on skin glands; and seasonal changes in skin glands.

CHARACTERISTIC PATTERN OF BEHAVIOR. Observation on the golden hamster indicates that the lateral glands of the males are as a rule examined by estrous females and the odor obviously excites them sexually, stimulating secretion from their clitoral glands. The secretion from clitoral glands, but not from the lateral glands, affects the sexual behavior of males examining females. Experimental removal of the lateral glands from males delays or actually prevents copulation (Lipkow, 1954).

In the golden hamster, Lipkow (as well as other authors writing about other animals) observed that there is not uniform excitement of receptive females to the smell of a particular male. Some females avoid contact with one male while accepting another immediately. This is also true of male preferences for females. The above observations provide direct evidence of the importance of skin glands in the selection of partners.

Buechner and Schloeth (1965), in describing the mating behavior of the Uganda kob, paid special attention to the behavior of males in relation to the female's inguinal glands. In two series of observations involving 48 and 100 separate copulations, nuzzling of the glands by the male occurred on 63 and 68 percent of occasions, respectively. There are few reports like the one quoted above which give quantitative data on the involvement of skin glands which would allow for a clear comparison of their role in relation to other components of reproductive behavior.

SEXUAL DIMORPHISM. Sexual dimorphism of skin glands has been observed in many species of mammals but only in some, including the rat (Freinkel, 1963), has it been studied specifically. In the European wild rabbit the weight and secretory activity of the anal, inguinal, submandibular, and orbital glands have been measured in animals of known ages. It has been shown that sexual dimorphism of the glands develops at the age when the animals attain sexual maturity (Mykytowycz, 1965, 1966a, 1966b). Glands were larger and more active in adult males than in adult females. There was no sexual dimorphism in the case of the anal gland of the European hare *Lepus europaeus* and the size of the inguinal glands was greater in adult females. These and other observations indicate that the same skin glands may not be of equal importance to both sexes nor of equal importance to different species.

Sexual differences have also often been found in the odor from skin glands. Thus, the odor of the houses of male wood rats, *Neotoma fuscipes*, is reported to be stronger than that of adult females, especially during the breeding season (Vestal, 1938). Herreid (1960) has referred to the stronger smell of male Mexican free-tailed bats, *Tadarida brasiliensis mexicana*, and our own observations indicate that the secretions from the anal and inguinal glands of male wild rabbits smell more strongly than those of females (Hesterman and Mykytowycz, 1968). It may be worthwhile emphasizing that the recognition of sex is of importance not only directly for reproduction but also in other aspects of intersexual behavior. For instance in many species of mammals, males have an *absolute inhibition* against fighting females (Lorenz, 1966).

EFFECT OF CASTRATION.   The effect of castration provides further evidence that skin glands are associated with reproductive activity. Experiments were carried out with many species of mammals including the possum (Bolliger, 1944b; Bolliger and Hardy, 1944), rabbit (Coujard, 1947; Mykytowycz, 1965, 1966a, 1966b), rat (Beaver, 1960), Mongolian gerbil (Mitchell, 1965), musk shrew (Dryden and Conaway, 1967), short-tailed shrew (Pearson, 1946), dromedary (Mimram, 1962), hamster (Hamilton and Montagna, 1950), and guinea pig (Martan, 1962).

In all these studies, changes in the skin glands of males castrated before and after puberty were obvious. Hypoplasia, hypotrophy, decrease of secretion, as well as reduction of odor and rate of application of the secretion of the glands were recorded. Not all of the studies mentioned above involved the castration of females, but in those that did the changes recorded were not so uniform as in the males.

While in some species, e.g., musk shrew, positive regression of the glands in castrated females is reported (Dryden and Conaway, 1967), in others the reverse occurs. For example, in the rabbit, stimulation of the secretion and of gland growth in castrated females has been seen (Mykytowycz, 1965, 1966a, 1966b).

THE EFFECT OF INJECTION OF SEX HORMONES.   As in castration experiments, the control of skin glands by androgens has been clearly established in all species examined. Injections of male hormone preparations produce hypertrophy, hyperplasia, enhance secretion, and increase the rate of application of the secretion. Thiessen et al. (1968) have pointed out the suitability of skin gland measurements as indices of androgen level in the Mongolian gerbil. Observations, collected on preputial glands of the rat (Montagna and Noback, 1946) and on wild rabbit glands, lend support to this idea. Sink (1967) suggested a relationship between androstenals and sex pheromones—including secretion from preputial glands in swine. The stimulating effects of estrogens were not as general. For instance, Coujard (1947) reported the stimulating effect of folliculin on the anal and inguinal glands of the rabbit, and Dryden and Conaway (1967) found a similar effect of estradiol on the apocrine glands of the musk shrew. However, Beaver (1960) could not demonstrate the stimulation of preputial glands of the rat by estrogens, and Pearson (1946) actually emphasized the *inhibiting* effect of estrogens on the lateral glands of the short-tailed shrew, referring also to similar reports for the rat and hamster. The inhibitory effect of estrogen on sebaceous glands has also been reported by Ebling (1963).

The effect of estrogen has been studied particularly with respect to sebaceous glands. The general findings have been summarized by Strauss and Pochi (1963), who discussed the effect of hormones on the sebaceous glands in man. The action of estrogen on the sebaceous glands is complex, and indications of individual variations also exist. The mechanisms of the action of estrogen remain obscure and cannot be explained at present. There is no evidence from either experimental animals or man which would indicate that estrogen in physiologic amounts could be responsible for increased sebum production in females.

Further research is needed, and attention should be centered on the possibility that different skin glands may function for different purposes in animals of different sexes and species. Hence, they may be affected by the same hormone to varying degrees. Thus, the administration of estrogen to castrated female brush-tailed possums seems to stimulate secretions from the skin glands of the pouch area while the sternal glands remain unaffected (Bolliger, 1944b).

SEASONAL CHANGES.   Seasonal changes in the secretory activity of the skin glands as measured by various methods have been reported in many species.

Thus Quay (1953) described variations in glandular activity in five species of *Dipodomys*, and analyzed the functional significance of these variations with reference to available information on the life histories and habits of the species involved. He used as criteria the area of the dorsal glands and the percentage of occurrence of the secretion. Different patterns of glandular activity were found in different species. Seasonal and sexual differences were present in some species but not in others. Certain phases of activity could be associated with the breeding season or annual moult in some species but not in others. The fact that some species which were sampled had broadly overlapping ranges and habitats suggests that physiologic rather than environmental events were responsible for the variations. The behavioral patterns associated with the glandular secretions are broadly different in at least some of the species since glandular activity is subject to great seasonal and sexual modifications.

Linsdale and Tevis (1951), while studying the seasonal changes in the activity of the ventral glands of the wood rat, used the presence or absence of secretion and coloration of the fur overlying the gland as an indication of glandular activity. They reported that the highest activity coincided with the breeding season and the time when territorial marking is most pronounced because of the importance of shelter. In June, a nonbreeding month, none of the adult males nor the females showed signs of secretion from the ventral gland. There were also great individual contrasts in the duration of glandular activity.

The weights of the odoriferous glands from approximately 1,300 European wild rabbits from five different geographic regions of Australia have been determined (Mykytowycz, 1966a, 1966b, 1966c). The results indicated that fluctuations which occurred were correlated with the breeding season. Histologic examination of material collected from 180 animals from one locality over a period of two years also showed higher secretory activity associated with the breeding season.

In another study (Hesterman and Mykytowycz, 1968) the glands from animals killed in the field were used to prepare aqueous dilution series. The dilutions at which the odor from the various samples could just be detected by a panel of human judges gave a measurement of the intensity of the odor produced by the animals at different seasons. For the anal gland, the highest odor intensity indices were obtained from both sexes during the breeding season.

Territorial marking

The existence of territoriality was first shown in birds but has subsequently been demonstrated in free-living mammals.

This is not the place to discuss the concept, precise meaning, and misconceptions of the terms "territoriality" and "home range"; this has been done on many occasions and recently by Schenkel (1966) and Jewell (1966). It is true, as Schenkel suggests, that in many publications, the territorial function of skin glands has been accepted prior to establishing evidence that the species in question really possesses "territory"— an area in which an individual, a pair, or a group does not tolerate the presence of another conspecific. The word "territoriality" in most publications dealing with scent glands has been applied rather to home ranges—areas within which the animal moves.

Recently, Kleinman (1966), in discussing the scent marking in Canidae, gives a definition of marking which is also shared by many other authors. In her view, scent marking serves to maintain the animal's familiarity with its environment. Odor is added to specific visual landmarks both to familiarize the animal with new territory and to refamiliarize it with old terrain. The term "identification" rather than "familiarization" has been used by others.

Marking, in my view, seems to become territorial by virtue of the behavioral characteristics of the species. In territorial species the marking activity of an individual is often correlated with its social rank. Under these circumstances, it is not difficult to imagine that the presence of the odor will show not only an animal's presence but also its level of influence and readiness to defend the marked area.

Geist (1964), referring to the behavior of the mountain goat, *Oreamnos americanus*, suggests that the marking of a territory acts as an extension of the animal. If one assumes that the scent of a conspecific may intimidate, a stranger entering the area marked is at a disadvantage prior to meeting the territorial owner. The importance of territory and the role of boundary marking within it have been discussed very frequently in relation to population dynamics (Wynne-Edwards, 1962) and most recently by Lockie (1966) with special reference to weasels, *Mustella nivalis*, and stoats, *M. erminea*.

Reports referring to the territorial function of skin glands are very numerous. In practically all cases where other functions were not apparent, this role has been suggested and justified, because in the sense of identification-familiarization with a given area, any scent derived from the animal's own body would be of importance. Noticeably lacking from such reports are data on the application of scent marks in relation to the territory, to the individuals involved, and on the reactions of other animals to the marking sites.

The existence of permanent, special marking sites within an animal's territory is generally known. For instance, there are "signposts" on which the European red deer, *Cervus elaphus*, and other Cervidae deposit secretion from their antorbital glands (Graf, 1956); special feces-deposition sites in many species of animals (Hediger, 1944, 1949; Wynne-Edwards, 1962; Tarasov, 1960).

The non-random distribution of rabbit dung-hills and their concentration in relation to the presence of other rabbit colonies which emphasize their role as a a territorial marker, has been referred to eariler (see section dealing with the application of scent gland secretions). The warning effects of strange dung-hills have been seen in nature and have also been tested under experimental laboratory conditions.

Indirect evidence of the involvement of skin glands in territorial marking comes also from results of studies which showed the larger size and more intense secretion of the glands in individuals most concerned with the defense and maintenance of territories.

Warning

In discussions of forms of expression in mammals, frequent mention is made of the use of odors, including those produced by skin glands, as warning signals by many species. It has been pointed out that marking generally can be regarded as a warning (Eibl-Eibesfeldt, 1957). Dogs seem to be able to control voluntarily the flow of secretions from the anal sacs. This happens when they are frightened. Avoidance has been observed in dogs experimentally confronted with the odor of anal-sac secretion from another dog (Donovan, 1967).

Apart from the type of warning aimed at repelling an antagonist, animals use signals aimed at raising alarm among conspecifics or members of other species; this form of behavior, useful for survival of the species, may not always be beneficial to the individual eliciting it.

The use of the tarsal glands by mule deer in relation to warning and alarm has been observed by Nichol (1938). He found that does discharge the scent to warn and attract fawns, and all deer do it when dogs are nearby or when other unusual disturbances take place. Deer lost from corrals and trying to return, signal to the herd by this means and the penned animals respond when the scent reaches them. It is always the most "nervous" animals which emit the odor first.

Aggression

The relation between olfactory stimulation and aggressive behavior has been studied in mice by Ropartz (1968). His results suggest that olfactory stimuli are involved in the release of aggressive behavior in males and that the olfactory system includes a mechanism, apparently involving the olfactory bulbs, necessary to elicit fighting. However, the extent to which the odor from preputial glands was involved in these observations is not known.

Linsdale and Tomich (1953) indicate the use of scent from the tarsal glands by mule deer as a threat, particularly by bucks in rut. They also describe the appearance of the tarsal glands in frightened and alerted animals. The tarsal gland tufts are flared into rosettes. These are closed after a few steps and then again spread to expose the slit-like opening, thus increasing the exposure of the basal parts to the air.

Geist (1964) emphasized the significance of scent in the agonistic displays of mountain goats. The supra-occipital gland is used for marking purposes by these animals. Marking is most conspicuous during agonistic encounters. He also emphasizes the significance of scent in agonistic displays of the musk-ox and black-tailed deer.

Marking is a regluar component of aggressive behavior in the golden hamster (Dieterlen, 1959), guinea pig, and other rodents (Kunkel and Kunkel, 1964). It is also regularly observed as part of the agonistic behavior of the European wild rabbit. Males engaged in fighting, during frequent interruptions in the actual physical combat, start digging and subsequently mark the freshly turned earth or other objects within their reach with the submandibular gland secretion. The deposition of fecal pellets has also frequently been observed during pauses in the fighting, but no quantitative measurements have been taken. The lifting up of the hind quarters and frequent flagging of the tail also suggest that the discharge of odor from the inguinal and anal glands may take place.

The discharge of odor from scent glands in fright is found in a wide variety of species including the black-tailed prairie dog, *Cynomys ludovicianus* (King, 1955). It is, however, among the civets (Viverridae) and skunks (Mustelidae) that the role of the skin glands reaches the highest degree of specialization for the purpose of defense (Blackman, 1911; Bourlière, 1954; Dücker, 1965; Wynne-Edwards, 1962; Goethe, 1964). Species of the genera Viverra, Viverricula, and Herpestes, such as the striped-necked mongoose and the crab-eating mongoose, empty their perianal glands when danger threatens.

The tigrine genet, *Genetta tigrina*, also secretes a malodorous, yellowish secretion from its anal glands when excited (Dücker, 1965).

According to Goethe (1964), the odor from the pregenital gland is used by mustelids for marking, while the supra-anal organ produces a secretion used in defense.

It is not clear from the available information whether the venom-producing crural gland of the platypus is also odoriferous but it is believed that this gland's chief function is in combat between males for territory and females (Calaby, 1968). Echidnas (family Tachyglossodae) also have a venom apparatus similar to that of the platypus. It is often difficult to decide which pattern of behavior and which olfactory signal should be accepted as an indication of warning, fright, or aggression.

COMPARATIVE STUDIES OF SKIN GLANDS

Comparative studies of skin glands were carried out by various workers. Recently, Montagna (1963) reported on the comparative aspects of sebaceous glands in mammals —in particular their histology, histochemistry, and problems of growth and differentiation.

Major contributions to comparative studies of skin glands have been made by Quay, who in a series of papers has reported on the dorsal glands in six species of Dipodomys (1954); the glands of the angulus oris in 79 species of rodents (1962, 1965b); the caudal glands of 15 species of pocket mice (1965a); and the integumentary modifications in over 20 species of North American desert rodents belonging to five families (1965c).

Despite the awareness of the possible behavioral functioning of all these glands, it was not possible to derive any conclusions related to behavior from the results obtained in all studies mentioned above, due to the lack of recorded observations.

The aim of the studies quoted above, and of others to which reference has not been made, can be expressed by the ideas of Kratochvil (1960). In referring to the results of a study of the preputial glands in nine species of Microtids he suggested that the skin glands are less exposed to environmental influences than are external body markings or skeletal characteristics. Due to this, glandular characteristics are a good tool in the taxonomy, particularly of Microtidae, among which it is often impossible to determine which characteristics are phylogenetically older and which have been induced by adaptation. He maintains that the structure of the glands reflects the physiological relationship of different species and their behavioral similarities.

Better grounds for speculation on the connection between the development of the skin glands and the behavior in related species were provided by comparative studies of the skin glands of the European wild rabbit and the European hare. The anal glands, which it is suggested function for territorial marking, are only one-tenth the size in adult male hares as compared with rabbits, despite the fact that the body size of hares is more than twice that of rabbits (Mykytowycz, 1966a). In the hare, sexual dimorphism in the anal gland is not obvious, as in the rabbit, nor is the increase in the size of the gland with increasing age as marked. Differences exist between the two species also in relation to the chin glands where the glands from male rabbits are approximately four times larger than those of male hares (Mykytowycz, 1965).

The territorial behavior of the two species is different. The gregarious rabbit possesses a small home range and a sharply defined territory with which it must constantly identify itself. This it does using the secretions from the skin glands.

The density of populations of the solitary European hare never reaches the level at which acute competition for space would occur. Hares occupy large home ranges. The demarcation of territory with scent under these circumstances is not so important and would also be physically difficult to execute. Maintenance of the appropriate distance between individual animals probably involves other communication methods.

In the case of the inguinal glands, a different relationship is found between the two species. Unlike the chin and anal glands, the inguinal glands are larger in hares than in rabbits. In rabbits, the inguinal glands are larger in males than in females, while in hares the contrary applies (Mykytowycz, 1966b).

Information on the reproductive behavior of hares suggests that olfactory cues play an important role in the finding of an estrous female by a male. It seems that in its search for an estrous female the male hare is guided by the scent of secretion from her inguinal glands which, unlike the rabbit's, cover the inside of the thighs with a thin yellowish oily layer.

Further indication of the reflection of the behavioral functioning of the skin glands in different species of animals emerged during a preliminary examination of the swamp rabbit, *Sylvilagus aquaticus*, and the cottontail rabbit, *Sylvilagus floridanus*. A comparative study of the behavior of these two North American lagomorphs, carried out by Marsden and Holler (1964), revealed that there was strong territorialism in the swamp rabbit while the cottontail was never observed to defend a permanent territory. In the swamp rabbit, "chinning" was observed while the cottontails were seen sometimes to rub twigs or vegetation with the corners of their eyes. Examinations revealed that the chin gland is well developed in the swamp rabbit but not in the cottontail (Mykytowycz, unpublished). It would be very interesting to study the skin glands in all American species of lagomorphs in relation to their behavior.

SKIN GLANDS IN PRIMATES

With regard to the role of olfaction in man's behavior, the problem has not been studied seriously as yet, although there are indications that it may be of more importance than is generally assumed and may vary between cultural groups. Sporadic observations are available which show the involvement of skin glands. For instance, Schultze-Westrum (1968) has drawn attention to the behavior of members of the Kanum-irebe tribe of southern New Guinea, which was originally observed by H. Nevermann in 1934. To demonstrate friendship for a departing visitor, the host wipes sweat from the visitor's armpit with his fingers, smells them and then rubs the fingers on his own chest. In connection with this observation, one has to think about the origin of the London Cockney's saying "he gets up my nose," and similar expressions in other European countries. Odor from skin glands is obviously also important in sex attraction or repulsion. The role of skin glands in odor production is well documented and well recognized at least by cosmeticians,* and the part played by bacterial decomposition in this process has been mentioned above.

Moncrieff (1966), using the subjective judgments of a panel of human judges of different ages and sex, recently carried out a study of preferences for a number of odors of non-human derivation. Interesting results could be obtained by using similar methods to study odors produced by human skin glands.

There seems to be a general shift away from olfaction as an important means of communication in higher primates. The anatomic structure of the olfactory organs alone supports this view. However, despite this there are indications that monkeys and apes may make more use of olfaction than is generally believed. Thus, it appears that odor can aid in individual recognition, especially between mother and child, and may indicate the presence of estrus in females. It has been suggested that the strong odor of semen in the rhesus monkey may also serve as a sexual attractant (Marler, 1965). The role of pheromones in behavior of primates has been reviewed recently by Michael and Keverne (1968). They demonstrated also that male rhesus monkeys become aware of sexual receptivity of females through olfactory stimuli derived apparently from vaginal glands.

Prosimians seem to rely more on olfaction than the other primates. There are reports on the occurrence of skin glands and on behavior patterns in some prosimians which leave little doubt as to the functional significance of their secretions. Thus, Hill (1953), in the first volume of his monograph on primates, states that sexual attraction in the lemur is affected by the secretion of specialized glands in the perineal region. Members of the genus repeatedly rub their perineal glandular skin upon each other and upon the branches of trees and other objects. This behavior is not confined to any particular season and may therefore have more than just a sexual significance. It probably serves to keep members of a troop or family together.

Petter (1962) and Jolly (1966) have reported that certain species of lemurs scent-mark their territory. For instance, male *Propittecus* have a scent gland on their throat the secretion from which when rubbed onto branches attracts the attention of other individuals.

The brachial glands of the ring-tailed lemur, *Lemur catta,* and the gentle lemur, *Hapalemur griseus,* were described by Montagna (1962, 1963). Evans and Goy (1968) deal at length with the function of the cutaneous glands—antebrachial, scrotal, shoulder—in the behavior of *L. catta.* Epigastric glands were recorded from

---

* Goodhart (1960) in discussing the significance of human hair patterns suggested that the retention of pubic and axillary hair in man may be connected with the concentration of apocrine glands in these areas.

the tarsiers, *Tarsius carbonerius* and *T. borneanus* by Hill (1951). Poglaven-Neuwall (1962, 1966), referring to the mandibular, throat, and abdominal glands of the kinkajou, *Potos flavus*, suggested their importance in range and trail marking and sex identification. Sprankel (1962) described marking with the secretion from the gular gland in the tree shrew, *Tupaia glis*, and Martin (1968) refers also to the abdominal gland, suggesting that both of these, in combination with urine and faeces and possibly saliva, function for general marking of the territory. However, Martin questions the taxonomic status of the genus.

Specialized glandular skin areas have also been reported in some simian primates. Thus, sternal glands occur in the drill, *Mandrillus leucophoeus* (Hill, 1944), the spider monkey, *Ateles geoffroyi* (Wislocki and Schultz, 1925) and the orangutan, *Pongo pygmaeus* (Schultz, 1921; Wislocki and Schultz, 1925). Circumanal, perineal, and pubic glands and glands in the labia majora have been reported in three species of marmosets by Wislocki (1930).

Sporadic information on the occurrence of skin glands in other primates and speculations on their role are scattered through Hill's monograph on the anatomy and taxonomy of primates.

CHEMICAL STUDIES OF SKIN GLANDS

Information on the chemistry of skin gland secretions has come mainly from industrial chemists who were interested in their commercial uses in perfumery rather than in demonstrating their communicative power. Products of animal skin glands are used in perfumery as fixatives, the addition of a small amount of them improving the quality of scents.

Relatively few animal products are used in perfumes. The most common ones—apart from ambergris which is formed from undigested remnants of squids and octopuses in the intestines of the sperm whale, *Physeter catodon*—are civet, castoreum and musk (Moncrieff, 1951). Civet is obtained mainly from an Abyssinian species of civet cat, *Viverra civetta*. The secretion from males kept in captivity is collected periodically by scraping the perineal pouches.

Castoreum is a product of the large anal glands of both sexes of the Siberian beaver, *Castor fiber*, and the Canadian beaver, *Castor canadensis*. Beavers are believed to mark their territory with the secretion of these glands.

The best known of the odorous animal secretions is that obtained from the male musk-deer, *Moschus moschiferus*, which occurs mainly in the Himalayas of South China. The musk-producing gland is located on the abdomen between the umbilicus and the preputium. The gland secretion is intensified during the breeding season which suggests that it is important in reproductive behavior.

Musk is considered to have a hormonal-like activity in stimulating male sexual behavior. Moncrieff (1951), quoting Sano (1938), reports that application of musk to the comb of a capon stimulated the growth of the comb. In North America, the muskrat, *Fiber zibethicus*, of Louisiana is harvested for the secretion of its preputial gland.

Kingston (1965) while discussing present knowledge of the chemistry of scent glands, emphasizes that information in this field is very scanty.

Studies have been confined mainly to those of the civet cat, beaver, and musk deer, as well as some related species, although Hardy (1949) has described the chemical composition of the odor secretions from 20 species of mammals.

The chemical components of castoreum are remarkable in their variety and nature. The presence of more than 40 compounds has been established. These compounds are believed to be derived partly from the diet and partly from breakdown products normally found in the urine of the vertebrates (Lederer, 1949). To determine their importance in animal behavior, each of them should be tested separately for its communicative power; this is not a simple task if one considers the highly complicated pattern of behavior in mammals.

More recently, ethologists became interested in the chemistry of skin gland secretions. Thus, Müller and Lemperle (1964) produced a brief communication in which they demonstrated the usefulness of gas-liquid chromatography in the analysis of skin gland secretions from five species of mammals (chamois, red deer, roe deer, dog, and fox). This tentative report indicates the presence of at least 12 components in the secretion of the antorbital gland of the male red deer. Differences in the chromatograms of the anal secretions from male and female Alsatian dogs were found. Several components present in the secretions from the male glands were missing from the female secretions.

Müller-Schwarze (1967a, 1967b, 1969) extended this work onto black-tailed deer. The odorous components from several hundred tarsal glands were extracted with petrol-ether, distilled and separated by means of gas-liquid chromatography. In applying the fractions experimentally to the tarsal tufts of animals kept in captivity, it was possible to demonstrate that the various fractions affect the behavior of the animals to a different degree. Approach licking, sniffing, following, tongue-protrusion, and so forth were used as criteria. It was shown that individual recognition by different tarsal odors is possible. Experiments with different concentrations revealed that two micrograms of the major component of a male tarsal gland extract in 0.25 ml of petroleum ether provided the most effective stimulus. Brownlee et al. (1969) isolated the major component of the male tarsal scent by preparative gas-liquid chromatography and identified it as cis-4-hydroxydodec-6-enoic acid lactone by spectro- and chromatographic methods.

Unsaturated γ-lactone, containing twelve carbon atoms, occurring naturally in butterfat, and synthesized material release the same behavioral responses as the deer lactone. While γ-lactones with ten, eleven, and twelve carbon atoms containing saturated side-chains do not possess the same stimulating power, three related unsaturated δ-lactones with ten and twelve carbon atoms give rise to similar but not so frequent responses (Müller-Schwarze, 1969).

Chemical studies of the secretions from the chin, anal, and inguinal glands of the rabbit are in progress. Preliminary tests revealed that the basic compositions of these secretions vary. The inguinal gland secretion is composed predominantly of lipids, while the secretion from the chin gland is mainly proteinaceous with some carbohydrate components. The anal secretion is a mixture of lipids and proteins. (Goodrich, in preparation). This observation reflects the diversity of behavioral functions of the different skin glands in the same animal. The presence of carbohydrates in the chin gland secretions suggests the possibility that gustation is involved in communication through this particular organ. Separate chemical components of rabbit's skin gland secretions are currently being tested to determine their behavioral importance.

Conclusion

The wide occurrence of specialized skin glands of itself suggests their importance in the life of mammals. This has been realized generally in the past but numerous early studies of skin glands rarely went beyond the scope of merely recording their presence and suggesting possible functions. They did not provide a clear understanding of the role of the glands. The reason for this was the lack of information on the behavioral patterns of animals as well as the failure to collect systematically data on the function of the glands and the application of their secretions. Furthermore, the lack of techniques suitable for odor and olfaction studies inhibited interest in skin gland secretions.

Now the situation has changed. With progress in ethology it has been demonstrated that it is possible, at least in some species, to back the study of the function of skin glands with behavioral observations. New techniques suitable for the study of odors have also been developed, although these are still not always accessible to the nonspecialist. There is a clear need for specialists in various fields of biology—physiologists, especially electrophysiologists, biochemists, histologists, and others—to cooperate in studies of the function of skin glands.

In studying the communicative powers of skin gland secretions consideration must be given to the fact that a response to odor may not always be demonstrable by visible changes in behavior. This was emphasized recently by Wenzel (1965) with regard to olfactory perception in birds. Often these responses are on a physiologic level in the form of changes in breathing rate, heart beat, blood pressure, widening of the pupils or as long-term effects as is best illustrated by pregnancy-block in mice. During the experiments in which rabbit kittens were confronted with the odors of skin gland secretions, changes in the breathing rate and nostril movements were recorded. Their increase reflected the interest of kittens in the odor to which they were exposed (Mykytowycz and Ward, in press).

In the same way as the ability of individuals to produce a signal varies from species to species, from season to season, depending on sex, social status, age, density of population and other factors, so the perception of a signal by an individual animal may vary. Ottoson (1963), reviewing some aspects of the functioning of olfactory systems, referred to the changing acuity in odor detection in man and other mammals. The olfactory capacity varies greatly and it is possible that local changes in the nasal mucosa or changes in the central nervous system, are involved. This changing ability to perceive signals should be taken into consideration when studying the communicative powers of skin gland secretions.

On the information available it is apparent that our knowledge of the role of skin glands in mammalian communication is still very limited. One can hardly talk about recent advances, and much will have to be done before our knowledge in this field of mammalogy equals that of the entomologist's on the role of pheromones in the insect's life. The highly complicated pattern of behavior of mammals makes the task more difficult, but at least one can claim at this stage that there is a clear picture of the problems involved and a better understanding of the methods which could be employed to solve them.

ACKNOWLEDGMENTS

I wish to thank Mr. E. R. Hesterman of the Division of Wildlife Research, CSIRO, for his assistance in the preparation of the manuscript, and Dr. C. Mims of the Australian National University for helpful criticism.

REFERENCES

Andrew, R. J. 1964. The displays of the primates. *In* Buettner-Janusch, J., ed. Evolutionary and Genetic Biology of Primates. Vol. 2. New York, Academic Press, Inc. pp. 277-309.

Beaver, D. L. 1960. A re-evaluation of the rat preputial gland as a "dicrine" organ from the standpoint of its morphology, histochemistry, and physiology. *J. Exp. Zool.*, 143:153-173.

Benjamin, R. M., Halpren, B. P., Moulton, D. G., and Mozell, M. M. 1965. The chemical senses. *Ann. Rev. Psychol.*, 16:381-416.

Blackman, M. W. 1911. The anal glands of *Mephitis mephitica*. *Anat. Rec.*, 5:497-523.

Bolliger, A. 1944a. On the fluorescence of the skin and the hairs of *Trichosurus vulpecula*. *Aust. J. Sci.*, 7:35.

——— 1944b. The response of the sternal integument of *Trichosurus vulpecula* to castration and sex hormones. *J. Proc. Roy. Soc. N.S.W.*, 78:234-238.

——— and Hardy, M. H. 1944. The sternal integument of *Trichosurus vulpecula*. *J. Proc. Roy. Soc. N.S.W.*, 78:122-133.

Bourlière, F. 1954. The Natural History of Mammals. New York, Alfred A. Knopf, Inc.

Bowers, J. M., and Alexander, B. K. 1967. Mice: Individual recognition by olfactory cues. *Science*, 158:1208-1210.

Brownlee, R. G., Silverstein, R. G., Müller-Schwarze, D., and Singer, A. G. 1969. Isolation, identification and function of the chief component of the male tarsal scent in Black-tailed deer. *Nature* (London), 221:284-285.

Bruce, H. M. 1965. Olfactory pheromones and reproduction in mice. *In Proc. 2nd Int. Congr. Endocrin., London, 1964.* pp. 193-297.

Buechner, H. K., and Schloeth, R. 1965. Ceremonial mating behaviour in Uganda Kob (*Adenota kob thomasi* Neumann). *Z. Tierpsychol.*, 22:209-225.

Calaby, J. H. 1968. The platypus (*Ornithorhynchus anatinus*) and its venomous characteristics. In Bücherl, W., et al., eds., Venomous Animals and Their Venoms, Vol. 1. New York, Academic Press, Inc. pp. 15-29.

Carr, W. J., and Caul, W F. 1962. The effect of castration in the rat upon the discrimination of sex odors. *Anim. Behav.*, 10:20-27.

—— Loeb, L. S., and Dissinger, M. L. 1965. Responses of rats to sex odors. *J. Comp. Physiol. Psychol.*, 59:370-377.

Chitty, D. 1964. Animal numbers and behaviour. *In* Dymond, J. R., ed. Fish and Wildlife: A Memorial to W.J.K. Harkness. Don Mills, Ont., Longmans Canada Ltd., pp. 41-53.

Christian, J. J., and Davis, D. E. 1964. Endocrines, behaviour and population. *Science*, 146:1550-1560.

—— Lloyd, J. A., and Davis, D. E. 1965. The role of endocrines in the self-regulation of mammalian populations. *Recent Progr. Hormone Res.*, 21:501-578.

Corbet, G. B. 1966. The terrestrial mammals of western Europe. London, G. T. Foulis & Co. Ltd.

Coujard, R. 1947. Etude des glandes odorantes du lapin et de leur influencement par les hormones sexuelles. *Revue Canad. Biol.*, 6:3-14.

De Maud, J. W. 1940. The effects of olfactory cues on the maze learning of white rats. *Trans. Kans. Acad. Sci.*, 43:337-338.

Dieterlen, F. 1959. Das Verhalten des syrischen Goldhamsters (*Mesocricetus auratus* Waterhouse). *Z. Tierpsychol.*, 16:47-103.

Dobroruka, L. J. 1960. Einige Beobachtungen an Ameisenigeln, *Echidna aculeata* Shaw (1792). *Z. Tierpsychol.*, 17:178-181.

Dominic, C. J. 1964. Source of the male odour causing pregnancy-block in mice. *J. Reprod. Fert.*, 8:266-267.

Donovan, B. T., and Kopriva, P. C. 1965. Effect of removal or stimulation of the olfactory bulbs on the oestrous cycle of the guinea pig. *Endocrinology*, 77:213-217.

Donovan, C. A. 1967. Some clinical observations on sexual attraction and deterrence in dogs and cattle. *Practitioner's Notebook*, 62:1047-1048.

Dryden, G. L., and Conaway, C. H. 1967. The origin and hormonal control of scent production in *Suncus murinus*. *J. Mammalogy*, 48:420-428.

Dücker, G. 1965. Das Verhalten der Viverriden. *Handb. Zool.*, 8:38, 10(20a): 1-48.

Dutt, R. H., Simpson, E. C., Christian, J. C., and Bornhorst, C. E. 1959. Identification of preputial glands as the site of production of sexual odour in the boar. *J. Anim. Sci.*, 18:1557.

Eadie, W. R. 1954. Skin gland activity and pelage descriptions in moles. *J. Mammalogy*, 35:186-196.

Ebling, F. J. 1963. Hormonal control of sebaceous glands in experimental animals. *In* Montagna, W. et al., eds. Advances in Biology of Skin. Elmsford, N.Y. Pergamon Press, Inc., Vol. 4, pp. 200-219.

Eibl-Eibesfeldt, I. 1957. Ausdruckformen der Säugetiere. *Handb. d. Zool.*, 8:12,10 (6):1-26.

—— 1964. Das Duftmarkieren des Igeltanrec (*Echinops telfairi* Martin). *Z. Tierpsychol.*, 22:810-812.

Eisenberg, J. F. 1966. The social organizations of mammals. *Handb. d. Zool.*, 8:39, 10(7):1-92

Evans, C. S., and Goy, R. W. 1968. Social behavior and reproductive cycles in captive Ring-tailed Lemurs (*Lemur catta*). *J. Zool. (London)*, 156:181-187.

Fiedler, W. 1964. Die Haut der Säugetiere als Ausdrucksorgan. *Studium Gen.*, 17:362-390.

Frädrich, H. 1967. Das Verhalten der Schweine (Suidae Tayassuidae) and Flusspferde (Hippopotamidae). *Handb. d. Zool.*, 8:42, 10(26):1-44.

Frank, F. 1956. Das Duftmarkieren der Grossen Wühlmaus, *Arvicola terrestris* (L.). *Z. Säugetierk.*, 21:172-175.

Freinkel, R. K. 1963. The effect of age and sex on the metabolism of the preputial gland of the rat. *In* Montagna, W., et al., eds. Advances in Biology of Skin. Elmsford, N. Y., Pergamon Press, Inc., Vol. 4, pp. 125-134.

Geist, V. 1964. On the rutting behavior of the mountain goat. *J. Mammalogy*, 45:551-568.

Godfrey, J. 1958. The origin of sexual isolation between bank voles. *Proc. Roy. Phys. Soc. Edinburgh*, 27:47-55.

Goethe, F. 1964. Das Verhalten der Musteliden. *Handb. d. Zool.*, 8:37, 10(19):1-80.

Goodhart, C. B. 1960. The evolutionary significance of human hair patterns and skin colouring. *Adv. Sci.*, 17:53-59.

Graf, W. 1956. Territorialism in deer. *J. Mammalogy*, 37:165-170.

Hamilton, J. B., and Montagna, W., 1950. The sebaceous glands of the hamster. I. Morphological effects of androgens on integumentary structures. *Amer. J. Anat.*, 86:191-234.

Hardy, E. 1949. Lesser known sources of musk *Perf. Essent. Oil Rec.*, 40:93.

Harvey, E. B., and Rosenberg, L. E. 1960. An apocrine gland complex of the pika. *J. Mammalogy*, 41:213-219.

Hediger, H. 1944. Die Bedeutung von Miktion und Defäkation bei Wildtieren. *Schweiz. Z. Psychol.*, 3:170-182.

——— 1949. Säugetier-Territorien und ihre Markierung. *Bijdragen tot de Dierkunde*, 28:172-184.

Henkin, R. I. 1967. The definition of primary and accessory areas of olfaction as the basis for a classification of decreased olfactory acuity. *In* Hayashi, T., ed. *Proc. 2nd Int. Sympos. Olfaction and Taste 2, Tokyo.* London, Pergamon Press. pp. 235-252.

Herreid, C. F., II. 1960. Comments on the odors of bats. *J. Mammalogy*, 41:396.

Hesterman, E. R., and Mykytowycz, R. 1968. Some observations on the intensities of odors of anal gland secretions from the rabbit *Oryctolagus cuniculus* (L). CSIRO Wildlife Res. 13:71-81.

Hill, W. C. O. 1944. An undescribed feature in the Drill (*Mandrillus leucophoeus*). *Nature* (*London*), 153:199.

——— 1951. Epigastric gland of *Tarsius*. *Nature* (*London*), 167:994.

——— 1953. Primates, comparative anatomy and taxonomy I. Strepsirhini. Edinburgh, Edinburgh University Press.

Jacobson, M., and Beroza, M. 1964. Insect attractants. *Sci. Amer.*, 211:20-27.

Jewell, P. A. 1966. The concept of home range in mammals. *Symp. Zool. Soc. Lond.*, 18:85-109.

Jolly, A. 1966. Lemur Behavior: A Madagascar Field Study. Chicago, University of Chicago Press.

King, J. A. 1955. Social behavior, social organization, and population dynamics in a black-tailed prairie dog town in the Black Hills of South Dakota. *Contr. Lab. Vert. Biol. Univ. Michigan.* No. 67.

Kingston, B. H. 1965. The chemistry and olfactory properties of musk, civet and castoreum. *In Proc. 2nd Int. Congr. of Endocrin, London, 1964.* pp. 209-214.

Kleinman, D. 1966. Scent marking in the Canidae. *Symp. Zool Soc. London,* 18:167-177.

Kligman, A. M. 1963. The uses of sebum? *In* Montagna, W. et al., eds. Advances in Biology of Skin. Vol. 4: The Sebaceous Glands. Elmsford, N. Y., Pergamon Press, Inc. pp. 110-124.

Koenig, L. 1957. Beobachtungen über Reviermarkierung sowie Droh-, Kampf- und Abwehrverhalten des Murmeltieres (*Marmota marmota* L.). *Z. Tierpsychol.,* 14:510-521.

——— 1960. Das Aktionssystem des Siebenschläfers (*Glis glis* L.) *Z. Tierpsychol.,* 17: 427-505.

Kratochvil, J. 1960. Sexualdrüsen bei den Säugetieren mit Rücksicht auf Taxonomie. *In* Kratochvil, J., and Pelikan, J., eds. Symp. Theriologicum Brno. Praha, Czechoslovak Akademy of Sciences. pp. 175-187.

——— 1964. Das männliche Genitalsystem des europäischen Bergmurmeltieres, *Marmota marmota latirostris* Krat. 1961. *Z. Säugetierk.,* 29:290-304.

Kulzer, E. 1961. Über die Biologie der Nil-Flughunde (*Rousettus aegyptiacus*). *Natur Volk (Frankfort),* 91:219-228.

Kunkel, P., and Kunkel, I. 1964. Beiträge zur Ethologie des Hausmeerschweinchens, *Cavia aperea f. porcellus* (L.). *Z. Tierpsychol.,* 21:602-641.

Lane-Petter, W. 1967. Odour in mice. *Nature (London),* 216:794.

Lederer, E. 1949. Chemistry and biochemistry of some mammalian secretions and excretions. *J. Chem. Soc.,* pp. 2115-2125.

Le Magnen, J. 1952. Les phénomènes olfacto-sexuals chez le rat blanc. *Arch. Sci. Physiol.,* 6:295-331.

Linsdale, J. M., and Tevis, L. P., Jr. 1951. The dusky-footed wood-rat. Berkeley, University of California Press.

——— and Tomich, P. Q. 1953. A herd of mule deer. Berkeley, University of California Press.

Linsday, D. R. 1965. The importance of olfactory stimuli in the mating behaviour of the ram. *Anim. Behav.,* 13:75-78

Lipkow, J. 1954. Uber das Seitenorgan des Goldhamsters (*Mesocricetus auratus auratus* Waterh.). *Z. Morphol. Ökol. Tiere,* 42:333-372.

Lockie, J. D. 1966. Territory in small carnivores. *Symp. Zool. Soc. London,* 18:143-165.

Lorenz, K. 1966. On Aggression. London, Methuen and Co. Ltd.

Mainardi, D., Marsan, M., and Pasquali, A. 1965. Causation of sexual preferences of the house mouse: The behaviour of mice reared by parents whose odour was artificially altered. Societa Italiana di Scienze Naturali e del Museo Civico di Storia Naturale di Milano, 104:825-838.

Marler, P. R. 1965. Communication in monkeys and apes. *In* DeVore, I., ed. Primate Behaviour. New York, Holt, Rinehart & Winston, Inc., pp. 544-584.

——— and Hamilton, W. J. 1966. Mechanisms of animal behavior. New York, John Wiley & Sons, Inc.

Marsden, H. M., and Bronson, F. H. 1964. Estrous synchrony in mice: alteration by exposure to male urine. *Science,* 144 (1469).

——— and Holler, N. R. 1964. Social behaviour in confined populations of the cottontail and swamp rabbit. *Wildlife Monogr.,* No. 13.

Martan, J. 1962. Effect of castration and androgen replacement on the supracaudal gland of the male guinea pig. *J. Morphology*, 110:285-297.

Martin, R. D. 1968. Reproduction and the ontogeny in tree-shrews, (Tupaia belangeri), with reference to their general behavior and taxonomic relationship. *Z. Tierpsychol.*, 25:409-495, 505-532.

McBride, G., Arnold, G. W., Alexander, G., and Lynch, J. J. 1967. Ecological aspects of the behaviour of domestic animals. *Proc. Ecol. Soc. Aust.*, 2:133-165.

Michael, R. P., and Keverne, E. B. 1968. Pheromones in the communication of sexual status in primates. *Nature* (London), 218:746-749.

Mimram, R. 1962. Les glandes occipitales du dromadaire. *Cahiers de la Faculté des Sciences, Rabat. Sér. Biol. Anima.*, 1:7-42.

Mitchell, O. G. 1965. Effect of castration and transplantation on ventral gland of the gerbil. *Proc. Soc. Exp. Biol., Med.*, 119:953-955.

——— 1967. The supposed role of the gerbil ventral gland in reproduction. *J. Mammalogy*, 48:142.

Moncrieff, R. W. 1951. The Chemical Senses. London, Leonard Hill.

——— 1966. Odour Preferences. London, Leonard Hill.

Montagna, W. 1962. The skin of lemurs. *Ann. N.Y. Acad. Sci.*, 102:190-209.

——— 1963. Comparative aspects of sebaceous glands. *In* Montagna, W., et al., eds. Advances in Biology of Skin. Elmsford, N.Y., Pergamon Press, Inc., Vol. 4, pp. 32-45.

——— and Noback, C. R. 1946. The histology of the preputial gland of the rat. *Anat. Rec.*, 96:41-54.

Moore, R. E. 1965. Olfactory discrimination as an isolating mechanism between *Peromyscus maniculatus* and *Peromyscus polionotus*. *Amer. Midland Nature*, 73:85-100.

Müller, D., and Lemperle, E. 1964. Objektivierung und Analyse olfaktorischer Signale der Säugetiere mit Hilfe der Gaschromatographie. *Naturwissenschaften*, 51:346-347.

Müller-Schwarze, D. 1967a. Social odors in young mule deer. Paper presented to AAAS New York Meeting, Dec. 26-31, 1967.

——— 1967b. Social odors in young mule deer. *Amer. Zool.*, 7:430.

——— 1969. Complexity and relative specificity in a mammalian pheromone. *Nature* (*London*), 233:525-526.

Müller-Using, D., and Schloeth, R. 1967. Das Verhalten der Hirsche (Cervidae). *Handb. d. Zool.*, 8:43, 10(28):1-60.

Myers, K. 1966. The effects of density on sociality and health in mammals. *Proc. Ecol. Soc. Aust.*, 1:40-64.

——— and Poole, W. E. 1959. A study of the biology of the wild rabbit, *Oryctolagus cuniculus* (L.), in confined populations. I. The effects of density on home range and the formation of breeding groups. *CSIRO Wildlife Res.*, 4:14-26.

——— and Poole, W. E. 1961. A study of the biology of the wild rabbit, *Oryctolagus cuniculus* (L.), in confined populations. II. The effects of season and population increase on behaviour. *CSIRO Wildlife Res.*, 6:1-41.

Mykytowycz, R. 1958. Social behaviour of an experimental colony of wild rabbits, *Oryctolagus cuniculus* (L.) I. Establishment of the colony. *CSIRO Wildlife Res.*, 3:7-25.

——— 1959. Social behaviour of an experimental colony of wild rabbits, *Oryctolagus cuniculus* (L.) II. First breeding season. *CSIRO Wildlife Res.*, 4:1-13.

———— 1965. Further observations on the territorial function and histology of the submandibular cutaneous (chin) glands in the rabbit, *Oryctolagus cuniculus* (L.). *Anim. Behav.*, 13:400-412.

———— 1966a. Observations on odoriferous and other glands in the Australian wild rabbit, *Oryctolagus cuniculus* (L.) and the hare, *Lepus europaeus* P. I. The anal gland. *CSIRO Wildlife Res.*, 11:11-29.

———— 1966b. Observations on odoriferous and other glands in the Australian wild rabbit, *Oryctolagus cuniculus* (L.) and the hare, *Lepus europaeus* P. II. The inguinal glands. *CSIRO Wildlife Res.*, 11:49-64.

———— 1966c. Observation on odoriferous and other glands in the Australian wild rabbit, *Oryctolagus cuniculus* (L.) and the hare, *Lepus europaeus* P. III. Harder's lachrymal, and submandibular glands. *CSIRO Wildlife Res.*, 11:65-90.

———— 1968. Territorial marking by rabbits. *Sci. Amer.*, 218:116-126.

———— and Dudzinski, M. L. 1966. A study of the weight of odoriferous and other glands in relation to social status and degree of sexual activity in the wild rabbit, *Oryctolagus cuniculus* (L.). *CSIRO Wildlife Res.*, 11:31-47.

———— and Gambale, S. 1969. The distribution of dung-hills and the behavior of free-living rabbits, *Oryctolagus cuniculus* (L.), on them. *Forma et Functio*, 1:333-349.

———— and Hesterman, E. R. 1970. The behavior of captive wild rabbits, *Oryctolagus cuniculus* (L.), in response to strange dung-hills. *Forma et Functio*, 2:1-12.

———— and Nay, J. 1964. Studies of the cutaneous glands and hair follicles of some species of Macropodidae. *CSIRO Wildlife Res.*, 9:200-217.

———— and Ward, Margaret M. Some reactions of nestlings of the wild rabbit, *Oryctolagus cuniculus* (L.), when exposed to natural rabbit odours. *Forma et Functio*. (in press)

Nelson, J. E. 1967. Communication in mammalian population ecology. *Proc. Ecol. Soc. Aust.*, 2: 117-123.

Nichol, A. A. 1938. Experimental feeding of deer. *Univ. Ariz. Coll. Agr. Exp. Sta. Tech. Bull.*, 75.

Nichols, D. G. 1944. Further consideration of American house mice. *J. Mammalogy*, 25:82-84.

Ortmann, R. 1960. Die Analregion der Säugetiere *Handb. d. Zool.*, 8:26, 3(7):1-68.

Ottoson, D. 1963. Some aspects of the function of the olfactory system. *Pharmacol. Rev.*, 15:1-42.

Parkes, A. S. 1963. Discussion. Some observations on prolactin secretion. *In* Nalbandov, A. V., ed. Advances in Neuroendocrinology. Urbana, University of Illinois Press.

———— and Bruce, H. M. 1961. Olfactory stimuli in mammalian reproduction. *Science*, 134:1049-1054.

Pearson, O. P. 1946. Scent glands of the short-tailed shrew. *Anat. Rec.*, 94:615-625.

Petter, J. J. 1962. *In* DeVore, I., ed. Primate Behavior. New York, Holt, Rinehart & Winston.

Pocock, R. J. Numerous publications. *Proc. Zool. Soc. Lond.*, 1910-1927.

Poglayen-Neuwall, I. 1962. Beiträge zu einem Ethogramm des Wickelbären (*Potos flavus* Schreber). *Z. Säugetierk.*, 27:1-44.

———— 1966. On the marking behaviour of the Kinkajou (*Potos flavus* Schreber). *Zoologica*, 51:137-41.

Quay, W. B. 1953. Seasonal and sexual differences in the dorsal skin gland of the kangaroo rat (*Dipodomys*). *J. Mammalogy*, 34:1-14.

———— 1954. The dorsal holocrine skin gland of the kangaroo rat (*Dipodomys*). *Anat. Rec.*, 119:161-176.

———— 1955. Histology of cytochemistry of skin gland areas in the caribou, *Rangifer*. *J. Mammalogy*, 36:187-201.

———— 1959. Microscopic structure and variation in the cutaneous glands of the deer, *Odocoileus virginianus*. *J. Mammalogy*, 40:114-128.

———— 1962. Apocrine sweat glands in the angulus oris of microtine rodents. *J. Mammalogy*, 43:303-310.

———— 1965a. Variation and taxonomic significance in the sebaceous caudal glands of pocket mice. (Rodentia: Heteromyidae.) *Southwest Natur.*, 10:282-287.

———— 1965b. Comparative survey of the sebaceous and sudoriferous glands of the oral lips and angle in rodents. *J. Mammalogy*, 46:23-37.

———— 1965c. Integumentary modifications of North American desert rodents. *In* Lyne, A. G., and Shost, B. F., eds. Biology of the Skin and Hair Growth. Sydney, Angus and Robertson, pp. 59-74.

Reiff, M. 1956. Untersuchungen über natürliche und synthetische Geruchstoffe, die bei Ratten und Mäusen eine stimulierende Wirkung auslösen. *Acta Tropica*, 13:289-318.

Ropartz, P. 1968. The relation between olfactory stimulation and aggressive behaviour in mice. *Anim. Behav.*, 16:97-100.

Roth-Kolar, H. 1957. Beiträge zu einem Aktionsystem des Aguti (*Dasyprocta aguti aguti* L.). *Z. Tierpsychol.*, 14:362-375.

Rue, L. L., III. 1965. Cottontail. New York, Thomas Y. Crowell Company.

Schaffer, J. 1940. Die Hautdrüsenorgane der Säugetiere. Berlin, Urban und Schwarzenberg.

Schenkel, R. 1966. Zum Problem der Territorialität und des Markierens bei Säugern —am Beispiel des schwarzen Nashorns und des Löwens. *Z. Tierpsychol.*, 23:593-626.

Schloeth, R. 1956. Zur Psychologie der Begegnung zwischen Tieren. *Behaviour*, 10:1-79.

Schultz, A. H. 1921. The occurrence of a sternal gland in orang-utan. *J. Mammalogy*, 2:194-196.

Schultze-Westrum, T. 1965. Innerartliche Verständigung durch Düfte beim Gleitbeutler, *Petaurus breviceps papuanus* Thomas (Marsupialia, Phalangeridae). *Z. Vergl. Physiol.*, 50:151-220.

———— 1968. Ergebnisse einer zoologisch-völkerkundlichen Expedition zu den Papuas. *Umschau*, 10:295-300.

Sebeok, T. A. 1965. Animal communication. *Science*, 147:1006-1014.

Shelley, W. B. 1956. The role of apocrine sweat in the production of axillary odor. *J. Soc. Cosmet. Chem.*, 7:171-175.

Shorey, H. H., and Gaston, L. K. 1967. Pheromones. *In* Kilgore, W. W., and Doutt, R. L., eds. Pest Control. New York, Academic Press, Inc., pp. 241-265.

Signoret, J. P., and du Mesnil du Buisson, F. 1961. Étude du Comportement de la truie en oestrus. *In Proc. Int. Congr. Animal Prod. IV* (Hague), 2:171-175.

Sink, J. D. 1967. Theoretical aspects of sex odor in swine. *J. Theoret. Biol.*, 17:174-180.

Sleggs, G. 1926. The adult anatomy and histology of the anal glands of the Richardson ground-squirrel, *Citellus richardsonii* Sabine. *Anat. Rec.*, 32:1-43.

Sprankel, H. 1962. Histologie und biologische Bedeutung eines jugulo-sternalen Duftdrüsenfeldes bei *Tupaia glis* Diard 1820. *Verh. Deut. Zool. Ges.*, 1961:198-206.

Stanley, A. J., and Powell, R. A. 1941. Studies on the preputial gland of the white rat. *Proc. Louisiana Acad. Sci.*, 5:28-29.

Strauss, J. S., and Pochi, P. E. 1963. The hormonal control of human sebaceous glands. *In* Montagna, W., et al., eds. Advances in Biology of Skin. Elmsford, N.Y. Pergamon Press, Inc., Vol. 4, pp. 226-254.

Takaki, S., and Tagawa, M. 1961. Hard and soft faeces of the rabbit and their relation to the anal gland secretion and the contents of the caecum. *Jap. J. Zool.*, 70:248-252. (in Japanese.)

Talbot, L. M., and Talbot, M. H. 1963. The wildebeest in western Masailand, East Africa. *Wildlife Monogr.*, No. 12.

Tarasov, P. P. 1960. Biological significance of scent glands in mammals. *Zool. Zh.*, 49:1062-1068 (in Russian).

Thiessen, D. D., Friend, H. C., and Lindzey, G. 1968. Androgen control of territorial marking in the Mongolian gerbil (*Meriones unguiculatus*). Science, 160:432-434.

Thomson, J. A., and Pears, F. N. 1962. The functions of the anal glands of the brushtail possum. *Victorian Natur.*, 78:306-308.

Vestal, E. H. 1938. Biotic relations of the wood-rat (*Neotoma fuscipes*) in the Berkeley Hills. *J. Mammalogy*, 19:1-36.

Wenzel, B. M. 1967. Olfactory perception in birds. *In* Hayashi, T., ed. *Proc. 2nd Int. Sympos. Olfaction and Taste 2, Tokyo.* London, Pergamon Press. pp. 203-217.

Whitten, W. K. 1966. Pheromones and mammalian reproduction. *Adv. Reprod. Physiol.*, 1:155-157.

Whitten, W. K. Personal Communication.

Wilson, E. O. 1965. Chemical communication in the social insects. *Science*, 149:1064-1071.

Wislocki, G. B. 1930. A study of scent glands in the marmosets, especially *Oedipomidas geoffroyi*. *J. Mammalogy*, 11:475-483.

—— and Schultz, A. H. 1925. On the nature of modifications of the skin in the sternal region of certain primates. *J. Mammalogy*, 6:236-244.

Wynne-Edwards, V. C. 1962. Animal dispersion in relation to social behaviour. Edinburgh, Oliver & Boyd Ltd.

Zannier, F. 1965. Verhaltensuntersuchungen an der Zwergmanguste. *Thelogale undulata sufula*, in Zoologischem Garten Franfurt am Main. *Z. Tierpsychol.*, 22:672-695.

12

Man's Ability to Perceive Odors*

TRYGG ENGEN

Walter S. Hunter Laboratory of Psychology
Brown University
Providence, R. I. 02912

DETECTION AND THE NATURE OF THRESHOLDS 363
OLFACTORY DISCRIMINATION AND RESOLVING POWER 367
PSYCHOPHYSICAL SCALING AND THE POWER FUNCTION 369
THE NOSE AS AN INFORMATION CHANNEL 372
CONCLUSION 381

A man is watching television. Downstairs in a closet his coat is smoldering from a hot pipe ash left in the pocket when he walked his dog earlier in the evening. Of course he cannot see, hear, or feel the smoldering coat. Will he be able to smell it before a serious fire breaks out? He will be able to smell it eventually as the smoke intensity increases. Suppose he falls asleep; will he be able to smell it and wake up before he becomes asphyxiated? This is the kind of information the nose provides to supplement the ears, eyes, and other sensory channels.

Our television viewer now thinks he detects something. This is the first and classic question of *detection* which concerns whether or not man judges that something is present. Did he smell something or was it just his imagination? By definition, this is a very difficult decision for him to make in the case of weak stimuli in any sense modality.

Let us suppose that he decides that he detects the smell of something. The observer is now aroused and he sniffs at the air in an attempt to determine whether the odor is getting stronger. This is the problem of *discrimination*. How much greater must the concentration of the odorant become before one is likely to judge that the smell is stronger? Two beliefs about man's ability to perceive odors seem common. One of these is that as contrasted with his ability to deal with qualitative differences, man's ability to detect intensity differences is relatively poor.

* This paper was prepared while the author was supported by a grant from the United States Public Health Service (HD-02358).

362 / MAN'S ABILITY TO PERCEIVE ODORS

The discrimination and recognition problems concern only ordinal differences in perceived intensities which may be described quantitatively by Weber's Law in terms of concentration. *Psychophysical scaling* takes the quantitative problem further and attempts to determine the mathematical form of the function relating perceived intensity to stimulus intensity. In other words, how does the subjective intensity of smell "grow" as the concentration of the odorant is increased? The evidence shows that this function is a negatively accelerated function. It, like the functions in other sense modalities, conforms to the power law such that pairs of stimuli which form equal physical ratios of intensity also form equal ratios of subjective intensity. The value of the ratio tends to distinguish between the various sense modalities and for that reason is important in evaluating sensory sensitivity.

The observer must also determine what the smell is, that is, he must classify it. This is the problem of *recognition*. Another belief about the sense of smell is that man can recognize or identify thousands of odor qualities. At best, this is not a very precise statement and evidence tends to suggest that it is wrong. In the case of vision we usually recognize objects easily and without much deliberation and so it is in the case of smell, providing we know the source of the odor. This restriction may be very important in the case of smell. For example, we easily recognize the smell of toasting bread and brewing coffee in the kitchen, especially at breakfast time. However, remove the familiar context and leave man to decide on the basis of his nose alone, and he easily becomes confused and often fails to identify even familiar odors.

Indeed, our television viewer may not be able to identify the stimulus as smoke until it becomes very strong because he does not expect it, for expectation is a powerful influence on his sensory channels of communication. Once he knows that it is smoke, he must try to locate the source. There is a dearth of information about man's ability in this task which is the specialty of the bloodhound.

The problems of detection, discrimination, scaling, and recognition are the main problems of psychophysics and must be considered basic to any objective understanding of man. The possession of an olfactory system makes man capable of responding to certain energies in the environment and thereby provides information which would otherwise not be available to him. This chapter is concerned with the problem of how man's behavior might be modified, or how well one might communicate with him by manipulating olfactory stimuli. It concerns itself with the nose as an informer although only one section deals with the term "information" in the more precise meaning of information theory (Attneave, 1959). The terms "olfaction" and "smell" are used in a molar or psychological sense and without any precise physiological connotation.

DETECTION AND THE NATURE OF THRESHOLDS

In the hundred years since Fechner's work (1860) psychophysical thresholds have been used as one of the main indices of sensitivity. A high threshold thus indicates low sensitivity and vice versa. The nose is apparently a sensitive instrument as compared with physical instruments (Stuiver, 1958); however, different investigators have obtained quite different thresholds for the same compounds, and for that reason it has never been possible to describe reliably the sensitivity of the nose (e.g., Allison and Katz, 1919; Bach, 1937). Difficulties have even been encountered in reaching the objective decision that dogs have a keener sense of smell than men (Moulton et al., 1960). It has long been obvious that such problems were at least partly caused by differences in experimental procedures, but the basic problem probably was Fechner's assumption that there is a threshold in the first place. The main idea of classical psychophysics was that the sensory system had a fixed cutoff or an absolute threshold for stimuli that would discharge the system. However, due to a complex set of factors, such as the efficiency of the connecting paths and the background level of activity of the system, it was assumed that the effect of a threshold stimulus would produce a random distribution. If it is assumed that this distribution is normal, a measure of central tendency would then represent the stimulus threshold. Different psychophysical methods were considered as only different ways of obtaining and analyzing different data pertaining to the same threshold. Empirical evidence, much of it accumulated recently in audition and vision, has generally failed to support the classical theory. The theory of a fixed cutoff is losing support. In its place a detection theory based on a decision analysis model of psychophysical threshold has been proposed (Green and Swets, 1966). In essence, detection theory proposes that there is no fixed cutoff and no stimulus threshold, and therefore no sensory threshold. In general, the problem of detecting a stimulus may be considered as a problem of signal-to-noise ratio. Whenever an experimenter presents a stimulus there will be noise presented because of external uncontrollable events, variability in the stimulus, spontaneous neural firing, or because it has been deliberately introduced by the experimenter. It is assumed that the noise has an effect on the sensory system of the same sensory quality as the stimulus, and that there is no fixed criterion the psychophysical observer can apply to it to classify his experiences in terms of "Yes, I smelled something" versus "No, I did not smell anything." Instead of considering the psychophysical task as one of categorizing experiences into two classes, detection theory considers it to be analogous to statistical sampling and deciding whether the observation sampled was obtained from a trial during which a stimulus was presented or a so-called catch-trial during which a "blank," e.g., a diluent, was presented. It is important to note that the diluent, which may represent noise in this example, and a stimulus (a weak concentration of an odorant) could produce the same sensory effect such as "I smelled something." There is in any trial a sensory event, but the observer's problem is to decide whether stimulus or noise produced it. His response depends on (1) the effect of the stimulus relative to the effect of noise, because the sensory effect of a stimulus is continuous and not discrete as assumed in classical theory, and (2) on the observer's conception of the situation, that is, the criterion by which he decides whether he smelled something or not.

Investigators in food science and smell have long been concerned with the basic effect of practice or sophistication of the observer, training of panels, and the like. These, according to detection theory, are problems primarily of the observer's criterion and not of his sensitivity. For example, amyl acetate can be detected *and* identified at one concentration because it smells like "banana oil" but it may be detected at a much lower level where it is more difficult to characterize in terms of past association, but where it may still be detected or distinguished from the effect of a diluent. Engen (1960) obtained such results with several odorants and also a decrease in conventional thresholds as a function of practice. Recent experiments in the Brown laboratory were designed (1) to extend and improve the study of ability to detect odors as a function of practice and (2) to illustrate that changes in threshold response are easily produced by changes in the observer's criterion, depending on his motivational state (Semb, 1968).

The method employed was designed to reduce response biases of the classic method of limits (Guilford, 1954) which has typically been used in measuring olfactory threshold. Pfaffmann (1951) should be consulted in a general review of the literature on olfactory sensitivity. The method of limits (or serial exploration) is the most direct of the classic methods of locating a threshold. The experimenter varies the stimulus in small steps in consecutive ascending and descending series, and the observer is asked to report for each stimulus value whether or not he can detect it. The data are the physical stimulus values corresponding to a shift in the observer's responses from "yes" to "no." Absolute threshold is defined as the stimulus value, I, which the observer detects on half of the trials but fails to detect on the remaining half of the trials. This value is considered a psychological zero, and will be designated as I_0. Semb used a modification of this method called the method of double random, "yes-no" staircases first proposed by Cornsweet (1962). This method employs an ascending and a descending series simultaneously. The concentrations are presented in discrete steps and are identical in the two series, and the series used on any trial is determined at random. A trial is presented and the observer must report either "Yes, I smell it" or "No, I did not smell it." If he responds "Yes," the next trial sampled from that series will contain the concentration just below it; if he responds "No," the next trial sampled from that series will contain the concentration just above it. That is, the procedure follows that of the method of limits, but the observer cannot predict whether the concentration on each trial will be weaker or stronger than that of the preceding trial. At the same time, the observer's response determines which stimulus is presented within each series when it is sampled next. This procedure tends to counteract the most common problems encountered in the method of limits, which are biases due to expectation and habituation, that is, a tendency of the observer to respond in terms of what he has learned about the experiment rather than experienced sensations.

Two women about 20 years old served as observers. Both were college students with previous experience in psychophysical experiments, although not in olfaction. Both were paid $1.50 per hour.

The experiment was performed in an air-conditioned and ventilated room with a temperature of close to 68° F and humidity of about 50 percent.

n-Butyl alcohol was used as the test odorant, and it was diluted in diethyl phthalate by pipette. The observer sniffed the solution from cotton wrapped around a glass rod.

Each trial consisted of the presentation of 100 percent diethyl phthalate as a standard stimulus followed in 2 seconds by a comparison stimulus which was either a certain concentration of the butanol or 100 percent diethyl phthalate. The observer's task was to judge whether butanol was present in the comparison stimulus or whether the comparison stimulus was a duplicate of the standard stimulus. Fifty-four trials

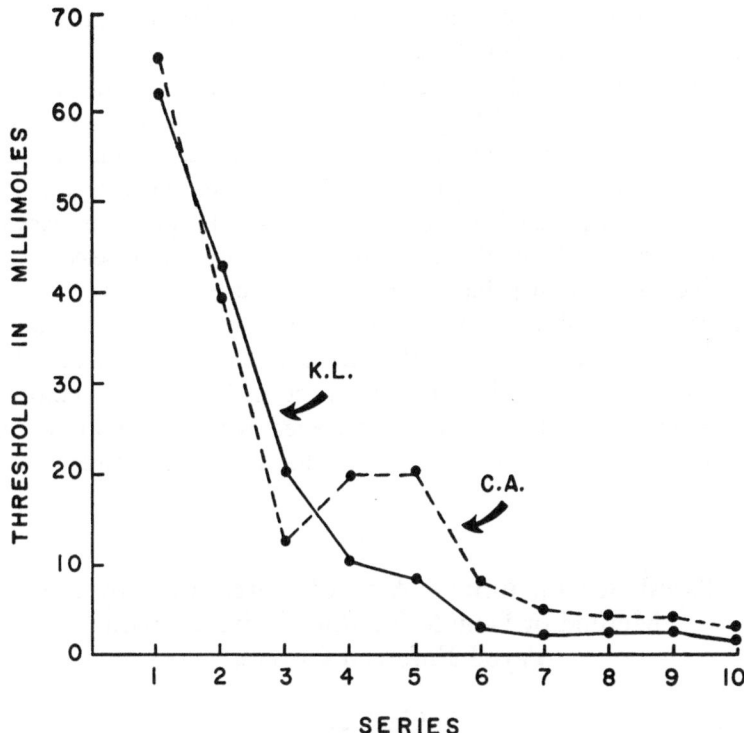

FIG. 1. Mean thresholds in millimoles of butanol as a function of consecutive series (practice sessions) for 2 observers, KL and CA. (*From* Semb. 1968. *Perc. Psychophys.*, 4:335-340.)

were run during each experimental session with a 10-second intertrial interval. On four of those trials the second stimulus was identical to the standard stimulus, i.e., 100 percent diethyl phthalate. If the observer said yes on those occasions, she was fined two cents. The first 20 trials were used for practice, and threshold was based on the remaining 30 trials.

The mean concentration in millimoles to which a "yes" response was obtained was computed for each session and this value corresponded to the stimulus threshold obtained with the classical method of limits. These means are shown for both observers (CA and KL) as a function of practice in successive sessions in Figure 1. The effect of practice was unmistakable and followed the same general curve of both observers. It may be worth noting that the "intertrial interval" (10 seconds) seemed short enough that decreased sensitivity (and thus increase in threshold) was considered possible before performing the experiment. If such adaptation did have an affect, it was minimal compared with the effect of practice.

The decrease of the threshold as a result of training can be understood as a change of the observer's response criterion following experience in the experiment. It is clearly at odds with the theory that thresholds reflect a fixed sensory response which is initiated automatically when a stimulus is increased above a certain value. The hypothesis proposed here is that the observer's detection performance improves and stabilizes, in olfaction as in other perceptual tasks, as a function of the criterion by which he attempts to distinguish between the effect of the odor and the qualitatively similar sensory effect of the diluent.

If an observer does indeed use a criterion of this sort to decide whether to call it an odor or not, then it should be possible to show that the observer is capable of changing this criterion. In the next experiment a so-called "payoff matrix" was introduced, so that rather than being paid by the hour the observer was now to be paid strictly in terms of his performance. The experimental procedure was the same, except that as the standard stimulus the experimenter arbitrarily selected a different stimulus variable (a criterion point) from the series used previously in each of 10 sessions. The observer's task was now to respond "Yes" when the comparison stimulus was stronger than the standard and "No" when it was weaker. He received two cents for a correct response and was fined two cents for an incorrect response. Threshold values were computed as above and are compared with the arbitrary criterion in Table 1. In addition standard deviations and the difference between the arbitrary

TABLE 1

Thresholds for n-Butyl Alcohol Determined by the Method of Double Random Staircases with Payoff Enforced Criteria*

Subject CA

SESSION	ENFORCED CRITERION	MEAN THRESHOLD	OBTAINED t†
11	2.732	2.951	.982
12	13.660	13.661	.001
13	2.732	2.804	.318
14	54.641	58.373	−.885
15	13.660	15.435	1.749‡
16	54.641	51.784	−.894
17	13.660	13.296	−.322
18	2.732	2.659	−.321
19	54.641	56.151	.352

Subject KL

SESSION	ENFORCED CRITERION	MEAN THRESHOLD	OBTAINED t†
11	54.641	53.236	−.281
12	2.732	2.731	−.003
13	13.660	14.753	.979
14	2.732	2.805	.269
15	13.660	13.296	−.321
16	54.641	54.694	.014
17	13.660	14.571	.836
18	54.641	53.183	−.635
19	2.732	2.659	−.322

* From Semb. 1968.
† $t_{.05} = 1.699$.; $df = 29$.
‡ Significant at the .05 level.

criterion and the observer's criterion were evaluated by a *t* test. Only one reliable difference was obtained, and the results show in general that the observers were quite able to change their criteria to conform to one selected by the experimenter. The important point is that there seems little doubt that the observer is able to establish his own criterion and to judge the stimuli accordingly, although the present data are not sufficient to prove the point.

These data are consistent with a crucial assumption in detection theory to the effect that the threshold value obtained will depend upon the judgmental criterion adopted by the observer. The two psychological variables which have been especially important in determining the observer's criterion are (1) the consequences of the observer's decision in terms of monetary losses and gains, as in the present experiment, and (2) the probability that the stimulus will be presented because this probability determines the observer's expectation, set, or attitude. This idea is not new. Slosson reports a classroom demonstration on hallucinations in the Psychological Review in 1899 as follows:

> I had prepared a bottle filled with distilled water carefully wrapped in cotton and packed in a box. After some other experiments I stated that I wished to see how rapidly an odor would be diffused through the air, and requested that as soon as anyone perceived the odor he should raise his hand. I then unpacked the bottle in the front of the hall, poured the water over the cotton, holding my head away during the operation and started a stopwatch. While awaiting results I explained that I was quite sure that no one in the audience had ever smelled the chemical compound which I had poured out, and expressed the hope that, while they might find the odor strong and peculiar, it would not be too disagreeable to anyone. In fifteen seconds most of those in the front row had raised their hands, and in forty seconds the "odor" had spread to the back of the hall, keeping a pretty regular "wave front" as it passed on. About three-fourths of the audience claimed to perceive the smell . . . More would probably have succumbed to the suggestion, but at the end of a minute I was obliged to stop the experiment, for some of the front seats were being unpleasantly affected and were about to leave the room . . ." (p. 407)

No research has been done on this important variable in olfactory detection. These psychological variables are as important as sensory variables in understanding human performance in a detection situation. Auditory detection theory has reached a particularly advanced stage and has been able to develop a measure of sensitivity, d', which is independent of these psychological variables (Green and Swets, 1966). Advancement of such a theory in olfaction depends on research on the "noise" in the olfactory system.

OLFACTORY DISCRIMINATION AND RESOLVING POWER

If the assumption of detection theory is correct, that the sensory continuum is continuous rather than discrete and that any magnitude of stimulus plus the added effect of sensory noise can be detected with a certain probability depending on the observer's criterion, the classical distinction between absolute and differential threshold disappears. From a similar theoretical basis Ekman (1959) has also shown that absolute and differential sensitivity may be essentially similar measures. Increases in the magnitude of a stimulus once it becomes detectable 100 percent of the time will be reported as increases in subjective intensity. The difference threshold or limen ($\triangle I$) is measured by presenting the observer with a constant standard concentration which is compared with a series of variable comparison stimuli, ranging in concentration from a value judged definitely weaker than the standard, to one judged definitely stronger than the standard. The value of $\triangle I$ is determined statistically as that difference which can be detected 50 percent of the time, just as the absolute threshold (I_0) is that value which can be detected 50 percent of the time. The reason that they may be considered two aspects of the same sensory problem is that, although no standard or criterion is prescribed by the experimenter in determining I_0, the observer is likely to adopt a certain criterion on his own. The value of this criterion may correspond to the effect of the value of the ever-present noise on the sensory continuum, or to a higher value corresponding to the combined effect of a weak stimulus and noise. Absolute threshold is in that sense a special case of differential threshold, but differential olfactory sensitivity has barely been studied. Gamble (1898) and Zigler and Holway (1935) used Zwaardmaker's olfactometer, and Wenzel (1949) used an Elsberg apparatus and then no work is reported until Stone presented his work on differential sensitivity with an air-dilution olfactometer. Stone's olfactometer provides a better method for presentation of the stimulus by delivering purified air with a calculated concentration of odorant into a Plexiglas hood surrounding the observer's head.

All of the above investigators were concerned with Weber's Law which originally was written as

$$\frac{\triangle I}{I} = k$$

This relation provides a means for measuring relative sensitivity with a fraction proposed to be constant within a modality. Weber's fraction, or k, tends to be constant for moderate values of I, but usually increases greatly as detection of I becomes a problem. For that reason a modified Weber's Law has long since been proposed that states

$$\frac{\triangle I}{I + a} = k, \text{ or } \triangle I = k(I + a)$$

where a is a small value of the stimulus dimension, I, and may be considered the value of sensory noise. A threshold value of I measured with conventional psychophysical methods would be an adequate definition of this noise for most purposes, for a is only a significant factor at low values and it may be omitted for high values of I without affecting the results appreciably. Figure 2 shows a test of Weber's modified law by Stone and Bosley (1965) with $\triangle I$ plotted as a function of $I + I_t$, where I_t is an estimated stimulus threshold. $\triangle I$ was obtained by a modified method of constant stimuli (Guilford, 1954). On each discrimination trial the observer was asked to select which sample, one of which was always the standard or I, was not intense. Absolute threshold was also estimated for each observer at the end of the discrimination test. Data were pooled from nine observers and evaluated by regression analysis with the

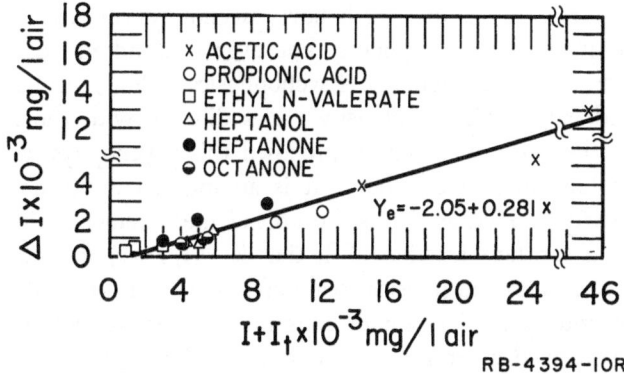

Fig. 2. The difference limen, △I, as a function of the standard, I, corrected for threshold, I_t. The values plotted are averages for 9 observers. (Reprinted with permission of author and publisher: Stone, H. & Bosley, J. J. *Psychological Reports*, 1965. 20:657-665, Figure No. 5, page 664.)

results shown in Figure 2. Weber's fraction, or k, is 0.281. Accordingly, a change of approximately 28 percent would be required in stimulus magnitude before the average observer was able to detect it 50 percent of the time. Measured in terms of relative sensitivity, the sense of smell is the dullest of the senses although taste discrimination in only slightly keener with a Weber fraction of about 0.20. This compares with fractions of less than about 0.10 for audition, touch, and vision. For example, for pitch of a pure tone it is 0.003 (Woodworth and Schlosberg, 1954).

It must be borne in mind that these Weber fractions depend on the psychophysical methodology by which they were obtained. It is an arbitrary index, subject to the same biases involved in detecting weak stimuli and equally likely to benefit from the developments in detection theory mentioned above.

PSYCHOPHYSICAL SCALING AND THE POWER FUNCTION

The ability of man to adjust to his environment depends on his ability to perceive the environment. When, for example, he sniffs and judges that a certain smell is getting stronger, he is making a kind of "measurement" that no physical measuring instrument can. Psychophysical scaling refers to comparison of such subjective measurement and physical measurement. It is concerned with the basic question of how subjective (perceived or judged) intensity grows as a function of a physical measure of intensity. Weber's Law deals only with the physical side of this psychophysical problem. It would have been useful to start the discusion of the present topic with that problem, because the psychophysical relation may be the proper background for consideration of detection and discrimination, rather than to pursue the practical example of the detection of smoke.

If the concentration of an odorant is halved, will perceived odor intensity also be halved? The answer is definitely no. Until recently very little consideration has been given to this kind of question partly because so little is known about the stimulus and partly because researchers did not believe psychophysical observers were capable of making such quantitative judgments. For over a hundred years since Fechner published his work, The Elements of Psychophysics (1860), it has been assumed that the perceived magnitude of intensity varies directly with the logarithm of the physical value of the stimulus. Fechner proposed this psychophysical relation on the basis of Weber's Law mentioned above. It is an indirect approach which uses the differential sensitivity defined as $\triangle I$ as the unit of measurement. Since Weber's Law states that $\triangle I$ is a constant fraction of I, that is $\triangle I/I = k$, the physical unit must be increased as we ascend the physical scale of intensity in order that the subjective increment corresponding to $\triangle I$ remain constant or "just noticeably different." If it is assumed that perceived magnitude is the sum of all the just noticeable differences which come before it on the scale, logarithms are especially convenient when dealing with the relative magnitude of perceived intensity.

Few tests were made of Fechner's "Law" with the indirect method of differential sensitivity, and more direct questioning of psychophysical observers has completely failed to support it. A direct approach entails that the subjective or perceived numerical property desired by the experimenter is stated in the instructions he gives to the observer; for example, "how much stronger is odor A than odor B?" Instead of Fechner's rather complex statistical discriminability assumption, these direct methods make only the assumption that the observer is able to quantify his perceptual observations in simple arithmetical terms. Several experiments which asked such questions in different ways and which used quite different stimuli, from pure compounds to coffee odor, have shown that subjective odor intensity is a negatively accelerated function of concentration. All these findings conform to Stevens' power function of the form $R = cS^n$, where R is subjective intensity, c is a constant reflecting the arbitrary numerical unit used by the observers, and n is the exponent (slope) of the function (Stevens, 1961). Expressed in logarithmic terms, $\log R = n(\log S) + \log c$, which is a simple linear equation suitable for graphical presentation of the data (Jones, 1958a, 1958b; Reese and Stevens, 1960; Engen and Lindstrom, 1963; Engen, Cain and Rovee, 1968).

The most recent research in psychophysical scaling and probably the best as far as methodology is concerned is reported by Cain (1968). The stimuli were presented by an olfactometer in which the odorant was diluted with air and sniffed through a Teflon nose piece. Fifteen observers judged each of seven concentrations twice in irregular order under the method of magnitude estimation. Their instructions were to judge how strong each odor was by assigning a number to it. The observer was to strive to make the ratios between the numbers assigned to different concentrations match the ratios between the subjective intensity of the odors. The results were analyzed by a computer program which eliminates all the variance of the data except that due to individual differences in the exponents. Table 2 presents geometric means, log geometric means, and standard deviations of the logs obtained for butanol, which provide a representative psychophysical scale in olfaction. The geometric means define the psychological scale values which may be compared with the physical scale values of concentration, but the data will first be considered from a simpler approach in order to answer the question raised above about halving intensity. For example, changing the physical concentration of the odorant by reducing it from

TABLE 2

Geometric Means and Standard Deviations of the Magnitude Estimate Scale Values of Butanol*

	CONCENTRATION (%)						
	100.00	50.00	25.00	12.50	6.25	3.13	1.56
Geometric mean	34.0	19.9	10.7	7.4	4.9	2.8	2.6
Log geometric mean	1.53	1.30	1.03	0.87	0.69	0.45	0.41
S. D. of logs	0.23	0.13	0.16	0.15	0.15	0.24	0.26

* From Cain. 1968.

100 percent to 50 percent reduces the subjective intensity to only 60 percent of the psychological value corresponding to the 100 percent concentration, that is, $19.9/34.0 = .59$ according to Table 2. This relation illustrated by 50 percent versus 60 percent represents a fairly reliable and general rule describing how subjective odor intensity varies as a function of physical concentration. This simple relation is important in evaluating man's ability to perceive odors and ought to be considered in practical applications of psychophysics. It indicates that the olfactory system tends to compress the sensory stimulation such that subjective output is a decelerating function of the physical input. By comparison, the heaviness of weights and the magnitude of electric shock are accelerating functions of their respective stimulus intensities, but deceleration has been observed in most such experiments on a variety of sensory dimensions.

When the present results (Table 2) are plotted in log-log coordinates, a linear function is indicated as shown in Figure 3. This is the power function proposed by Stevens (1961) as a general psychophysical law. What distinguishes one sense modality from another is the exponent of the function, which in the present case is

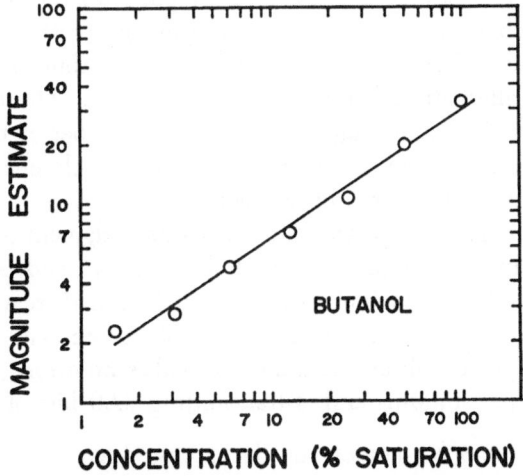

FIG. 3. Log magnitude estimation scale values of odor intensity as a function of log concentration of butanol. The values are geometric means for 15 observers. *Courtesy of Dr. W. S. Cain.*

.64. It is interesting to note that the exponent in olfaction seems to vary as a function of water solubility (Engen, 1965). Cain (1968) determined that the rank-order correlation between the exponent and the water solubility of the compounds was +.94 for the compounds scaled by Jones (1958b) and Engen (1961), and in the case of his own data the correlation was perfect.

THE NOSE AS AN INFORMATION CHANNEL

If, instead of different concentrations of the same odorant, one were to present the observer with qualitatively different odorants or compounds, then how well would he be able to discriminate? Man would probably do very well, some would think excellently (Behnke, 1954), but this problem must be defined precisely.

How would one measure the performance? Discrimination in the classical sense discussed above has been successful for the reason that psychological or behavioral ability can be expressed in physical units such as concentration. On what basis would one select odorants for comparison from the enormous number of compounds available? How would this ability in olfaction compare with the ability of a human observer to discriminate between pairs of different objects? There are no simple psychophysical answers to these problems, for this is essentially the psychophysiological problem of the "afferent code for sensory quality" (Pfaffmann, 1959).

The most fruitful psychophysical approach to this problem has been the application of information theory, which makes it possible to describe psychological events in probabilistic terms. The basic unit of this theory can be applied to any set of stimulus categories and provides a common unit, the bit, whereby one can determine the number of different stimuli an observer can identify correctly by rank-order. The unit of measurement is simply the logarithm to the base two of this number. It is not necessary to assume that sense modalities operate in a binary manner, and information theory as applied here provides only a simple model for describing dimensions of input and output. For the present purposes the number of categories correctly identified, obtained by taking the antilogarithm of the number of bits, will be considered. That is, information is defined as the logarithm to the base two of the number of stimulus alternatives being judged in a session. The unit of information measurement, the bit (for binary digit), specifies the number of two-choice decisions which the observer must make in order to identify which one of the alternative stimuli is presented at a certain trial or occasion.

According to this method of analyzing absolute judgments, if four stimuli are used in one session and one of them is selected at random and presented to the observer, he must decide from which of two groups it is, and then which of two stimuli it is. If he makes no errors in such a situation the observer is said to transmit 2.0 bits of information. If eight stimuli are used and he makes no errors, he is making three two-choice decisions per stimulus and is transmitting 3.00 bits of information. Since $\log \frac{1}{x} = -\log x$, this can be mathematically expressed as

$$T = \log_2 \left(\frac{1}{p} \right) = -\log_2 p$$

where T is information transmitted per stimulus and p is the probability of occurrence of each stimulus, or .125 in the last example. Therefore

$$T = \log_2\left(\frac{1}{p}\right) = -\log_2 p = 3.00$$

In the present context psychologists are interested not in the particular stimulus but in the average amount of information or uncertainty the observer can transmit with respect to a defined sensory attribute. For that reason the average amount transmitted per stimulus is computed and weighted in terms of the probability of occurrence of each stimulus such that

$$T = -\sum_{}^{i} p_i \log p_i$$

where p_i is any alternative stimulus. This is known as the Shannon-Wiener measure of informaton, and the antilogarithm of this measure is what has been used in psychology as a measure of the number of stimuli that the human observer can discriminate perfectly. The maximum amount of information which can be transmitted is limited by the number of alternative stimuli presented. In simple terms, the more errors the observer makes the more T is reduced, but it is generally more useful to talk about T than the number of errors. Basically the present problem was to estimate (1) the number of intensity levels of one odorant and (2) the number of qualities (different odorants) that the average observer can discriminate without making errors when the stimuli are presented singly and without any other cues or comparisons available to him.

Judgments were obtained from groups of 5 to 8 paid observers. The odorants were presented in test tubes. For the intensity judgments a geometric dilution series of 100, 50, 25, 12.5, and 6.25 concentration of the odorant provided a large enough step between adjacent stimuli to prevent any problems of discrimination, in the sense discussed above. Such dilution series were prepared for amyl acetate, heptanal, heptane, and β-phenylethyl alcohol. First, the observer was required to learn to identify the rank-order of these concentrations as the "strongest," "next strongest," etc. by comparing them as often as he wished, in order to avoid confounding verbal coding and odor discrimination (see discussion following the paper by Dravnieks, 1968). This procedure was repeated at the beginning of each session. Then, when the observer was satisfied that he had learned the ordering, the experimenter would represent one stimulus at a time in a random order; that is, the observer now has the task of identifying its rank-order as before but without the aid of any comparison. The correct rank-order was indicated by the experimenter once the observer had made his judgment. Each stimulus was presented 50 times in different random orders to each observer. About one week was required by the observer to complete each set of 5 stimuli in several daily sessions of about 12 minutes each. Table 3 presents the results for each odorant separately and shows that the results for the average observer are very close to 1.5 bits/stimulus for all odorants. Taking the antilogarithm of 1.5 bits yields the value of approximately 3, which is the amount of information the average observer is able to transmit (recall and identify). There is a small but reliable improvement in accuracy of identification with an increase in the size of the

TABLE 3

Average Amount of Information in Bits Obtained for Five Alternative Stimuli with 100, 50, 25, 12.5, and 6.25 Percent Dilution Concentration Values for Each of Four Odorants*

ODORANT	N	MEAN	S.D.
n-amyl acetate	5	1.52	.37
n-Heptanal	4	1.53	.16
n-Heptane	5	1.51	.33
β-phenylethyl alcohol	5	1.58	.27

* From Engen and Pfaffmann. 1959. *J. Exp. Psychol.*, 58:24.

physical step between the stimuli; identification is also better with strong stimuli than with weak stimuli. Amount of information in bits/stimulus varied from 1.4 to 1.7 as a function of size of step between stimuli and concentration of stimuli.

A more important variable than these variables is the amount of practice given the observers. Each observer judged a total of 12 different sets of five stimuli in a different random order in order to evaluate the effect of practice regardless of odorant, size step and concentration. Figure 4 shows that the average performance of the observers improved from just over 1 bit/stimulus, or two categories, to almost 2 bits/stimulus, or four categories. This, as in the case of the detection data above, shows that the ability to perceive odors depends on experience. Man's relatively poor reputation in odor perception as compared with other sense modalities undoubtedly reflects the effect of disuse. Civilized man, it is often suggested, does not depend on smell for survival and uses his nose mostly for pleasure.

Although performance was not perfect with five alternative stimuli even after

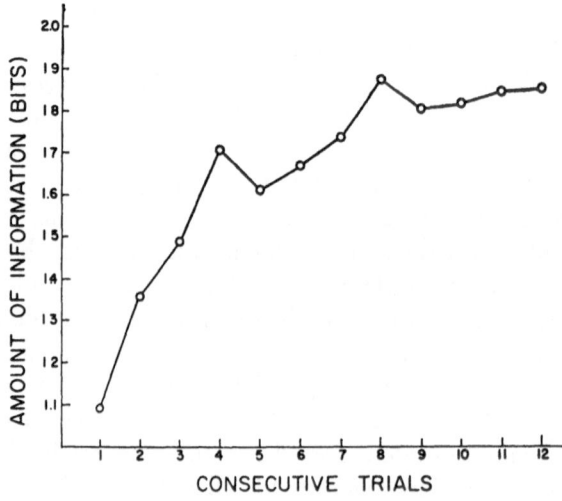

FIG. 4. Amount of information in bits/stimulus as a function of practice trials. The values plotted are averages for 8 observers. (*From* Engen and Pfaffmann. 1959. *J. Exp. Psychol.*, 58:25.)

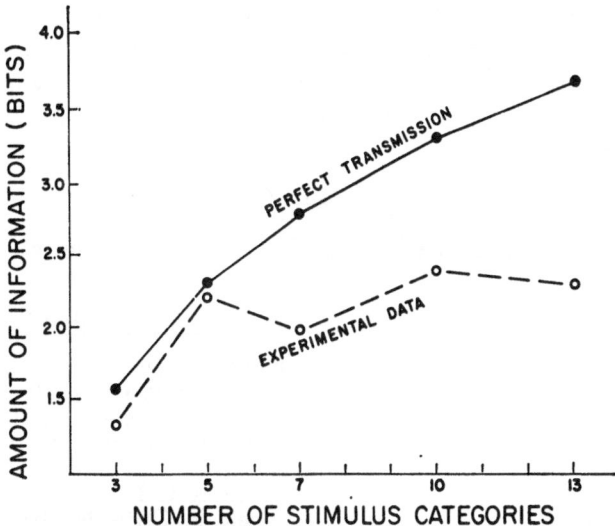

Fig. 5. Amount of information in bits/stimulus, for the best of 8 observers as a function of the number of stimulus intensities presented in a recognition task. (*From* Engen and Pfaffmann. 1959. *J. Exp. Psychol.*, 58:25.)

this long practice, it is possible that constraining the observers to this relatively small number of categories, or "input information" had put a ceiling on the best possible performance or "channel capacity." To assess the extent to which channel capacity may depend on input information, the best of the observers, who had reached almost perfect performance with five alternative stimuli at the end of the 12 practice sessions shown in Figure 4, was presented consecutively, 3, 5, 7, 10, and 13 alternative stimulus categories of amyl acetate. A geometric dilution series was again prepared by progressively halving the concentration of undiluted amyl acetate. The results in Figure 5 show that for this well-practiced observer the judgment efficiency is nearly perfect, or matches the input information, for up to five alternative stimulus categories, but no appreciable improvement results from further increments in the number of alternative stimuli. This was a very conscientious observer, but all his efforts to improve were in vain at this stage. There is no evidence that a "chemist's nose" would do any better under these conditions. It is not unlikely that performance would vary with the selection of odorant and method of presentation of the odorant, but there can be little doubt that the ability to make such unidimensional judgments is limited.

These results agree with findings from similar experiments in other sense modalities. In general, Miller (1956) has concluded that man's ability to identify values of unidimensional stimuli are limited to approximately seven. Once again the higher scores have been obtained in vision (e.g., brightness) and hearing (e.g., loudness) and the lower scores in taste (e.g., concentration of salt) and smell. (See Corso, 1967 for further information.)

The poor performance with odor intensities has often been contrasted with a supposedly great ability to discriminate and remember odor qualities. The advantage of an informational analysis of this problem is that it makes possible this kind of comparison both between and within sense modalities. In the experiment on odor quality Engen and Pfaffmann (1960) used the two sets of odorants shown in Table 4. The sets of "different" odorants were selected to sample a variety of qualities, e.g., fragrant, spicy, fruity, etc., and the "similar" odorants to sample only sweet-fruity.

TABLE 4

Compounds Used in the Quality Experiments*

DIFFERENT	SIMILAR
Acetanisol—16	n-Octyl acetate—24, 10
Acetic acid—24, 15, 10, 7, 5	n-Nonyl acetate—24, 10, 7, 5
Acetone—24, 15	n-Decyl acetate—24, 7, 15
n-Amyl acetate—24, 16, 15	n-Undecyl acetate—24, 10
Allyl Caproate—24, 16, 15, 10, 7	Aldehyde C-7
Benzaldehyde, N.F.—16	Aldehyde C-8
Benzene—24, 16, 10, 7	Aldehyde C-10—24, 10
Benzyl Cinnamate—7, 5	Aldehyde C-14 (peach)—24, 15
n-Butanol—24, 7	Aldehyde C-14 (pure)—24, 10
n-Butyric acid—10	Aldehyde C-16—24, 15
Camphor (lt. syn.)—24, 16	Aldehyde C-18—24, 10
n-Caprylic acid	Allyl caproate—24, 15, 7
Citral—24, 15	Amyl acetate—24
Clove oil—24, 15	Amyl phenylacetate—24, 15, 10, 5
Coumarin—24, 15	Amyl propionate
Diacetone alcohol—24	Benzyl acetate "Coeur"—10
Ethyl acetate—7	Benzyl butyrate—24, 15, 10
Eugenol—24, 16, 15, 10, 7	Cinnamyl acetate—24, 15, 10, 5
Guaiacol—24, 16, 15	Cinnamyl butyrate—24, 15
1-Heptanol—24, 16, 15	Cinnamyl propionate—24
n-Heptane—24, 15, 7	Citral—24, 10
n-Hexane—24, 16, 10	Citronellyl acetate—24, 5
Indol—24, 15, 10	Citronellyl butyrate—24, 15, 7
Isopropanol—16, 15, 10, 5	Dimethyl anthranilate
Linalool—24	Geranyl butyrate
Menthol—24, 16, 10	Geranyl propionate
Methanol—16, 10, 5	Linalool—24, 15, 7
Methyl salicylate—24, 16, 5	Menthol
Musk—24, 16	Methyl anthranilate Standard
α-Novoviol—15	Methyl salicylate—24, 15, 7
β-Phenylethyl alcohol—24, 16, 15	β-Phenylethyl acetate—24, 15, 5
Pyridine—24	β-Phenylethyl alcohol—24, 15
Safrole—24	β-Phenylethyl isovalerate—24, 15, 10, 7
Vanillin—24	Raspberry aldehyde
n-Butyl acetate—15	Tolyl acetate—24, 15
Nitrobenzene	Vanillin

* From Engen and Pfaffmann. 1960. *J. Exp. Psychol.*, 59:215.

The first experiment on quality used 24 different odorants matched in subjective intensity against a 6.25 solution of amyl acetate, which had been judged by a group of other observers to be of moderate intensity. These odorants are identified by the numeral 24 in Table 4.

Next, the matched odorants were compared with information transmitted for

TABLE 5

Means and S.D.s of Bits of Information Transmitted by 5 Ss for Various Samples of Odorants*

SAMPLE OF ODORANT	MEAN	S.D.
Different odorants of equal and medium intensity	4.00	.17
Different odorants, unequal intensities	4.03	.19
Similar odorants, unequal intensities	3.86	.43

* From Engen and Pfaffmann. 1960. *J. Exp. Psychol.*, 59:215.

24 different and 24 similar undiluted odorants and therefore unmatched in subjective intensity.

The procedure followed for quality judgments was the same as that for intensity with one exception. Instead of first learning the rank-order of the stimuli, in the initial phase of the quality experiments, each observer was required to label every odorant used in a particular experiment with his subjective association to that odorant; for example, amyl acetate was usually called "banana oil." This is a relatively easy task, for it avoids the restriction of an arbitrary response code and requires a minimum of verbal ability. After each odorant had been named and checked carefully for duplications, the observers were required to identify the odorants by these labels when the odorants were presented singly and in random orders. The experimenter indicated the correct label after the observer had made his judgment. Each odorant was presented 24 times.

The results are shown in Table 5 and show that the average observer transmitted four bits of information about quality; that is, he could use 16 categories (the antilog$_2$ of 4.00) without error. Whether or not the subjective intensity was controlled was of little consequence because matched stimuli of moderate intensity and stimuli unmatched in intensity show almost the same results (4.03 bits); however, the similarity of the quality of the stimuli played a role, because the amount of information transmitted for the sweet-fruity sample was relatively low, 3.86, or 14 categories.

The purpose of the second experiment was to determine whether or not the information transmitted for intensity (1.5 bits or 3 categories) would add to the information transmitted for quality (4 bits or 16 categories). Sixteen different odorants (identified by "16" in Table 3) were presented at three concentration values, 100,

TABLE 6

Means and S.D.s of Bits of Information Transmitted by Five Ss for 3 Levels of Intensity for 16 Odorants*

TRANSMISSION	MEAN	S.D.
Quality–Responses	3.51	.14
Intensity–Responses	.27	.05
Interaction	.09	.01

* From Engen and Pfaffmann. 1960. *J. Exp. Psychol.*, 59:216.

25, and 6.25 percent for a total of 48 different stimuli or input information of 5.5 bits. The prediction was that the observer should be able to identify all of these stimuli without making errors. The procedure was the same as above, beginning with familiarization with the rank-order of intensity for each of 16 odorants labeled by the observer himself. The judgments were analyzed by a multivariate method (McGill, 1954) which resulted in Table 6. The prediction was not confirmed, but the results are very interesting, and perhaps revealing with regard to how man perceives and uses information about odors. Performance was close to prediction for quality, although errors were made, but performance on intensity, indicated by the low .27 bits, was almost no better than chance. In other words, this suggests that almost all the information is contained in quality. It is possible that intensity plays a role only at extreme values, although the observer is capable of identifying at least three categories on this dimension when there is no information on quality available. The small interaction of .09 bits is consistent with this intepretation. If it is added to the 3.51 bits for quality, the results indicate that 3.60 bits were transmitted about quality for a specified intensity; likewise, .09 bits added to the .27 bits for intensity indicate that .36 bits of information

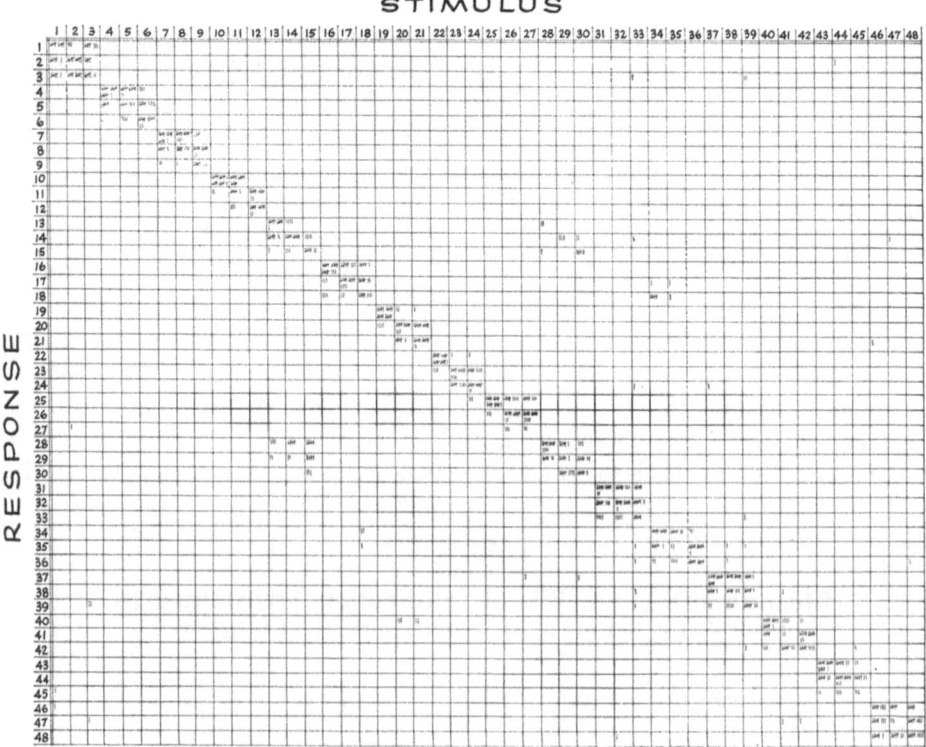

FIG. 6. Stimulus-response matrix for a typical observer. See text for details. (*From* Engen and Pfaffmann. 1960. *J. Exp. Psychol.*, 59:217.)

was transmitted about intensity for a specified quality. Figure 6 illustrates the results of this experiment with data from a single observer. The stimuli are shown in an arbitrary code where numbers 1-3, 4-6, etc. in successive triplets are odors of the same quality but of different concentrations. Figure 6 shows that the typical kind of error made was to confuse these concentrations while accuracy on quality was very high.

The last experiment of quality investigated maximum channel capacity for both different and similar odorants with 5, 7, 10, 15, 24, and 36 alternative stimuli at 100 percent concentration. The results are shown in Figure 7 and are consistent with those obtained in the other experiments on quality. Performance improves up to 24 alternatives and then seems to reach an asymptote. It is interesting to note that performance was never perfect even when fewer than 16 alternatives were presented. In fact, it tends to emphasize that, contrary to popular belief, man's ability to identify odors by quality is not outstanding but falls within the range of ability to identify multidimensional stimuli in other sense modalities, for example, brightness and hue in vision (Corso, 1967). It should be emphasized that a common and always surprising experience in this situation is that even the most familiar odors may be mislabeled, and this kind of confusion increases with the number of alternatives (or surprises) presented by the experimenter.

No evidence of practice was obtained in these experiments on quality, and this may indicate that observers are generally more practiced in judging the qualities as compared with the intensities of odors. The present experiments require hundreds of judgments over a period of weeks, but perhaps improvement in such a task requires extremely long training. Jones (1968) addressed himself to this problem in a study of

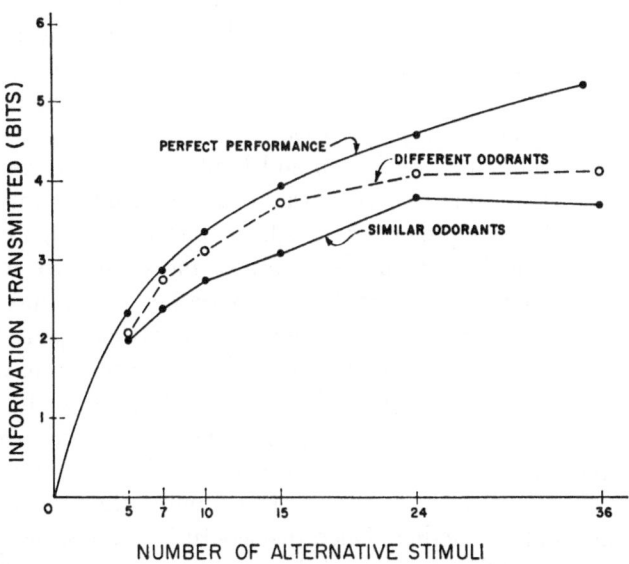

FIG. 7. Amount of information in bits/stimulus as a function of the number of stimulus qualities presented from two sets of odorants in a recognition task. The values plotted are averages for 5 observers. (*From* Engen and Pfaffmann. 1960. *J. Exp. Psychol.*, 59:217.)

the ability of chemists and perfumers in order to resolve the conflict between the present kind of evidence and the popular belief about ability to identify odors. Jones discusses the claim that a successful perfumer would be able to identify absolutely at least 2,000 odors, which in terms of information theory corresponds to more than 11 bits/stimulus. He was able to secure the cooperation of "an experienced individual who had presumably a genuine "chemist's nose," who was a chemist by training and profession and had indeed been a perfumer" (p. 134). This observer and Jones together assembled a collection of 45 diverse odorants. According to Jones, "When these substances were presented in random order, the subject correctly identified 16" (p. 135).

Jones's next experiment investigated the effect of learning on a few unpracticed observers' ability to identify odors in a procedure apparently similar to but more extensive than the one followed by Engen and Pfaffman (1960) wherein the observers familiarized themselves with the odors before identifying them one at a time. The results for four observers, including Jones himself, ranged from slightly less than 4 to 4.6 bits.

In search of the limit of man's capacity in this task Jones's third experiment used two perfumers (who apparently wished to remain anonymous) already trained in identifying odors. One of them had about 25 years and the other five years of experience, and together they proposed a list of "192 odors which they were fairly certain they could readily identify" (p. 136). These odorants were presented in groups of 16 along with a list containing the identification of the odorants, and the task of the perfumer was to smell each odorant and check its name against that list. Compared to the procedure followed by Engen and Pfaffmann (1960), this is a somewhat easier task of recognition which eliminates response coding altogether. Also, the observers selected the odorants used in the experiment. The results did not support the perfumers' expectation, because they made 4 and 5 errors on the first group of 16 odorants. Overall, the more experienced perfumer mislabeled 28 and the other 42 of the odorants. These data do not lend themselves to calculation of information transmitted in bits, but they do suggest at least a performance superior to that of the liberal arts college students used by Engen and Pfaffmann. These results still fall far short of recognizing 2,000 odorants—or transmitting eleven bits of information by the nose. That level of ability represents a belief completely unsupported by objective tests. Jones's (1968) data show again that training and experience improves sensory judgments, but, as he points out, "it is difficult to see how one could provide specific fresh training from any extensive list of odors within a time span short enough to make the necessary responses sufficiently available to affect the results very much. While one is learning a new set the old ones are interfered with" (p. 137). The basis for the belief in man's, or at least the perfumer's, great capacity to identify odors does not refer to identification or recall and recognition of single odorants on the basis of smell alone. Identification has been confused with discrimination based on a comparison of one kind or another. Once that confusion is resolved, the conclusion seems clear that man's ability to judge odor qualities is not phenomenal but falls within the range of his ability to judge tones and colors.

CONCLUSION

It seems reasonable to conclude that as an instrument even the nose of a trained perfumer is not outstanding when compared with the other senses. It is more sensitive than physical instruments in detection tasks but is subject to the attitude and experience of the observer. The resolving power of the sense of smell is poor, and relatively little variation in behavior can be expected from variation of odorant concentration. Even in the ability to identify and classify qualities the nose is only a mediocre instrument, and there is no evidence available to support the hypothesis that training, as in the case of perfumers, will yield remarkable improvements in performance. Side-by-side comparison of qualities must not be confused with ability to recognize; it is a multidimensional discrimination analogous to judging by sight whether or not two automobiles are the same or different or by ear whether or not musical compositions are the same or different.

The most important function of the nose may be not in transmitting messages about the outside world but in motivating the organism after the message has been received. Once a human observer has recognized an odor, or thinks he has, he more or less automatically judges whether the odor indicates something good, bad, or indifferent; that is, will he approach, avoid, or ignore it? Having judged the odor as smoke, our television viewer would attempt to locate a fire. From now on he is alerted to visual, auditory, and cutaneous cues. Precise localization by smell alone would seem an inherently difficult task in view of the nature of the stimulus, although it would get stronger as one approaches the source. The important function of the nose is as an informer initiating the activity in the first place.

The possession of an olfactory system makes man capable of responding to certain aspects of his environment. This chapter has taken a quantitative approach to man's ability in adjusting to his environment by the sense of smell. However, the study of odor perception from a more qualitative, motivational, and hedonic approach remains a relatively unexplored and potentially rewarding area. This volume demonstrates that this is the case in animal behavior. Many people have observed that odors trigger memories. However, sometimes the effect is unconscious in the sense that people may not be able to pinpoint the event that produces the reaction. For example, one woman invariably felt nauseated while using a certain soap. The explanation of the discomfort became clear once she remembered that this was the soap she used during the time she was experiencing morning sickness. Another woman "admitted" to the author how embarrassed she felt for liking the odor of skunk. She had pondered this problem and it came to her gradually that when she was a child and had found herself in the presence of skunk odor her "overprotective" mother had invariably made some remark to the effect that "Isn't that a lovely smell, dear," apparently in an effort to spare her child the unpleasant feelings *she* experienced in the situation. In experiments the author has done on the dimension of odor pleasantness (e.g., Engen and McBurney, 1964) individual responses deviating greatly from the average can often be explained on a similar basis. Sometimes they cannot be accounted for, and the person may not even be able to associate his present psychological state with the

382 / MAN'S ABILITY TO PERCEIVE ODORS

effect of odor. Often people become aware of the importance of smell only after they have lost it, and this too serves as an indication of the subtlety of odors. Of course, these are only anecdotes, but they may well point to the right direction for the psychological investigation of the function of olfaction in man. Smell may be more important to him if it is considered as something much more than a conscious or verbalized experience.

Interest in the importance of the sense of smell in man has increased to the extent that it has now begun to be considered in the study of personality and psychosexual development of children. One analytic notion bandied about is that human pleasures of smell are important not only in taste but in sex as well, but their enjoyment has been repressed by both patient and doctor alike. In any case, it has become a truism to some that man's ability to perceive odors is not necessary for survival in modern civilization, but it is also true that in Western Civilization man has come to associate odor or smell with something bad. The very small range of odor qualities promoted by the perfume industry compared to the enormous range available to them from modern chemistry is indicative of the small range of odor perferences permitted.

REFERENCES

Allison, V. C., and Katz, S. H. An investigation of stenches and odors for industrial purposes. *J. Indust. Eng. Chem.*, 1919. 11:336-338.
Attneave, F. Applications of Information Theory to Psychology: A Summary of Basic Concepts, Methods, and Results. New York, Henry Holt and Company, 1959.
Bach, H. Paper in Gsundheits Ing., 1937. 60:222-225 discussed in R. W. Moncrieff. The Chemical Senses, London, Leonard Hill Limited, 1951.
Behnke, A. R. Basic odor research correlation: Introduction. *Ann. N. Y. Acad. Sci.*, 1954. 58:15-21.
Cain, W. S. Olfactory Adaptation and Direct Scaling of Odor Intensity. Unpublished Ph.D. Thesis, Brown University, 1968.
Cornsweet, T. N. The staircase-method in psychophysics. *Amer. J. Psychol.*, 1962. 75:485-491.
Corso, J. F. The Experimental Psychology of Sensory Behavior. New York, Holt, Rinehart, and Winston, 1967.
Dravnieks, A. A. Approaches to objective olfactometry. *In* Tanyolac, N. ed. Theories of Odors and Odor Measurement. Robert College Research Center, Bebek, Istanbul, Turkey, 1968.
Ekman, G. Weber's law and related functions. *J. Exp. Psychol.*, 1959. 47:343-352.
Engen, T. Effect of practice and instruction on olfactory thresholds. *Percept. Motor Skills*, 1960. 10:195-198.
————— Direct scaling of odor intensity Report No. 106, University of Stockholm Psychological Laboratories, 1961.
————— Psychophysical analysis of the odor intensity of homologous alcohols. *J. Exp. Psychol.*, 1965. 70:611-616.
————— Cain, W. S., and Rovee, C. K. Comparison of olfaction in the newborn infant and the adult human observer. *In* Tanyolac, N. ed. Theories of Odors and Odor Measurement. Robert College Research Center, Bebek, Istanbul, Turkey, 1968.

——— and Lindström, C. A. Psychophysical scales of the odor intensity of amyl acetate. *Scand. J. Psychol.*, 1963. 4:23-28.

——— and McBurney, D. H. Magnitude and category scales of the pleasantness of odors. *J. Exp. Psychol.*, 1964. 68:435-440.

——— and Pfaffmann, C. Absolute judgments of odor intensity. *J. Exp. Psychol.*, 1959. 58:23-26.

——— and Pfaffmann, C. Absolute judgments of odor quality. *J. Exp. Psychol.*, 1960. 59:214-219.

Fechner, G. T. Elemente der Psychophysik. Leipzig, Breitkopf and Härterl, 1860. English translation of Vol. 1 by H. E. Adler (D. H. Howes and E. G Boring, eds.) New York, Holt, Rinehart, and Winston, 1966.

Gamble, E. M. The applicability of Weber's Law to smell. *Amer. J. Psychol.*, 1898. 10:82-142.

Green, D. M., and Swets, J. A. Signal Detection Theory and Psychophysics. New York, Wiley, 1966.

Guilford, J. P. Psychometric Methods (2nd Ed.) New York, McGraw-Hill, 1954.

Jones, F. N. Scales of subjective intensity of odors of diverse chemical nature. *Amer. J. Psychol.*, 1958a. 71:305-310.

Jones, F. N. Subjective intensity of scales for three odors. *Amer. J. Psychol.*, 1958b. 71:423-425.

——— Information content of olfactory quality. *In* Tanyolac, N. ed. Theories of Odors and Odor Measurement. Robert College Research Center, Bebek, Istanbul, Turkey, 1968.

McGill, W. J. Multivariate information transmission. *Psychometrika*, 1954. 19:97-116.

Miller, G. A. The magical number seven, plus or minus two: Some limitations on our capacity for processing information. *Psychol. Rev.*, 1956. 63:81-97.

Moulton, D. G., Ashton, E. H., and Eayrs, J. T. Studies in olfactory acuity. 4. Relative detectability of n-aliphatic acids by the dog. *Anim. Behav.*, 1960. 8:117-128.

Pfaffmann, C. Taste and smell. *In* Stevens, S. S. ed. Handbook of Experimental Psychology. New York, Wiley, 1951.

——— The afferent code for sensory quality. *Amer. Psychol.*, 1959. 14:226-232.

Reese, T. S., and Stevens, S. S. Subjective intensity of coffee odor. *Amer. J. Psychol.*, 1960. 73:424-428.

Semb, G. The detectability of the odor of butanol. *Perc. Psychophys.*, 1968. 4:335-340.

Slosson, E. E. A Lecture Experiment in Hallucinations. *Psychol. Rev.*, 1899. 6:407-408.

Stevens, S. S. The psychophysics of sensory function. *In* Rosenblith, W. A. ed. Sensory Communication. New York, Wiley, 1961.

Stone, H, and Bosley, J. J. Olfactory discrimination and Weber's Law. *Percept. Motor Skills*, 1965. 20:657-665.

Stuiver, M. Biophysics of the Sense of Smell. 'S-Gravenhage (Netherlands) Excelsior, 1958.

Wenzel, B. M. Differential sensitivity in olfaction. *J. Exp. Psychol.*, 1949. 39-124-143.

Woodworth, R. S., and Schlosberg, H. Experimental Psychology. New York, Henry Holt, 1954.

Zigler, M. J., and Holway, A. H. Differential sensitivity as determined by the amount of olfactory substance. *J. Gen. Psychol.*, 1935. 12:372-382.

General Discussion
(Chapters 8 through 12)

BRONSON: We have many things to discuss: we've heard papers on the induction of sexual behavior in a generalized mammal, the mouse; we've heard about sexual attraction in infra-human primates; and we've heard a good discussion of some of the concepts of olfactory perception in humans. Now, we're all mammals, we're all primates, we're all humans, and I assume most of us are interested in sex, so we should have many concepts to consider. Just on the basis of semantics alone, there have been papers in which the word "pheromone" has been used routinely, and we've heard papers in which this term has been avoided. We've heard Dr. Whitten, for example, say how he and I have argued whether or not the term "releasing pheromone" has any business at all in mammalian research, and we've even heard a couple of snide comments about the term "communication" itself.

BARDACH: To start things off, I have two points to make: to supplement the mammalian skin gland paper of Dr. Mykytowycz, I would like to mention a group of fish—the so-called "glandulo-caudine fish"—in which the males have a gland at the caudal peduncle. They are being investigated now, but it seems that this gland is active in courtship. Perhaps this would be an unusual thing in fish. The pheromone emitter in this case would be concentrated in a more distinct fashion than it is for other signalling substances that modify behavior.

What interests me more is this problem of social interactions, the problem of status and individual recognition. I'm having difficulty in resolving, in the experiments that I described to you, whether one should postulate the existence of a status system apart from individual recognition, or whether individual recognition leads to status recognition—whether the two can be used interchangeably or otherwise. I would like to ask my colleagues, who work with mammalian social behavior and smell, to say something about this.

BRONSON: I'll make a comment on this because your question falls into my field of interest. I think that most of us who work with mice tend to think that social status is initially determined by a physical encounter. From this point on, I have no doubt that individual odors function strongly as social reinforcers. For example, recent papers have shown that you can evoke fighting in mice by removing one mouse from the group, putting urine from another male on it, and then returning it. The result is almost as though its social status had not been determined previously.

BURGHARDT: I'd like to ask Drs. Whitten and Bronson why they feel that the terms "releasing" or "releaser" are not appropriate for mammalian forms. The way I understood Dr. Whitten is that just because a "releaser" may not be innate, we shouldn't use the term "release"—well, I don't see the connection. Schleidt has discussed in some detail the various types of releasers, of which an innate releaser is only one type, so I don't see why it must be innate before one can use the term "releaser."

BRONSON: I don't like the term itself—let's start out by saying that I get the impression from this term that there is an almost ritualized pattern of behavior that is, in a sense, held back like water behind a dam. If the appropriate stimulus occurs, the gates go down, and the water or the behavior pours forth. Now, mammalian behavior—whether it is aggression, sex, or whatever—is simply too modifiable to be considered in this manner. For example, the response to a sex attractant may be positive, negative, or no response at all, depending upon the social status of the responder. You say that there can be both innate and learned releasers, but there are many people who are not going to keep that possibility in mind. If I then use the term "releaser" someone immediately says, "Aha! I know exactly what he means"—but maybe he doesn't. The term simply retains too much of its original meaning for me; for example, it implies a degree of stereotypic response that is not accurate in terms of most mammalian behavior.

STÜRCKOW: I think that Dr. Wilson should speak about it because he distinguishes between two types of pheromones in insects: "releaser" pheromones which work rapidly and bring immediate response, and "primer" pheromones which work on a long-term basis. This might help in distinguishing "releasers," or in deciding whether or not you have a releaser.

Secondly, I was very interested in the remark by Dr. Mykytowycz that injection of a pheromone-containing solution did not result in a response, while ingestion and olfactory reception or perception brought an effect. Where, then, does the release moment occur? If the pheromone enters the blood level through injection, nothing occurs; but if it is received through receptive membranes as they exist in the olfactory mucosa and intestines, a release effect is obtained. Does the phenomenon of reception through a membrane cause a change which is perhaps neurally transmitted to the brain to bring about a release effect, while the pheromone in the blood does not effect the essential point? I would like to hear more concerning the different ways in which a releaser can become effective.

WILSON: The dichotomy suggested by Bossert and myself five years ago seemed to be very applicable to the existing knowledge of that time, which was almost entirely entomologic. You will recall that very little had been done of a systematic nature on mammals, although a good deal of our knowledge of primer pheromones came from the work on the reproductive pheromones. Therefore, we didn't worry very much about the exact meaning of the term "releaser." In insects, where responses appear to be truly innate, screening often takes place at the level of the sensory receptors themselves, and there are few complications other than those affecting threshold which could be in the category of learning. Now when we come to mammals, I agree with Dr. Bronson that the use of the term "releaser" (particularly in primates) should be purged of the old Lorenz/Tinbergen meaning, which involved the notion of stored energy and other features of an outdated neurophysiologic model. We should not allow its usage to imply that similar responses and similar physiologic mediating phenomena are involved in primates and insects. I suppose that as knowledge increases in the primates we will want to develop a new terminology and a new classification. Meanwhile, the distinction between releasers and primers remains at least pedagogically useful.

LE MAGNEN: Concerning the mammalian body odors, I want to report the recent work of a French worker, Dr. Ropartz, who has found in mice that in addition to the already known body odors of these mice, there exist two odors, one emanating from the coagulating gland which is responsible for the group effect (i.e., augmentation of the weight of the adrenal gland), and the other emanating from the feet of the mice. The latter is an individual odor responsible for the augmentation of the general activity of the mice and for some modification of the social hierarchy of the colony of mice.

BURGHARDT: To come to my interests, and defend myself (since you mentioned my "snide" comments on communication), I would like the term "chemical communication" clarified, as well as "communication" in general. To get some discussion going, I will take the devil's advocate position that it is a meaningless—or at least unnecessary— term.

UNIDENTIFIED VOICE: I understand communication by chemical signals to mean communication in the sense of transfer of information, and the term chemical signals means that information is contained somehow in the chemical substance—that is, in their structure, in their bonding, in their shape, and something about them that we call chemical. This is where the information is; to transfer these substances means that these substances in going from one place to another can carry the information.

BURGHARDT: I would not say that it is communication.

BARDACH: Do you not suppose that transfer of information would fit well with Burghardt's work as well as that of many others. Communication implies two individuals either of the same or different species, while in his case it is information from the environment either transferred or received olfaction or gustation.

WERTHESSEN: May we ask the disputant to define communication in his terms? What is he doing now except to communicate?

BURGHARDT: Well, I think there are many places where information is transferred which most people would not call communication or even consider animal communication. Don't I get some information from a tree when I detour around it because it is an obstacle? There is a transfer of information involved in this case which I would call perception rather than communication. If you feel that mutuality or some reciprocal action is involved, then what about symbiotic relationships? Would we want to consider these truly communicative in function? I think that the danger lies in the word then becoming so broad that it is essentially meaningless. Maybe we need, secondly, to order symbolic activity before we have communication. But isn't this limited to language? If so, we need only the term language, and none of our chemical work would apply. Conceptions of communication based on information are inadequate, that it is our much less empirical notion of "intent" on the part of "a" which is implicit in examples we feel comfortable in calling communication.

WERTHESSEN: If an animal is signalling that it is in heat to another animal (a female to a male), then by whatever means it is doing it, it is communicating to a potential sexual partner. The message is that the female is fertile and willing to be fertilized. This is most important from the point of view of reproduction. I think the critical thing we are missing here is a bit of the physiology of reproduction. In all species, females are fertile for only a very short time. Sexual intercourse at any other time is useless. Outside our own particular species copulation does not occur unless the female is fertile. There is a tremendous evolutionary advantage in having intercourse restricted to the fertile period of the female. The unique behavior pattern of man may be a secondary development related to communal living.

The female has, therefore, the problem of informing the male to "do it now, it is worth while." I think any system—such as sex skin on the baboon and many other primates, or (as shown today) an odor by which this information is transferred to a sexual partner—can be defined as communication.

UNIDENTIFIED VOICE: Could you define it more abstractly, with an example?

WERTHESSEN: All you need here is a message transmitted by a chemical.

UNIDENTIFIED VOICE: This, then, is down to an information factor again.

WERTHESSEN: All right then, that is communication. It happens when animal "a" uses a system or device (odor, color) for information transfer to inform "b" that a particular situation exists.

UNIDENTIFIED VOICE: So it has to be two ways: "a" has to give a signal; "b" then, has to respond before you can have communication.

WERTHESSEN: Yes, you always have to imply a response. The response potential is there; otherwise there is no communication. I think this is where you are slipping up in recognizing what is the basic reason behind it—at least in the sexual performance. This is honest-to-goodness communication. I have seen it function in the bush and it works very well in getting males and females together at just the right time.

STUART: I think Dr. Burghardt's problem is really the subdivisions of communication. To have communication, one must have an emitter, a channel carrying the message, and a receiver, so that the many factors in the environment can obviously, in the same sense, impart information. Furthermore, perhaps Dr. Burghardt is not making a distinction between intraspecific communication and interspecific communication, and he is perhaps wondering whether communication ought to be a two-way relation— a dialogue as it were. Usually, as in human communication, once the message has been received, a message is sent back: in other words, a dialogue. I think that intraspecific chemical communication by definition implies pheromones, and perhaps the term "pheromone" should be restricted to this. Anything else can be conceived of as a chemical signal or a chemical stimulus setting off a behavioral response.

WILSON: Communication is an alteration of the response probabilities which is adaptive in nature, meaning that the rules have been programmed by natural selection.

ENGEN: I just want to make a silly comment. Information means the same as uncertainty in communication theory. If there are no alternatives or surprises, there is no uncertainty or information. So we could get hung up on this semantic problem without really solving it. The only solution would be to make some specific definition of the term. That would require some model, preferably a mathematical model such as communication theory. I do not think we are advanced enough in this field to get anywhere with that kind of elegance so I would suggest we drop the topic and agree that we use "communication" in a broad, general common-sense meaning. I think we all understand each other when we use the term.

STÜRCKOW: We have another term which is difficult to comprehend: inter- or intraspecific. I think you mean inter- or intra-individual specific but we also have inter-species specific, inter-genera specific, inter-cell specific. We need to define which different concepts are being distinguished. The term "interspecific" has to have a context, if I am to understand its meaning.

STUART: Since I used the terms, I will try to say what I mean by them. "Intraspecific" is a normal zoologic term meaning "between members of a species," and "interspecific" means "between members of different species."

WILSON: May I return to biology for a moment? (And it takes some courage because I want to deal with a subject that is really outside my area of competence). Dr. Whitten called for hypotheses concerning the adaptive significance of the Bruce effect. This effect, to remind you, is the abolition of pregnancy by the odor of strange males. The one hypothesis that has been raised, is that the effect serves as a density-dependent population control mechanism. This idea is hard for me to conceive. But I would like to raise another hypothesis, namely that the Bruce effect is really not an adaptive phenomenon having to do with intraspecific communication, but that it is an epiphenomenon that has to do with isolation among species of rodents. It is possible that one of the mechanisms by which rodent species isolate themselves, where more than one species occur together, is by odor differences. To use the terminology of the evolutionary biologist, the odor differences to which the different strains of *Mus* and *Peromyscus* are so sensitive may be prezygotic isolating mechanisms, or the beginnings of prezygotic isolating mechanisms. In other words, there may be a very sensitive response on the part of rodents to curtail reproductive activity in the presence of a strange rodent odor, *that odor normally being the odor of strange species*. It is well known that it is very non-adaptive to mate with strange species and to produce hybrids. In support of this hypothesis we can recall the work of Godfrey on the vole populations on islands off England. Different populations tend to produce different odors, in the pattern generally regarded as being part of early geographic speciation. The mice prefer mates with the same odor. To recapitulate, it may be that in the course of rodent evolution, odors play an important role in speciation. And if that is true, then the rodents may be very sensitive to slight differences in odors, a phenomenon which would be manifested in laboratory experiments when different genetic strains were brought together.

BRONSON: Could I say one thing with regard to Dr. Wilson's comment? I am not sure that, when considering either the induction of estrus or the block to pregnancy in *Mus* or *Peromyscus*, you really have to postulate an epiphenomenon. It seems to me that Dr. Whitten now knows quite a bit about the genetics of these phenomena, and they seem to be fairly simple. He can correct me if I am wrong on this, but it would seem to me that there are obvious selection advantages for the ability of a male to inhibit a pregnancy that would have developed from another male's insemination, induce estrus again, and then copulate with the female himself. There may, of course, be additional functions such as preventing inbreeding and ensuring isolation. The other point I would like to make is that there is a better example than the *Clethrionomys* work, namely, Moore's doctoral dissertation at the University of Texas in which he tested sexual attraction in males of two types of *Peromyscus,* one of which occurred in nature sympatrically with a variety of other types of *Peromyscus,* one of which was geographically isolated. The geographically isolated type of male showed little ability in detecting species differences in odor, whereas the type of male taken from the sympatric situation detected its own species of female easily.

MASON: Dr. Burghardt, you saw quite different responses in snakes taken from different localities. How do you explain the difference within a species, or the difference as being only environmental?

BURGHARDT: I do not think that I meant to imply that the differences in my four litters from three geographic areas were environmental. What I hope to have shown in that study was that even within the same species, but from different geographic areas, there can be different effectiveness in the "releasing values" (or whatever term you want to use) of the same stimuli. Somehow, geographic separation has led them to respond somewhat differently to extracts which are still effective. In other words, it is not so much the type of extract attached which was different, but the relative effectiveness of the various extracts.

MASON: Yes, it is a quantitative difference. You wouldn't care to explain that?

BURGHARDT: All I can say is that in these various locations, there were different habitats.

MASON: Were these snakes hatched in captivity?

BURGHARDT: The females were collected in Wisconsin, Iowa, and two from northern Illinois. I kept the snakes in captivity until they gave birth and I tested 21 of each of these 4.

MASON: Does this have to be explained as a genetic difference then? It can't be, can it?

BURGHARDT: Why can't it be? They are the same species.

MASON: You say they are mutations within a species relative to these particular likes—a quantitative difference?

BURGHARDT: Well, I think within many species there are morphologic differences between various localities, so I think we should expect behavioral differences too. Recently, it has been shown that populations of the same species of lizards can have different push-up displays. Intraspecifically you have the same genetic processes going on; this is what evolution is all about. You do expect, of course, more differences between species than within species.

I might comment on the idea that there may be some environmental effect on the native snake's behavior. For instance, the females were feeding on different foods, so you could say that my results are all due to what the female has been eating. I am trying to do genetic studies now, but they are very difficult with snakes. I do know what the females were feeding on in captivity while I had them. I know that if I fed the female minnows all the time during gestation, the young could still respond much more to worms than to fish. Two of the species I have discussed were oviparous forms, and yet the young had a response profile that was very specific also, so that I am inclined to believe that there is a strong genetic factor.

BRONSON: All you are trying to say is that all populations of the same species do not necessarily have the same genetic make-up.

JOHNSTON: Dr. Whitten, I was very interested in your question about a fixative during your discourse. Could a pheromone really be a melange of a few components held together by a fixative? Did you mean that this melange would be a product of a given gland?

WHITTEN: That was a comment about the odors in general, in particular reference to the odors which individuals identify. This has considerable meaning when it comes to recognition of a strange male which would induce the Bruce effect and there must be some recognition factor there. I was not referring to the fixative function with respect to the specific pheromone action.

JOHNSTON: Mr. Chairman, I am going to venture a timid speculation. In this context—recognition of the individual—I am more than ever enthused about the notion that the scent is a melange of three or four components held together by a fixative (in the sense that their various rates of volatility are approximately equalized). This would be the way the individual differences of a body scent of a given species could be communicated. Dravnieks has an apparatus for collecting human odors at the Illinois Institute for Technology in Chicago. It is essentially a large chamber into which a person of around 5 feet 6 inches height can be rolled on a litter. An air stream is passed through to collect the volatiles, which are fed into a gas-liquid chromatograph and an odor signal obtained. (Of course, the donors have been processed before they are put in there.) In other words, there are end components in variable, individualistic proportions, according to the chromatogram.

Communication by Chemical Signals: Conclusion

J. LE MAGNEN

Laboratoire de Physiologie des
Sensibilités Chimiques et
Régulations Alimentaires
de l'E.P.H.E.
Collège de France—Paris, France

The present knowledge of chemical systems of communication among animals and of their characteristic features has been reviewed in various chapters of this book. In the overall view of this field we may ask what is the survival value of the chemosensory apparatus and chemical signals? In other words: which essential requirements of animal species are satisfied by these systems of communication through their properties, and what are these peculiar properties?

To answer the question, a preliminary general consideration has to be made.

The differentiated chemosensory exteroceptors represent a sensory system parallel to the internal chemoreception operating in the relations between cells and tissues in the "milieu intérieur." Hypothalamic gluco- and osmoreceptors, carotid bodies and possibly neurons responding to chemical stressors, correspond to the "sweet, salt, acid, and bitter" analysis of gustatory nerve discharges. The chemo-selectivity of every cell, and among them, of free blood cells (for example, lymphocytes), is an homolog of the olfactory-like sensitivity of bacteria and protozoa (Adler, 1965; Metalnikov, 1913). The response of olfactory cells in the differentiated organs of multicellular organisms has been classically compared with the response either of postsynaptic neuronal membranes to chemical transmitters, or of receptive tissues to hormonal messages.

Thus, two analogous chemosensory systems operate: one inside the body for the relations between cells and organs, the other for the relations of free organisms both among themselves and with their external environments. These two systems are intermingled and—as shown below—are correlated in their respective functions. The reason for this coexistence and this correlation is to be found in the following general

393

fact: the interochemoreceptors are mostly involved in the pattern of physiologic regulations insuring the constancy of the "milieu intérieur." This concept of Claude Bernard may be extended to behavioral regulations and to the role of chemical messages in relation to their regulations. Through regulatory behaviors, animals build up, in average space and time, a kind of constancy of their environment: constant temperature and respiratory medium, constant supplies of water and food. Likewise, the self-preservation of individuals by defensive reactions, and of species by selective mating, implies a general constancy of social relationships. As an extension, and as a requirement of the constancy of the "milieu intérieur," this building up through the behavior of a mean constant environment needs the use of a chemical system of communications in addition to other sensory modalities.

Initially, the survival value of this system of communication was "chemical" in nature, that is, it defended the organism against stresses related to the state and composition of its chemical environment. All behaviors designed to provide the above constancy of the environment through relations and exchanges with the medium, requires a selective appraisal of this chemically differentiated environment. The selection of food of a constant caloric and specific value is only possible after a chemosensory analysis of the biochemical properties of the food sources.

Individuals, species, and faunistic families are biochemical entities. Inasmuch as these entities are reflected in an emitted chemical signal, this signal, better than other sensory cues, will be the sensory basis of individuality, mother-young relationships, selective mating, defensive reaction against enemies, and so on. Hence the preservation of these chemical individualities will be insured.

The second advantage of chemical communication is represented by the high level of specificity of messages and by the accuracy of sensory discriminations. The possibility of differentiating and identifying various sources depends both on the emission by these sources of specific chemical signals and on the separative power or discriminative performances of sensory organs. These two specificities have been tested separately or together in numerous wonderful works bearing both on the chemical identification of active stimuli and on animal responses.

Many examples of this high degree of specificity have been given in preceding chapters. The active material of insect sexual pheromone is so species-specific that it has been possible to name this substance, when chemically identified, by the name of the species (for example, "Bombikol"). A great number of specific responses to such insect attractants, phagostimulants, alarm, and trail substances, are now known to depend on a single chemical structure. The smallest modification of these structures, cis-trans or optical isomers for example, leads to the disappearance of the responses (Bates and Sigel, 1963; Butenandt et al., 1959). When the active material is not known, the discrimination of species and individual odors (by fish, mice, and others for example), demonstrates the existence of individual and species scents. Simultaneously, their discriminability by olfactory receptors is demonstrated. Many of these responses have been shown to be highly specific, as, for instance, the trail-following of fire-ants (Wilson, 1962a, 1962b), the responses of monophagous or oligophagous insects to food stimuli (Schoonhoven, 1967-1968).

All these data on behavioral responses to natural and mixed odors, and to pure chemical substances have confirmed the existence (already known in man) of an olfactory discriminative accuracy reaching as far as the level of molecular differentiation. Electrophysiologic investigations, looking for the mechanism of this discrimination of molecular units by different patterns of discharges from nonspecialized receptor cells, have shown in insects (Boeckh, 1968; Schneider, 1962, 1963) that some of these olfactory cells are highly specialized. They respond only to the sex attractant or food stimuli of the species. Moreover, only males, and sometimes males of one species, possess these specific receptivities.

Thus the specificity and accuracy of responses are based not only upon selective reactions to a particular signal, discriminated from a background of other active signals, but also upon "all or none" efficiencies of some of these chemicals at the receptor level.

Without a supplementary specificity through a temporal or spatial modulation of signals (which is, as shown below, limited), and without consideration of the informative capacity due to the transmission properties of the channel, the signal-to-noise ratio depends on the level of these two specificities of both emission and reception. This ratio is as high as are these specificities and, as shown above, is very high indeed. In the most favorable cases the insulation of the channel of transmission is remarkable. This is the case, for example, when an insect attractant is emitted only by the female of a particular species and is perceived only by the male of the same species. The equivalent in acoustic signals would be a highly specific call of an emitter (which is common), but heard only by a given and appropriate receptor, others being deaf to the same signal.

The most original characteristic of this chemical system, recently substantiated by a variety of works, is the neuroexocrine-endocrine modulation of both emitted signals and responses.

Since body odors are sources, there is much evidence that emitted messages are directly controlled by, or are consequences of the neuroendocrine-exocrine mechanisms. Thus, when discriminated by receivers, they provide information about the various physiologic states involved in such mechanisms.

The existence of individual and species odors is shown by the ability of various animals to discriminate between two congeners as well as between congeners and members of other species such as their predators (Bowers and Alexander, 1967; Goz, 1942; Todd, Atema and Bardach, 1967; and others). The secretion of sebaceous glands are the probable sources of these individual and specific chemical signals in mammals. This supposition has been verified for the alarm reaction of salmons to L-serine, one of the products of this secretion of mammalian skin (Alderdice, Brett, Idler and Fagerlund, 1954; Idler, Fagerlund and Mayoh, 1956). It is also probable that the individual and species specificity of these signaling secretions is related to the biochemical individuality demonstrated by immunologic compatibility tests. The inability of dogs to discriminate between the odors of human twins tends to confirm this (Kalmus, 1955). Thus, an olfactory "signature" exists which conveys, via an exocrine mechanism, a genetic, fundamental, biochemical pattern.

It has been repeatedly demonstrated in various species that the secretion of more specialized exocrine glands acting as sexual messages are directly controlled by sexual neuroendocrine mechanisms (Barth, 1965 in female insects; Le Magnen, 1951a, 1951b, 1951c, 1951d, in rats; Hamilton, 1949-1950 in hamsters; Dutt, Simpson,

Christian and Barnhart, 1959, in swine). According to the sex, estrogen or androgen hormones modulate the specific secretion. These hormones per se, or their break-down products excreted in urine, may provide the specific olfactory message acting upon the appropriate receivers and eliciting the specific response. This fact has been recently demonstrated in the onset of estrus synchrony in female mice (Whitten effect) in which the testosterone content of male urine seems the effective agent (Bronson and Whitten, 1968; Marsden and Bronson, 1964). Other neuroendocrine mechanisms are sources of emanating odors and they are therefore directly involved in the genesis of a carrier of a particular physiologic information. The alarm substance of fish, for example, responsible for the well known "Schreckreaktion" is probably due to skin secretions and their modification by emotional stress (Todd, Atema and Bardach, 1967).

Thus, in the same animal a number of emanating substances, each derived from a particular neuroglandular system, form a true olfactory language. This language possesses its own semantics carried by a specific chemical material or a pattern of chemicals present in various secretions. In mice, for example, five or six different odors of different sources and physiologic significance, and effects upon receivers, have already been identified: the individual odor (Bowers and Alexander, 1967), the alarm odor (Muller-Velten, 1966), the male urine odor (Bronson and Whitten, 1968; Marsden and Whitten, 1964) and the alien male odor (responsible for the Bruce effect: Bruce, 1967; Parks and Bruce, 1961), to which must be added two odors recently identified by Ropartz (1966a, 1966b, 1968): an individual and a group odor emanating respectively from sweat glands of pads and from the coagulating glands of the male genital tract.

Thus, a chemical patterning of information exists in such systems with unlimited possibilities for coding quantitative and qualitative aspects of physiologic events. The study of these neuroendocrine-emitted stimuli relationships, completing the classical stimuli-responses relationships in receivers, is a new, fascinating field open to further research.

The neuroendocrine modulation of responses to chemical stimuli occurs in two types. Either the behavioral response depends in its quality and intensity on the endocrine state of the receiver, or the response itself is a neuroendocrine one, that may or may not be associated with a behavioral reaction.

In the first type the neuroendocrine control of responses may take place by way of an hormonal action interfering with chemosensitivity itself. In insects, a difference of sensitivity of the two sexes to various odors and particularly to the sex attractant has been recognized. The fact has now been electrophysiologically confirmed by the demonstration, for example, of the lack of a single unit responding in female *Bombix mori* to its own pheromone (Schneider, Block, Boeckh, and Priesner, 1967). Such sex-linked all-or-none, or differential sensitivity to certain types of odors, has not yet been extensively verified in vertebrates, but it seems plausible. However, in these species the modulation of the sensitivity level under the influence of neuroendocrine states has been verified. The first evidence of such an hormonal modulation of sensitivity has been the demonstration of the systemic fluctuation of olfactory acuity with menstrual cycle in women (Köster, 1965; Le Magnen, 1948-1950; Meixner,

1955; Vierling, 1967). A peak of sensitivity occurs at the time of ovulation or estrus This lowering of threshold is particularly accentuated for musky odors like that of the C15 lactone Exaltolide. The presence of androsten possessing a strong musky odor (Prelog and Ruzicka, 1944) in male human urine has been subsequently demonstrated (Brooksbank and Haslewood, 1961). The general sensitivity of women is highly affected by ovariectomy and restored by estrogenic treatment (Le Magnen, 1950; Schneider, Costiloe, Howard, and Wolf, 1958). Recently this hormonal dependency of olfactory sensitivity has also been demonstrated in fish of both sexes (Hara, 1967). Its occurrence in mammals other than man, and particularly its variation with estrus cycle, is very probable. Along the same lines, adrenal deficiencies lead in man to an apparent increase of olfactory and gustatory acuity reversible after hormonal therapy (Henkin et al., 1962-1966).

Such hormonal modulation of sensory mechanisms are unknown, and, of course, inconceivable in other sensory modalities, thus representing a typical feature of chemosensory systems.

Electrophysiologic data concerning the role of centrifugal fibers, and their facilitatory or inhibitory effects on afferent olfactory pathways demonstrate the mechanism of these hormonal influences. At the peripheral level, stimulation of sympathetic fibers both in olfactory (Tucker and Beidler, 1956) and gustatory (Kimura, 1961) systems modifies the thresholds. At bulbar and prepiriform levels centrifugal fibers coming from thalamus and mesencephalic reticular formation, and particularly from the hypothalamus (Powell, Cowan, and Raisman, 1965) probably exert such a selective control (amplification or inhibition) on afferent pattern discharges under hormonal influences.

Whether or not they have this primary basis in sensory mechanisms, variations of behavioral responsiveness to olfactory and gustatory cues with hormonal status are very common and now well known. It is not useful to list them. It is, for example, the case in the rats ability to discriminate between odors of estrus and diestrus females only present in mature male rats (Carr and Caul, 1962; Le Magnen, 1951b, 1951c). Variations under the influence of endocrine (Insuline) or, in general, humoral factors in relation to the nutrition status, hunger and satiety are well documented (Le Magnen, 1963-1969; Long and Tapp, 1967; Pfaffmann and Hagstrom, 1955). In insects, the reciprocal relationships between satiety, its neural and humoral background, and contact chemoreceptors thresholds have been fully demonstrated (Dethier, 1967; Gelperin, 1966).

One of the most original features of chemosensory systems of communication is represented by the direct releasing action of chemical stimuli upon endocrine systems. Through short neuron pathways, true somato-visceral reflexes participate in these effects. Details have been given in various chapters of this book on experimental results showing that specific olfactory messages may stimulate or inhibit ovarian maturation and functioning. As animal pheromones, plant phagostimulants may directly affect ovogenesis and oviposition in parasitic phytophagous insects (Barton-Browne, 1960; Cadeilhan, 1965; Hudson, 1967; Robert, 1964, and others). In mammals, prolactin release may be, as in mice (Bruce effect), inhibited by specific olfactory stimulation. These external chemical releasers are as effective in directing hypothalamohypophyseal functions as sucking and vaginal stimulations are effective in releasing LH and oxytocin. These latter responses have been classically considered as reflexes.

Such acquired or innate reflexes are seen in action in the short-term regulation of food and water intake. Recent works have clearly shown that oro-pharyngeal stimulation by food elicits anticipated insulin release, of adrenal medullary and probably neurohypophyseal discharges, according to the specific gustatory or olfactory acting stimuli (Nicolaidis, 1969). In this phenomenon (see below) the overlap between external and internal chemosensory systems of homeostasic regulation is evident.

Electrophysiologic investigation has shown that hypothalamic neurons, activated or inhibited after an intracarotid infusion of hypertonic saline solutions, are also activated or inhibited by this solution applied to the tongue (Nicolaidis, 1968). Single unit activity in both feeding (lateral) and satiety (medio-ventral) hypothalamic centers is modified by olfactory stimulation (Oomura et al., 1967). Other hypothalamic sites in the female rat are activated by olfactory afferents more in the proestrus than in the diestrus phase of the menstrual cycle (Barraclough and Cross, 1963).

Such a neuroendocrine control of emitted signals, and of sensitivities and responsiveness in both emitters and receivers, form a very typical loop of positive and negative feed-back mechanisms. These mechanisms, acting through chemical systems of communication, are involved in sexual behavior and more generally in the social relations of animal species.

Another peculiar characteristic of chemosensory communication among animals— one which contributes in some situations and types of behaviors to its survival value— appears in the special conditions of transmission of chemical messages.

Chemicals diluted in air are either transmitted signals from an emitter to a receptor, or are a stationary, fixed, or recorded signal resulting from the deposition of chemical materials on various substrates. Owing to this latter property the topographic organization of signals may involve a special behavior (territorial marking, trail-laying). The exploration of these spatial patterns is another typical olfactory behavior (trail-following, for example). Of course all intermediaries exist between these two types of transmitted, and fixed, activity-explored signals.

The conditions of the channel of transmission in air of chemical stimuli as a source of directional and distance information are very disadvantageous compared with light and sound propagation. The spreading of chemical signals in still air depends on slow and nondirectional diffusion. The ratio of given concentration to threshold concentration defines an active sphere, and, taking into account the diffusion coefficient, the time of efficacy in each point (Wilson, 1965). From these parameters it is computed that some alarm substances may act as brief signals, effective, in *Pogonomyrmex badius* (Latreille), during less than 35 seconds in an "active space" attaining a maximum radius of about 6 cm. Inside this active space (which, in the case of insect attractants, for example, is very large), the directional and distance information in the olfactory field in still air is theoretically nil. At a large distance, the intensity gradient may not provide this information. As demonstrated by Kettlewell, 1946, Hannes (1965), and Martin (1964) in insects, bilateral localization, and therefore directional and distance information, is only effective at a short distance and on the front of an active space. These still-air conditions are, in practice, only realized in small active spaces, and particularly at a short distance from the source—for example, a fixed adsorbed source in the ground. In that situation (trail-following in dogs for instance) the von Bekesy (1964) mechanism of localization seems fully effective.

In moving air, the spreading of chemical signals depends entirely on the wind. But this dependency per se yields a source of directional information. Upwind and guided by their anemotaxis, insects progress in the olfactory field in the direction of the female. A comparable rheotaxis is probably effective in the progression of the spawning salmon up to its parental river (Ueda, Hara, and Gorbman, 1967).

It should be pointed out that such a chemical system of transmission in air cannot be temporally modulated. In acoustic and luminous signals this modulation carries the maximal quantity of the transmitted information. A temporal patterning of emissions is sometimes realized, for example, by female insects emitting rhythmically their sexual attractant. But, as shown recently by Bossert (1968), this temporal modulation is very rapidly smoothed out by diffusion and might be active in the best circumstances only at a distance of one meter.

As noted above, the essential carrier of the specific information allowing the receiver to identify a source is the chemical structure itself and its decoding by the chemosensory apparatus.

The laying or the natural occurrence of stationary sources of chemical stimuli allow spatial modulation in this system of communication. It strongly resembles the visual signalling system used by man with road-signals, arrows, lights, and buoys. The territorial marking behavior has been described in rabbits (Mykytowycz, 1965-1967), and the trail-following in fire ants. In dogs, the information content of the trail is very high. In addition to its very accurate directional and distance information, the trail carries vectorial information. A dog, upon intersecting a trail, can immediately find the direction taken by the layer. The origin of this dromic-antidromic discrimination is unknown. The differential sensitivity associated with the binarinal mechanism of localization also operates to enhance the accuracy of the trail-following. When service dogs are tested in a Y-shaped trail, two branches having been laid by the same layer at 60-minute, 30-minute, or 15-minute intervals, the dogs upon reaching the crossing point always choose the more recent branch. The threshold in some subjects is less than 15 minutes (Le Magnen, and Dugas du Villard, unpublished).

The last and perhaps the most important original characteristic of chemical stimuli is their ability to acquire signifying property in consummatory acts—to become in the proper sense, *signals.*

The attractiveness or repellency of these signals in general, their feeding or sexual stimulatory or inhibitory effects, have often been demonstrated to be the result of innate reactions. This is probably the case in the response to insect sexual attractants. Various species (fish: Goz, 1942; Kleerekoper, 1967; Todd, Atema, and Bardach, 1967; rat: Reiff, 1958; Vernet-Maury, 1968) respond to body odors of specific prey or predators they have not previously experienced. The stimulatory and inhibitory effects on feeding of sweet and bitter substances, respectively, are present in neonates (Pfaffmann, 1936). Food preferences or selective feeding based upon olfactory cues are also manifest in inexperienced young—in snakes for example (Burghardt, 1967). The phylogenetic origin of such innate responses—their inheritance—raises many unanswered and important questions.

Whatever may be this origin and mode of transmission of innate reactions, the close association of chemosensory stimuli with various consummatory acts (feeding, mating, and so on) offers the best conditions to acquire by experience during the life span of animals the releasing efficacy of these behavioral responses. These conditions are the contiguity between the sensory stimulation and the consummatory effect. In this situation, analogous to those of experimental operant conditioning, the sensory stimulus easily becomes the feedback regulator of the consummatory response. The role of such a learning process in feeding regulation has been the subject of extensive experimentation. It is easy to show in rats, for example, that the repetitive association of an olfactory cue with a required food (Harris, Clay, Hargreaves, and Ward, 1933; Le Magnen, 1959-1968; Scott and Quint, 1946) makes this olfactory cue active in regulating either selective or caloric appetites. Through this alimentary conditioning, innate reactions may be positively or negatively reinforced or substituted. At least in vertebrates, this oral chemosensory function in feeding acquired through this learning process is one of the major roles played by these sensory systems in behaviour. Similar facts might be described in mating behavior.

Finally, the most common cause of the place occupied by chemosensory messages, and of their preferential use by many species must not be neglected. Sight depends on light, and hearing on sounds. Chemical stimuli are always present or available. All nocturnal species (and, among them, most of mammals) hunt and seek for their food or their sexual partners, in darkness. Thus, odors become the dominant guide for their behavior. The same is true in the underground life of other species living inside their burrows and galleries, and among fishes living below a certain depth. Often hearing cannot be used. Nature is more a world of scents than a source of noise.

The reason, among others, for the modest role of chemical sensation in communication in our species is perhaps that man is diurnal, has discovered artificial illumination, and is also a very noisy and chattering animal.

REFERENCES

Adler, J. 1965. Chemotaxis in *Escherichia coli*. *Cold Spring Harbor Symp. Quant. Biol.*, 30:289-292.

Alderdice, D. F., Brett, J. R., Idler, D. R., and Fagerlund, U. 1954. Further observations on olfactory perception in migrating adult coho and spring salmon: Properties of the repellent in mammalian skin. *Prog. Rep. Pacific Coast Stat. Fish Res. Board (Canada)*, 98:10-12.

Barraclough, C. A., and Cross, B. A. 1963. Unit activity in the hypothalamus of the cyclic female rat: Effect of genital stimuli and progesterone. *J. Endocrinol.*, 26:339-359.

Barth, R. H. 1965. Insect mating behavior: Endocrine control of a chemical communication system. *Science*, 149:882-883.

Barton-Browne, L. 1960. The role of olfaction in the stimulation of oviposition in the blowfly, *Phormia regina*. *J. Insect Physiol.*, 5:16-22.

Bates, R. B. and Sigel, C. W., 1963. Terpenoides: Cis-, trans- and trans-cis-nepetalactones. *Experientia,* 19:11, 564-565.

Bekésy, G. von. 1964. Olfactory analogue to directional hearing. *J. Applied Physiol.,* 19 (3):369-73.

Boeckh, J. 1968. Qualitative and quantitative reaction range of single insect olfactory receptor cells. *In* Tanyolaç, N., ed. NATO Institute of advanced studies on the theories of odor and odor measurement, Istanbul, 1966, pp. 213-224.

Bossert, W. H. 1968. Temporal patterning in olfactory communication. *J. Theoret. Biol.,* 18 (2):157-170.

Bowers, J. M., and Alexander, B. K. 1967. Mice: Individual recognition by olfactory cues. *Science,* 158 (3805):1208-1210.

Bronson, F. H., and Whitten, W. K. 1968. Oestrus-accelerating pheromone of mice: Assay, androgen-dependency and presence in bladder urine. *J. Reprod. Fert.* 15:131-134.

Brooksbank, B. W. L., and Haslewood, G. A. D. 1961. The estimation of Androst-16-en-3α-ol in human urine. *Biochem. J.,* 80:488.

Bruce, H. M. 1967. Effects of olfactory stimuli on reproduction in mammals. *In* Holtensholme, G. E. W. and O'Connor, M., eds. Effects of External Stimuli on Reproduction. Ciba Found. Study Group no. 26, London, Churchill, pp. 29-42.

Burghardt, G. M. 1967. Chemical-cue preferences of inexperienced snakes: Comparative aspects. *Science,* 157:718-721.

Butenandt, A. R., Beckmann, D., Stamm, D., and Hecker, E. 1959. Uber den Sexual-Lockstoff des Seidenspinners *Bombyx mori.* Reindarstellung und Konstitution. *Z. Naturforsch.,* 14b:283-284.

Cadeilhan, L. 1965. Stimulation de la ponte d'Acrolepia assectella (Zell). par la présence de la plante hôte. *C. R. Acad. Sci.,* 261:1106

Carr, W. J., and Caul, W. F. 1962. The effect of castration in rat upon the discrimination of sex odours. *Anim. Behav.,* 10:20-27.

Dethier, V. G. 1967. Feeding and drinking behavior of invertebrates. *In* W. Herdel et al., eds. Handbook of Physiology. Washington, D.C., Amer. Psychol. Society sect. 6, vol. 1:79-96.

Dutt, R. H., Simpson, E. C., Christian, J. C., and Barnhart, C. E. 1959. Identification of preputial glands as the site of production of sexual odor in the boar. *J. Anim. Sci.,* 18:1557.

Gelperin, A. 1966. Investigations of a foregut receptor essential to taste threshold regulation in the blowfly. *J. Insect. Physiol.,* 12: 829-841.

Göz, H. 1942. Uber den Art-und individualgeruch bei Fischen. *Z. Vergl. Physiol.,* 29:1-45.

Hamilton, J. B., and Montagna, W. 1950. The sebaceous glands of the Hamster. I. Morphological effects of androgens on integumentary structures. *Amer. J. Anat.,* 86:191-233.

Hannes, F. 1965. Uber den sinnesphysiologischen Mechanismus beim Aufsuchen und Auffinden entfernter Duftquellen in strömender Duftquellen in strömender Luft durch fleigende Insekten im Gegensatz zu kriechenden. *Biol. Zentralbl.,* 84 (2):191-203.

Hara, T. J. 1967. Electrophysiological studies of the olfactory system of the goldfish, *Carassius auratus* L. III. Effects of sex hormones on olfactory activity. *Comp. Biochem. Physiol.,* 22:209-225.

Harris, L. J., Clay, J., Hargreaves, F. J., and Ward, A. 1933. Appetite and choice of diet: the ability of the vitamin B deficient rat to discriminate between diets containing and lacking the vitamin. *Proc. Roy. Soc. London*, Ser. B, 113:161-190.

―――― and Bartter, F. C. 1966. Studies on olfactory thresholds in normal man and in patients with adrenal cortical steroids and of serum sodium concentration. *J. Clin. Invest.*, 45(10):1631-1639.

―――― Gill, J. R., Bartter, F. C., and Solomon, D. H. 1962. On the presence and character of the increased ability of the Addisonian patient to taste salt. *J. Clin. Invest.*, 41:1364-1365.

Hudson, A., and McLintock, J. 1967. A chemical factor that stimulates oviposition by *culex tarsalis coquillet* (Diptera, Culicidae) *Anim. Behav.*, 15:336-341.

Idler, D. R., Fagerlund, U., and Mayoh, H. 1956. Olfactory perception in migrating salmon. I. L-serine, a salmon repellent in mammalian skin. *J. Gen. Physiol.*, 39(6):889-892.

Kalmus, H. 1955. The discrimination by the nose of the dog of individual human odors and in particular the odours of twins. *Brit. J. Animal Behav.*, 3:25-31.

Kettlewell, H. B. D. 1946. Female assembling scents with reference to an important paper of the subject. *Entomologist*, 79:8-14.

Kimura, K. 1961. Factors affecting the response of taste receptors of rat. *Kumamoto Med. J.*, 14:95-99.

Kleerekoper, H. 1967. Some effects of olfactory stimulation on locomotory patterns in fish. *In* Hayashi, T., ed. Olfaction and Taste II, 2nd Int. Symp. Tokyo 1965, Oxford, Pergamon Press, p. 625-645.

Koster, E. P. 1965. Olfactory sensitivity and the menstrual cycle. *Rhinol. Int. III*, (1)57-64.

Le Magnen, J. 1948. Un cas de sensibilité olfactive se présentant comme un caractère sexuel secondaire féminin. *C. R. Acad. Sci.*, 226:694-695.

―――― 1950. Nouvelle données sur le phénomène de l'exaltolide. *C. R. Acad. Sci.*, 230:1103-1105.

―――― 1951a. Etude des phénomènes olfacto-sexuels chez le rat blanc. Méthode de détermination de la réponse de l'animal aux odeurs biologiques du mâle et de la femelle. *C. R. Soc. Biol.*, 145:851.

―――― 1951b. Etude des phénomènes olfacto-sexuels chez le rat blanc. Variations de l'odeur biologique de la femelle avec son état sexuel et la discrimination de ces odeurs par le mâle adulte. *C. R. Soc. Biol.*, 145:854.

―――― 1951c. Etude des phénomènes olfacto-sexuels chez le rat blanc. Variations avec leur état sexuel de la réponse des mâles à l'odeur de la femelle et réponse des femelles à l'odeur du mâle. *C. R. Soc. Biol.*, 145:857.

―――― 1951d. Etude des phénomènes olfacto-sexuels chez le rat blanc. Variation de sensibilité olfactive absolue du mâle et de la femelle en fonction de leur état hormonal sexuel. *C. R. Soc. Biol.*, 145:1636.

―――― 1959. Effets des administrations post-prandiales de glucose sur l'établissement des appétits. *C. R. Soc. Biol.*, 153:212-215.

―――― 1963. Le contrôle sensoriel dans la régulation de l'apport alimentaire. in: L'Obésité, Paris, Expansion Scientifique, p. 147-171.

―――― 1969. Peripheral and systemic actions of food in the caloric regulation of intake. *Ann. N.Y. Acad. Sci.*, 157:1126-1156.

Long, C. J., and Tapp, J. T. 1967. Reinforcing properties of odors for the albino rat. *Psychon. Sci.*, 7(1):17-18.

Marsden, H. M., and Bronson, F. H. 1964. Estrous synchrony in mice: alteration by exposure to male urine. *Science*, 144:1469.

Martin, H. 1964. Zur Nahorientierung der Biene um Duftfeld zugleich ein Nachweis für die Osmotropotaxis bei Insekten. *Z. Vergl Physiol.*, 48(5):481-533.

Meixner, C. H. Changes in olfactory sensitivity during the menstrual cycle. M.A. Thesis, Providence, R.I., Brown University, June 1955.

Metalnikov, S. 1913. Sur la faculté des infusoires "d'apprendre" à choisir la nourriture. *C. R. Soc. Biol.*, 74:701-703.

Montagna, W., and Hamilton, J. B. 1949. The sebaceous glands of the hamster: II some cytochemical studies in normal and experimental animals. *Am. J. Anat.*, 84:365-388.

Muller-Velten, H. 1966. Uber den Angstgeruch bei der Hausmaus (*Mus musculus* L.). *Z. Vergl. Physiol.*, 52(4):401-429.

Mykytowycz, R. 1965. Further observations on the territorial function and histology of the submandibular cutaneous (chin) glands in the rabbit, *Oryctolagues cuniculus* (L.). *Anim. Behav.*, 13(4):400-412.

——— 1967. Communication by smell in the wild rabbit. *Proc. Ecol. Soc. Aust.*, 2:125-131.

Nicolaidis, S. 1968. Réponses des unités osmosensibles hypothalamiques aux stimulations salines et aqueuses de la langue. *C. R. Acad. Sci.*, 267:2352-2355.

——— 1969. Early systemic responses to oro-gastric stimulation in the regulation of food and water balance. Functional and electrophysiological data. *Ann. N.Y. Acad. Sci.*, 157:1176-1200.

Oomura, Y., Ooyama, H., Naka, F., and Yamamoto, T. 1967. Microelectrode positioners for chronic animals. *Physiol. Behav.*, 2(1):89-91.

Parkes, A. S., and Bruce, H. M. 1961. Olfactory stimuli in mammalian reproduction. *Science*, 134:1049-1054.

Pfaffmann, C. 1936. Differential responses of the newborn cat to gustatory stimuli. *J. Genet. Psychol.*, 49:61-67.

——— and Hagstrom, E. C. 1955. Factors influencing taste sensitivity to sugar. *Am. J. Physiol.*, 183:651.

Powell, T. P. S., Cowan, W. M., and Raisman, G. 1965. The central olfactory connexions. *J. Anat.*, 99:791-813.

Prelog, V., and Ruzicka, L. 1944. Unteruschungen über Organextrakte. Uber zwei moschusartig riechende Steroide aus Schweinetestes-Extrakten *Helv. Chim. Acta*, 27:61.

Reiff, M. 1958. Untersuchungen über naturiche und synthetische Geruchstoffe, die bei Ratten und Maüsen eine stimulierende Wirkung auslösen. *Acta Trop.* (Suisse), 13(4):289-318.

Robert, P. 1965. Influence de la plante-hôte sur l'activité reproductrice de la teigne de la betterave *Scrobipalpa* (Phthorimaea) *Ocellatella boyd* (Lepidoptere Plutellidae). *Proc. XII Int. Congr. Entomol.* London 1964.

Ropartz, P. 1966a. Mise en évidence d'une odeur de groupe chez les souris par la mesure de l'activité locomotrice. *C. R. Acad. Sci.*, 262:507-510.

——— 1966b. Mise en évidence du rôle d'une sécrétion odorante des glandes sudoripares dans la régulation de l'activité locomotrice chez la souris. *C. R. Acad. Sci.*, 263:525-528.

——— 1968. The relation between olfactory stimulation and aggressive behavior in mice. *Animal Behaviour*, 16(1):97-100.

Schneider, D. 1962. Electrophysiological investigation on the olfactory specificity of sexual attracting substances in different species of moths. *J. Ins. Physiol.*, 8:15-30.

———— 1963. Electrophysiological investigation of insect olfaction. *In* Zotterman, Y., ed. Olfaction and Taste. Oxford, Pergamon Press, p. 85-103.

———— Block, B. C., Boeckh, J., and Priesner, E. 1967. Die Reaktion der männlichen Seidenspinner auf Bombykol und seine Isomeren: Elektroantennogramm und Verhalten. *Z. Vergl. Physiol.*, 54(2):192-209.

Schneider, R. A. Costiloe, J. P., Howard, R. P., and Wolf, S. 1958. Olfactory perception thresholds in hypogonadal women: changes accompanying administration of androgen and estrogen. *J. Clin. Endocrinol. Metabol.*, 18:379-390.

Schoonhoven, L. M. 1967. Chemoreception of mustard oil glucosides in larvae of *Pieris brassicae*. Proc. Konink. Nederl. Akad. Wetensch., Ser. C, *Biol. Med. Sci.*, 70(5):556.

———— 1968. Chemosensory bases of host plant selection. *Annual Rev. Entomol.*, 13:115-136.

Scott, E. M., and Quint, E. 1946. Self-selection of diet: II. The effect of flavor. *J. Nutr.*, 32:113-119.

Todd, J. H., Atema, J., and Bardach, J. E. 1967. Chemical communication in social behavior of a fish, the Yellow Bullhead (*Ictalurus natalis*). *Science*, 158(3801): 672-673.

Tucker, D., and Beidler, L. M. 1956. Autonomic nervous system influence on olfactory receptors. *Am. J. Physiol.*, 187:637.

Ueda, K., Hara, T. J., and Gorbman, A. 1967. Electroencephalographic studies on olfactory discrimination in adult spawning salmon. *Comp. Biochem. Physiol.*, 21(1):133-144.

Vernet-Maury, E., Chanel, J., and Le Magnen, J. 1968. Comportement émotif chez le rat; influence de l'odeur d'un prédateur et d'un nonprédateur. *C. R. Acad. Sci.*, 267:331-334.

Vierling, J. S., and Rock, J. 1967. Variations in olfactory sensitivity to exaltolide during the menstrual cycle. *J. Applied Physiol.*, 22(2):311-15

Wilson, E. O. 1962. Chemical communication among workers of the fire ant *Solenopsis saevissima* (Fr. Smith): I. The organization of mass-foraging. *Anim. Behav.*, 10(1-2):134-147.

———— 1962. Chemical communication among workers of the fire ant *Solenopsis saevissima* (Fr. Smith): II. An information analysis of the odour trail. *Anim. Behav.*, 10(1-2):148-158.

———— 1965. Chemical communication in the social insects. *Science*, 149(3688): 1064-1071.

INDEX

Page numbers in *italics* refer to tables and figures

ACTH, 320
Adenohypophysis, 321
Adrenals, 314, 317
Aggregation drive, 253
Aggressive behavior, 314, 319
  in fish, 216, 224-226
Aeoliscus, 232
African toad (Xenopus), affection for opposite
        sex, 202
Alarm reaction
  cloacal fluid
    alarm substance of King snake, 250
    defensive role of, 250
    obnoxious odor of, 249
  of skunk oil, in rattlesnakes, 256
  substances eliciting, 303
Alarm substance, 176-189
  characteristics of, 184-185
  in the Dolichoderinae, 178-180
    chemistry of, 178, 179
    effects of 2-heptanone, 179-180
    function of, 178
    point of origin of, 178-179, 180
    specificity of, 179
  in the Dorylinae, 177-178
  in the Formicidae, 176-177
  in the Formicinae, 180-182
    chemistry of, 181-182
    dendrolasin, 181
    point of origin of, 181
  in the Myrmicinae, 182-183
    chemistry of, 182-183
    point of origin of, 183
  in the Ponerinae, 177
    chemistry of, 177
    point of origin of, 177
  specificity of, 185
Alarm transmission system, characteristics of,
        183-184
  biological, 184
  chemical, 183
  physical, 183
Allomone, 3, 80, 81
Alosa alosa, 217
Ambergris, 350
Amino acids, 235
Amitermes herbertensis (Mjöb), 92, 93
Amitermes vitiosus (Hill), 93

Ammonium hydroxide, 253
Amphiprion (clownfish), 232
Amyl acetate, 131, 257, 363
Androconia
  of the butterfly, 43
  as scent glands, 43
Androgens, 312, 316, 317, 343
  index of, 343
Anemotaxis, positive, 50
Anestrous, 313, 314
Anguilla anguilla (European eel), 212, 215
Anoplura. SEE *Pediculus*
Antennae, 108, 109, 117, 162
  as used in alarm reactions, 182
  in ants, 162
  combs, 202
  long, 202
Antherea (saturnid silk moth), 134
Antherinops affiniis (top smelt), 231
Antorbital gland, 351
Aphrodisiacs, 42-44
  as used in courtship behavior, 42
  secreted by the dorsal abdominal gland, 43
  secreted by the gustatory organ, 42
  secreted by the olfactory organ, 42
Apis mellifica, 117. SEE ALSO Bees
Appetitive behavior, 259
Astyanax (blind cavefish), 212
Attractancy of chemical compounds, 110, 111,
        115, 130
Audition, human
  identification of stimuli, 363
  lack of absolute threshold, 375
Avertin, 317

Bark beetle. SEE *Ips*
Bathygobius soporator (goby), 211, 216
Bees, 108, 109, 111, 112, 113, 114, 115, 117,
        120, 134, 139
Beetles. SEE Coleoptera
Benzaldehyde, 43, 114
Benzene, 114
Benzyl mercaptan, 128
Blennies, 217
Blennius pavo, 217
Blood, 100
Boars, 313
Body odor, 113

Bombus spp., 99
Bombyx mori (silkmoth), 94, 117, 130-132, 134, 139-142, 396
Bombykol, 2, 50-51, 129, 139, 140, 141
Bullheads. SEE Catfish
Butterflies, 109, 115, 132

"Calling" behavior, 88, 89
Calliphora erythrocephala (blowfly), 121, 122, 123, 132, 145
Calliphora vomitoria (blowfly), 128
"Canalization," 289
Campaniform sensilla. SEE Sensilla
Cannibalism, 8
n-Caproic acid, 95, 310
  in insects, 95, 142
  in mammals, 310
Carassius carassius (crucian carp), 231
Carausius morosus (Phasmidae), 117
Carbon atoms, number of and stimulating effectiveness of, 235, 311
Carbon dioxide, 139
Carbon monoxide, 116
Carnegiella strigata (hatchet fish), 231
β-Carotene, 145
Carp, 213, 231
Carvone, isomers of, 29
Castoreum, 311
Castrates, castration, 315, 316
Caterpillars, 114, 116
Catfish, SEE ALSO *Ictalurus* spp., 209-211, 213, 214, 217, 220-229, 231, 233, 234, 236
Cats, 317
Cavefish. SEE *Astyanax*
Ceratitis capitata, 27
Channel capacity, 375
Chemoreception
  chemical factors in, 30-31
  physical factors in, 31-33
Chemosensory systems, 393-394
Chemical communication, in insects, behavioral and ecologic aspects, 35-65
Chemical perception, in newborn snakes, 277-303
  chemical analysis of the effective substances, 302-303
  the comparative perspective, 282-291
  concentration of chemicals in, 293-295
  modifying the "innate schema," 296-302
  role of various senses in, 279-282
  stimulus and response deprivation and, 291-293
Chemical senses, in reptiles
  evolution and morphology of, 242-245
    Chelonia, 243
    Crocodilia, 243
    Rhynchocephalia, 243-244
    Squamata, 244-245
  role of, 245-277
    alarm reaction, 249-250

courtship, 251-254
  Chelonia, 251
  Crocodilia, 251
  Sauria, 253
  Serpentes, 253-254
feeding behavior, 256-277
  Chelonia, 256-257
  Squamata, 257-277
    research after 1935, 269-277
    research before 1935, 258-269
  intraspecific aggregation behavior, 247-249
  maternal behavior, 254-255
  method of investigation, 245-246
  orientation, 246-247
  predator recognition, 255-256
  territorial markings, 250-251
"Chemist's nose," 377, 380
"Chinning," 334, 348
Chin pressing, 253
Chorda tympani, 139
Chromatography, 235
Cichlasoma nigrofasciatum, 219
Cichlids, 219
Cilia, 214
Citral, 56, 57, 181
Citronellal, 28, 57, 181
Civetone, 311
cis-Civetone, 24
Circadian rhythms, 230. SEE ALSO Diurnal rhythms
Clevelandia ios (goby), 232
Cloaca, in snakes, 249
Cloacal rubbing, 253
Cockroach. SEE *Periplanata*
Codfish, 213, 214
Coding of information in insect chemoreceptors, 132, 143
Cold receptor cells, 116, 133, 137
Coleoptera, 114-118, 122-124, 127, 128, 132, 133, 136, 138, 139
  bark beetles, 117, 132
  dung beetles, 114
  potato beetles, 111, 115, 117, 132, 133, 138, 139
  sexton beetles, 115, 122, 123, 127, 128, 134
  water beetles, 115
Colonies, of termites
  aggregation, 81, 82, 84
  caste determination, 87
  caste recognition, 85-87
  cohesion, 81, 84
Colony odor, 98-101
Communication, chemical
  neuroendocrine control of, 396-398
  spatial features of, 398-399
  specificity of, 394-395
  survival value of, 394
Communication, in mammals, systems of, 328
Conditioning experiments, 110
  with fish, 212, 221
Cooperative behavior in fish, 224, 225

Copulin, 217
Corpora allata, 58
  and attractive odor, 42
Corticosterone, 314
Corynopoma spp., (glandulocaudine fish), 236
Courtship behavior in fish, 216-218
Cryptotermes cynocephalus, 89
Cryptotermes havilandi (Sjöst), 89
Cyclopentanone, 129

Deermouse, 313, 318. SEE ALSO *Peromyscus*
Defense secretions, 80
Dendrolasin, 57, 180, 181
Detection theory, 367-368
Diurnal rhythm, 321. SEE ALSO Circadian
  rhythm
cis-3-*cis*-6-*trans*-8-Dodecatriene-1-ol, 96
Dogs, 212, 312
Drepanotermes rubriceps (Froggatt), 92, 93
Drosophila ampilophila (pomace fly), 108, 109

Eels, 211, 213, 214, 215
Elasmobranchs, 213
Electroantennogram (EAG), 109, 129-131, 139-143, 145
Electron microscopy, 109, 119, 122, 124, 125, 126
  olfactory peg, 124
  olfactory sensillum, 109, 119
  taste hair, 109, 122, 124-126
Endocrine status, 311, 312
Environmental factors, 312, 313, 321
Esomus lineatus, 231
Esox (pike), 233
Estradiol, 314
Estrogen, 310, 314
Estrous cycle, 310-314, 316, 319, 320
Ethanol, 114, 137
Euthynnus pelamis (tuna fish), 208
Evolution, 110, 234, 237, 318
Exaltone, 310
"Extract-response profile," 301

Fatty acids, 95, 114, 127, 134, 142, 310, 311
Feces, 309, 312
Fechner's law, 370
Feeding reaction, in reptiles, 271
Fierasfer, 232
Flies, 108, 109, 114-116, 121-123, 125, 128, 137
  blowflies, 116, 119, 121-123, 125, 126, 128, 134, 136, 137, 144, 145
  muscid, 109, 137
  pomace, 108, 109
  stable, 109, 123
Flow capillary for liquid stimulus delivery in experiments with insects, 145
Free nerve endings, 214. SEE ALSO Trigeminal nerves

Gambusia, 233
Gasterosteus aculeatus (stickleback), 213, 214, 217, 236
Generator potential in insect chemoreceptors, 109, 116, 127-130, 142
Genetic factors in controlling ability to recognize odors and in controlling pheromone effects, 312, 313, 314, 316
Geotrupes, 114, 116
Geraniol, 23, 49
Glands
  accessory sex, 315
  anal, 53, 217, 329, 333, 334, 347
  aphrodisiac, 44
  axillary, 337
  brachial, 349
  caudal, 217, 236
  cephalic, 91
  cloacal, 249
  cutaneous, 217, 309, 349
  dorsal abdominal, 49
  Dufour's, 172, 174, 175-176, 179, 181, 190
  epidermal follicular, 250-251
  inguinal, 336, 348, 351
  lateral, 342
  mandibular, 46, 49, 54, 56, 62, 63, 177, 180, 181, 182, 183
  metanotal, 44
  metatarsal, 303
  Nassanoff's scent, 46, 49, 53-54, 56
  odoriferous, 341, 344, 347
  Pavan's, 165-166, 167, 174, 190
  pedal, 332
  perianol, 347
  postnasal, 249
  pregenital, 347
  preputial, 315, 334, 341, 343
  scent, 311
  sebaceous, 343
  slime, 229, 232, 233, 234, 235
  sternal, 46, 94-97, 99, 100, 329
  submandibular, 337, 341, 346
  supra-anal, 178, 180, 190
  supra-occipital, 346
  tarsal, 330, 332, 346, 351
  ventral, 329, 344
Glandulocaudine fish, 217, 236
Glucose, 134
Glycosides (alkaloid), 117, 132, 134, 136, 139
Glucuronates, 315
β-Glucuronidase, 315
Glycol, 113
Goats, 317
Gobiosoma bosci (naked goby), 235
Gobio fluviatilis, 231
Goby, 231, 232, 235
  blind, 211, 215, 216
Goldfish, 214
Gonadotropin, 318, 322
Grasshopper, 169. SEE ALSO *Melanoplus* ssp.
Guinea pig, 311, 317

Guppy, 217
Gustation, in reptiles, 242, 243
Gypsy moth, 32

Habitat conditioning, 249
Habrobracon (parasitic wasp), 114
Hair pencils, 43
  of the *Danaus* butterfly, 43
  of the *Lycorea ceres ceres*, 43
Haplochromis multicolor, 217
Hatchet fish. SEE *Carnegiella*
Hemichromis bimaculatus (jewel fish), 218,
    229, 233
Hepsetia stipes (an antherinid fish), 231
Heptanol, 128
2-Heptanone, 56, 57, 179, 180
  as an alarm substance, 174, 180
  as a defense substance, 179
  as a repellant, 180
n-Hexanoic acid. SEE *n*-Caproic acid
trans-2-Hexenal, 42
3-Hexen-1-ol, 83-85
History of research on insect chemosensory
    systems, 108
Homing, 206, 207
  in turtles, 246-247
Homeostasis, 102
Host plant selection, 109, 110, 114
Housefly. SEE *Musca domestica*
Humidity, 116, 137
Hydrous (water beetle), 115-116
9-Hydroxy-*trans*-2-decenoic acid, 49, 55
Hypothalamus, 311, 313, 317, 321, 322
Hypsoblennius spp. (blennies), 217

Ictalurus natalis (yellow bullhead), 209-211,
    213, 220-228, 233, 234, 236
Ictalurus nebulosus (brown bullhead), 217,
    218, 229, 233
Imprinting, 312
Individual recognition, 312
  in fish, 220-224, 226, 228
Inhibitory scent, 62
Induction of the estrous cycle, in mice, 313,
    315, 319
Information transfer, 7
"Innate perceptual schema," 291
"Innate" responses, in newborn snakes, 277
Input information, 375
Insecticides, 137, 138, 139
"Instinctive regression," 291
Isoamyl acetate, 131
Isopentyl acetate, 57
Interspecific chemical communication, in fish,
    232-234
Ionone, 24
Ips confusus (Scolytidae), 117, 132
Isolating mechanism, 236
Isomerisms
  geometric, 23-27
  optical, 28-29

structural, 22
Isotopes, 30
Isovaleraldehyde, 111

Jacobson's organ, 242-243, 244, 245, 247, 254,
    255, 261, 262, 263, 265, 271,
    272, 273, 275, 277, 280, 291, 309,
    311
Jewel fish. SEE *Hemichromis*

Kalotermes spp., 82-89
Keto-enol equilibrium, 29-30

Lactone, 310, 311
Lampreys, 213, 217
"Learned" responses, in newborn snakes, 277
Lebistes reticulatus (guppy), 217
Lenzites trabea (Pers. ex Fr.), 100, 101
Lepomis megalotis (longear sunfish), 207, 233
Leptine, 117, 139
Leptotyphlops, 10
Limb pads, as a source of odor, 312
Limonene, 92, 93
Limulus, 137
Linalool, 28
Loaches (*Cobitidae*), 214
Locust, 134, 142
Locusta migratoria, 142
Lucilia spp. (fly), 116

Malpighian tubules, and attractive odor, 41
Mandibular gland secretion (bee), 117
Mating stance, 341
Maze for fish, 228
Meat, decaying, odor of, 127, 128
Median effective concentration, 115
Melanoplus differentialis (Acrididae), 117, 122,
    124
Menthol, 26-27, 29
Menthone isomers of, 26-27
Methodology, 145-147. SEE ALSO Olfactometer
4-Methyl-3-heptanone, 56
Mice, 311-320
Microtermes edentatus, 85
Minnow (Cyprinidae), 207, 211-215, 220, 229,
    230-233
Molecular weight, 311
Morays, 214
Mouthpart, 108, 109
Mus, 318
Mus musculus (house mouse), 318, 319
Musca domestica (housefly), 137
Musks, musk-like odors, 310, 350
  civetone, 311
  exaltone, 310
  muskone, 311

Nannacara anomala (dwarf cichlid), 219
Nasutitermes spp., 80, 90, 91, 94, 95, 97-100
Necrophorus (sexton beetle), 115, 116, 123,
    127, 128, 134

Neotermes spp., 99
Nerol, 23, 114
Nerolic acid, 49
Nerve impulses (single unit activity) in insect chemoreceptors, 109, 127-129, 132-134, 136-139, 142, 143
  site of initiation of, 107, 110
Nomeus, 232
Nonbiologic potentials, 131, 132. SEE ALSO Receptor potentials, and Generator potentials
Nose, the, as an informational channel, 372-380

Odor
  human response to, 112, 114, 115, 116
  mammalian, sources of, 328-329
  man's ability to perceive, 361-380
Odor discrimination, bees, 113, 117
Odor gradients, 50
Odor perception, in mammals, 335
  gustatory, 335
  nasal, 335
Odor presentation apparatus, for fish, 219, 220, 221, 222, 224. SEE ALSO Olfactometer
Odor spots, 46
Odor streaks, 46
Odor theories, 143
Odor trails, 44, 45-46, 47, 50
  life of, 45
  physical appearance of, 45-46
Olfaction, in reptiles, 242, 243
Olfactometer, 82, 108, 111, 114, 115
Olfactory bulb
  of fish, electrical activity in, 206
  in mammals, 310, 313, 314, 316
Olfactory communication, characteristics of, 175-176
Olfactory cues, in mice, 339
Olfactory discrimination, 312, 366-367
  in fish, 206, 207, 226
Olfactory epithelium, 310, 311
  elimination of, in fish, 227-229
  regeneration of, in fish, 227
Olfactory receptors, in mammals, 310, 311
Olfactory response, factors influencing, in fish, 230
Olfactory system, in fish, anatomy of, 213, 214
Olfactory threshold, measurement of, 364
Orientation
  in fish, 206-211, 213, 229
  in insects, 82, 94-98
Orthokinesis, 82
Orthoptera, 117
9-Oxodecenoic acid, 42, 60, 63, 65
9-Oxo-*trans*-2-decenoic acid, 42, 60, 62
Oysters, 235

Paraffin oil, 113
Parent-young relations, in fish, 218, 219

"Payoff matrix" experiment, 366
Pediculus humanus corporis (human louse), 116, 117, 132
Perch, 233
Perfumes, 114, 311
Periplanata americana (cockroach), 131, 140, 142
Peromyscus maniculatus bairdii (prairie deermouse), 313, 318, 319, 320, 321
pH, effects of, 131, 133, 137
Pheromones, 2-3
  and aggression, 52-55
  alarm (alerting), 53, 55-57
  chemical nature of, 83-85, 91-93, 95, 96, 98, 232, 235, 311, 315
  classification of, 36, 80, 81, 88, 94, 102
    olfactory acting, 36
    orally acting, 36
    primer, 36
    release, 36
  concentrations of, 31-33
  definition of, 80, 88, 90, 94, 230, 310
  ecologic significance, 318, 320
  gustatory, 36, 43-44
  olfactory, 36, 43-44, 54
  olfactory sex-attractant, 37-42, 50, 52, 65
    species-nonspecific, 40
    species-specific, 39-40, 43
  primer, 80, 81, 310, 311, 313-316, 318, 320-322
  relations between, 317
  releaser, 310
  signalling, 310
  source and production of, 231, 316
  surface, 47-49
    action of, 47
    composition of, 47
    uses of, 48-49
  trailmarking, 36-37
    aerial, 37
    terrestrial, 37
Pheromone cycle, hypothesis of, 63
Phenylethyl alcohol, purity of, 21
Philosamia cynthia, 132
Philtrum, 311
Phormia regina (blowfly), 125, 126, 134, 136, 137
Porthetria dispar, 32
Photinus, 16
Photuris, 16
Phoxinus laevis (minnow), 212, 229, 230, 232, 233
Phoxinus phoxinus (Elritze minnow), 212, 214, 215
Pieris, 132, 133
Pigs, 313, 317, 318
Pike. SEE *Esox*.
Pimephalas notatus (blunt nose minnow), 207
Pinene, 91
Pituitary, pituitary hormones, 317, 322
Plants, in relation to homing, in fish, 207

Plasma, 314
Pogonomyrmex badius, 398
Podura (collembola), 144
Polistes, 93
Polyethism, 83
Porthetria dispar (gypsy moth), 131, 140, 141
Portugese man-of-war, 232
Precopulatory behavior, role of olfaction in, 341
Pregnancy block, in mice, 312, 313, 316-321, 341
Prey-attack response, in reptiles, 271, 291
"Primary drive reduction," 299
Proboscis, 109
Prolactin, 317
Propionic acid, 127, 134
Propyl isobutyl ketone, 57
Prostoglandins, 311
Protoparce, 132
Pseudopregnancy, 313, 314
Psychophysical scaling, 362, 363, 369-372
Psychophysical threshold, 363
Pterine-like substances, 232
Purine, 232
Purity, 20-21

Queen excluder, 63
Queen rearing, control of in social insects, 61-64
  accepting newly mated queens, 61
  loss of a queen, 61-62
  pheromones, 62-63, 64
  9-oxodecenoic acid and, 63
  production of few queens, 61
  production of many queens, 61
  queens, excrements and, 64
  reproductive swarming and, 63
Queen substance, 62-63

Rabbits, 317
Rasbora heteromorpha, 230
Rats, 313, 317
Rays, 213
Receptor mechanisms, 143-145
Receptor potentials, in insect chemoreceptors, 127, 128, 145. SEE ALSO Generator potentials
Recognition, problem of, 361, 362
Releasers, social, 80, 81, 236
"Releasing mechanism," 291
Repellancy of chemical substances, 110, 111, 115, 130, 137
Reserpine, 317
Resolving power, 365-367
Responses, behavioral and neural, compared, 139-142
Reticulitermes flavices (Kollar), 89, 94, 96-98, 100, 101
Reticulitermes lucifugus, 84, 86, 88
Rhamdia (catfish), 228
Roach. SEE *Rutilus*

Rodents, 317, 318
Rudd. SEE *Scardinius*
Rutilus rutilus (Roach), 229

Salmon, 206, 208, 211, 217, 235
Salt, 132, 137, 138, 139, 148. SEE ALSO Sodium chloride
Salt receptor cell, 136-138
Scardinius erythrophthalmus (Rudd), 229
Scent folds, 41
Scent glands, 41-42, 311
  anal, 41
  sternal, 41
  tergal, 41
Scent glands, mammalian
  approaches to the study of, 329-330
    use of gas-liquid chromatography in, 330
    use of psychometric methods in, 330
  occurence of, 330-331
    cervidge, 331
    lagomorphs, 331
    nocturnal species, 331
Scent-producing organ (bee), 113, 114
Scent rings, 41
Scent sacs, 41
Schooling behavior, in fish, 228, 229, 233, 235, 236
Schreckreaction, 396
Sea anenome, 232, 233
Sea cucumbers, 232
Sebum, 336-337
Sensillum, 107-110, 119-121, 124-126, 132, 133
  accessory cells, 107, 119, 120
  basiconicum, 118
  campaniform, 94, 96
  maxillary taste, 132
  placodeum, 109, 120
  trichodeum, 118
  structure of, 118-122
Sensitivity of a receptor cell, change in, 110
Serine, 235
Sexual maturity, control of development of, 57-61
  effect of pheromones on, 57-58, 59, 60
  egg-laying potential and, 58-59
  oogenesis and, 60
  ovary size and, 59, 61
Shad, 217
Shannon-Wiener measure of information, 373
Sharks, 213
Sheep, 317
"Signposts," 345
Sign stimuli, 236, 246
Silkmoth. SEE *Bombyx*
Siluriformes, 209, 211, 212, 214, 229
Skatole, 32, 114
Skin glands, mammalian
  apocrine sudoriferous, 335
  chemical studies of, 350-352
    castoreum, 350
    musk, 350

Skin glands (cont.)
  comparative studies of, 347-348
  distribution of secretions of, 332-335
    active marking, 332-334
    frequency of, 334-335
    passive marking, 332
    gustatory perception and, 335
  holocrine sebaceous, 335
  primate, 348-349
  sebaceous, 336
  types of messages conveyed by odor from, 336-347
    age appraisal, 340
    aggression, 346-347
    characteristic pattern of behavior, 342
    effects of castration on, 343
    effects of injection of sex hormones on, 343-344
    group appraisal, 339
    individual appraisal, 338-339
    messages related to reproduction, 341-342
    olfaction, 337
    seasonal changes, 344
    sexual dimorphism, 342
    social status appraisal, 340-341
    territorial marking, 344-345
    warning, 346
Smelt. SEE Antherinops
Sniffing, differential, in fish, 214
Social behavior, in fish, 227-229, 235-237
Sodium chloride, 139. SEE ALSO Salt
Sows, 313
Speciation, 234
Sperm, 311
Sphyrna (hammerhead shark), 213
Spontaneous activity, in insect chemoreceptors, 133
Statistical evaluation, 111
Steroids, 315
Stevens' power function, 370
Stickleback. SEE Gasterosteus
Stimulus-response (dose-response) relations, 111
Stoichactis (giant sea anemone), 232
Stomoxys (stable fly), 123
Strain differences, in mice, 314-316, 318
Stress, effects of, in fish, 226, 314
Structure-activity relations, 110
Sucrose, 117, 137
Sugar, 109, 110, 113, 114, 127, 132, 148
Sugar receptor cells, in insects, 127, 137
Sunfish, 233
Sunfish, longear. SEE Lepomis
Supplementary reproductives, 61, 63, 64
Suppression of the estrous cycle, in mice, 313, 315, 320
Symbiosis, 232, 233
Synchronisation of the estrous cycle, in mice, 314, 315, 317

Tandem behavior, 89
Taste, 109, 116, 117, 132, 212, 213, 311

anatomy of, in fish, 213, 215
Taste buds, in fish, 209, 210, 214, 215
Taste thresholds, absolute sensitivity, in fish, 212, 213, 215
Telemone, 80
Temperature, effects of, 137
Tench. SEE Tinca
Tenodera angustipennis (Mantidae), 117
Termes bellicosus, 84
Terpenoids, 95, 98
Terpinolene, 92
Territorial behavior
  in fish, 216, 224-226, 228
  in mammals, 312
Territorial marking with odors, 312. SEE ALSO Colony odors
"Territoriality," 345
Testosterone, 312
Thresholds
  detection and nature of, 362-369
  in fish, 212
  of odor, 110, 114-116
  of taste, 117, 137
Tinca vulgaris (tench), 231
Tomatine, 133, 136, 137, 139
Tongue, Jacobson's-organ system, 272, 281
Trail and alarm substances, in ants, 161-190
Trail pheromones, 276
  as a chemical communicator, 175-176
Trail-laying phenomenon, 162
  in insects, 94-98, 100
  in mammals, 312
Trail substance, 162-175
  in the Dolichoderine, 165-166
    interspecificity of, 166
    point of origin of, 165
  in the Dorylinae, 166-168
    interspecificity of, 167-168
    point of origin of, 166, 167
  in the Formicidae, 162-163, 164
  in the Formicinae, 163-165
    chemical analysis of, 163
    interspecificity of, 165
    point of origin of, 163-165
  in the Myrmicinae, 168-175
    appearance of, 168
    life span of, 168-171
    point of origin of, 168, 171, 172, 173, 174
    specificity of, 173, 174
Training of insects, 108, 109, 114, 115. SEE ALSO Conditioning
Trigeminal nerves, chemosensitivity of, 311
Trophallaxy, 83-84
Trout, 208, 209, 211
Trypodendron lineatum (Scolytidae), 117, 118
Tumors, 313
Tuna, 208, 209, 211
Types of receptor cells ("specialists" and "generalists"), 134
Typhlogobius californiensis (blind goby), 211, 215, 216, 218

Uexküll's "functional circle" approach, 260-261
Urechis (burrowing worm), 232
Urine, 309-312, 314-316, 318-319

Vaginal smears, 315
Valeric acid, 310, 311
Vanessa, 132
Vanillin, 32
Ventral organ, 165
trans-Verbenol, 53
"Viscous" substance (in taste hair of blowfly), 123, 125, 134, 234, 236
Vitamin A, 145
Vomeronasal epithelium, 243, 244, 245

Vomeronasal organ. SEE Jacobson's organ

Wasp, 114, 116
Water, 132-133
Water receptor cells, in insects, 127, 137
Weber fractions, 368, 369
Weber's law, 362, 368, 370
Wilson nest, 91
Wood, odor of, 101

Zootermopsis angusticollis (termite), 131
Zootermopsis spp., 84, 86, 89-94, 96, 97, 99
Zwaardmaker's olfactometer, 368